Statistical Analysis and Modelling
of Spatial Point Patterns

STATISTICS IN PRACTICE

Advisory Editors

Stephen Senn
University of Glasgow, UK
Marion Scott
University of Glasgow, UK

Founding Editor

Vic Barnett
Nottingham Trent University, UK

Statistics in Practice is an important international series of texts which provide detailed coverage of statistical concepts, methods and worked case studies in specific fields of investigation and study.

With sound motivation and many worked practical examples, the books show in down-to-earth terms how to select and use an appropriate range of statistical techniques in a particular practical field within each title's special topic area.

The books provide statistical support for professionals and research workers across a range of employment fields and research environments. Subject areas covered include medicine and pharmaceutics; industry, finance and commerce; public services; the earth and environmental sciences, and so on.

The books also provide support to students studying statistical courses applied to the above areas. The demand for graduates to be equipped for the work environment has led to such courses becoming increasingly prevalent at universities and colleges.

It is our aim to present judiciously chosen and well-written workbooks to meet everyday practical needs. Feedback of views from readers will be most valuable to monitor the success of this aim.

A complete list of titles in this series appears at the end of the volume.

Statistical Analysis and Modelling of Spatial Point Patterns

Janine Illian
*School of Mathematics and Statistics, University of St Andrews,
Scotland, UK*

Antti Penttinen
*Department of Mathematics and Statistics,
University of Jyväskylä, Finland*

Helga Stoyan
*Institut für Stochastik, TU Bergakademie Freiberg,
Germany*

Dietrich Stoyan
*Institut für Stochastik, TU Bergakademie Freiberg,
Germany*

John Wiley & Sons, Ltd

Other Wiley Editorial Offices

John Wiley & Sons Inc., 111 River Street, Hoboken, NJ 07030, USA

Jossey-Bass, 989 Market Street, San Francisco, CA 94103-1741, USA

Wiley-VCH Verlag GmbH, Boschstr. 12, D-69469 Weinheim, Germany

John Wiley & Sons Australia Ltd, 42 McDougall Street, Milton, Queensland 4064, Australia

John Wiley & Sons (Asia) Pte Ltd, 2 Clementi Loop #02-01, Jin Xing Distripark, Singapore 129809

John Wiley & Sons Canada Ltd, 6045 Freemont Blvd, Mississauga, ONT, L5R 4J3

Wiley also publishes its books in a variety of electronic formats. Some content that appears in print
may not be available in electronic books.

Library of Congress Cataloging in Publication Data

Statistical analysis and modelling of spatial point patterns / Janine Illian . . . [et al].
    p. cm. — (Statistics in practice)
   Includes bibliographical references and index.
   ISBN 978-0-470-01491-2 (cloth : acid-free paper)
   1. Spatial analysis (Statistics)   I. Illian, Janine.
   QA278.2.S72 2008
   519.5—dc22

                         2007045547

British Library Cataloguing in Publication Data

A catalogue record for this book is available from the British Library

ISBN 978-0-470-01491-2 (HB)

Typeset in 10/12pt Times by Integra Software Services Pvt. Ltd, Pondicherry, India

This book is printed on acid-free paper responsibly manufactured from sustainable forestry
in which at least two trees are planted for each one used for paper production.

Contents

Preface

Spatial point processes are mathematical models that describe the arrangement of objects that are irregularly or randomly distributed in the plane or in space. The patterns formed by the objects are analysed in many scientific disciplines; hence a great variety of objects may be considered such as atoms, molecules, biological cells, animals, plants, trees, particles, pores, or stars and galaxies. At a basic level the data simply consist of point coordinates describing the locations of the objects, but additional characteristics of the objects can also be included in the analysis. These additional characteristics may, for instance, describe the size or type of an object and are usually referred to as 'marks'. Point process analysis is in many ways distinct from the classical statistical methodologies presented in undergraduate textbooks. However, some of the more fundamental classical statistical issues remain influential; for example, sampling, exploratory data analysis, parameter estimation, model fitting, testing of hypotheses and separating signal from noise may all form part of a point process analysis.

Point process statistics is perhaps the most developed and beautiful branch of the modern field of spatial statistics; this is perhaps because points are the most elementary of geometrical objects and lead to data structures that are particularly clear and useful. Sometimes, however, point data have to be analysed in combination with other data from variables that vary continuously in space. This requires an application of spatial statistical methods that fall outside the realm of spatial point processes.

Recent decades have seen a strong increase in the development of point process methodology, based on a profound theoretical development and driven by applications from many different fields of science. In addition to the classical fields of application such as archaeology, astronomy, particle physics and forestry, today other fields such as ecology, biology, medicine and materials science extensively apply point process methods. This development is facilitated by the advent of new and improved technologies that may be used to collect point pattern data. Whereas the first point patterns were small and collected manually, modern data sets tend to be much larger and are collected using automated methods. Ecologists have found fascinating relations in plant communities by considering plants as points marked by characteristics such as size, species or genotype. For instance, they have shown that spatial structure determines ecological processes in the short term and ecological processes modify spatial structure in the long term. Physicists, on the other hand, have used point process methods to study physical structures, for example, packings of hard spheres or other objects, where phase transitions appear. Astronomers have analysed the spatial distribution of galaxies in the universe with particularly powerful statistical methods based on point processes.

The aim of this book is to present statistical methods that are relevant in practice to readers from all these areas. Indeed, there is no point process methodology specific to ecology or physics; the methods are universal, and ideas developed in one field of application may be of value in another. Consequently, not every example in this book is of an ecological or physical nature. Ecologists and physicists are encouraged to translate the examples into their own language. In a few cases this may be difficult. For ecologists it may be impossible to apply the idea of packings of hard spheres and for physicists cluster processes rarely are suitable models for physical phenomena. Knowledge of such structures in one discipline may eventually turn out to be equally useful in the other; for example, solutions applied to the physical problem of packings of hard spheres have the potential to be informative in spatial studies on the swarming behaviour of birds and fish or, in the planar case, on the distribution of communities of plants.

Readers are encouraged to study all the examples even if these are from outside their specific field of interest, taking into account that it is really the geometrical structure that is being analysed. A pattern originating from an entirely different area of science may well be geometrically similar to patterns formed by more familiar objects. Consider, for instance, the pattern of gold particles, which is frequently discussed in this book. A pattern with similar geometrical properties might, on a different spatial scale and with its own interpretation, also appear in quite different contexts. The results of the statistical analysis should then be translated into the terminology relevant to the reader – this might even generate surprising new ideas.

Readers from fields of applications where only planar patterns are analysed are asked for their forbearance when spheres and even the d-dimensional case are discussed, as this is sometimes necessary for the sake of brevity and elegance.

This book, which may be regarded as a successor to Stoyan and Stoyan (1994), is intended for an audience of readers with widely varying knowledge of mathematics, statistics and computer science. The authors hope that this book will prove useful to students on a variety of courses, as well as to scientists both within and outside the field of mathematical statistics, who may be interested in the underlying principles and theoretical ideas. Some readers will write their own programs for the statistical procedures, many will work with open source libraries such as spatstat (http://spatstat.org) in R (R Development Core Team, 2007; Baddeley and Turner, 2005, 2006) or with commercial software, whereas others may simply want to understand the capabilities of point process statistics or the output generated by point process software.

The authors hope that readers will enjoy the large number of examples. They are encouraged to use the data files provided in `http://www.wiley.com/go/penttinen` and to analyse these in more detail using their own software. Some of the methods are presented without examples, mainly due to lack of space, or when further explanation seems unnecessary, as for example with the numerous indices. For some methods that have been developed recently, convincing examples could not be found. These methods are nevertheless presented here in the expectation of potential future applications.

The book mainly presents mathematical-statistical facts. Proofs of these are only provided when they are considered helpful in understanding the ideas underlying the statistical methods. The mathematically inclined reader may use this book as an introductory text, as a source of examples and ideas and as a motivation for further, more detailed study of these topics in other literature such as Stoyan et al. (1995), van Lieshout (2000) and Møller and Waagepetersen (2004).

The authors have tried to present many different methods developed in different fields of point process statistics that merit communication to a broader audience. This leads to an extensive presentation of non-parametric statistical methods. But it turned out to be impossible to present all the existing knowledge of point process statistics. In general, this book focuses on traditional and proven methods, which are preferred over mathematically complex methods and, therefore, such developing areas as spatio-temporal and Bayesian point pattern modelling are only briefly discussed.

This book is not intended to be read from cover to cover. Of course the chapters and sections have a logical order and the book can be read in this way. But the reader is perhaps more likely to regard some of the material as less important at a first reading. Hence, initially some sections may be ignored to provide a general understanding of the methods. These are marked by *. The reader is encouraged to jump from one chapter to another, and this is facilitated by a comprehensive index and a notation index. In particular, Chapter 3 may be ignored at a first reading unless a reader is specifically interested in finite point processes. Some basic knowledge of the ideas in Chapter 4 is helpful for an understanding of Chapter 3. Before reading Section 6.6 the reader should have read Section 3.6.

Chapter 1 presents fundamentals of the underlying theory, motivating examples, sampling methods and historical remarks. Chapter 2 studies a particular fundamental model, the Poisson process and, closely related to this, tests of the hypothesis of complete spatial randomness. Chapter 3 considers finite point processes – processes that exist only within a bounded window, which influences the distribution of the points. The important particular case of finite Gibbs processes (or Markov point processes) is discussed in much detail. The pivotal Chapter 4 presents the statistical theory for stationary point processes. It is this theory that many scientists refer to as 'point process statistics', as it comprises the K-function and second-order methods in general. Towards the end of the chapter some methods for clearly inhomogeneous patterns are also discussed. Chapter 5 discusses an analogous theory for marked point processes and presents a wealth of statistical characteristics and methods. Chapter 6 introduces a suite of stationary point process models, after initially discussing general principles of model building. It considers classical models such as Cox processes, cluster processes, hard-core processes and stationary Gibbs processes. Additionally, spatial-temporal processes are considered, an area that is still in its infancy and currently undergoing rapid development. Furthermore, statistical methods are presented which enable the analysis of correlations between point processes and random fields or fibre processes. Finally, Chapter 7 presents important approaches to parameter estimation for point process models, and explains various, typically simulation-based tests for point process models.

All the methods are illustrated by many examples with data from various areas of application. These examples aim to show the reader the wide range and potential applications of the statistical methodology. They are listed for readers' convenience on pp. xvii–xix.

This book differs from others on point process statistics not only in its application-oriented approach but also in some technical aspects. Densities play a central role, in particular the pair correlation function, since these functions are easier to interpret than cumulative functions such as Ripley's K-function. Furthermore, many nonparametric methods and many new characteristics for marked point processes are presented. However, some of the more recent Markov chain Monte Carlo methods are discussed in less detail.

It is a pleasure for the authors to thank all those friends and colleagues who helped by providing data, suggesting references, allowing us to view unpublished manuscripts, and reading and commenting on preliminary drafts of sections or chapters. In alphabetical order these include: A. J. Baddeley, F. Ballani, S. Barot, A. Bellmann, U. Berger, D. Burslem, R. Capobianco, O. Davies, P. Diggle, D. J. Daley, A. Elsner, C. Geiss, U. Hahn, J. Heikkinen, L. Heinrich, H. Hildenbrandt, A. Holroyd, V. Isham, U. Jansen, A. Järvinen, S. Kärkkäinen, G. Last, M. N. M. van Lieshout, K. Lochmann, K. Mäkisara, V. J. Martinez, T. Mattfeldt, J. Mecke, A. Mikhail, M. Myllymäki, M.-A. Moravie, M. Nummelin, Y. Ogata, J. Ohser, J. Pfänder, A. Pommerening, T. Rajala, E. Renshaw, B. D. Ripley, A. Särkkä, K. Schladitz, M. Sonntag, S. Soubeyrand, U. Tanaka, E. Tomppo, S. Torquato, M. Vihola, A. Wade, K. Wälder, A. Weiss, R. van de Weygaert and S. Wolf. We would also like to thank the students of the point process seminar at the University of Jyväskylä for helpful discussions. T. Rajala undertook the hard job of producing the many figures for this book.

In the summer of 2006 D. J. Daley showed a draft of Chapter 15 of Daley and Vere-Jones (2008) to one of the authors along with Vere-Jones (2008), which aided the writing of Section 6.10 on space–time processes. The book also benefited from fruitful discussions with U. Hahn on the statistics of non-stationary point processes. V. J. Martínez supported the authors in writing the text in Section 1.3.4. A. J. Baddeley informed the authors about the calculation of set-geometrical quantities. And M. N. M. van Lieshout discussed with the authors questions of the theory of finite point processes, leading to an improved presentation of Chapter 3.

This work was partially supported by the Academy of Finland (projects 111156 and 208284).

The authors are also grateful for the assistance of M. Robakowski, who did a large part of the technical work. Finally, they wish to express their gratitude to K. Sharples of John Wiley & Sons, Ltd for her continuous support, to Richard Leigh for his careful work and constructive suggestions as copy editor.

The authors   November 2007

List of examples

1

Introduction

The aim of this first chapter is to provide both an introduction to point process statistics and a short overview of the theory of point processes. Section 1.1 describes the aims of point process statistics and Section 1.2 introduces five empirical point patterns from various scientific disciplines that differ with regard to their statistical structure. These patterns are discussed in later chapters in a variety of contexts. Section 1.3 consists of some historical notes which describe early approaches to statistical methods for the analysis of point patterns. Section 1.4 discusses some basic technical sampling methods and point coordinate measurement.

Section 1.5 introduces basic notation and fundamental theoretical ideas such as intensity and intensity function, moments and product densities. Section 1.6 then discusses the notions of stationarity and isotropy, properties which are assumed by many of the most popular methods of point process statistics. Section 1.7 discusses the general concept of summary characteristics. As in classical statistics, these are numerical values or functions that describe the distributional behaviour of point patterns in a concise way. Section 1.8 aims to contribute to a systematic description of point patterns and discusses various approaches to constructing secondary structures for point processes. The idea here is to construct other geometrical objects based on an empirical point pattern with the aim of finding interesting properties of the pattern

Statistical Analysis and Modelling of Spatial Point Patterns   J. Illian, A. Penttinen, H. Stoyan and D. Stoyan
© 2008 John Wiley & Sons, Ltd

reflected in the secondary structures, which include tessellations, graphs and random fields.

Simulations are an important tool in point process statistics, and thus Section 1.9 introduces the main ideas relevant to simulation approaches as a preparation for more extensive discussions in later chapters.

1.1   Point process statistics

The aim of point process statistics is to analyse the geometrical structure of patterns formed by objects that are distributed randomly in one-, two- or three-dimensional space. Examples include locations of trees in a forest stand, blood particles on a glass plate, galaxies in the universe, and particle centres in samples of materials. These objects are represented in a natural and elegant way by points and marks. The *points* describe the locations of the objects, and the *marks* provide additional information, thus characterising the objects further, e.g. through their type, size or shape. Being based on this data structure, point process statistics forms (perhaps the most efficient and central) part of spatial statistics.

So how is the analysis of spatial pattern data usually approached? A simple graphical representation of the pattern of objects as a point map is a very useful preliminary step towards understanding its properties: visual inspection provides a qualitative characterisation of the type of the pattern even if rather vague terms are used in the initial description (clustered, aggregated, clumped, patchy, regular, inhibited, uniform, even); see Figure 1.1.

It may also indicate correlations among marks or between the point density and spatial covariates, i.e. other random structures which influence the point distribution such as soil property or physical influences. However, for a precise quantification and a more standardised description and finer distinction between types of spatial behaviour, appropriate statistical methodology has to be applied. These methods

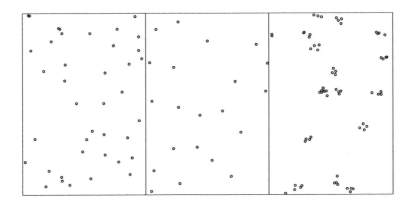

Figure 1.1   Three simulated point patterns: (left) random, (centre) regular, (right) clustered.

provide information on more subtle differences in spatial structures that are not apparent to the naked eye.

Spatial point process statistics differs fundamentally from classical statistics, which typically analyses independent observations and applies concepts such as the mean $\bar{x}$, the sample variance s^2 and the distribution function $F(x)$. Much statistical theory is based on the assumption of a Gaussian or normal distribution or is a consequence of the central limit theorem. Based on this, statistical tests and confidence intervals have been derived that are in common use in all areas of statistics. Point process statistics, however, is confronted with various types of correlation in the patterns. The relative inter-point distances are correlated as well as the numbers of points in adjacent regions. In addition, the characteristics of the objects represented by the points may be (spatially) correlated. Hence the statistical analysis is very much concerned with detecting and describing these correlations.

Many different aspects of the nature of a specific spatial point pattern may be described using the appropriate statistical methods. The simplest of these is the point intensity, i.e. the average number of objects per unit area or volume, if point density can be considered constant across space. Note that this resembles the use of the sample mean $\bar{x}$ in classical statistics. If the point density is variable, intensity maps may be constructed and these may be related to maps describing the values of covariates across space, perhaps in the context of geographical information systems. However, characteristics that are more typical of point process statistics describe correlations among the points in the pattern relative to their distances, e.g. distances to nearest neighbours or numbers of neighbours within given distances. Large parts of this book discuss different ways of describing and characterising point patterns in this way.

In other words, the main aim of point process statistics is to understand and describe the short-range interaction among the points, which explains the mutual positions of the points. Quite often this concerns the degree of clustering or repulsion (inhibition) among points and the spatial scale at which these operate. The analysis of a point pattern also provides information on underlying processes that have caused the patterns, as well as on the geometrical properties of the structure represented by the points. Point process statistics may help to model these structures and to find suitable model parameters, which may be used for classification and to identify structural changes in point patterns depending on time or physical parameters. On the whole, point process statistics can do both – characterise an entire pattern by a small number of interpretable numbers (e.g. indices) or curves (e.g. the K-function) and characterise the individual points by natural or artificial marks.

The analysis of marked point patterns is more interesting and often provides deeper insight into the processes that are causing the pattern than an analysis of unmarked point patterns. Often the marks are 'qualitative', i.e. the pattern is multivariate and consists of several types of points, e.g. different species, ages and size classes. This kind of point pattern data may be regarded either as a superposition or union of single-type point patterns, or alternatively as a single point pattern with different types of labels where the labels indicate the type of points. In this situation the aim is to explore correlations among the different types of points and the spatial scale and range of these correlations. A data set with different types of points has a higher degree of complexity as opposed to non-labelled data. For example, some data sets describing tree locations in tropical forests labelled according to species can consist of several hundred species in a 50 ha plot!

Even more general information may be assigned to the individual points. The marks can be continuous variates, vectors of variates or even stochastic processes. Examples from forestry include the diameter at breast height (dbh) of a tree, last year's growth increment, and a time series of dbh over several consecutive years. Other examples are particle diameter, size and shape. A common feature of these measurements is that they describe the status or property of the object associated with a point location. Data sets with 'quantitative' (or real-valued) marks are highly complex as they reflect various correlations among the objects represented by the points and contain an abundance of information on the system of objects. Point process statistics may be used to detect these correlations and hence provide information contained in these data sets.

In addition to correlations among (i) the point locations, (ii) marks and (iii) marks and point locations, also (iv) correlations with covariates are of interest. Indeed, the degree of inhomogeneity of the point locations within the observation window often reflects the influence of covariates, which may be regionalised variables, e.g. elevation, precipitation or soil property or discrete structures such as geological fault lines or river courses. Including both mark and covariate information in the analysis of a point pattern has become increasingly relevant. An increase in the scientific interest in the association between point patterns and covariates, coupled with an increase in the amount of detailed spatial data including covariates as a result of modern technology, makes this type of analysis more and more important. In particular, rapid development in areas such as geographical information systems and image analysis has broadened the interest in the influence of covariates on spatial patterns. These types of issues are usually addressed by initially constructing maps of both the intensity and the marks and investigating whether they are dependent on or independent of covariates. The resulting point pattern data can be very extensive and complex. Extracting information from this type of data remains a challenge in point process statistics.

Note that in some cases the observation window itself substantially influences the results of the statistical analysis. For example, if the observation window is too small, a spurious trend in the point density may be observed. Using a larger window, however, can reveal that density fluctuations on a spatial scale similar to the window size are normal and the point distribution is globally homogeneous. For this reason, the issue of choosing an appropriate window is discussed repeatedly in this book, in particular in Chapters 3 and 4.

In addition to the statistical description of point patterns and often in combination with it, suitable point process models can be defined and fitted to data. This is very similar to the approaches taken in classical statistics, where models play an important role. They lead to specific distributions for data, such as the Gaussian distribution, and simplify and systematise thinking and further work. Point process statistics uses its own types of models, which are of a different nature but often not more complicated than the models in classical statistics. These models may be used to formulate scientific hypotheses in terms of model parameters. Statistical approaches may then be used to test whether properties of the pattern derived from these hypotheses are reflected in the pattern, and hence whether the patterns support or disprove the hypotheses. In particular, these models enable the simulation of point patterns, which may be a helpful way to understand the underlying natural processes that have formed the pattern.

This book treats point pattern *analysis,* i.e. it discusses methods that may be used to extract information hidden in data using modern statistical methods, using various summary characteristics. It also deals with *synthesis,* i.e. the construction of point process models. The book aims to make point process methods accessible to applied researchers such that these methods become an everyday tool within applied statistics and that users are encouraged to use modern ideas and methods of statistics in the analysis of point patterns.

The summary characteristics as well as the models discussed in this book may be used to describe and analyse point patterns from any applied discipline. The examples referred to throughout this book are from a wide range of scientific fields. The methodology may be applied in any other area of research where similar data structures occur to answer the reader's original question in physics, biology, ecology, etc.

1.2 Examples of point process data

1.2.1 A pattern of amacrine cells

In biology and medicine many data sets consist of spatial point patterns which are most suitably analysed with point process methods. A classical example is the pattern formed by the amacrine cells in the retina of a rabbit, which is perhaps best described in Diggle et al. (2006). Figure 1.2 shows the pattern, which consists of the locations of the 'on' and 'off' cells in planar projection within a 1060 by 662 μm rectangular section of the retina. The retina is a neural structure at the back of the eye which converts light into electrical impulses. It is a three-dimensional structure consisting of several types of cells arranged in different layers. The 10 μm cholinergic amacrine cells are among these cells. They are either 'on' or 'off' cells, depending on whether they are excited by an increase or a decrease in illumination.

These data represent a bivariate point pattern as the points are marked with two discrete marks (○) and (●). The combined pattern looks rather random and homogeneous. It is only a small part of the retina, taken from an area where the pattern continues in a similar way outside the observation window. The overall distribution of cells varies with eccentricity, becoming less dense in the parafoveal regions and periphery of the retina. The pattern is a good example of data that may be analysed with statistical methods for stationary marked point processes. These

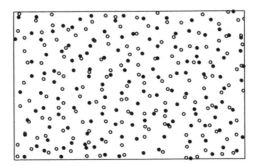

Figure 1.2   The amacrine cell data as in Diggle et al. (2006), with 152 on (o) and 142 off (•) cells. The cell locations lie in different layers and are projected onto a plane, and so cells of different types can partially overlap, while cells of the same type can never overlap. Courtesy of P.J. Diggle. With kind permission of Springer Science and Business Media.

are discussed in Chapter 5 and aim to study the short-range interaction among the cells. Visual inspection indicates that both subpatterns show some tendency towards regularity with repulsion within a distance that exceeds the size of the cells. On the other hand, cells of different types appear close together and there appears to be only weak repulsion.

This book analyses the data of the amacrine cells in various contexts revealing some further properties of the patterns beyond the findings in Diggle et al. (2006), where a suitable model was fitted to the data.

1.2.2   Gold particles

Point process statistics have been applied in many areas of the health sciences and cell biology. As an example, Figure 1.3 shows an ultrathin section of a pellet of purified tobacco rattle virus after immunogold labelling (IGL) by particles of colloidal gold. IGL is a powerful method for the detection of antigens in samples of embedded tissues depending on the exposure of antigen sites on the surface of the thin section. The gold particles, which are visible through an electron microscope, are coupled to the antigen sites yielding information on the spatial structure of these. The locations of 218 gold particles in a window of 1064.7×676 nm (rescaled as 630×400 lu, lu = length unit, 1 pixel in image analysis) are shown in the figure. The cell centres form a point pattern in a rectangular sampling window. The observation window was chosen independently of the point pattern. Hence the data may be regarded as a sample from a larger ultrathin section and again methods for stationary point processes may be applied (see Chapter 4). The data originate from Roberts (1994) and have been analysed previously with point process methodology in Glasbey and Roberts (1997). These data represent a point pattern marked with quantitative marks, the gold particle diameters. They are used here both as an

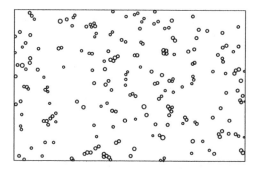

Figure 1.3   Ultrathin section of a pellet of purified tobacco rattle virus after immunogold labelling with a goat antirabbit gold (size 15 nm) probe in a rectangular window of size 1064.7 × 676 nm. The 218 gold particles are identifiable as dark spots in the electron-microscopic image. The diameters of the small circles are proportional to the gold particle diameters. Data courtesy of C. Glasbey.

example of the application of statistical methods for stationary point processes as explained in Chapter 4 and of marked point processes in Chapter 5. Visual inspection indicates slight regularity (repulsion), whereas the points are clustered at larger distances.

1.2.3   A pattern of Western Australian plants

An important area of applied point process statistics is the analysis of planar point patterns formed by plant communities. Throughout this book patterns of trees in forests are studied in various contexts, but here a pattern of herbaceous plants is introduced. Figure 1.4 shows the 207 locations of plants of the species *Phlebocarya filifolia* in a 22 × 22 m square. The coordinates were measured on a 10 cm grid, such that the minimum inter-plant distance is 10 cm. This discretisation causes some difficulties in the statistical analysis, since some smoothing is necessary to obtain reasonable results.

Phlebocarya is endemic to Western Australia and known from scattered locations from Eneabba to Perth, Western Australia, where it grows in shrublands and eucalyptus woodlands in sandy, extremely nutrient-poor soil. It is a tufted herb with short rhizomes, i.e. it propagates by horizontal underground stems that send out roots and shoots from its nodes. The leaves are basal, 250 to 400 mm long and up to 0.6 to 1.8 mm wide. The plant produces numerous flowers along multiple branching leafless stems. The data originate from an extensive ecological study of many plants in the same plot (Armstrong, 1991; Illian et al., 2008). The issue of analysing a multi-type pattern is considered in Example 4.19. The species *P. filifolia* was selected as an example for this book because it provides a good example

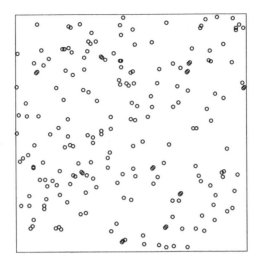

Figure 1.4   Positions of 207 *Phlebocarya filifolia* plants in a 22×22 m square at Cooljarloo near Perth, West Australia. Data courtesy of P. Armstrong.

of a homogeneous clustered pattern; the clustering behaviour can be explained as resulting from propagation mechanisms.

1.2.4 Waterstriders

The application of point process statistics in biology is not restricted to non-motile organisms such as plants; the methods are also suitable for cross-sectional data on positions of animals at a specific time point. Figure 1.5, which is based on a photograph, shows the positions of Palaearctic waterstriders (*Limnoporus rufoscutellatus*) on a water surface. Waterstriders are arthropods that live on the water surface and move at high speed from time to time. Individuals communicate by sending signals along the water surface by vibrating their front legs. The spatial patterns formed by individuals contain information on the animals' behaviour. Various aspects of behaviour, such as habitat selection at different stages in the life cycle (larval stages, juvenile, adult), territoriality and cannibalism, are of ecological interest and have been studied in the literature, both in experiments and in the natural environment. The figure shows the last larval stage (stage 5) in a sub-rectangle of a water surface of irregular shape. The rectangle was chosen such that methods for stationary point processes can be applied even though the pattern is very small, i.e. the pattern is treated as if it were a small rectangular area in a very large water surface with many waterstriders. Modern statisticians would probably prefer to observe the entire finite pattern and apply the methods of Chapter 3 for finite point processes. As part of the analysis, graphs are constructed the vertices of which are the points of the pattern (see Section 1.8.5), and which reflect the geometrical structure of the pattern.

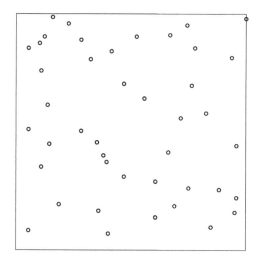

Figure 1.5   Positions of 43 Palaearctic waterstriders in a vessel. The window size is 53.6×53.6 cm. The data arose from a series of ecological experiments. Data courtesy of M. Nummelin.

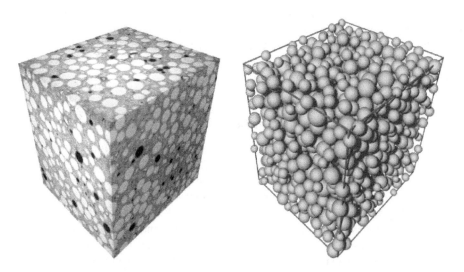

Figure 1.6   (left) A sample of concrete (self-flowing refractory castables) of size $10 \times 10 \times 10$ mm, obtained by computerised tomography. The white circles are cut corundum grains, the black ones air pores. (right) Result of 3D image analysis: the grains are approximated by spheres of variable radii. Courtesy of F. Ballani. With kind permission of Springer Science and Business Media.

1.2.5   A sample of concrete

Materials science is another important field of application for point process statistics. In this context, the points are typically constructed, e.g. centres of pores, particles or grains. Size and shape parameters of the objects are used as marks.

A well-known building material, which has both pores and grains, is concrete. Figure 1.6 shows a three-dimensional sample of a special type of concrete, so-called self-flowing refractory castables; see Ballani (2006) for details. It was produced for research purposes with spherical refractory aggregates shown in white. The black objects are air pores. Using techniques from image analysis, a system of spheres was constructed which approximates the set of all (white) particles. In this way, marked point process data were obtained, where the sphere centres are the 'points' and their radii the 'marks'.

1.3   Historical notes

The following historical notes describe early statistical approaches to the analysis of spatial point processes in the context of forestry, medicine, ecology and astronomy, which are to this day important fields of application of point process statistics. Refer to Daley and Vere-Jones (2003) for more details on the history of point processes, but note that in that book the discussion focuses on the one-dimensional case, i.e. processes over time, the origin of the term 'point process'.

Note that this section assumes some prior knowledge on the basic concepts of point process statistics as some terms are used which are discussed much later in the book. Readers who are new to the field may prefer to initially skip this section and return to it later, once they have gained some knowledge of Poisson processes, intensities and pair correlation functions.

1.3.1   Determination of number of trees in a forest

In his book *Die Forst-Mathematik* ('Forestry Mathematics'), written for foresters and published as early as in 1835 in Gotha, Germany, Gottlob König considered the issue of estimating the mean number of trees per unit area (stems per hectare) in a forest, denoted by λ and referred to as 'intensity' in this book. This remains a difficult problem, because the necessary field work is often very laborious and costly, even the simple counting of trees in larger stands. König's idea was to measure *distances* from n randomly chosen reference trees to their nearest neighbours. It is obvious that he chose this measuring strategy for practical reasons, as it can be done by simply using a measuring rod or tape rather than by determining the location of every single tree. König suggested the estimator

$$\hat{\lambda}_K = \frac{1}{d^2} \qquad (1.3.1)$$

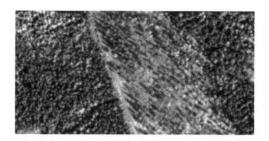

Figure 1.7 Radar photograph showing the variability of tree positions in a stand. Data courtesy of E. Tomppo and K. Mäkisara of the Finnish Forest Research Institute.

where

$$\overline{d^2} = \frac{1}{n} \sum_{i=1}^{n} d_i^2$$

is the empirical second moment of the distances $d_1, \ldots, d_n$ from n sample trees to their nearest neighbours. Equation (1.3.1) is based on the assumption that the trees form a regular square lattice. This may make sense if the forest was regenerated by planting the saplings in this form and may otherwise be an acceptable approximation. Indeed, based on this assumption, the nearest-neighbour distance is a constant (the lattice spacing δ), and the area per point is δ^2, resulting in $\lambda = 1/\delta^2$. König's estimator is an empirical version of this relation.

This estimator and the idea of estimating the mean number of trees based on tree-to-nearest-tree observations have turned out to be particularly sensitive to the nature of the pattern of trees, which is in nature rather variable, see Figure 1.7. For example, if the trees are 'randomly' located (Chapter 2 discusses the notion of randomness in detail), the corresponding maximum likelihood estimator is

$$\hat{\lambda}_P = \frac{1}{4(\overline{d})^2} \tag{1.3.2}$$

with

$$\overline{d} = \frac{1}{n} \sum_{i=1}^{n} d_i,$$

which usually yields different results than (1.3.1). This shows that results of intensity estimation obtained with nearest-neighbour methods vary with the nature of the point distribution, a problem that was first noted by German foresters in the late nineteenth century. In modern terminology, König's estimator is not a robust estimator of the intensity as it varies with the spatial distribution of the points, and it is therefore no longer in use today.

Whereas the sensitivity to pattern formation of the tree-to-nearest-tree distance methods introduced by König is a drawback for intensity estimation, it is useful in the detection of different patterns formations. Today, the idea is widely used in the study of patterns of points and is discussed in several contexts in this book.

König was not alone in his approach. The Swedish Royal Forester Israel af Ström published in 1830 his *Handbok för skogshushållare* ('Handbook for Foresters'), in which he suggested a random-sampling-based method for the inventory of forest resources. This idea is the basis of what came to be called 'strip surveys' or transect sampling, where parallel strips of width 5 alns (approximately 3 m) were chosen and trees were sampled within the strips and then measured. The corresponding probabilistic background was worked out later by Matérn (1960), who also developed specific stochastic models for point patterns.

The interest in developing point process methods for forestry has continued since Matérn (1960) – see, for example, Warren (1972), Stoyan and Penttinen (2000) and Pretzsch (2002) – with applications in forest inventory and forest ecology. Recent developments in forestry research focus on models in which the individual trees play an important role, which makes the point process approach even more relevant.

1.3.2 Number of blood particles in a sample

The German physicist and pioneer of modern scientific optical technology Ernst Abbe described and discussed how to determine the number of blood particles per unit volume in a blood sample (see Figure 1.8) in a paper of 1879 (see Seneta, 1983). The device was called a 'counting chamber' and manufactured by Carl Zeiss, Jena, where Abbe served as director at the time.

The counting of blood particles was carried out as follows. A diluted blood sample with known volume was spread on a thin plate of glass as a layer of 0.1 mm and analysed under a microscope. Then the counting was carried out in a net of

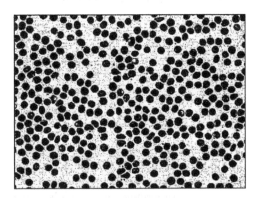

Figure 1.8   A blood sample. The red blood particles are displayed in black. The size of the rectangular window is about $225 \times 182\,\mu$m. Data courtesy of T. Mattfeldt.

adjacent squares of side length 0.05 mm on the bottom of the plate corresponding to a volume of 0.005 ml. This method is obsolete today; flow cytometry is now used, where the blood is sucked through a capillary and physically analysed.

In his paper, Abbe deals with the statistical error due to sampling rather than to errors caused by sample preparation and the equipment, and deals with a statistical model: he assumes that the probability of observing n particles in a sample is

$$p(n) = \frac{e^{-\nu} \nu^n}{n!} \qquad \text{for } n = 0, 1, 2, \ldots, \tag{1.3.3}$$

known as the Poisson distribution. Here ν is the expected number of blood particles in a sample of fixed volume. Further, if ν is large, and this is the case in practice, then Abbe suggested the use of the Gaussian distribution with mean and variance ν. Today this is well known as the normal approximation of the Poisson distribution.

An important issue concerns providing a definition of the statistical error that has a useful interpretation in applications. Abbe's choice was the concept of 'probable error'. This is an interval for which, in a large number of replications, half the observations exceed or fall below it (related to the 50% confidence interval) when the observations are from the model. Abbe noted that when using a volume of 0.001 mm^3, or counts in four adjacent cells on the plate, typically with $\nu \approx 50$, the relative error is around 10 %. His conclusion was that the result derived from a single sample of this size is measured with a large uncertainty.

1.3.3   Patterns of points in plant communities

In 1922 The Svedberg, a Swedish chemist (the 'The' originates from his first name Theodor) and Nobel Laureate in 1926, studied how individual plants were distributed in a plant community. As he mentions in this work, his observations should be treated as preliminary and his only purpose was 'to draw plant ecologists' attention to a hitherto unnoticed form of statistical vegetation analysis'.

Svedberg's reasoning is based on observations of counts in squares where a large number of squares of fixed size were randomly located in the community. The locations of the squares are chosen independently of the plant positions. Within each square, the number of plants was counted. The side length of the squares varied in the range 2–50 cm depending on the species, and the number of squares was between 50 and 261, being typically around 100. Note that this type of observation is still in common use.

Svedberg introduced a 'reference' model which, according to him, marks the 'normal' condition in a plant community where no 'forces' are acting between the individuals. This reference model may be used to distinguish normal, clustered and regular plant communities.

The reference model is based on counts of individuals in squares of area q and the distribution of this number is derived from the 'Poisson series' (1.3.3). In other

Table 1.1   Distribution of counts of *Viola tricolor* (Wild pansy) samples in 105 squares of side length 0.5 m.

Number of plants	Number of squares	Probability (%)	
		observed	expected
0	26	25	22
1	32	31	33
2	24	23	25
3	16	15	13
4	4	4	5
5	3	3	2

words, if the plot is virtually divided into a large number of squares of area q, then $100 \cdot p(n)\%$ of the squares will have n individuals. Note that the point process model underlying this approach is the *Poisson process*. The Poisson process is still used today as a reference model describing a 'normal' case, a situation without interaction. Svedberg studied data for seven species and found 'normal' dispersion as well as clustering and regularity. In his study, Svedberg applied the variance-to-mean ratio of the counts. This is known as the *dispersion index* and was used by Fisher et al. (1922) in the context of bacteria on a microscope plate. Since then, this index has been discussed in the literature in many variations; see also Section 4.2.4.

Consider the example data set from Svedberg (1922) in Table 1.1. The frequency table describes the distribution of plant counts of *Viola tricolor* in squares of side length 0.5 m. In this example, according to Svedberg, only weak deviation from normal dispersion towards clustering was observed.

It is known today that the size of the sampling unit (square of size q) is crucial, to the extent that using two different values of q for the same plant community may yield inconsistent results. For example, consider a plant community where the plants form small-scale clusters, which in turn are randomly located. The use of a small square correctly reveals this clustering, whereas the application of larger squares may indicate normal randomness and not detect clustering. The method based on single-sized squares is not flexible enough to detect structures in a spatial pattern on two or more different spatial scales. Svedberg seemed to be aware of the size-sensitivity of his approach because he applied different square sizes depending on the species.

Svedberg interpreted the case of 'normal' dispersion as the situation where no interaction between the individuals exists in the community. This may be true for physical systems and for many biological systems, but the latter are often much more complex. In this sense, the conclusions derived by Svedberg are daring. Today it is known in ecology that patterns which follow the Poisson distribution can also result from evolution processes such as self-thinning of originally clustered patterns by a competition process.

1.3.4   Formulating the power law for the pair correlation function for galaxies

Astrostatistics is an old and classical field of spatial statistics; see the books by Peebles (1980), Babu and Feigelson (1996) and Martínez and Saar (2002). Astronomers study both spatial random fields in space (the density field, the velocity field and others) and spatial point patterns, in modern times preferably the positions of galaxies (or rather galaxy centres) in three-dimensional space. (Note that our sun is just one amongst hundreds of thousands of stars in our galaxy, the Milky Way.) Figure 1.9 shows a sample from a galaxy redshift catalogue.

Although the cosmological principle assumed by Einstein in 1917 states that the universe is isotropic and homogeneous on large scales, early observational

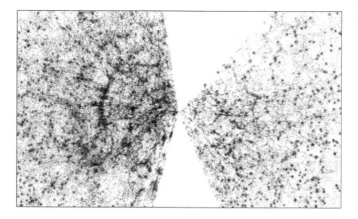

Figure 1.9   A sample of galaxies from the SDSS galaxy redshift catalogue. The image shows a composition of two slices through the third data release of the SDSS survey, offset by around 100 million light-years from the centre (located in our galaxy). The slice is oriented such that an impression may be obtained of the intricate pattern that has become known as the Cosmic Web: filaments and walls draped around large near-empty void regions, with dense and compact clusters of galaxies residing near the nodes of the network. The clusters are easily discernable as conspicuous stretched objects. The stretching, known as 'fingers of God', is a pure artefact: due to internal motions of galaxies within dense clusters (with velocities in excess of 1000 km/s), their redshift is not only determined by the distance of the cluster but also by the corresponding thermal velocity along the line of sight. The image was prepared by Rien van de Weygaert using the interactive COSMOS SDSS3 planetarium program. Data courtesy of R. van de Weygaert.

studies had already found that matter in the universe is distributed in clusters on smaller scales. For example, Hubble (1934)[1] found that counts of galaxies in telescope fields have a strongly skewed distribution, which is similar to a lognormal distribution. Bok (1934) and Mowbray (1938)[1] found that the variance of these counts is considerably larger than the variance of a random distribution following a Poisson process.

An early cluster point process model was introduced by Neyman, Scott and Shane (Neyman and Scott 1952, 1958; Neyman et al. 1953), assuming that galaxy patterns can be interpreted as samples from a statistically homogeneous and isotropic point process. These authors used a specific Neyman–Scott process that is now called the Thomas process and consists of randomly distributed isotropic clusters of points. In the statistical analysis they applied the pair correlation function $g(r)$ which will be introduced in Section 4.3. This characterises the frequency of interpoint distances; if $g(r)$ is large for some r then the inter-point distance r occurs frequently in the process. The function $g(r)$ used for the model looks similar to that in Figure 6.2 on p. 377, but it resembles the right-hand side of a Gaussian bell curve; in particular $g(0)$ takes a finite value.

Neyman, Scott and Shane had statistical difficulties as they obtained different model parameters for different galaxy catalogues (files of coordinates of galaxies). Therefore, they modelled the distribution of galaxy clusters as a random superposition of groups of galaxies of varying size, including superclusters.

The data analysed in Totsuji and Kihara (1969) were of better quality than those analysed previously, as they used the 1967 catalogue by Shane and Wirtanen of the Lick Observatory. According to Totsuji and Kihara (1969), Shane and Wirtanen 'published the distribution of galaxies with ... apparent magnitude brighter than 19^m on the celestial sphere. Photographic plates ... $6° \times 6°$ [in size] covering the sky north of declination $-23°$ were divided into $10' \times 10'$, and images of galaxies in each of these squares were counted. The results are tabulated as counts in solid angles $1° \times 1°$; the counts in the original solid angles $10' \times 10'$ have not been published.' The authors used a linear integral equation which makes it possible to use projected two-dimensional data for the estimation of the three-dimensional pair correlation function. Modern galaxy catalogues contain three-dimensional coordinates, which result from the use of the redshift of each individual galaxy as a distance indicator by means of the Hubble law and a given cosmological model. Nevertheless, since peculiar velocities of galaxies contaminate the measured redshifts, astronomers distinguish between real space and redshift space, the second being a distorted version of the first.

The assumption of a power law simplified the calculations, and Totsuji and Kihara (1969) estimate the parameters r_0 and s in

$$1 + g(r) = \left(\frac{r_0}{r}\right)^s$$

[1] See Peebles (1980) or Martínez and Saar (2002).

as $r_0 = 15.3 \cdot 10^6$ light-years and $s = 1.8$. They found s graphically by inserting empirical values into a band of curves corresponding to $s = 1.7$, 1.8, 1.9 and 2.

Today the power law of the pair correlation function is generally accepted; see Martínez and Saar (2002). The interpretation is that there are no characteristic scales or quantities; see Peebles (1980). Modern estimates of s (today the symbol γ is used) are around 1.7–1.8. Nevertheless, very recently Zehavi et al. (2004) found deviations in the power-law behaviour of the real space correlation function that are interpreted as different clustering regimes.

These days, statisticians seek to develop realistic and simple point process models which have a pair correlation function $g(r)$ with a pole at $r = 0$ of order γ. It is not difficult to provide examples of non-realistic models; see Buryak and Doroshkevich (1996) and Snethlage et al. (2002). Incidentally, these examples show that a power law for the pair correlation function alone is not necessarily an indicator of any 'fractal' behaviour. The parameter γ is now estimated based on least-squares techniques, for which the simple models are training objects.

The books by Peebles (1980), Martínez and Saar (2002) and the paper by Jones et al. (2004) are recommended as further reading.

1.4 Sampling and data collection

1.4.1 General remarks

In point pattern statistics it is important to identify a suitable strategy for converting a specific real-world point pattern, e.g. a pattern formed by trees in a forest or cells in a tissue sample, into statistical data. This issue has to be addressed prior to any point pattern analysis, and the chosen strategy influences the choice of suitable statistical methods and models.

Data collection methods are strongly dependent on the objects represented by the points, the objectives of the study and the available resources, and all these are different for different applications. Generally, the best choice is only the best choice for a very specific example, due to the specific situation and study aims. However, some general guidelines for data collection can be given. Central aspects that need to be considered include the spatial scale, the relationship to the environment, the morphology, size and density of the objects, and available methods. From a statistical point of view, the aim of data collection methods is to optimise unbiasedness and representativeness and to control sampling errors.

The two main aspects in a strategy for data collection are appropriate *sampling* and appropriate *measurement* methods as these are both sources of uncertainty. Sampling concerns the strategy of extracting information on point patterns in a real-life situation, whereas measurement is the technical realisation of the data collection approach. Sampling is a traditional area of statistics and many useful methods have been developed in the literature. Recently, measurement technologies have undergone rapid development such that more and more extensive and high-quality point pattern data have become available.

In general, two different approaches are used to sample point patterns, *field methods* and *mapping*. Field methods have traditionally been used in forestry and ecology as convenient data collection methods. These only measure small fractions of the study area, usually as counts of numbers of points in randomly located rectangles or spheres ('area-based' sampling), or as distance measurements from sampled points ('point-related' sampling).

Area-based sampling is used mainly to determine the total number or mean number of points per unit area and to detect deviations from complete randomness (see Chapter 2).

Point-related sampling approaches use random locations in the study area or randomly chosen points in the point pattern or both. They are useful for the detection of interesting small-scale structural properties. In addition to two- or three-dimensional sampling windows or sampling points, other types of sampling units such as randomly or systematically located strips and lines (transects) are also common.

Field methods often produce *sparse* data, exploiting the fact that the data were observed at locations that are far enough apart to be considered almost independent. This is particularly useful when the number of sampling units is small compared to the study area, which is always a bounded set W with finite area or volume $v(W)$. Sparseness is an advantage in the estimation of interesting parameters and construction of statistical tests. Indeed, if sparse sampling is done properly, spatial correlation, typical of spatial data, is often negligible and hence the distributional theory and statistics are straightforward, close to classical statistics. However, this book focuses on spatial statistics and thus does not cover sparse data any further.

Despite these advantages, field methods are tied to pre-fixed scales and allow only limited statistical modelling. Extensive literature exists for this traditional area of spatial statistics; see, for example, Ripley (1981), Krebs (1998), Diggle (2003) and Upton and Fingleton (1985).

Today mapping is the most commonly used data collection method for point patterns. Its popularity is at least partly due to advanced statistical methodology and modelling, which can analyse the point pattern simultaneously at a number of different spatial scales. In this data collection approach, all locations of the objects within a specific subset of the space are recorded. This set is called the *observation window*. In some cases, the window is predetermined by the application, e.g. in experimental studies, where the data are a census of the objects. In these cases, it is not necessary to choose a sampling strategy.

More common are situations where the researcher determines the size, shape and positioning of the observation window. Mapping and the choice of the observation window are discussed in more detail below.

As noted, in addition to the point locations, information on the objects represented by the points or on the sampling window may be useful for an appropriate analysis or may even be necessary. If the point pattern is not unlikely to be heterogeneous within the sampling window, observations on other spatial variables that explain the heterogeneity may be very important and improve modelling and interpretation.

These variables are called (environmental) *covariates*. The spatial pattern may also be affected by differences in the properties of the objects represented by the points such as their size, shape, age and species. As mentioned above, these are generally described by marks.

Mapped point pattern data may consist of the following elements:

1. Specification of the observation window W.

2. The point record of each point in the sampling window in the form $\{x, m_1, m_2, \ldots, m_p\}$, where x represents the d-dimensional coordinates of the point location and $m_1, \ldots, m_p$ are marks collected for the point at x.

3. Covariate information as spatial measurements of type $Z(y_j)$, $y_j \in W$, for $j = 1, \ldots, l$, where the points y_j form a lattice and do not coincide with the points of the point pattern. The $Z(y_j)$ usually describe a continuous regionalised variable, which is defined at every point in W, whereas the marks are only given and defined for the points in the point pattern.

Not all of these elements are always available for a specific point pattern. An appropriate analysis of the point pattern requires at least information on the window W and the point locations $\{x\}$.

A further type of data collection is *repetitive mapping*, commonly used in the context of space–time analyses, where the same sampling window is measured at two or more points in time. This type of data can be represented by a marked point pattern (using the observation time instants as marks). In some cases it is informative to study the movements of the points as well as their lifetimes (refer to Chapter 6 for more details on space–time processes).

Field methods and mapping are rather different approaches. However, some of the data summaries originally developed for field sampling, such as the distance of a point to its nearest neighbour, may also be applied to mapped data. In modern point process statistics, where often very large point patterns are analysed as a result of improved data collections methods, field methods are still relevant in the preliminary analysis.

1.4.2 Choosing an appropriate study area

An important aspect of point pattern sampling is the *choice of the observation window W* or the study area. In some cases W is given a priori, in particular when a local phenomenon (a finite point process as in Chapter 3) is studied, such as pores of a metallic foam in a pipe, fungal spots on leaves, and cell centres in a small biological organ. Then W is a *window of existence*.

In many other cases, the point pattern of interest is larger than any feasible samples that may be used in a statistical analysis, e.g. in studies of forests, materials or the universe (in galaxy statistics). In these cases, the choice of the window is closely related to the scientific question. If the local interaction among points (relationships among neighbours) is of interest, a window with (nearly) homogeneous point distribution may be suitable, to avoid influences of larger-scale inhomogeneity. If the study focuses on global fluctuations of points, larger windows are more suitable. Here, two rather different scenarios occur: the density fluctuations may be regarded as a unique inhomogeneous phenomenon (e.g. the influence of two mountains on the spatial pattern of a forest) or as a part of a large homogeneous phenomenon.

As far as the shape of the window is concerned, convenient choices include rectangles and circles for planar patterns; sometimes polygons are used. For three-dimensional data parallelepipeds are often used, but sometimes the observation conditions dictate more complicated windows, as in astronomy; see Martínez and Saar (2002). The issue of choosing the *size* of the window is discussed in Section 4.8.

1.4.3   Data collection

Field methods

Once the window has been chosen, data can be collected. Clearly, the choice of the data collection method has a pronounced impact on the choice of the statistical analysis methods. The most informative type of point pattern data consists of the Cartesian coordinates of all points. Often their measurement is too labourious, or partial or condensed information might be sufficient for a particular study. For example, measurements are sometimes taken in the field and are directly summarised in indices that describe the spatial behaviour of the pattern.

An old field technique is *quadrat counting*, i.e. counting points in subwindows that may or may not form a lattice. Here, the exact point coordinates are not collected, which simplifies data acquisition but limits the statistical analysis. Nevertheless, valuable results can be obtained; see Section 2.7.2 and the ecological literature such as Krebs (1998), who also discusses the choice of shape and size of the 'quadrats', which do not necessarily have to be quadratic.

Another old field technique is *distance sampling*, where distances from test points (which form part of the pattern) or from test locations (which are not part of the pattern, perhaps points on a measurement lattice) are measured and analysed statistically, as discussed in Section 1.3.1 above. Statistical methods that are based upon this approach are discussed at several points throughout this book. A special case is *line transect sampling*; see Buckland et al. (2001) or Krebs (1998) for details. Here a line is used as a reference to collect data on the point pattern of interest in a strip of fixed or variable width (which often has to be estimated statistically). An observer moving along the line determines the coordinates of the points $x_i = (\xi_i, \eta_i)$ within the strip by their sighting distances r_i and sighting angles θ_i, where the ξ-axis is parallel to the line and the η-axis is vertical (see Figure 1.10),

$$\xi_i = r_i \cos \theta_i, \qquad \eta_i = r_i \sin \theta_i. \tag{1.4.1}$$

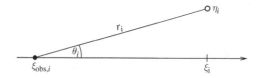

Figure 1.10   Sighting distances and angles in line transect sampling.

Also, the direct measurement of specific inter-point distances yields many of the *indices* discussed in Sections 4.2.4 and 5.2.4.

Mapping

The *Cartesian coordinates* of points are determined either for all points in the whole window, resulting in mapped data, or only within specific (circular) sample plots with centres on a systematic sampling grid; the latter approach is common in forestry. A number of different methods, both simple and advanced, can be used to determine these coordinates.

In forestry and ecological studies, the coordinates are often determined *in situ*. This can be done very efficiently by using *polar coordinates*. The distances r_i to the points $x_i = (\xi_i, \eta_i)$ from some central position (which may be one of the points of the pattern or another suitable location) are measured and the azimuths φ_i, the angles between the north (or another natural) direction and the rays towards the points, are measured clockwise; equation (1.4.1) yields the Cartesian coordinates. This procedure can be repeated for further central points to cover the whole pattern. Moeur (1993) describes this method in detail. A similar approach is the *measurement of point networks* or *triangulation* where, in triangles of points, angles are measured and, for some of the pairs of points, the distances. Formulas of planar trigonometry yield the (ξ, η) coordinates. For distance and angle measurements classical (measuring tape, theodolite) or modern (laser-based takymeter) equipment can be used. Total mapping of a forest area of 1 ha can take 1–3 days when all trees with some minimum height (e.g. 1.3 m) are mapped. The time is even longer if all plant species are mapped, e.g. for ecological studies.

Sometimes photographs are used to construct point pattern data by scanning the image and clicking on points to identify and save the coordinates. An analogous method on a larger scale is described in Stein and Georgiadis (2006). The aim of the study was to measure the coordinates of centres of herds of herbivores in Kenya based on data derived from aircraft flying over the savannah at heights between 70 m and 130 m above ground level, following transects spaced 1 km apart. Whenever a herd was spotted, the aircraft deviated from its flight-line to circle the herd until the number of its members was counted. The geographical coordinates of the centres of herds were recorded using a Trimble GPS receiver. Overlaps and double counts at the boundaries of the blocks were identified and subtracted. Refer to Krebs (1998) for further aerial sampling methods.

Methods from photogrammetry, the geometrical measurement of objects in analogous digital images (Konecny, 2003), are very helpful. Problems resulting from perspective and lens distortion in photographs are overcome by rectification, but the point pattern can be determined much more accurately when photographs are taken vertically rather than at an angle. Applications in forestry and ecology are described in Bai et al. (2005) and Dean (2003).

Whenever possible, the original image should be directly transferred from the camera into a computer. This technique is common in microscopy, and digital photography facilitates this approach as well. Experience shows that the subsequent coordinate determination by mouse click is affected by subjective factors. It is thus advisable that the same person performs this analysis step on all images if more than one image is to be analysed.

Image data can also be obtained by remote sensing (see Kerle et al., 2004), i.e. by methods that may be used to derive information on the objects through devices that are not in direct contact with the objects. For very large distances aerial cameras, scanners or radar are used. The photographs or other images resulting from this are scanned and converted into a digital format, usually a pixel image.

Another source of image data are techniques such as computerised tomography. These methods yield three-dimensional pixel structures, which are the starting point for computations that construct the point coordinates of centres of objects numerically.

For the further processing of digital image data, methods from image analysis are often used. In the simplest case the point coordinates are computed numerically, e.g. as centres of gravity or of surrounding circles or rectangles if the objects in the image are clearly defined. Sometimes the 'objects' are not given a priori but must be determined computationally. A large number of approaches to this have been considered within image analysis, such as the determination of connected components, watershed algorithms (Vincent and Soille, 1991) or Bayesian image analysis (see Van Lieshout, 1995; Winkler, 2003). The approaches regard the images as noisy images of spatial particle systems, e.g. of spherical particles. The aim is to calculate a sharp image, i.e. to find a particle system which best matches the noisy data.

In this context, optimisation techniques such as simulated annealing are commonly used. Figure 1.6 shows both the CT data and the system of spheres for a sample of concrete; see also the discussion in Ballani (2006). Imaging of objects is an important problem in optics research (Ovryn and Izen, 2000; Kvarnström and Glasbey, 2007). The technique is simplified when the particles are spherical and uniform in appearance. Then each particle appears as a bright circular set of pixels and its centre is the geometric centre of the brightness-weighted centroid (Kvarnström and Glasbey, 2007).

Sometimes noise generated by noisy points can be an issue; statistical approaches to this problem are discussed in Lund and Rudemo (2000).

Note that reducing an image to a marked point pattern is often a very useful and appropriate data reduction step, as it allows the application of the powerful methods of point process statistics to the image. However, it often also results in an undesirable loss of information. In these cases, it may be better to stick with image analysis methods, i.e. statistical methods for random fields and sets as described in Cressie (1993), Stoyan et al. (1995), Ohser and Mücklich (2000) and Ohser and Schladitz (2008).

The collection of spatial point pattern data is often an expensive and laborious part of a study. The methodology has improved rapidly in recent years and, as a consequence, automatised data collection methods can often produce extensive data sets at minimal costs. However, the objects are often measured imperfectly or indirectly. Transforming this type of data to point pattern data requires careful pre-handling and the use of pattern recognition methodology.

1.5   Fundamentals of the theory of point processes

Point processes are *stochastic models* of irregular point patterns. Within mathematical theory, point processes can be defined and studied in abstract spaces, but this book mainly considers planar and spatial (i.e. two- and three-dimensional) point processes as these are the most relevant processes in applications. In accordance with the literature, the perhaps confusing expression 'point *process*' is used here. The term 'process' implies that some development over time is considered, but in most cases time-independent phenomena are studied. Physicists and engineers might prefer the term 'point field', which was also used in Stoyan and Stoyan (1994).

A *point pattern* is a collection of points in some area or set and is typically interpreted as a sample from (or realisation of) a point process. In the notation the points are often numbered. This is done only for convenience and does not imply any meaningful order of the points.

There is an extensive literature on point processes, ranging from rather theoretical to more applied texts. Mathematical introductions to the fundamental theory include Cox and Isham (1980), Daley and Vere-Jones (1988, 2003, 2008), Van Lieshout (2000), Stoyan et al. (1995), and Møller and Waagepetersen (2004).

This section, along with Sections 4.1 and 5.1, provides those aspects of the fundamental theory that are relevant for this book. In this way the book is self-contained and makes the reader familiar with notation and terminology and eventually the statistical methods. The exposition is frequently based on heuristic explanations avoiding mathematical details and theory, which can be found in the references. Nevertheless, formulas and difficult mathematical concepts cannot be avoided, but the authors encourage the reader to follow the text since its study is also indispensable for a thorough understanding of the statistical methods.

A point process is denoted by N. Note that this symbol is used here for two different mathematical descriptions of point processes.

- N may denote a function operating on sets or, in more mathematical terms, N is a random *counting measure*. For a subset B of $\mathbb{R}^d$, $N(B)$ is the random number of points in B, i.e. the set B is assigned the number $N(B)$.[2] It is assumed that $N(B) < \infty$ for all bounded sets B, i.e. that N is 'locally finite'. Clearly, $N(B)$ regarded as a function of B has the fundamental property of additivity, i.e.

$$N(B_1 \cup B_2) = N(B_1) + N(B_2)$$

for disjoint B_1 and B_2, and similarly for countably many sets.

- N may also denote a *random set*, i.e. the set of all points $x_1, x_2, \ldots$ in the process. In other words,

$$N = \{x_i\} \quad \text{or} \quad N = \{x_1, x_2, \ldots\};$$

$x \in N$ means that the point x is in the set N. The set N can be finite or infinite. If it is finite the total number of points may be deterministic or random.

The set-theoretical notation can be used without any problems since throughout this book all point processes are assumed to have a property referred to as *simplicity*, i.e. all points are different, and do not coincide, i.e. $x_i \neq x_j$ if $i \neq j$. Note that the theory of point processes also considers models with multiple points.

Remarks

(1) Note that some mathematicians do not like to use the same symbol N for the two different concepts as above. They base the definition on the counting measure N and refer to the corresponding point set as the 'support of N' and use a different symbol for it.

(2) In this book, random points in N are always referred to as 'points'. In the literature, these are sometimes called 'events', 'trees' or 'sites'. The points in $\mathbb{R}^d$ that may or may not coincide with points in N are called 'locations' or 'positions'. Hence one may say 'point x_i is at location x', meaning $x_i = x$. The x_i are dummy variables used in order to emphasise the nature of N as a point sequence. Thus, for example, x_1 is *not* a special point with a special property such as the point of N closest to the origin o of the space, but just any point in the process.

---

[2] Technically B is a Borel set.

Point process sums

Often sums of the form

$$S_f = f(x_1) + f(x_2) + \ldots = \sum_{(i)} f(x_i) = \sum_{x \in N} f(x) \qquad (1.5.1)$$

have to be considered, where f is some real-valued function. (In mathematical terms, f has to be a 'measurable' mapping.) The symbol $\sum_{(i)}$ denotes 'summation over all points of N' and denotes both $\sum_{i=1}^{\infty}$ for an infinite process and $\sum_{i=1}^{n}$ for a finite process.

Example 1.1.   Point process sums

(1) *Seed density.* Assume that the points represent the locations of plants and each plant produces seeds independently of the other plants. The seeds are dispersed randomly around the parent plants, in the same way for all plants. The density of seeds from a plant at position x at location y depends on the distance r of y from x denoted as $md(r)$, for instance $d(r) = \exp(-cr)$. Here m is a parameter proportional to the mean number of seeds per tree, which is assumed to be identical for all trees and c is some positive parameter. Given all parent points $x_1, \ldots, x_n$, the total seed density at the position y may be determined by considering the superposition or the sum of the plant-related densities. Denote this total density by $S_f = S_f(y)$ with reference to (1.5.1) with

$$f(x_i) = md(\|y - x_i\|)$$

(2) *Counting of birds by distance sampling.* Let the points be positions of birds in a forest. A bird singing at position x is still audible at a point with distance r from x with probability $p(r) = 1 - ar$ for $r \leq 1/a$. Given the bird locations, what is the mean number of birds that can be heard by a bird-watcher at position y? It is simply the sum of the different probabilities and can be denoted by S_f as in (1.5.1) with

$$f(x_i) = p(\|y - x_i\|)$$

Note that in the mathematical literature sums as in (1.5.1) are also written as integrals,

$$\sum_{x \in N} f(x) = \int f(x) N(dx) \qquad (1.5.2)$$

In the context of sums of functions the *indicator function* $\mathbf{1}_B(x)$ often appears as it may be used to calculate sums of functions for points in a subset B of $\mathbb{R}^d$. The indicator function is defined as

$$\mathbf{1}_B(x) = \begin{cases} 1 & \text{for } x \in B, \\ 0 & \text{otherwise,} \end{cases}$$

where B is some subset of the space $\mathbb{R}^d$. Then

$$\sum_{x \in N \cap B} f(x) = \sum_{x \in N} \mathbf{1}_B(x) f(x) \tag{1.5.3}$$

is the sum of the $f(x_i)$ restricted to the x_i in B. Furthermore, we have

$$N(B) = \sum_{x \in N} \mathbf{1}_B(x).$$

In order to avoid clumsy notation in the context of statistical estimators, point sums referring to the observation window W are described by the following simplified notation:

$$\sum_{x \in N} \mathbf{1}_W(x) f(x) = \sum_{x \in W} f(x) \tag{1.5.4}$$

and

$$\sum_{x,y \in N} \mathbf{1}_W(x) \mathbf{1}_W(y) f(x, y) = \sum_{x,y \in W} f(x, y). \tag{1.5.5}$$

Number distributions of a point process

Classical statistics usually uses a single distribution to describe a specific phenomenon. This distribution may be discrete if a random integer-valued variable is analysed, or continuous if a continuous random variable is analysed. A point process, however, can be described by infinitely many random variables. These are discussed in detail in the following.

The most fundamental of these functions are the *number distributions* given by

$$\mathbf{P}(N(B) = n) \qquad \text{for } n = 0, 1, \ldots,$$

and

$$\mathbf{P}(N(B_1) = n_1, \ldots, N(B_k) = n_k) \qquad \text{for } n_1 = 0, 1, \ldots, \ n_k = 0, 1, \ldots.$$

The first term describes the probability that there are exactly n points in the set B and the second that there are exactly $n_1, \ldots, n_k$ points in k sets $B_1, \ldots, B_k$. The first probability is described by a univariate distribution, the second by a multivariate distribution. There are infinitely many sets B and the probabilities can differ among these sets. Hence, there are infinitely many number distributions that describe a point process.

A specific type of these probabilities are the emptiness or *void probabilities* describing the probability that there are no points in a specific subset B, i.e.

$$\mathbf{P}(N(B) = 0).$$

For example, if $B = b(x, r)$ is the sphere (or disc) of radius r centred at x, then $\mathbf{P}(N(b(x, r)) = 0)$ is the probability that this disc does not contain any points. This probability may also be interpreted as the probability that the distance between the nearest point of N to the position x is larger than r. Using this interpretation and fixing x, this probability can be regarded as a function of r, leading to the *spherical contact distribution function* or location-to-nearest-point distance d.f.

$$H_{s,x}(r) = 1 - \mathbf{P}(N(b(x, r)) = 0) \qquad \text{for } r \geq 0. \qquad (1.5.6)$$

It is discussed in detail in Section 4.2.5.

Point process distribution

At some points in this book even more general probabilities for point processes are considered that have the form

$$\mathbf{P}(N \in \mathcal{A}).$$

This describes the probability that the point process N is in the set $\mathcal{A}$, where $\mathcal{A}$ is a set of point patterns with a specific property. If, for example, $\mathcal{A}$ is the set of all point patterns with no points in the set B, then

$$\mathbf{P}(N \in \mathcal{A}) = \mathbf{P}(N(B) = 0).$$

Mean numbers for point processes

In classical statistics, mean values are a fundamental concept. This is similar in the context of point processes, where mean numbers in fixed sets are particularly important. The value $N(B)$ for a set B is a random variable and, if B is bounded, it makes sense to consider the mean $\mathbf{E}(N(B))$, where $\mathbf{E}$ is the symbol for *expectation*; see Appendix A. Thus

$$\mathbf{E}(N(B)) = \text{mean number of points of } N \text{ in } B.$$

Clearly, this mean depends on the set B, and it is therefore a (deterministic) function operating on sets (more precisely, a measure). Therefore the notation

$$\Lambda(B) = \mathbf{E}(N(B)) \tag{1.5.7}$$

is used and Λ is called the *intensity measure*.

Under some continuity conditions, which are usually satisfied in practical applications of point process statistics, a density function $\lambda(x)$ exists that is called the *intensity function* with

$$\Lambda(B) = \int_B \lambda(x)\mathrm{d}x. \tag{1.5.8}$$

Remarks

(1) The continuity condition is violated if, for example, the points are arranged on a lattice.

(2) The integral in (1.5.8) should be interpreted as a volume integral (in $\mathbb{R}^d$), where $\mathrm{d}x$ is the volume element. Using more traditional notation, the integral could be written as

$$\int_B f(x)\mathrm{d}A, \int_B f(x)\mathrm{d}V \quad \text{or even} \quad \int_B f\mathrm{d}A, \int_B f\mathrm{d}V.$$

Statistically estimated intensity functions will be discussed on pp. 116 and 289.

It is clear that $\lambda(x)$ is proportional to the point density around a location x. If $\mathrm{d}x$ is the volume of an infinitesimal sphere centred at x, then $\lambda(x)\mathrm{d}x$ is the probability that there is a point in this sphere.

Proof. Let $b(x)$ be this small sphere. It is so small that $\mathbf{P}(N(b(x)) \geq 2)$ can be ignored. Then

$$\int_{b(x)} \lambda(x)\mathrm{d}x \approx \lambda(x)\mathrm{d}x = \mathbf{E}(N(b(x))) = 0 \cdot p_0 + 1 \cdot p_1 = p_1$$

with $p_i = \mathbf{P}(N(b(x)) = i)$.

Conditional intensity

Many important point process models are defined in terms of a refined version of the intensity function $\lambda(x)$, the so-called *Papangelou conditional intensity* $\lambda(x|\mathcal{X})$, where x is a deterministic location and $\mathcal{X}$ a point pattern. The conditional intensity is fundamental for example for Gibbs processes (see Sections 3.6 and 6.6), where explicit formulas for $\lambda(x|\mathcal{X})$ can be given. It is also an important tool in the context of simulations of these processes (see Section 3.6).

Loosely speaking, $\lambda(x|\mathcal{X})dx$ is the conditional probability that there is a point of N in an infinitesimal sphere $b(x)$ of volume dx containing x, given the realisation $\mathcal{X}$ of N outside of $b(x)$. The quantity $\lambda(x|N)$ is a random variable with mean

$$\mathbf{E}(\lambda(x|N)) = \lambda(x) \qquad \text{for } x \in \mathbb{R}^d.$$

Point density distribution function

Sometimes the intensity function $\lambda(x)$ is rather irregular such that it may be useful to consider the *point density distribution function* $G(t)$ (Ghorbani et al., 2006) defined as

$$G(t) = \frac{\nu(W_t)}{\nu(W)}, \tag{1.5.9}$$

where ν is area (volume), W is the window of observation for the point process and W_t the subset of W defined as

$$W_t = \{x \in W : \lambda(x) \le t\},$$

i.e. the set of locations where the intensity is below the threshold t. In other words, $G(t)$ is simply the fraction of W where the intensity function is smaller than t. Section 3.3.3 presents an example where the point density d.f. is applied.

The Campbell theorem

The mean value of a sum S_f with non-negative $f(x)$ as in (1.5.1) can be calculated in a very elegant way using the *Campbell theorem,* which states that

$$\mathbf{E}S_f = \mathbf{E}\left(\sum_{x \in N} f(x)\right) = \int f(x)\lambda(x)dx. \tag{1.5.10}$$

In other words, it suffices to know $f(x)$ and the intensity function $\lambda(x)$ for the calculation of $\mathbf{E}S_f$.

Remarks. The formula is clearly true for all indicator functions $f(x) = \mathbf{1}_B(x)$. For these functions the integral in (1.5.10) is

$$\int \mathbf{1}_B(x)\lambda(x)dx = \Lambda(B),$$

and $ES_{1_B} = \Lambda(B)$ by definition of the intensity measure Λ because of

$$S_{1_B} = \sum_{x \in N} 1_B(x) = N(B).$$

Those readers familiar with measure theory will know that there are theorems that may be used to show the same for general non-negative measurable function f.

In Chapters 4 and 5 the Campbell theorem is applied repeatedly in the simple case of a constant $\lambda(x)$.

For Example 1.1 the mean values are

(1) $ES_f = \int md(\|x\|)\lambda(x)dx,$

(2) $ES_f = \int p(\|y - x\|)\lambda(x)dx,$

for seed density and bird counting respectively.

Variances and higher-order moments

In classical statistics, variances are very important and fundamental distributional parameters in addition to means. This is similar for point processes. Thus the variance $\mathbf{var}N(B)$ of the random variable $N(B)$ may be considered, given by

$$\mathbf{var}N(B) = \mathbf{E}(N(B) - \mathbf{E}N(B))^2$$

or

$$\mathbf{var}N(B) = \mathbf{E}(N(B) - \Lambda(B))^2, \tag{1.5.11}$$

$$\mathbf{var}N(B) = \mathbf{E}(N(B))^2 - \Lambda(B)^2. \tag{1.5.12}$$

Clearly, the numerical value of $\mathbf{var}N(B)$ depends on the set B, and thus a point process is associated with infinitely many variances (for the infinitely many subsets B of $\mathbb{R}^d$). However, this set function is not a measure as it is not additive.

Higher-order moments may also be considered, i.e. moments of the form

$$\mathbf{E}(N(B))^k \qquad \text{for } k = 2, 3, \ldots.$$

Furthermore, the correlations of the numbers of points for different sets can be studied. Consider two sets B_1 and B_2. For the two random variables $N(B_1)$ and $N(B_2)$, the covariance and correlation coefficient can be defined:

$$\mathbf{cov}(N(B_1), N(B_2)) = \mathbf{E}((N(B_1) - \Lambda(B_1))(N(B_2) - \Lambda(B_2))) \tag{1.5.13}$$

and

$$\mathbf{corr}(N(B_1), N(B_2)) = \frac{\mathbf{cov}(N(B_1), N(B_2))}{\sqrt{\mathbf{var}N(B_1)\mathbf{var}N(B_2)}}. \tag{1.5.14}$$

These characteristics can be calculated if the so-called moment measures are known.

There are two families of moment measures, the *(normal) moment measures* $\mu^{(k)}$ and the *factorial moment measures* $\alpha^{(k)}$.

Moment measures

The kth-order moment measure of a point process N is the measure $\mu^{(k)}$ defined by

$$\int_{\mathbb{R}^{nd}} f(x_1, \ldots, x_n)\mu^{(n)}(\mathrm{d}(x_1, \ldots, x_n)) = \mathbf{E}\left(\sum_{x_1, \ldots, x_n \in N} f(x_1, \ldots, x_n)\right), \tag{1.5.15}$$

where $f(x_1, \ldots x_k)$ is any non-negative measurable function on $\mathbb{R}^{nd}$. In particular,

$$\mu^{(k)}(B_1 \times \cdots \times B_k) = \mathbf{E}\left(N(B_1)\cdots N(B_k)\right)$$

and, if $B_1 = \cdots = B_k = B$,

$$\mu^{(k)}(B^k) = \mathbf{E}\left(N(B)^k\right).$$

Thus $\mu^{(k)}$ yields the kth moment of the real-valued random variable $N(B)$, which is the number of points in B. Special cases are

$$k = 1: \ \mu^{(1)}(B) = \mathbf{E}\left(N(B)\right) = \Lambda(B),$$
$$k = 2: \ \mu^{(2)}(B_1 \times B_2) = \mathbf{E}\left(N(B_1)N(B_2)\right),$$
$$\mathbf{var}(N(B)) = \mu^{(2)}(B \times B) - (\Lambda(B))^2.$$

The covariance of the random variables $N(B_1)$ and $N(B_2)$ can also be expressed in terms of the moment measure:

$$\mathbf{cov}(N(B_1), N(B_2)) = \mathbf{E}\left(N(B_1)N(B_2)\right) - \mathbf{E}\left(N(B_1)\right)\mathbf{E}(N(B_2))$$
$$= \mu^{(2)}(B_1 \times B_2) - \Lambda(B_1)\Lambda(B_2).$$

Factorial moment measures

The kth-order factorial moment measure $\alpha^{(k)}$ of the point process N is defined by

$$\int f(x_1, \ldots, x_k)\alpha^{(k)}(\mathrm{d}(x_1, \ldots, x_k)) = \mathbf{E}\left(\sum_{x_1, \ldots, x_k \in N}^{\neq} f(x_1, \ldots, x_k)\right). \tag{1.5.16}$$

Here f is any non-negative function on $\mathbb{R}^{kd}$, and the sum $\sum^{\neq}$ is a sum of all k-tuples of *distinct* points in N including all permutations of given points. This is where $\alpha^{(k)}$ and $\mu^{(k)}$ differ: in (1.5.16) the sum omits all k-tuples in which two or more entries are the same. Hence if $B_1, \ldots, B_k$ are pairwise disjoint sets then

$$\mu^{(k)}(B_1 \times \cdots \times B_k) = \alpha^{(k)}(B_1 \times \cdots \times B_k). \tag{1.5.17}$$

For $k = 2$,

$$\mu^{(2)}(B_1 \times B_2) = \Lambda(B_1 \cap B_2) + \alpha^{(2)}(B_1 \times B_2) \tag{1.5.18}$$

by the definitions of $\mu^{(2)}$, Λ and $\alpha^{(2)}$, and using

$$\sum_{x_1, x_2 \in N} \mathbf{1}_{B_1}(x_1) \mathbf{1}_{B_2}(x_2) = \sum_{x_1 \in N} \mathbf{1}_{B_1 \cap B_2}(x_1) + \sum_{x_1, x_2 \in N}^{\neq} \mathbf{1}_{B_1}(x_1) \mathbf{1}_{B_2}(x_2).$$

Since

$$\alpha^{(k)}(B^k) = \mathbf{E}(N(B)(N(B) - 1) \cdots (N(B) - n + 1)),$$

the quantity $\alpha^{(k)}(B^k)$ is the so-called kth-order factorial moment of the random variable $N(B)$; this is the reason for the name 'factorial moment measure'.

Product densities

The product density $\varrho^{(k)}$ describes the frequency of possible configurations of k points. Suppose that $b_1, \ldots, b_k$ are pairwise disjoint spheres with centres $x_1, \ldots, x_k$ and infinitesimal volumes $dV_1, \ldots, dV_k$. Then $\varrho^{(k)}(x_1, \ldots, x_k)dV_1 \cdots dV_k$ is the probability that there is a point of N in each of the $b_1, \ldots, b_k$. Technically, $\varrho^{(k)}$ is defined if some continuity properties are satisfied for $\alpha^{(k)}$ that usually hold in applications, local finiteness and absolute continuity with respect to the Lebesgue measure ν_{kd}. Then $\alpha^{(k)}$ has a density $\varrho^{(k)}(x_1, \ldots, x_k)$, the kth *product density*:

$$\alpha^{(k)}(B_1 \times \cdots \times B_k) = \int_{B_1} \cdots \int_{B_k} \varrho^{(k)}(x_1, \ldots, x_k)dx_1 \cdots dx_k. \tag{1.5.19}$$

Moreover, for any non-negative bounded function $f(x_1, \ldots x_k)$,

$$\mathbf{E}\left(\sum_{x_1, \ldots x_k \in N}^{\neq} f(x_1, \ldots, x_k) \right)$$
$$= \int \cdots \int f(x_1, \ldots, x_k) \varrho^{(k)}(x_1, \ldots, x_k) dx_1 \cdots dx_n. \tag{1.5.20}$$

Historically, product densities were introduced earlier than moment measures, probably because of the intuitive interpretation above.

For $k = 1$ the product density coincides with the intensity function $\lambda(x)$. In the important case of $k = 2$ the exponent '(2)' is often omitted. In this case it is possible that $\varrho^{(2)}(x_1, x_2)$ depends only on the distance r of x_1 and x_2, and then the simple symbol $\varrho(r)$ is used.

For a general point process, it is difficult to work with moment measures. The situation changes if the point process is stationary and isotropic. In this case the second-order moments can be expressed by functions such as Ripley's K-function and the pair correlation function; see Section 4.3.

Further distributional characteristics

The theory of point processes discusses further distributional characteristics such as the characteristic function, the Laplace functional and the Bartlett spectrum. None of these is covered in this book: the first two have rarely been used in spatial statistics, while a thorough discussion of the third would take up too much space. The Bartlett spectrum is of value in particular when many large patterns have to be analysed and when the data have been derived from image analysis. The Bartlett spectrum contains the same information as the pair correlation function, but in a different form. It is popular among physicists since it can be measured physically, e.g. by X-ray small-angle scattering. A classical reference is Bartlett (1964) and recent ones are Renshaw (1997, 2002), Mugglestone and Renshaw (1996a, 1996b, 2001) and Ohser and Schladitz (2008).

Marked point processes

Marked point processes are generalisations of point processes and are highly relevant in practical applications. Each point x_i is assigned a further quantity $m(x_i)$, which provides additional information on the object represented by the point, as discussed in the examples in Section 1.2.1 (qualitative marks, cell types) and 1.2.2 (quantitative marks, particle diameters). In this book the $m(x_i)$ are usually integers or real numbers, but much more general marks may also be considered. A marked point process is denoted by M. In the same way as for N, there are two different mathematical interpretations of M:

- M is a function operating on sets or a random counting measure. For a subset B of $\mathbb{R}^d$ and a subset C of $\mathbb{R}$, $M(B \times C)$ denotes the random number of marked points $[x; m(x)]$ with $x \in B$ and $m(x) \in C$. If $C = \mathbb{R}$ the measure M counts all points and ignores the marks, i.e.

$$M(B \times \mathbb{R}) = N(B),$$

where N is the point process of the points in M without the marks.

- *M* denotes a set of marked points,

$$M = \{[x_i; m(x_i)]\} \quad \text{or} \quad M = \{[x_1; m(x_1)], [x_2; m(x_2)], \dots \}.$$

The notation $[x; m(x)] \in M$ means that the marked point $[x; m(x)]$ is in M.

For marked point processes sums similar to those in (1.5.1) may be considered:

$$\begin{aligned}
S_f &= f(x_1, m(x_1)) + f(x_2, m(x_2)) + \cdots \\
&= \sum_{(i)} f(x_i, m(x_i)) = \sum_{[x; m(x)] \in M} f(x, m(x)),
\end{aligned} \quad (1.5.21)$$

where $f(x, m(x))$ is a real number assigned to $[x; m(x)]$. By including marks in the analysis of a point pattern, many spatial phenomena can be studied and described much more clearly and realistically. Consider the following examples in comparison to Example 1.1 above and note how the description of the situations benefits from the additional information contained in the marks.

Example 1.2. Marked point process sums

(1) *Seed density.* Assume that the points x_i represent the locations of plants and each plant produces seeds independently of the other plants. The seeds are dispersed randomly around the parent plants, in the same way for all plants. Seed dispersal depends on tree-dependent parameters, $m(x_i)$, which can be different for different plants. The density of seeds from a plant at position x at a location y depends on the distance r of y from x and is denoted by $m(x)d(r)$ with, for example, $d(r) = \exp(-cr)$. Clearly, the total seed density at position y can be calculated as

$$S_f = \sum_{(i)} m(x_i)d(\|y - x_i\|).$$

Note that the constant m in Example 1.1 has now been replaced by the variable $m(x_i)$.

(2) *Counting of birds by distance sampling.* Let the points x_i be locations of birds in a forest and the marks $m(x_i)$ the volumes of their song. A bird singing at position x at volume $m(x)$ is still audible at a point y of distance r from x with probability $p(r) = 1 - ar/m(x)$ for $r \le m(x)/a$. The mean number of birds heard by a bird-watcher at position y is given by

$$S_f = \sum_{(i)} (1 - a\|y - x_i\|/m(x_i)).$$

The mean behaviour of marked point processes can also be described by an intensity function $\lambda(x, m)$: the mean number of marked points of M located in the set B and with marks in C is given by

$$\mathbf{E}(M(B \times C)) = \int_{B} \int_{C} \lambda(x, m) \mathrm{d}m\mathrm{d}x. \qquad (1.5.22)$$

Second-order moments for marked point processes can be defined in a similar way as for the non-marked processes. This is discussed in detail for stationary point processes in Chapter 5.

1.6 Stationarity and isotropy

1.6.1 Model approach and design approach

In classical probability theory a random variable describes all possible outcomes of an experiment; in mathematical terms, it is a mapping from some probability space into some state space. In a rather abstract sense, a point process is also a random variable.

A simple example of a random variable describes the outcome of a roll of a die. Here the random variable takes on the values 1 to 6, and if the die is fair the probability for each of the six values is $\frac{1}{6}$. Now assume that it is possible to construct a die with infinitely many sides, with a two-dimensional point pattern on each of the sides. Every time the die is rolled, a point pattern is generated and the die represents the point process. Now assume somebody is observing the rolling of the die. This person sees different point patterns in the plane and, in particular, fluctuating values of the number of points in some fixed, deterministic subset B of the plane. In other words, every roll of the 'point pattern die' produces a different realisation of the random variable $N(B)$. This interpretation of a point process is called the *model approach:* the observer's position is deterministic, while the point process is random, producing different patterns in different observations or experiments.

The *design approach* takes a different perspective. Here the point pattern is deterministic and fixed, while the observer's location changes randomly. That is to say, if the observer's location is x and if the observation area is the disc $b(x; r)$ of radius r centred at x, 'the number of points in a disc of radius r' for a random x is also a random variable $N(b(x, r))$ as in the model approach.

This book is mainly written from the point of view of the model approach. However, experience of teaching non-mathematicians shows that the design approach is more natural than the model approach to many of these. This might be because they typically analyse one specific point pattern (a specific forest, some part of the sky, the whole universe, ...) or a small number of patterns (as in materials science or medicine) and usually the system of Cartesian coordinates is chosen arbitrarily.

However, when stationarity and ergodicity are assumed (see below), both approaches are equivalent and using the model approach is more than justified as it lends itself very well to explaining the mathematical details. Prior to a detailed discussion of these properties a few words on finite and infinite point processes may be useful.

1.6.2   Finite and infinite point processes

In the real world, all point patterns are finite. Nevertheless, it makes sense to interpret point patterns as parts of infinite patterns and to use *models* that assume infinite point processes. Whether an infinite point process model is used or not depends on the specific situation. Some point patterns have to be regarded as samples from a finite point process, because they represent phenomena that are strictly locally limited. Examples are:

- the bullet marks on a target;
- the location of seeds dispersed around a single plant;
- the centres of air pores in a piece of Swiss cheese.

In other cases, a point pattern may be regarded as a part of a much larger pattern in which the points are distributed according to the same laws as in the observation window. Examples include:

- the positions of trees in a forest;
- the grain or pore centres in homogeneous probes of materials such as metals or ceramics;
- the location of seeds dispersed by a large community of plants.

In the example of the forest the observation window may be 'biologically homogeneous', i.e. within the window the growth conditions are the same for all trees. The window might be surrounded by a larger forest area. In other words, the edge of the observation window is to some extent arbitrary and has no biological meaning. The trees close to the edge of the observation window do not behave differently from those closer to the centre of the window. Also, given equal environmental conditions, the pattern in the observation window might be continued infinitely in all directions. This means that the trees interact in the same way everywhere and generate similar fluctuations of local tree density and tree parameters (or marks).

In the seed example with a single plant that disperses seeds, the seed pattern is closely related to the specific parent plant. However, if there are many parent plants in a homogeneous pattern the combined seed pattern (i.e. the superposition of the patterns formed by the seeds from all trees) is also homogeneous. It may be interpreted as continuing infinitely beyond the observation window.

It might come as a surprise to some readers that the reason for considering infinite point processes is the fact that studying these processes is mathematically simpler than studying most finite point processes, if stationarity (and isotropy) is assumed. And, point process theory typically starts with infinite point processes, rather than considering infinite processes as limits of finite process; physicists would call these 'thermodynamic limits'.

1.6.3 Stationarity and isotropy

A point process N is called *stationary* if N and the translated point process N_x have the same distribution for all translations x. This is written as

$$N \stackrel{d}{=} N_x, \tag{1.6.1}$$

where N_x is the point process resulting from a shift of all points of N by the same vector x; if $N = \{x_1, x_2, \ldots\}$ then $N_x = \{x_1 + x, x_2 + x, \ldots\}$. The expression 'have the same distribution' means

$$\mathbf{P}(N(B_1) = n_1, \ldots, N(B_k) = n_k) = \mathbf{P}(N_x(B_1) = n_1, \ldots, N_x(B_k) = n_k)$$
$$= \mathbf{P}(N(B_1 - x) = n_1, \ldots, N(B_k - x) = n_k), \tag{1.6.2}$$

where the first equality in the second line holds because $N_x(B) = N(B - x)$ for all B and x, where $B - x = \{y - x : y \in B\}$ is the set B shifted by vector $-x$. (The number of points resulting from fixing the observation set B and counting the points of the shifted process N_x is equal to the number of points obtained when fixing the point process and shifting the observation set in the opposite direction.) Equation (1.6.1) implies

$$N(B) \stackrel{d}{=} N(B_x) \qquad \text{for all } B \text{ and } x, \tag{1.6.3}$$

i.e. the numbers of points in B and in the shifted set B_x have the same distribution. This has important consequences for the distributional characteristics of the point process. In particular, the intensity measure simplifies substantially (see Section 4.1):

$$\Lambda(B) = \lambda \nu(B), \tag{1.6.4}$$

i.e. the intensity measure is a multiple of the area or volume. The constant λ is called *intensity* or *point density* and may be interpreted as the mean number of points per unit volume.

Note that some readers may prefer the term 'homogeneous' to 'stationary'. Use of the term 'stationary' is linked to the use of the term 'point process' rather than 'point field'. The physicist Torquato (2002) uses the term 'statistically homogeneous', since many physicists and engineers use the word 'homogenous' in a different way. To them homogeneous means 'uniform' in a sense which is natural but difficult to quantify. In this context, the point patterns in Figures 1.2 and 1.3 would perhaps be interpreted as 'homogeneous', but those in Figures 1.4 and 6.3 would not. However, the term 'stationary' is used in this book in a sense such that all four patterns are likely to be regarded as samples from stationary point processes.

In terms of the design approach, stationarity means that the chance of observing some point configuration at a specific location is independent of the location.

A point pattern may deviate from stationarity in several ways:

- The intensity function $\lambda(x)$ or point density may not be constant but vary systematically. Consider, for example, the tree density in mountain forests, which decreases with increasing altitude, or see Figure 4.49.

- The local point configurations may be location-dependent. For example, in a cluster point process the cluster size may be location-dependent, or the points may be aggregated within one subregion and random in another. The marks may also be location-dependent. For example, in mountain forests some tree species may not grow in higher altitudes.

Note that it is impossible to prove rigorously that a specific point pattern is a sample from a stationary point process. Statistical tests can assess only some aspects of stationarity but never all. In other words, accepting the stationarity hypothesis for a given point pattern based on some test is only a necessary condition. Some point patterns look deceptively non-stationary even though they are samples from stationary processes, in particular for small observation windows. Consider, for example, a cluster process with large clusters observed in a small window such that it can contain at most one cluster. Similarly, samples from a non-stationary process may exhibit a local behaviour that is similar to the behaviour of a stationary process. For these reasons it is very helpful to justify stationarity based on non-statistical scientific arguments. In forestry, for example, an argument in favour of using methods for stationary processes might be that soil and climatic conditions do not vary within a research plot.

In many applications which are aimed at exploring the interaction among points the observation window may be chosen prior to the analysis. In this case, one may try to choose a homogeneous sample, in order to guarantee stationarity as an a priori property.

Isotropy is a concept that is analogous to stationarity. Rather than translations by vectors, rotations around the origin are considered here. In the planar case a rotation is described by an angle α between $0°$ and $360°$. If $x = (\xi, \eta)$ is a point in

$\mathbb{R}^2$ having the coordinates ξ and η then the rotated point $R_\alpha x$ has coordinates

$$\xi_\alpha = \xi \cos \alpha + \eta \sin \alpha \quad \text{and} \quad \eta_\alpha = -\xi \sin \alpha + \eta \cos \alpha.$$

A point process N is called *isotropic* if

$$N = \{x_1, x_2, \ldots\} \quad \text{and} \quad R_\alpha N = \{R_\alpha x_1, R_\alpha x_2, \ldots\}$$

have the same distribution for all α.

A point process that is both stationary and isotropic is called *motion-invariant*. The distribution of these processes is invariant also with respect to rotations around arbitrary points.

Note that there are processes which are non-stationary but isotropic with respect to a fixed location. A good example of this are the locations of fungi around a tree (see Byth, 1981; see also 'centred processes' in Section 3.3 below). Sections 4.5 and 5.4 present methods for analysing anisotropies in point patterns entitled 'orientation analysis'.

Weaker stationarity properties are sometimes assumed, e.g. certain forms of 'second-order stationarity'. This means that a process is not stationary but its second-order characteristics behave similarly to those of a stationary process; see Section 4.10.

1.6.4 Ergodicity

In addition to stationarity (or motion-invariance) ergodicity is also frequently assumed in point process statistics. If a point process is ergodic, it suffices to analyse one sample (i.e. one point pattern) of an appropriate size to obtain statistically meaningful results. Physicists would express this as 'the spatial average is equal to the time average'. In other words, the average over one large sample yields the same result as the average resulting from many (small) samples. For example, for ergodic point processes

$$\lim_{W \uparrow \mathbb{R}^d} \frac{N(W)}{\nu(W)} = \lambda \quad \text{with probability 1.} \tag{1.6.5}$$

Here λ is the intensity of the point process N, the mean number of points in the unit square or cube, which in this book is denoted by $\boxed{1}$. The expression $W \uparrow \mathbb{R}^d$ means that the window W converges in a reasonable sense towards the whole space; for example, W contains a sphere of radius r and r converges to ∞. A precise definition of ergodicity is beyond the scope of this book; see Stoyan et al. (1995). Note however that the model approach and the design approach yield the same results if a specific point process is ergodic.

An example of a non-ergodic point process is a process the samples of which are samples of stationary processes of intensity λ_i with probability p_i for $i = 1, 2$

and $p_1 + p_2 = 1$ and different intensities λ_1 and λ_2. This process is stationary, but (1.6.5) does not hold as the limit results in either λ_1 or λ_2. Lattice point processes where the points are at the nodes of a regular lattice, which may as a whole be randomly positioned in the space, are not ergodic either. A homogeneous Poisson process (see Chapter 2) is ergodic.

A sufficient condition for ergodicity is the following *mixing* property. A stationary point process N is mixing if, for all point process properties $\mathcal{A}$ and $\mathcal{B}$,

$\mathbf{P}$ ('N has the property $\mathcal{A}$ and N_x has the property $\mathcal{B}$')

$\rightarrow \mathbf{P}$ ('N has the property $\mathcal{A}$') $\mathbf{P}$ ('N_x has the property $\mathcal{B}$'),   as $\|x\| \rightarrow \infty$.

N_x is the point process N translated by the vector x (see p. 37). For example, $\mathcal{A}$ may mean that the sphere $b(o, r)$ of radius r centred at o does not contain any point and $\mathcal{B}$ may mean that the same sphere contains two points. Then, for a mixing N,

$$\mathbf{P}(N(b(o, r)) = 0, N(b(-x, r)) = 2) \rightarrow \mathbf{P}(N(b(o, r)) = 0) \cdot \mathbf{P}(N(b(o, r)) = 2).$$

In some sense, mixing means that distant parts of a point process are independent.

Gaussian distributions, central limit theorem

A famous theorem in probability theory says that the sums of many random variables are asymptotically normally distributed. Similar theorems have also been proved for the numbers of points in large windows (which can be regarded as numbers of points in many subwindows); see Ivanoff (1982), Heinrich and Schmidt (1985) and Heinrich (1986).

1.7   Summary characteristics for point processes

One of the most important objectives in all areas of statistics is to summarise data sets, i.e. to describe samples by a small number of numerical and functional characteristics. In classical univariate statistics, these characteristics include $\bar{x}$, s^2 and the empirical distribution function $\hat{F}_n(x)$. Summary characteristics have an important role in exploratory data analysis, since they reveal valuable information and describe important characteristics of the underlying distribution and may be used to identify suitable stochastic models. Typically, they are empirical analogues or estimators of theoretical characteristics such as the mean $\mathbf{E}X$, the variance $\mathbf{var}X$ and the distribution function $F(x)$. Not surprisingly, summary characteristics have also been defined and frequently used in point process statistics.

The summary characteristics discussed in the context of spatial point processes may be classified in two ways. One may consider *numerical* or *functional* but also *location-related* or *point-related* summary characteristics. In this chapter only a few examples of summary characteristics for stationary point processes are discussed in order to introduce general ideas. Throughout this book summary statistics are discussed in more detail and for different types of processes.

1.7.1 Numerical summary characteristics

The most important numerical summary characteristic for a stationary point process is the *intensity* λ, the mean number of points per unit area or volume. Intensity may be interpreted as a location-related characteristic: consider a test point or location x and the mean number of points in the sphere $b(x, r)$ with some radius r. This is $\lambda \nu(b(o, r)) = \lambda b_d r^d$, where b_d is the volume of unit sphere, i.e. this mean is essentially given by λ.

Many other numerical summary characteristics are point-related. If the points are marked by real numbers, the *mean mark* is an important point-related characteristic. Valuable characteristics may be defined based on constructed marks. For example, the distances $d(x)$ to the nearest neighbours of the points x may be used to calculate the mean nearest-neighbour distance m_D, a useful point-related numerical summary characteristic. The corresponding location-related characteristic is the mean distance from a test location to the nearest point process point.

A particularly important class of numerical summary characteristics are the *indices* that have typically been developed in biological contexts, and are often referred to as competition indices. They are usually point-related indices and are based on constructed marks using the points in some zone of influence or local neighbourhood of the reference point. This zone of influence can be determined either based on the k nearest neighbours or on a disc whose radius r determines the distance up to which the individual represented by the reference point interacts with neighbouring individuals. The constructed mark $c(x_i)$ for the reference point x_i is defined as

$$c(x_i) = \sum_{j=1}^{k} c_{ij} \quad \text{or} \quad c(x_i) = \sum_{j=1}^{n_i(r)} c_{ij},$$

where

$$c_{ij} = f(\|x_i - x_j\|, m(x_i), m(x_j)),$$

i.e. c_{ij} is calculated by means of a function f using the distance between the points x_i and x_j and their marks; $n_i(r)$ is the number of points in the disc $b(x_i, r)$. The index is then the mean mark corresponding to $c(x_i)$.

Consider, for example, the *mingling index* $\overline{M}_k$ based on

$$\overline{M}_k(x_i) = \frac{1}{k} \sum_{j=1}^{k} \mathbf{1}(m(x_i) \neq m(x_j))$$

for the case of discrete (integer-valued) marks (see Section 5.2.4 for more details). Here $\overline{M}_k(x_i)$ is simply the proportion of the k nearest neighbours of x_i having a mark different from that of x_i.

1.7.2 Functional summary characteristics

Probably the most important location-dependent functional summary characteristic is the spherical contact d.f. $H_s(r)$.[3] The 'location' x is used as the centre of a disc or sphere of radius r and the characteristic describes the probability that $b(x, r)$ is not empty, that it contains at least one point of N. When stationarity is assumed it suffices to consider $x = o$. (Or, in design-approach terminology, the observation point is used as the origin of the coordinate system.) Thus

$$F(r) = H_s(r) = 1 - \mathbf{P}(N(b(o, r)) = 0) \qquad \text{for } r \geq 0.$$

Important point-dependent functional summary characteristics are the nearest-neighbour distance d.f. $D(r)$[4] and Ripley's K-function $K(r)$. These describe the distribution of the distance from the points to their nearest neighbours and the mean number of points within distance r from points, respectively. A precise definition of these is based on the theory of Palm distributions; see Section 4.1. Mark distributions of constructed marks such as the $c(x_i)$ above may also be used as functional summary characteristics.

1.8   Secondary structures of point processes

1.8.1 Introduction

In the analysis of a spatial point pattern it can be very useful to apply methods that, strictly speaking, do not form part of point process statistics. To this end, new geometrical structures (referred to as 'secondary structures' in the following) are constructed based on the points in the pattern, and statistical methods suitable for the specific type of geometric structure are applied. The following secondary structures have been successfully used in this context:

- random sets,

- random fields (in particular, shot-noise fields),

- tessellations (in particular, the Voronoi tessellation),

- networks or graphs (in the graph-theoretic sense).

---

[3] In the literature $F(r)$ is often used for the spherical contact d.f., also called the empty-space function. The H in the symbol used here refers to 'hit'.

[4] The nearest-neighbour distance d.f. was originally denoted $G(r)$ by Diggle (1979), apparently simply because G follows F in the alphabet. The alert reader will recall that G denotes the point density d.f. in this book. The D used here refers to 'distance'.

The first two approaches assign a regionalised variable to a point process, i.e. a variable that has a value at every location in space. Applications include the analysis of correlations between point processes and covariates that are represented by regionalised variables, e.g. the correlation between the density of plant positions and other spatial variables. Refer to Example 6.7 for an ecological application. This section introduces each of the four secondary structures listed above in detail.

1.8.2 Random sets

A random set X is a set in the sense of mathematical set theory which depends on chance. Rigorous definitions of random (closed) sets can be found in Molchanov (2005) and Stoyan et al. (1995). The following is an important example, which is highly relevant in the context of point process statistics, is easy to understand and does not require reference to mathematical theory. Let N be a point process in $\mathbb{R}^2$. Take the points of N as centres of (closed) discs and consider the union of all these discs. This results in a pattern that is similar to the example shown in Figure 1.11.

The random set X_r may be described mathematically as

$$X_r = \bigcup_{x \in N} b(x, r) = N \oplus b(o, r), \tag{1.8.1}$$

where $\oplus$ denotes Minkowski addition; see Appendix B. In stochastic geometry this set is called a *germ–grain model*, where the 'germs' are the points x of N and the 'grains' are the discs $b(x, r)$. In the special case of a Poisson process the random

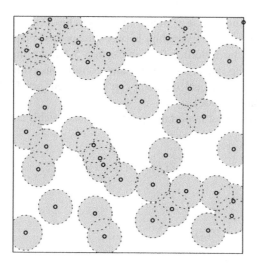

Figure 1.11   A set assigned to a point pattern. All points in Figure 1.5 are assigned discs of constant radius r, which can overlap, and the union is the set X_r shown in grey. This set provides interesting further information on the original point pattern.

set is called a *Boolean model* (with circular grains). Closed formulas are known for this model. Clearly, the model can be easily generalised to the three-dimensional and the general d-dimensional case.

But what additional information do these random sets provide to make them worth considering? Moving from a point process to a random set is useful since the random set describes aspects of the point distribution that other methods do not.

For example, returning to the random set shown in Figure 1.11, one may want to consider the area of the set X_r or its boundary length. The first approach results in a well-known functional summary characteristic, the spherical contact distribution function $H_s(r)$. Both this function and the second approach are discussed in Section 4.2.5. In addition, the topological properties of X_r may be analysed. For example, the *Euler number* of X_r or the *r-connectedness* of the points of N may be considered: two points of the point process are called r-connected if they are in the same component of X_r, i.e. if the points can be connected by a curve that lies entirely in the set X_r. It is particularly useful to consider statistical characteristics related to these properties, which depend on the radius r, as described in Mecke (2000).

1.8.3 Random fields

A random field $\{Z(x)\}$ is a family of random variables as explained in Appendix C and in books on geostatistics. In contrast to marked point processes, where the values $m(x_i)$ of the marks are given only for points x_i of the process, a random field has values $Z(x)$ in *all* $x \in \mathbb{R}^d$. Therefore random fields are also called *regionalised variables*. For a random set X one may put $Z(x)$ equal to 1 if $x \in X$ and 0 otherwise.

In point process statistics, random fields are secondary structures that result from regionalisation operations such as those described below. They are used to describe natural or technical spatial relationships between the points or have service functions in statistics such as the Bitterlich field in Section 5.2.2. An important application is the analysis of long-range correlations and correlations with covariates. In this context powerful geostatistical methods can be applied; see Chilès and Delfiner (1999) and Mandallaz (2000).

In the simplest and most important approach random fields are constructed based on point counts,

$$Z(x) = N(B + x), \qquad (1.8.2)$$

or mark sums,

$$Z(x) = \sum_{[x_i; m(x_i)] \in M} \mathbf{1}_B(x_i - x) m(x_i). \qquad (1.8.3)$$

In the first case the value of $Z(x)$ is equal to the number of points of N in the set $B + x$; if $B = b(o, r)$ then $Z(x)$ is the number of points in the sphere of radius r

centred at location x. In the second case a marked point process with quantitative marks $m(x)$ is considered. The value of $Z(x)$ is the sum of the marks of the points of M in the set $B + x$; if $B = b(o, r)$ then $Z(x)$ is the sum of the marks in the sphere of radius r centred at the location x.

More general and more interesting is what is called the *shot-noise field* $\{S(x)\}$. This random field describes a structure which results from the superposition of (random) *impulses,* also called *responses* or *effects,* which are related to the points of a (marked) point process N (M). These impulses are usually assumed to be homogeneous, i.e. they only depend on the mark of the point and the difference $x - x_i$ of the location x of interest and the point x_i. Then $s(x - x_i, m(x_i))$ is the contribution of point x_i to $S(x)$. Formally, $s(x, m)$ is the contribution of a point at origin o with mark m. In many cases a suitable name for the impulse function $s(x, m)$ is *attenuation function,* as its value decreases with the distance of x from o and really does describe an attenuation process. The $s(x, m)$ corresponding to (1.8.3) is $\mathbf{1}_B(-x)m$.

Superposition of the impulses of all points yields a shot-noise field:

$$S(x) = \sum_{[x_i; m(x_i)] \in M} s(x - x_i, m(x_i)) \qquad \text{for } x \in \mathbb{R}^d. \tag{1.8.4}$$

An important special case is

$$s(x - x_i, m(x_i)) = m(x_i) f(x - x_i),$$

where f is a probability density function. In this approach the 'point mass' $m(x_i)$ is continuously distributed around x_i.

However, there are also situations where the superposition follows the max-rule, i.e. the value at x is the maximum of the $s(x - x_i, m(x_i))$,

$$M(x) = \max_{[x_i; m(x_i)] \in M} s(x - x_i, m(x_i)) \qquad \text{for } x \in \mathbb{R}^d;$$

multiplicative superposition may also be considered, e.g. in the context of resource interference; see Wu et al. (1985).

Usually, the main aim is to determine characteristics of $\{S(x)\}$. Some formulas are given in Section 6.9 for the stationary case and independent marks. However, sometimes it is also interesting to know the value of $S(x)$ at a typical point of the point process N.

Example 1.3. Shot-noise fields

(1) *Seed density.* The points in N represent parent plants that disperse their seeds according to $s(x - x_i, m(x_i))$. This results in a random seed density field $\{S(x)\}$, where $S(x)$ is the seed density at the (deterministic) point x, e.g. the number of seeds in some quadrat centred at x.

(2) *Competition load* (Adler, 1996). The points in N again represent plants but now each plant competes with its neighbours. Clearly, the strength of the competition decreases with distance from the reference plant and a given point is influenced by the competition load from all its neighbours. The result is a random competition field $\{S(x)\}$, where $S(x)$ is the sum of the competition contributions from all plants in the vicinity of location x. In this context, the attenuation function $s(x, m)$ is called a 'local competition function'. In this example the value of the random field at a location x is probably less important than the total competition strength on the plants, i.e. the values $S(x_i)$ for $x_i \in N$.

(3) *Signal power in wireless communication* (Baccelli and Blaszczyszyn, 2001; Baccelli et al., 1997). The points x_i in N are fixed or mobile base-transceiver stations and the 'impulses' correspond to the signals emitted by antennas. A user of the communication network at location x is interested in the strength of the signals coming from the different antennas. The strength of the signals depends on the distances from the antennas and on noise. Of great interest is the existence of an antenna x_i with

$$s(x - x_i, m(x_i)) \left/ \sum_{k \neq i} s(x - x_k, m(x_k)) > \theta_i \right. ,$$

where θ_i is the so-called pilot-to-noise ratio of x_i.

The main benefit of referring to random fields is the fact that strong statistical methods have been developed for random fields which, by this approach, may also be exploited for point processes. See, for example, the remark on pair correlation function estimation on p. 233. Shot-noise fields are also used to construct point process models, the so-called shot-noise Cox processes; see Section 6.4.1.

1.8.4 Tessellations

A *tessellation* or *mosaic* divides the plane into non-overlapping polygons, or the space into polyhedra. Figure 1.12 shows an example of such a structure. Tessellations are often used as auxiliary structures, supporting the statistical analysis of point processes. In addition, tessellations are also relevant in the context of efficient computation algorithms and intensity estimation.

Perhaps the most important tessellation model is the *Voronoi tessellation*. Motivated by issues in number theory, Dirichlet (1850) and Voronoi (1908) considered regular tessellations of planes and higher-dimensional spaces. Dirichlet and Voronoi tessellations appear to have been applied independently in meteorology (Thiessen and Alter, 1911), metallurgy (Johnson and Mehl, 1939; Kolmogorov, 1937) and ecology (Matérn, 1960, 1986; Pielou, 1977); see Okabe et al. (2000), the standard

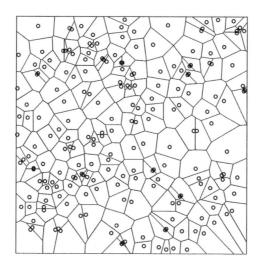

Figure 1.12 The Voronoi tessellation for the pattern of *Phlebocarya* (see Figure 1.4). All cells are convex polygons, and their area is small in regions of high point density.

'handbook' on the theory and application of tessellations,[5] for a historical sketch of the development of ideas.

The Voronoi tessellation is constructed with respect to a point process N in $\mathbb{R}^d$. Almost all x in $\mathbb{R}^d$ have a unique *nearest point* $n(x)$ in N. The *cell* $T(y)$ of a point y of N is defined by

$$T(y) = \left\{ x \in \mathbb{R}^d : n(x) = y \right\}.$$

The points on the boundary of cells have two or more nearest points in N. The cells $T(y)$ are all convex polygons but some can be unbounded. If all the polygons are bounded then the $T(y)$ constitute a tessellation of $\mathbb{R}^d$, the *Voronoi tessellation* relative to N. Some authors refer to the $d = 2$ case as the *Dirichlet* or *Thiessen* tessellation; in this book the term Voronoi tessellation is always used.

If N is a stationary point process with finite positive intensity λ, then almost surely all the $T(y)$ are bounded such that the corresponding Voronoi tessellation is indeed a random tessellation. In this case the tessellation is also stationary. The mean cell area or volume $\mathbf{E}(A)$ or $\mathbf{E}(V)$ is

$$\mathbf{E}(A) = \frac{1}{\lambda} \quad \text{or} \quad \mathbf{E}(V) = \frac{1}{\lambda}. \tag{1.8.5}$$

---

[5] Those references cited in this paragraph and not listed at the back of this book may be found in Okabe et al. (2000).

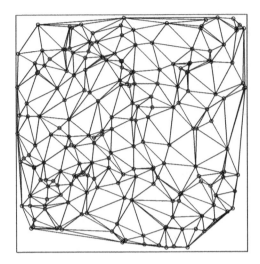

Figure 1.13   The Delaunay tessellation with respect to the pattern of *Phlebocarya* discussed in Section 1.2.3. Compare with Figure 1.12.

Voronoi tessellations are used in the data analysis of geometrical structures; see Chiu (2003). These tessellations have been successfully applied to packings of hard spheres. Finney (1979) uses the term 'polyhedral statistics'; see also Medvedev and Naberukhin (1987). Sibson (1980, 1981) describes the use of Voronoi tessellations derived from point patterns as a basis for 'natural neighbour interpolation' – interpolating a smooth function for data located at irregularly distributed points; see also Okabe et al. (2000) and Bernardeau and Van de Weygaert (1996). Thiessen and Alter had this in mind when they originally suggested this tessellation.

In point process statistics, distributional properties of the cells and of the tessellation as a whole are used. An important example is the cell area distribution, which, of course, depends on the distribution of the underlying point process and may characterise it. Unfortunately, this approach is rather problematic as almost all known formulas for Voronoi tessellations only apply if N is a stationary Poisson process. Some of these formulas are presented in Section 2.5.3.

Today, Voronoi tessellations can be straightforwardly constructed with software based on efficient iterative algorithms that is freely available through the internet.

If N is a point process where, as for the Poisson process, almost surely every node is touched by exactly three cells (in the planar case) or by exactly four cells (in the spatial case), another tessellation can be constructed, the *Delaunay tessellation* or *triangulation* (see Figure 1.13). In the planar case it is constructed from the triangles formed by the points of the point process whose cells share the same node. In this book, the Delaunay tessellation is regarded as a special type of point network and discussed below together with other networks.

Note that there are also tessellations of a quite different nature in which every point in a pattern is allocated a 'cell' of equal area or volume; see Hoffman et al. (2006) and Figure 1.14.

1.8.5 Neighbour networks or graphs

A *graph* (in the sense of graph theory) is a system of points, called *vertices*, which are connected by *edges* usually shown as straight-line segments. This book considers geometrical graphs the vertices of which are the points of a given point process *N*, with (undirected) edges that are either given a priori or constructed by some specified rule based on the local point configuration; see Marchette (2004), Penrose (2003, 2005) and Penrose and Yukich (2001). These graphs are called neighbour networks. Examples where edges are given a priori include (1) forests with trees with overlapping crowns, where the corresponding points are joined by edges, (2) hard spheres in direct contact and (3) neighbours in Gibbs processes.

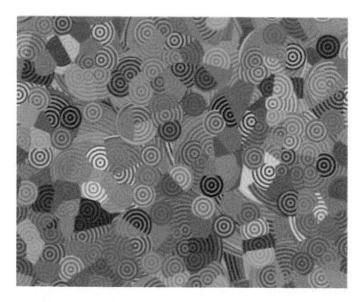

Figure 1.14 The 'stable tessellation' for a sample of a Poisson process with periodic boundary conditions. Every cell has the same area, but not all cells are connected. The figure shows the generating points with the allocated areas. Concentric circles around the points are used to aid the identification of the cells. Data courtesy of A. E. Holroyd.

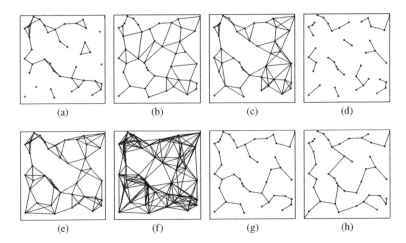

Figure 1.15   Various graphs for the point pattern of Figure 1.5 of waterstrider positions: (a) $G(N; r)$, (b) Gabriel graph, (c) sphere-of-influence graph, (d) 1-neighbour graph, (e) 4-neighbour graph (f) 8-neighbour graph, (g) minimal spanning tree and (h) radial spanning tree.

The latter is illustrated in Figure 3.21. Some of the graphs may be used to detect short-range correlations in point processes or help to detect clusters or anomalies in point patterns.

Figure 1.15 shows some graphs constructed for the point pattern of positions of waterstriders in Figure 1.5. These graphs show interesting differences and provide information on different properties of the pattern. For example, the 4-neighbour graph (Figure 1.15(e)) and sphere-of-influence graph (Figure 1.15(c)) indicate holes or gaps in the pattern. Graphs are also valuable tools in exploratory statistics of finite point processes; see Marchette (2004).

In the following, eight different graphs for a point process N are briefly described.

1. *Disc or sphere graph* $G(N; r)$. This graph inserts an edge between two points x and y in N whenever the distance between x and y is smaller than (or equal to) r, $\|x - y\| \leq r$.

2. *Delaunay tessellation.* The easiest way to understand this graph is to regard it as the dual tessellation of the Voronoi tessellation as explained at the end of Section 1.8.4.

3. *α-hull.* This graph inserts an edge between two points x and y in N whenever there is a disc of radius α centred at a suitable location such that x and y lie on the disc boundary, and there is no point of N in the interior of the disc. (In three dimensions the disc is replaced by a sphere; see Edelsbrunner and Mücke, 1994.)

4. *Gabriel graph.* This graph inserts an edge between two points x and y in N if the closed disc (sphere) centered at $(x+y)/2$ does not contain any other point in N. (Note that $(x+y)/2$ is the midpoint of the line connecting x and y.) The Gabriel graph is a subgraph of the Delaunay triangulation, i.e. every edge of the Gabriel graph is also an edge of the Delaunay graph.

5. *k-neighbour graph.* This graph inserts an edge between two points x and y in N whenever y is one of the k nearest neighbours of x (or x is one of the k nearest neighbours of y), for $k = 1, 2, \ldots$. If the kth nearest neighbour is not well defined (i.e. there is more than one point with the same distance), then some rule is used to define the neighbour, e.g. the lexicographic ordering of the point coordinates. A similar graph is considered in Chiu and Molchanov (2003). There each point in N is connected to its nearest neighbour, then to its second nearest neighbour, and so on until the point is contained in the interior of the convex hull of these nearest neighbours. The degree of the typical point in N is a useful concept for separation of clustering and repulsion behaviour of point processes.

6. *Sphere-of-influence graph.* The *sphere of influence* S_x of a point x of N is the sphere $b(x, d(x))$ of radius $d(x)$ centred at x, where $d(x)$ is the distance to the nearest neighbour of x. The sphere-of-influence graph inserts an edge between x and y if and only if the spheres S_x and S_y overlap.

7. *Radial spanning tree.* This graph is a tree in the graph-theoretic sense (non-directed, acyclic and connected), which is defined with respect to one particular location called the 'origin' o, which is not a point in N; see Baccelli and Bordenave (2007). N is assumed to be irregular, i.e. not to form a lattice. In this graph every point is linked with its nearest neighbour among those points which are nearer to o. The construction leading to the radial spanning tree is local and, unlike the graph in (8), does not minimise any global functional.

8. *Minimal spanning tree.* Assume first that N is finite. Then the minimal spanning tree is the connected graph with vertex set N of minimal total edge length. In the general case the minimal spanning tree is a graph with vertex set N and is defined as follows: an edge is inserted between the vertices x and y if and only if x and y are in different components of $G(N; \|x - y\|)$, and at least one of these components is finite.

If N is stationary (and isotropic), then graphs 1–6 are also stationary (isotropic) for suitable definitions of stationarity (and isotropy) for graphs. Graphs 1, 3 and 5 depend on parameters r, k and α. The behaviour of graph characteristics (e.g. the number of edges or components) dependent on these parameters provides valuable structural information about N.

Neighbour networks have been applied in image processing, for segmentation, classification and identification of clusters and isolated points as well as for the detection of clusters, isolated points and gaps in point patterns. The graphs discussed above can be constructed using the `spatgraphs` library in R; see `http://cran.r-project.org/doc/packages/spatgraphs.pdf`.

1.9 Simulation of point processes

Traditionally, simulation methods were only rarely used within classical statistics. Today this situation has changed mainly due to the increasing popularity of Bayesian methods and Markov chain Monte Carlo (MCMC) techniques. However, in point process statistics (stochastic) simulations have a long tradition and even simple questions require the use of simulations, i.e. the generation of random point patterns by means of random numbers. For this reason, simulation methods and algorithms are introduced here and discussed frequently throughout this book.

This section discusses general principles of simulation in the context of spatial point processes and specific aims of the approach. Simulation methods – or Monte Carlo (MC) methods – are used for a number of different aims:

- *Calculation of summary characteristics* for point process models. Many point process models are so complicated that explicit analytical formulas for even the most fundamental summary characteristics such as the intensity or second-order characteristics have not been found. Based on simulated point patterns derived from these models, summary statistics can be estimated statistically. For the applied statistician this has the benefit that these numerical approaches and algorithms are often much easier to understand than a complicated probabilistic proof. Unfortunately, it requires some programming but these days more and more MC software is becoming available, for example in R.

- *Goodness-of-fit tests* for point process models. Simulation tests replace the classical tests of mathematical statistics, which are not applicable in point process statistics, since the relevant test statistics do not have the classical distributions.

- *Investigation of the behaviour* of statistical methods. Simulation methods may be used to assess the performance of statistical methods. These are applied to simulated samples for point processes and the results are compared to known characteristics of the underlying model, in order to evaluate the methods and explore the sampling variation.

- *Visualisation* of point process models. Simulated samples from point process models can be displayed and so provide a better understanding of the spatial structures that may be generated from a particular model.

- *Reconstruction and extension.* In some applications, point pattern data have not been collected for the entire window and simulated data may be used to fill the 'gaps'. In other cases, one might want to simulate a continuation of a specific pattern outside the observation window, in particular in the context of edge correction.

In the following, some of the ideas underlying simulation methods in point process statistics are introduced, in particular the idea of statistical tests based on simulations.

In the classical approach, univariate or multivariate data are often analysed based on the Gaussian distribution. Its parameters can be estimated using simple unbiased estimators, and distributions such as the t-, F- and χ^2-distribution are used for significance tests. However, this becomes much more complicated in point process statistics. The following uses an analogy from classical statistics to explain the simulation approach.

Assume for the moment that you know:

- the formula for the p.d.f. of the Gaussian distribution,

$$f(x) = \frac{1}{\sqrt{2\pi}\sigma} \exp\left(-\frac{(x-\mu)^2}{2\sigma^2}\right);$$ (1.9.1)

- some method to generate random numbers from the distribution in (1.9.1);

- the ideas of parameter tests and goodness-of-fit tests, in particular the Kolmogorov–Smirnov test.

However, you do not know:

- the formulas for mean and variance of the normal distribution, i.e.

$$\mathbf{E}X = \mu,$$ (1.9.2)

$$\mathbf{var}X = \sigma^2;$$ (1.9.3)

- the critical values z_α of the Gauss test;

- the critical values k_α of the Kolmogorov–Smirnov test.

You can still carry out satisfactory statistics if you apply simulation methods, using a computer. The following explains how to proceed in the classical Gaussian case – and it will turn out that these basic ideas are essentially the same for point process statistics.

Probabilistic calculations

The calculation of means, moments and probabilities is an important problem, which can be solved analytically using well-known formulas. This is usually not possible in point process statistics. To understand the situation, assume that you want to determine $\mathbf{E}X$ for the Gaussian distribution (1.9.1) without knowing formula (1.9.2).

A simple approach is to generate k random numbers $x_1, \ldots, x_k$ corresponding to (1.9.1) with the values of μ and σ^2 of interest. Then you determine the sample mean $\bar{x}$. Due to the law of large numbers you expect that $\bar{x}$ approximates $\mathbf{E}X$. You will observe

$$\bar{x} \approx \mu$$

and might want to ask a mathematician to prove that (1.9.2) is indeed true.

In point process statistics, simple formulas for moments or probabilities do not exist in most cases and a friendly mathematician can provide help in particular by designing clever simulation algorithms; see Møller and Waagepetersen (2004). Here, simulation is often the only way to find numerical values for summary characteristics such as the intensity, the K-function, nearest-neighbour distance d.f. or estimation variances.

Testing hypotheses by Monte Carlo or simulation tests

Parametric hypotheses.   Assume you want to test the hypothesis $H_0 : \mu = \mu_0$ against $H_A : \mu \neq \mu_0$ with known σ^2. (Of course, you do know that the Gauss test solves this problem, but recall that for the moment this is assumed to be unknown.) Proceed as follows: calculate $\bar{x}$ from the data $x_1, \ldots, x_n$ and then determine the test statistic

$$z = \frac{\bar{x} - \mu_0}{\sigma} \sqrt{n}, \tag{1.9.4}$$

based on the idea that the difference $\bar{x} - \mu_0$ may be of interest for the test and that it may be useful to normalise by $\sigma/\sqrt{n}$, the standard deviation of the sample mean. Then you want to compare the empirical z in (1.9.4) with the corresponding values for a Gauss distribution with $\mu = \mu_0$ and σ^2.

To do this, n Gaussian random numbers are simulated k times with parameters μ_0 and σ^2 and the corresponding sample means $\bar{x}_i$ for $i = 1, \ldots, k$, and the z_i are calculated based on (1.9.4) using $\bar{x}_i$ instead of $\bar{x}$. If $H_0 : \mu = \mu_0$ is true, you expect the z_i to be 'similar' to z. The similarity is assessed by putting z and $z_1, \ldots, z_k$ in increasing order. $H_0 : \mu = \mu_0$ is not very likely to be true if z has an extreme position among the z_i. More specifically, H_0 is rejected if z is one of the $\frac{\alpha}{2}(k+1)$ smallest or largest values. Here α is the error probability of the test. If $k = 999$ and $\alpha = 0.05$ then the critical positions are 25 and 976, i.e. H_0 is rejected if z is in position 1–25 or 976–1000.

Goodness-of-fit tests.   Assume you want to test the hypothesis $H_0 : F = F_0$, i.e. that a d.f. of interest equals some d.f. F_0, such as a Gaussian distribution with parameters μ and σ^2. You may use the basic idea of the Kolmogorov–Smirnov test to compare two distributions based on the maximum difference. Assume that you do not have a formula for $F_0(x)$ but only know how to generate corresponding random numbers.

The first step is the estimation of $F_0(x)$. Generate a very large sample $y_1, \ldots, y_m$ of m random numbers from F_0 by simulation and calculate the empirical d.f.

$$\hat{F}_{0,m}(x) = \#\{y_i \le x\}/m.$$

Then calculate the empirical d.f. for the data $x_1, \ldots, x_n$,

$$\hat{F}_n(x) = \#\{x_i \le x\}/n,$$

and calculate the test statistic D that is the maximum distance between the two d.f.s. More formally, this difference is defined as

$$D = \max\{\max_{i=1,\ldots,n} |\hat{F}_{0,m}(x_i) - \hat{F}_n(x_i)|, \max_{i=1,\ldots,n} |\hat{F}_{0,m}(x_i) - \hat{F}_n(x_i - 0)|\}.$$

If D is 'small', $H_0 : F = F_0$ is likely to be accepted.

The decision whether D may be regarded as small is again made based on simulation methods. Generate k samples of n random numbers from $F_0(x)$ to obtain empirical d.f.s $\hat{F}_{n,l}(x)$ and calculate

$$D_l = \max\{\max_{i=1,\ldots,n} |\hat{F}_{0,m}(x_{i,l}) - \hat{F}_{n,l}(x_{i,l})|, \max_{i=1,\ldots,n} |\hat{F}_{0,m}(x_{i,l}) - \hat{F}_{n,l}(x_{i,l} - 0)|\}$$

for $l = 1, \ldots, k$. Then put D and $D_1, \ldots, D_k$ in increasing order. If D is one of the $\alpha(k + 1)$ largest values, you will conclude that D is not small and reject H_0. For $k = 999$ and $\alpha = 0.05$ the critical position in the ordered series is $(1 - \alpha)(k + 1) = 950$, i.e. H_0 is rejected if the value of D for the data is at a position that is more extreme than 950.

Even though the simulation approaches to classical statistical problems outlined above appear rather laborious and cumbersome, in the absence of closed-form expressions for many characteristics this is the standard approach in point process statistics. In other words, samples from point process models are generated and point process summary characteristics such as the L-function or the nearest-neighbour distance d.f. are used, instead of the d.f. F_0 above.

There are some issues that are specific to the simulation approach in the context of spatial point process statistics, however, and these are briefly outlined here. Usually, a point process is observed and simulated in some window W. If the point process of interest is *stationary,* i.e. the sample in W is considered to be a small part of a much larger pattern, then the simulation has to consider the impact of points outside W on the points within W. A common solution to this is to simulate the process in a large window W_{sim} which contains W and to use only the points in W as a sample.

Finite point processes can either have a random or a fixed number of points n in W. If random, n is often generated first and then, *conditional* on n, the corresponding point pattern is simulated. However, some models are simulated by MCMC methods where n fluctuates during the simulation.

Conditional simulation

In the context of probabilistic calculations it is necessary to simulate the model under the given distributional conditions, for example as a stationary process or with the prescribed distribution of the number of points. However, conditional simulations are often used for statistical tests, where the number of points n is fixed at the value observed in the original sample. In this case it is, of course, important that the right conditional model is simulated. This issue is discussed in Section 3.6 in the context of canonical and grand canonical Gibbs processes and for other processes in other parts of the book.

Conditional simulation is also used if some of the point locations are given and have to remain fixed. For example, a forester may measure tree positions and attributes in some circular sample plots in a stand and may then want to simulate a forest within the whole stand. Naturally, the locations of trees and their attributes within the circles have to be fixed and the forest is simulated only outside the circle. However, the relationships among the trees within and outside the circles have to follow the same general rules everywhere. For example, the minimum inter-point distance must not be violated.

Some general comments

In most cases, simulation methods can only be applied if a specific model can be assumed and if an algorithm is known that can be used to simulate this model. This book discusses in detail how an appropriate model may be identified for empirical data.

Simulation may seem to be a 'less mathematical' way of solving problems to the more mathematical reader. Nevertheless, it should be seen as a serious way to obtain results, as it is a specific type of numerical mathematics.

Results derived from simulation are often exact, as exact as the user wants them to be, depending perhaps on computing time. In contrast, approaches in classical statistics are often not as exact as they appear to be, since they are based on distributional assumptions which frequently do not hold, or use large-sample assumptions, but the meaning of 'large' is not always clear.

This book discusses simulations without any discussion of the algorithms used for random number generation. Within statistics, however, this is a topic of ongoing research and a number of reliable algorithms have been developed. The text assumes that the reader uses reliable software for this purpose. Refer to the literature on MC methods, and see Fishman (1996), Gentle (2003), Manly (2006) and Ripley (1987) for a detailed discussion.

2

The homogeneous Poisson point process

The Poisson process has a central role in point process statistics. It is fundamental to any successful analysis of point pattern data that the user is familiar with the basic properties of this process.

There are many situations in which the Poisson process is a suitable model, i.e. when the points are 'randomly' distributed in space. In addition, it serves as a basis for the construction of more complicated models. Perhaps even more importantly, the Poisson process is a null or benchmark model that may be used as a reference model to distinguish between point patterns exhibiting aggregation and repulsion. Finally, it admits analytical calculations of summary characteristics and thus provides some understanding of the theory of point processes.

Therefore, this chapter presents the theory and application of the Poisson process systematically and in detail. The exposition starts with the binomial point process, which is a finite variant of the Poisson process and provides an easy introduction into its theory. Then the Poisson process is defined and thoroughly investigated. A detailed justification is given as to why the Poisson process is the ideal model for

Statistical Analysis and Modelling of Spatial Point Patterns   J. Illian, A. Penttinen, H. Stoyan and D. Stoyan
© 2008 John Wiley & Sons, Ltd

complete spatial randomness. The simulation of the Poisson process is easy to understand, but the derivation of the summary characteristics is a little technical in places. However, the resulting formulas are valuable tools in applications as they allow a comparison of empirical characteristics with theoretical ones and facilitate the general understanding of the various summary characteristics used throughout this book.

Finally, issues of statistical inference are discussed – in particular, tests of the CSR hypothesis. Testing this hypothesis is a central problem in point process statistics. Indeed, when analysing a spatial point pattern one aims to detect and quantify spatial structure, with the final aim of relating it to underlying processes that have caused the observed pattern. Hence, it is necessary to verify that an observed aggregation or repulsion of points is really significant. Rejection of the CSR hypothesis for a given pattern can indicate this and may then be followed by the more complicated but also more interesting part of the point process analysis.

2.1   Introduction

This chapter introduces the simplest and most important infinite point process model: the homogeneous Poisson point process. More complicated models are much easier to understand once this null model has been discussed in detail. In addition, the Poisson process plays a central role in the theory of point processes as a reference or null model, leading to direct statistical applications.

A specific point pattern may well exhibit various types of interaction between its constituent points. For instance, the points may occur in clusters (see Figures 1.4, 1.9, 6.3 and 6.4) or may exhibit regularity (see Figures 1.2, 1.6, 6.7 and 6.23). In addition, there may be a *hard-core distance,* i.e. a disc or sphere of diameter r_0 around each point where no other points are located. In other words, a positive minimal inter-point distance may be determined for a given pattern. All of the above features may occur in combination in the same pattern. If neither of the above interactions is found in a point pattern, it may be regarded as completely spatially random. This property of a pattern is termed *complete spatial randomness* (CSR). A theoretical model for patterns with this property forms an important basis for comparison, as a *null model* and as a reference for the construction of summary characteristics.

Furthermore, calculations may be carried out of the extent and probability of fluctuations from the theoretical characteristics in samples. It is thus possible to determine objectively whether a specific fluctuation in an observed pattern is small enough to be considered insignificant, as it is a feature that might well be observed in a pattern that is a realisation from a completely random point process model.

By imposing axioms concerning stationarity and lack of interaction that are intuitively appealing, completely random processes can be characterised as Poisson point processes. As a result, it is possible to decide whether a given pattern exhibits spatial clustering or regularity or whether it may be considered CSR.

In addition, the Poisson process has a further role as a *basic building block for other more complicated models*; see Chapter 6. The definition of many cluster processes is based on Poisson processes, some hard-core processes result from Poisson processes by thinning and Gibbs processes are defined with direct reference to Poisson processes. Therefore, simulation procedures for point processes frequently comprise the construction of a Poisson point process, which is then modified to yield the required form.

For the interested reader, a history of the concept of the Poisson process may be found in Daley and Vere-Jones (1988, 2003). They report that the first recorded use of the Poisson process in spatial statistics appears to be that of Abbe (1879) as discussed in Section 1.3.2.

2.2   The binomial point process

2.2.1   Introduction

The *binomial point process* is the classical starting point for a discussion of CSR and hence of the Poisson process. However, while intuitively appealing, the binomial process does not suffice as a model for CSR. Nevertheless it is interesting as the most simple non-trivial example of a spatial point process model, and the Poisson process may then be considered to be a generalisation of the binomial process. Furthermore, it is a null model for finite point processes; see Chapter 3.

A binomial point process consists of n points, which are randomly scattered in a set W. The term 'randomly' means here that the n points $x_1, \ldots, x_n$ are uniformly and independently distributed in W. The set W is assumed to be bounded. Its area is denoted by $\nu(W)$, where the neutral symbol 'ν' may also denote volume, if W is a subset of $\mathbb{R}^3$, i.e. in the context of a three-dimensional spatial pattern.

Consider first a single point randomly distributed in space. Whereas this yields a very simple example of a trivial one-point pattern of no direct practical relevance, the union of several of these points yields the binomial point process. More specifically, a single random point x is *uniformly distributed* in W if

$$\mathbf{P}(x \in A) = \frac{\nu(A)}{\nu(W)} \tag{2.2.1}$$

for all subsets A of W. This equation means that the probability that x takes its position in the subset A of W is equal to the ratio of the areas of the sets A and W. This is the classical definition in the sense of geometrical probability.

Consider now a more interesting point process which consists of n points and apply formula (2.2.1) for each of these points. This yields the binomial point process for which

$$P(x_1 \in A_1, \ldots, x_n \in A_n) = P(x_1 \in A_1) \cdot \cdots \cdot P(x_n \in A_n)$$
$$= \frac{\nu(A_1) \cdot \cdots \cdot \nu(A_n)}{\nu(W)^n}, \qquad (2.2.2)$$

where $A_1, \ldots, A_n$ are subsets of W. This implies that all points are randomly scattered in W and are independent of each other's locations.

Figure 2.2 (on p. 64 below) shows the result of a simulation of a binomial point process in $W = \boxed{1}$ with $n = 50$ points.

If the points $x_1, \ldots, x_n$ form a binomial point process in W then the random pattern formed by these points is denoted $N_{W^{(n)}}$. The ordering of the points is ignored and $N_{W^{(n)}}$ can be regarded as a random set.

The binomial point process $N_{W^{(n)}}$ is the first non-trivial example of a point process in this book. It is close to but not in every sense *the* model for CSR, as will be shown below: in spite of the independence assumption in the construction there are some spatial correlations in the pattern, which result from the fact that the total number of points in W is fixed, i.e. equal to n. Only by assuming a random number of points and extending the approach to the whole space can the right definition of CSR be given.

2.2.2 Basic properties

The binomial point process $N_{W^{(n)}}$ owes its name to a distributional property. If A is a subset of W then the random number of points in A, denoted by $N_{W^{(n)}}(A)$, follows a binomial distribution, with parameters $n = N_{W^{(n)}}(W)$ and $p = p(A) = \nu(A)/\nu(W)$. More specifically:

$$P(N_{W^{(n)}}(A) = k) = \binom{n}{k} p^k (1 - p)^{n-k} \qquad \text{for } k = 0, \ldots, n. \qquad (2.2.3)$$

Since the mean (or expected value) of a binomial distribution is np, the mean number of points in A is

$$np = n \frac{\nu(A)}{\nu(W)} = \lambda \nu(A),$$

where λ is the mean number of points per unit area or volume, termed the *intensity* of the binomial point process,

$$\lambda = \frac{n}{\nu(W)}.$$

In general, void probabilities are very important characteristics for point processes. They are concerned with the event that a given set K is empty, i.e. that it does not contain any point, and are given for the binomial point process by

$$\mathbf{P}(N_{W^{(n)}}(K) = 0) = \frac{(\nu(W) - \nu(K))^n}{\nu(W)^n}, \tag{2.2.4}$$

where $\mathbf{P}(N_{W^{(n)}}(K) = 0)$ is the probability that there is no point in the subset K of W.

In addition, some of the finite-dimensional distributions of the binomial point process are given by a simple formula: if $A_1, \ldots, A_k$ are disjoint subsets of W with $A_1 \cup \cdots \cup A_k = W$ and if $n_1 + \cdots + n_k = n$, then

$$\mathbf{P}\left(N_{W^{(n)}}(A_1) = n_1, \ldots, N_{W^{(n)}}(A_k) = n_k\right)$$

$$= \frac{n!}{n_1! \cdots n_k!} \cdot \frac{\nu(A_1)^{n_1} \cdots \nu(A_k)^{n_k}}{\nu(W)^n}. \tag{2.2.5}$$

This distribution is the well-known multinomial distribution with parameters n, $p_1 = \frac{\nu(A_1)}{\nu(W)}, \ldots, p_n = \frac{\nu(A_n)}{\nu(W)}$, which is familiar from the analysis of count data using log-linear models. For overlapping sets $A_1, \ldots, A_k$ the formula is more complicated and will not be considered here.

Note that numbers of points in different subsets of W are *not* independent even if the subsets are disjoint. This is due to the fact that $N_{W^{(n)}}(A) = m$ directly implies $N_{W^{(n)}}(W \setminus A) = n - m$. Thus the number of points in one subset has influence on the number of points in another subset, and hence these are not independent. This demonstrates that the binomial point process is not sufficiently appropriate as a model of CSR since the number of points is fixed. Therefore, it is necessary to consider a suitable probability distribution of the random number of points, such that the number of points in one subset cannot be predicted from the number of points in another subset.

This distribution may be derived as follows. As noted, $N_{W^{(n)}}(A)$ follows a binomial distribution with parameters n and $p(A)$. The well-known Poisson limit theorem yields the following: if the total number of points n tends to infinity and the second parameter $p(A)$ tends to zero in such a way that the product remains fixed as $np(A) = \lambda \cdot \nu(A)$, then $N_{W^{(n)}}(A)$ is asymptotically Poisson distributed with mean $\lambda \cdot \nu(A)$. This limit can be obtained if the region W is enlarged to fill the whole of $\mathbb{R}^d$ while n is allowed to tend to infinity. If the ratio $n/\nu(W) = \lambda$ remains fixed as n increases and W is enlarged, then the Poisson limit will hold for $N_{W^{(n)}}(A)$ for any fixed bounded subset A of W. If there is a limiting point process N then it should have the property:

$N(A)$ is Poisson of mean $\lambda \cdot \nu(A)$ for each bounded set A.

As an implication of (2.2.5) such a limiting process should be the right model for CSR:

$N(A_1), \ldots, N(A_k)$ are independent if $A_1, \ldots, A_k$ are disjoint sets.

Hence the idea of assuming that the number of points in N is random and follows a Poisson distribution leads to the CSR model. The following describes a construction of this model that does not refer to limits; the limiting procedure above is used only to aid understanding.

2.2.3 The periodic binomial process

Assume that W is the rectangle with left lower vertex at the origin o and sides of lengths a and b parallel to x- and y-axis. Then the distributional properties do not change much if W is transformed to the unit square $\boxed{1}$ and the points $x_i = (\xi_{i,1}, \xi_{i,2})$ become $x_i = (\xi_{i,1}, \xi_{i,2})$. Now assume that the resulting pattern in the unit square is periodically continued in all other unit squares of the plane as shown in Figure 2.1, or, more technically, consider in addition to the original points $x_i = (\xi_{i,1}, \xi_{i,2})$ infinitely many copies $x_i = (\xi_{i,1}, \xi_{i,2})$ for all integers k and l. This pattern does not change under translations which transform points with integer-valued coordinates into points with integer-valued coordinates.

It makes sense to define functional summary characteristics for the periodic binomial process that originate in the theory of stationary point processes (see Chapter 4): $D(r)$, the distribution function of the distance from an arbitrary point in

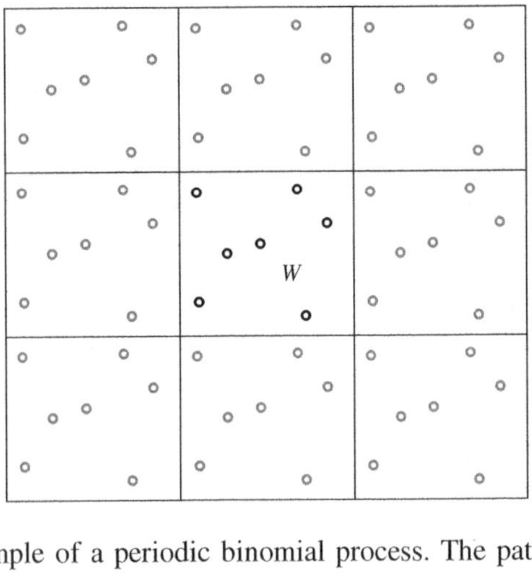

Figure 2.1 A sample of a periodic binomial process. The patterns in all squares are congruent.

the pattern to its nearest neighbour; $H_s(r)$, the distribution function of the distance from an arbitrary test point to its nearest neighbour in the pattern; and $K(r)$, the mean number of further points of the pattern in a disc of radius r centred at an arbitrary point of the pattern, divided by n. Note that the distances are measured with respect to the *torus metric* as defined on p. 184. For $\boxed{1}$ the functions can be calculated as follows:

$$D(r) = \begin{cases} 1 - \left(1 - \pi r^2\right)^{n-1} & \text{for } r \le \frac{1}{2}, \\ 1 - \left(1 - \sqrt{4r^2 - 1} + r^2 \left(4\arccos\frac{1}{2r} - \pi\right)\right)^{n-1} & \text{for } \frac{1}{2} < r \le \frac{\sqrt{2}}{2}, \\ 1 & \text{for } r > \frac{\sqrt{2}}{2}; \end{cases}$$

$$H_s(r) = \begin{cases} 1 - \left(1 - \pi r^2\right)^{n} & \text{for } r \le \frac{1}{2}, \\ 1 - \left(1 - \sqrt{4r^2 - 1} + r^2 \left(4\arccos\frac{1}{2r} - \pi\right)\right)^{n} & \text{for } \frac{1}{2} < r \le \frac{\sqrt{2}}{2}, \\ 1 & \text{for } r > \frac{\sqrt{2}}{2}; \end{cases}$$

$$K(r) = \begin{cases} \frac{n-1}{n}\pi r^2 & \text{for } r < 1, \\ 4 + \frac{n-1}{n}\pi r^2 & \text{for } 1 \le r < \sqrt{2}. \end{cases}$$

The proof of these formulas uses simple ideas of geometrical probability. Note that in the calculation of $D(r)$ and $K(r)$ the reference point is excluded; therefore, the exponent $n - 1$ and the factor $\frac{n-1}{n}$ appear; see the discussion at the end of Section 2.5.2.

2.2.4   Simulation of the binomial process

As explained in Section 1.9, simulation is an important method in point process statistics. This is due to the fact that for statistical inference it is often necessary to simulate from a point process model. Therefore, here the simulation of a binomial point process is described as a first simple example and as an introduction to general principles. Further examples will show that simulation approaches are often similar for more complex models and are based on the simple case considered here.

The simulation of a binomial point process is easily done by superpositioning random points independently in the required region. It is straightforward to simulate a random point that is uniformly distributed in $\boxed{1}$. If $\{u_n\}$ is a sequence of independent random numbers uniformly distributed in $[0, 1]$ then the points

$$x_i = (u_{2i-1}, u_{2i}) \qquad \text{for } i = 1, 2, \dots, \tag{2.2.6}$$

form a sequence of independent random points uniformly distributed in $\boxed{1}$. Figure 2.2 shows a sample of 50 points simulated in this way.

More generally, when patterns in a d-dimensional space are considered, a sequence of random points uniformly distributed in the d-dimensional hypercube

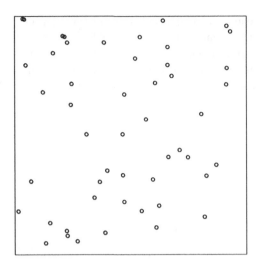

Figure 2.2   A simulation of 50 random points uniformly distributed in ⬜.

$[0, 1]^d$ is generated by the same mechanism, i.e.

$$x_i = (u_{(i-1)d+1}, \ldots, u_{id}) \qquad \text{for } i = 1, 2, \ldots . \qquad (2.2.7)$$

Translation and scale changes may be used to produce a sequence of points uniformly distributed in any fixed rectangle or hypercube.

The simulation procedure described above is the method of choice for a large number of cases in practice, as data have been collected on a rectangle or in a cube in many applications and the binomial point process will have to be simulated in a rectangle or cube. However, in some applications observation windows with complicated shapes make the simulation approach slightly more difficult. Simulation of a uniform random point in a bounded region W of arbitrary shape is tackled using one of three main techniques, of which only the planar case is considered here; all methods may be canonically generalised to higher dimensions.

(a) *Rejection sampling.* A rectangle R containing W is found and a sequence of independent uniform random points is simulated in R. These are rejected when they are outside W. The first point that falls in W is uniformly distributed in W. To obtain a binomial point process the whole procedure is repeated until n points have fallen in W, and these n points constitute the sample of the binomial point process. To maximise efficiency, R should be chosen to be as small as possible. Figure 2.3 illustrates the process. Note that the total number of generated points including those that were rejected is random.

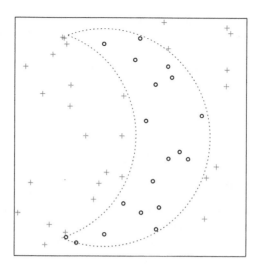

Figure 2.3   A simulated pattern of 19 random points in W based on the rectangle in Figure 2.2. The points o are uniformly distributed in W and the points + are outside W.

(b) *Approximation.* The region W is replaced by a disjoint union of k squares (of equal or different sizes) approximating W. A random point distributed uniformly in this union is simulated by choosing a square with probability proportional to its area and then simulating a random point uniformly distributed in this square. Exact simulations for complicated regions may be obtained by combining this technique with rejection sampling (a).

(c) *Transformation of coordinates.* If the region W exhibits some symmetry then a transformation of coordinates may be useful. For example, if W is the unit disc $b(o, 1)$, a uniform random point can be described in polar coordinates

$$x = (r, \theta) \qquad \text{for } r \text{ in } [0, 1] \text{ and } \theta \in (0, 2\pi].$$

The random variables r and θ are independent; θ is uniformly distributed in $(0, 2\pi]$ and r satisfies the law

$$\mathbf{P}(r \leqslant t) = t^2 \qquad \text{for } 0 \leq t \leq 1.$$

Thus, if u_1 and u_2 are independent random numbers uniformly distributed in $[0, 1]$ then the formulas

$$r = \sqrt{u_1} \quad \text{and} \quad \theta = 2\pi u_2$$

provide a method for simulating $x = (r, \theta)$.

Note that the binomial point process arises from the stationary Poisson point process by conditioning on n; see property (e) in Section 2.3.2 for details. Consequently, it is not necessary to discuss specific statistical methods for assessing the hypothesis that a given point pattern is a realisation of a binomial point process; one can apply the methods for stationary Poisson point processes described in Section 2.7.

2.3 The homogeneous Poisson point process

2.3.1 Introduction

This section presents the formal definition of the homogeneous Poisson point process, or homogeneous Poisson process for short. It follows straightforwardly from the discussions in the previous section and it also essentially explains the name of the process.

A homogeneous Poisson process N is characterised by two fundamental properties which have been already mentioned as asymptotic properties in Section 2.2:

(1) *Poisson distribution of point counts.* The number of points of N in any bounded set B follows a Poisson distribution with mean $\lambda \cdot \nu(B)$ for some constant λ; readers unfamiliar with the definition of the Poisson distribution may refer to formula (2.3.2) and p. 106.

(2) *Independent scattering.* The numbers of points of N in k disjoint sets form k independent random variables, for arbitrary k.

Property (2) is also known as the 'completely random' or 'purely random' property. Note that this property does not hold for the centres of hard objects that are randomly distributed in space. Models have been developed which may be suitable for these patterns but they are beyond the scope of the Poisson process theory; see Section 6.5.4 (the random sequential adsorption (RSA) model) and 6.6.1 (Gibbs hard-core process).

The number λ in (1) is the characteristic parameter, called the *intensity* or *point density*, of the homogeneous Poisson process. It describes the mean number of points to be found in a unit volume and is given by

$$\lambda \cdot \nu(B) = \mathbf{E}(N(B)) \qquad \text{for all bounded sets } B. \qquad (2.3.1)$$

In the following, λ is always positive and finite. (If $\lambda = 0$, the point pattern contains no points and an infinite λ corresponds to a pathological case.)

2.3.2 Basic properties

Let N be a homogeneous Poisson process with intensity λ. Once the intensity λ is known, the whole distribution of the homogeneous Poisson process can be determined from properties (1) and (2).

(a) *One-dimensional distributions.* Property (1) implies that

$$\mathbf{P}(N(B) = n) = \frac{\lambda^n (\nu(B))^n}{n!} \exp(-\lambda\nu(B)) \qquad \text{for } n = 0, 1, \dots . \tag{2.3.2}$$

This means that $N(B)$ has a Poisson distribution with parameter $\lambda\nu(B)$.

(b) *Finite-dimensional distributions.* Properties (1) and (2) imply that if $B_1, \dots, B_k$ are disjoint bounded sets then $N(B_1), \dots, N(B_k)$ are independent Poisson random variables with means $\lambda\nu(B_1), \dots, \lambda\nu(B_k)$. Thus

$$\mathbf{P}(N(B_1) = n_1, \dots, N(B_k) = n_k)$$

$$= \frac{\lambda^{n_1 + \dots + n_k} (\nu(B_1))^{n_1} \cdot \dots \cdot (\nu(B_k))^{n_k}}{n_1! \cdot \dots \cdot n_k!} \exp\left(-\sum_{i=1}^{k} \lambda\nu(B_i)\right). \tag{2.3.3}$$

From this formula the joint probabilities $\mathbf{P}(N(B_1) = n_1, \dots, N(B_k) = n_k)$ may be evaluated for general (possibly overlapping) sets $B_1, \dots, B_k$.

(c) *Stationarity and isotropy.* A point process $N = \{x_n\}$ is stationary if the translated process $N_x = \{x_n + x\}$ has the same distribution for all x in $\mathbb{R}^d$ (recall Section 1.6.3). A point process is isotropic if its distribution is invariant with respect to rotations, i.e. N and $\mathbf{r}N = \{\mathbf{r}x_n\}$ have the same distribution for every rotation $\mathbf{r}$ about the origin. A process is motion-invariant if it possesses both these properties. The homogeneous Poisson process N is defined by properties (1) and (2) above and the specification of the intensity λ. These properties and the characteristic λ are clearly invariant under rotation and translation. Therefore the homogeneous Poisson process N has to be stationary and isotropic, that is to say, motion-invariant. Typically, the older, more traditional name 'homogeneous' Poisson process is used rather than the term 'stationary' Poisson process.

That the stationarity and isotropy properties do indeed hold may be verified directly by establishing that the finite-dimensional distributions above remain the same no matter whether one uses a homogeneous Poisson process N or its translation N_x or its rotation $\mathbf{r}N$.

The Poisson process has a further 'conservation property' with respect to linear transformations, where the fundamental properties are retained but only the intensity changes. This property has a quite natural interpretation and application. Consider a pattern of trees observed as aerial images, where the images have been taken at different heights and angles. This results in different patterns, which are linear transformations of the same original pattern. If the true pattern is Poisson, then the same is true for all transformed patterns, but the intensities (mean numbers of trees per unit area on the images) do of course differ. The difference in intensity can easily be calculated such that the intensity of the original pattern can be determined.

More technically, the property is formulated as follows. Let $\mathbf{A}$ be a non-singular linear mapping from $\mathbb{R}^d$ to $\mathbb{R}^d$. If N is a homogeneous Poisson process with intensity λ, then $\mathbf{A}N = \{\mathbf{A}x : x \in N\}$ is also a homogeneous Poisson process and its intensity is $\lambda \cdot \det(\mathbf{A}^{-1})$, where $\det(\mathbf{A}^{-1})$ is the determinant of the inverse of $\mathbf{A}$. A simple special case is the linear transformation

$$x = (\xi_1, \xi_2) \rightarrow \left(\frac{\xi_1}{a}, \frac{\xi_2}{b} \right),$$

which transforms a rectangle of side lengths a and b into the unit square. Clearly, the intensity of the transformed process is $ab\lambda$.

(d) *Void probabilities.* The *void probabilities* v_K of a point process are the probabilities of there being no point of the process in given test sets K:

$$v_K = \mathbf{P}(N(K) = 0).$$

If N is a homogeneous Poisson process,

$$v_K = \exp(-\lambda\nu(K)). \tag{2.3.4}$$

The contact distribution functions are closely related to the void probabilities. An important special case is the *spherical contact distribution function* as introduced in (1.5.6),

$$H_s(r) = 1 - \mathbf{P}(N(b(o, r)) = 0). \tag{2.3.5}$$

By (2.3.4) this yields for the Poisson process case

$$H_s(r) = 1 - \exp(-\lambda b_d r^d) \qquad \text{for } r \geq 0. \tag{2.3.6}$$

This is the distribution function of the distance from o to the nearest point of N; see Section 4.2.5. The mean and variance of this distribution for $d = 2$ are given by

$$m_D = \frac{1}{2\sqrt{\lambda}} \tag{2.3.7}$$

and

$$\sigma_D^2 = \frac{1}{\pi\lambda} - \frac{1}{4\lambda}. \tag{2.3.8}$$

The idea can be generalised by considering an arbitrary set B with $\nu(B) > 0$ and $o \in B$. The *contact distribution function* $H_B(r)$ *(with respect to B)* is given by

$$H_B(r) = 1 - v_{rB} = 1 - \mathbf{P}(N(rB) = 0)) \qquad \text{for } r \geq 0.$$

It makes sense to consider, for example, the case of a rectangular B. The spherical contact d.f. is $H_B(r)$ with $B = b(o, 1)$.

(e) *Conditioning and binomial point processes.* Consider the restriction of the homogeneous Poisson process N to a bounded set W conditional on $N(W) = n$. This 'conditioning' yields a new point process which is actually the binomial point process in W with n points.

This assertion can be proved by showing that the finite-dimensional distributions of the two processes coincide. In fact, it suffices to consider only the void probabilities. If K is a bounded subset of W then the void probability for K of the conditioned homogeneous Poisson process is given by

$$
\begin{aligned}
\mathbf{P}(N(K) = 0 | N(W) = n) &= \frac{\mathbf{P}(N(K) = 0, N(W) = n)}{\mathbf{P}(N(W) = n)} \\
&= \frac{\mathbf{P}(N(K) = 0)\mathbf{P}(N(W \setminus K) = n)}{\mathbf{P}(N(W) = n)} = \frac{(\nu(W) - \nu(K))^n}{\nu(W)^n}
\end{aligned}
$$

by (a) after substitution and cancellation. This formula coincides with (2.2.4) for the void probabilities of the binomial point process.

2.3.3   Characterisations of the homogeneous Poisson process

One may ask whether properties (1) and (2) are both necessary to completely characterise a homogeneous Poisson process? Indeed, the properties are not logically independent; see Kingman (1993). Rényi (1967) showed that property (1) implies property (2); the Poisson distribution property forces the point process to have the independent scattering property as described in (e) above. One may thus suspect that property (2) is superfluous.

However, if property (1) is weakened, property (2) may be necessary for the characterisation of the homogeneous Poisson process. More specifically, property (2) does *not* follow from (1) if (1) holds only for the class of all connected subsets of $\mathbb{R}^d$. Moreover, Moran (1976) shows that it is not enough to assume the independence and Poisson distribution of counts of points in k arbitrary disjoint convex sets for some fixed k. In particular, one can construct point processes with Poisson counts in one size of quadrats which are nevertheless not Poisson processes (Dale, 1999).

Another set of properties also characterises the Poisson process, and should be borne in mind as it frequently provides a *prima facie* case for assuming that an empirical point pattern is a realisation of a homogeneous Poisson process. This characterisation asserts that a process must be a homogeneous Poisson process if the following three properties are satisfied:

(i) *Simplicity.* Two points never coincide, i.e. the process is a *simple* point process; see Section 1.5.

(ii) *Stationarity.* This has been defined above; see also Section 1.6.

(iii) *Independent scattering.* This is property (2) as given at the beginning of this section.

The most important condition is that of independent scattering. In effect, it asserts that there is no interaction between the points of the pattern, and it may be tested statistically. In some cases, independent scattering is suggested by underlying biological or physical theories. Whether this is plausible or not, it frequently provides the starting point for a statistical analysis, even if only as a null hypothesis.

Note that a Poisson process is sometimes used as a 'white noise' process, corresponding to the independent and identically distributed (i.i.d.) random variables in a times series context, which model completely random disturbance effects. A good explanation of the fact that point patterns often behave like samples of Poisson processes is the *Poisson convergence theorem* which says that the superposition of many thin point processes is a Poisson process (see Kallenberg, 2002, Theorem 16.18).

2.4 Simulation of a homogeneous Poisson process

As the above sections have shown, many formulas have been derived for the Poisson process and many important characteristics can therefore be calculated exactly. Nevertheless, it is still important to be able to also simulate this process. There are three main reasons for this:

- There are situations where even Poisson process characteristics are difficult to obtain, for example in the context of tests, as shown in Section 2.7.7 below.

- The Poisson process is a building block for more complex processes. These can often only be explored by simulation. The simulation of many more complex models is based on the simulation of a Poisson process.

- To motivate the use of simulations in point process statistics and to facilitate understanding, it might be helpful initially to use a simple example.

The starting point for the simulation of a homogeneous Poisson process in a bounded set W is property (e) in Section 2.3.2 that conditioning on the total number of points in W yields a binomial point process. As a result, once the total number of points has been determined, simulation methods for the binomial process as described in Section 2.2 may be applied.

Thus the simulation essentially consists of two steps. First, the number of points in W is determined by simulating a Poisson random variable, and then the positions of the points in W are determined by simulating a binomial point process in W with the number of points determined in the first step.

The following describes two possibilities for the first step, the simulation of a Poisson random variable. Which of the two methods is more appropriate depends on the mean $\lambda \cdot \nu(W)$ one intends the simulated pattern to have.

If $\lambda \cdot \nu(W)$ is small, a linear (or one-dimensional) Poisson process may be simulated, exploiting the fact that for such a process the (one-dimensional) inter-point distances are independent exponential random variables. For this purpose random numbers e_i from an exponential distribution with mean 1 are generated. This can be achieved by applying the inversion method to uniform random numbers. If u_i is uniform on $[0, 1]$ then $e_i = -\ln(u_i)$ is as required. The desired Poisson random variable is the smallest n for which

$$\sum_{i=1}^{n+1} e_i > \lambda \cdot \nu(W). \qquad (2.4.1)$$

Since addition of logarithms is equivalent to multiplication, the Poisson variable can be determined more elegantly by

$$\prod_{i=1}^{n+1} u_i < \exp(-\lambda \cdot \nu(W)). \qquad (2.4.2)$$

If $\lambda \cdot \nu(W)$ is large, some form of rejection technique can be used; see Devroye (1986) and Ripley (1987). Alternatively, the central limit theorem may simply be exploited. It states that for large μ, a Poisson random variable with mean μ approximately follows a Gaussian distribution with mean μ and variance μ. This is because the mean and variance coincide for a Poisson distribution. Thus a Gaussian random number may be generated by well-known methods and then rounded to an integer.

In the special case where W is a disc or a sphere, another method may be used based on polar coordinates, as described on p. 65. This method, developed by Quine and Watson (1984), is more elegant than the rejection method and is called 'radial generation'.

2.5 Model characteristics

2.5.1 Moments and moment measures*

Just as moments are important characteristics for 'standard' random variables, so are the corresponding entities for the Poisson process and for general point processes. Section 1.5 shows that the moments of a point process are given by measures. These *moment measures* are described here for the specific case of the homogeneous Poisson process, where explicit formulas are known. These may appear rather technical to the applied reader, who might prefer to skip this section. At a later stage, however, when it has become clearer that the theory is relevant, it might be useful to return to this section.

Consider the homogeneous Poisson process N and take a bounded subset B of $\mathbb{R}^d$. Then $N(B)$ is a random variable with first moment or mean

$$\lambda \nu(B) = \mathbf{E}(N(B)). \tag{2.5.1}$$

If B_1 and B_2 are two bounded sets, $N(B_1)$ and $N(B_2)$ are two random variables with a non-centred covariance

$$\mu^{(2)}(B_1 \times B_2) = \mathbf{E}(N(B_1)N(B_2)).$$

The second-order quantity $\mu^{(2)}(B_1 \times B_2)$ is evaluated by using properties (1) and (2) in Section 2.3.1 as follows. Note that both B_1 and B_2 may be decomposed into disjoint unions

$$B_1 = (B_1 \cap B_2) \cup (B_1 \setminus B_2),$$
$$B_2 = (B_1 \cap B_2) \cup (B_2 \setminus B_1).$$

Applying property (2) and the fact that N satisfies

$$N(A \cup B) = N(A) + N(B)$$

for arbitrary non-intersecting A and B, one can establish

$$
\begin{aligned}
\mu^{(2)}(B_1 \times B_2) &= \mathbf{E}(N(B_1)N(B_2)) \\
&= \mathbf{E}(N(B_1 \setminus B_2)) \cdot \mathbf{E}(N(B_2 \setminus B_1)) \\
&\quad + \mathbf{E}(N(B_1 \cap B_2)) \cdot \mathbf{E}(N(B_2 \setminus B_1)) \\
&\quad + \mathbf{E}(N(B_1 \setminus B_2)) \cdot \mathbf{E}(N(B_1 \cap B_2)) + \mathbf{E}((N(B_1 \cap B_2))^2) \\
&= \mathbf{E}(N(B_1)) \cdot \mathbf{E}(N(B_2)) + \mathbf{E}((N(B_1 \cap B_2))^2) \\
&\quad - (\mathbf{E}(N(B_1 \cap B_2)))^2.
\end{aligned}
$$

Property (1) ensures that $N(B_1 \cap B_2)$ is Poisson with mean and variance $\mathbf{E}(N(B_1 \cap B_2))$. Using (2.5.1) the final formula can be derived:

$$
\begin{aligned}
\mu^{(2)}(B_1 \times B_2) &= \mathbf{E}(N(B_1))\mathbf{E}(N(B_2)) + \mathbf{E}(N(B_1 \cap B_2)) \\
&= \lambda^2 \cdot \nu(B_1) \cdot \nu(B_2) + \lambda \cdot \nu(B_1 \cap B_2). \tag{2.5.2}
\end{aligned}
$$

Thus, the *second-order moment measure* $\mu^{(2)}$ can be expressed in terms of λ and the volume measure ν. Note that $\mu^{(2)}$ is a measure on $\mathbb{R}^d \times \mathbb{R}^d$.

Variances and covariances can be calculated directly from the second-order moment measure (this is also true for general point processes). The relevant formulas are

$$\textbf{var}(N(B)) = \mu^{(2)}(B \times B) - (\textbf{E}(N(B)))^2,$$

$$\textbf{cov } (N(B_1), N(B_2)) = \mu^{(2)}(B_1 \times B_2) - \textbf{E}(N(B_1)) \cdot \textbf{E}(N(B_2))$$

for all sets B, B_1 and B_2. These equations follow immediately from the definitions of variance and covariance. In the case of the homogeneous Poisson process one can use (2.5.2) or calculate directly from properties (1) and (2) to show that

$$\textbf{cov}(N(B_1), N(B_2)) = \lambda \cdot \nu(B_1 \cap B_2). \tag{2.5.3}$$

Note that $\mu^{(2)}(B_1 \times B_2)$ can also be expressed as the expectation of this sum:

$$\mu^{(2)}(B_1 \times B_2) = \textbf{E} \left(\# \{(x_1, x_2) : x_1 \in N \cap B_1, x_2 \in N \cap B_2\} \right)$$

$$= \textbf{E} \left(\sum_{x_1, x_2 \in N} \mathbf{1}_{B_1}(x_1) \mathbf{1}_{B_2}(x_2) \right).$$

That is, $\mu^{(2)}(B_1 \times B_2)$ is the mean number of pairs of points (x_1, x_2) in N with $x_1 \in B_1$ and $x_2 \in B_2$.

The two terms in (2.5.2) correspond to the dissection of this sum into the sum over distinct pairs of points $x_1, x_2 \in N$ and the sum over equal points $x_1 = x_2 \in N$. For some purposes it is convenient to subtract out the second of these terms. The result is the *second-order factorial moment measure* $\alpha^{(2)}$ given by

$$\alpha^{(2)}(B_1 \times B_2) = \textbf{E} \left(\# \{(x_1, x_2) : x_1 \in N \cap B_1, \ x_2 \in N \cap B_2, \ x_1 \neq x_2\} \right)$$

$$= \textbf{E} \left(\sum_{x_1, x_2 \in N}^{\neq} \mathbf{1}_{B_1}(x_1) \mathbf{1}_{B_2}(x_2) \right).$$

Here $\sum^{\neq}$ denotes the summation over all pairs (x_1, x_2) such that $x_1 \neq x_2$.

Since the terms $\mu^{(2)}(B_1 \times B_2)$ and $\alpha^{(2)}(B_1 \times B_2)$ differ only by the expectation of the sum

$$\sum_{x_1, x_2 \in N : x_1 = x_2} \mathbf{1}_{B_1}(x_1) \mathbf{1}_{B_2}(x_2) = \sum_{x \in N} \mathbf{1}_{B_1 \cap B_2}(x),$$

we have

$$\mu^{(2)}(B_1 \times B_2) = \alpha^{(2)}(B_1 \times B_2) + \Lambda(B_1 \cap B_2).$$

In the case of the homogeneous Poisson process,

$$\alpha^{(2)}(B_1 \times B_2) = \lambda^2 \cdot \nu(B_1) \cdot \nu(B_2) = \mathbf{E}(N(B_1)) \cdot \mathbf{E}(N(B_2)). \qquad (2.5.4)$$

Thus the *second-order product density* $\varrho^{(2)}$ has the very simple form

$$\varrho^{(2)}(x_1, x_2) = \lambda^2 \qquad \text{for } x_1, x_2 \text{ in } \mathbb{R}^d. \qquad (2.5.5)$$

By analogy with higher-order moments for random variables, higher-order moment measures may be defined for point processes. The *nth-order moment measure* $\mu^{(n)}$ is a measure on $\mathbb{R}^{nd}$ defined by

$$\mu^{(n)}(B_1 \times \cdots \times B_n) = \mathbf{E}\left(N(B_1) \cdot \cdots \cdot N(B_n)\right)$$

$$= \mathbf{E}\left(\sum_{x_1,\ldots,x_n \in N} \mathbf{1}_{B_1}(x_1) \cdot \cdots \cdot \mathbf{1}_{B_n}(x_n)\right)$$

for sets $B_1, \ldots, B_n$. The *nth-order factorial moment measure* $\alpha^{(n)}$ is a measure on $\mathbb{R}^{nd}$ given by

$$\alpha^{(n)}(B_1 \times \cdots \times B_n) = \mathbf{E}\left(\sum_{x_1,\ldots,x_n \in N}^{\neq} \mathbf{1}_{B_1}(x_1) \cdot \cdots \cdot \mathbf{1}_{B_n}(x_n)\right),$$

where $\sum^{\neq}$ denotes summation over n-tuples $(x_1, \ldots, x_n)$ of distinct points. The corresponding *nth-order product density* $\varrho^{(n)}$ is given by the simple formula

$$\varrho^{(n)}(x_1, \ldots, x_n) = \lambda^n \qquad (2.5.6)$$

and $\alpha^{(n)}$ has the form

$$\alpha^{(n)}(B_1 \times \cdots \times B_n) = \lambda^n \cdot \nu(B_1) \cdot \cdots \cdot \nu(B_n). \qquad (2.5.7)$$

2.5.2 The Palm distribution of a homogeneous Poisson process

In point process theory the so-called 'typical' point of a point process N is frequently considered. Informally, this is a point that has been chosen by a selection procedure in which every point of the process has the same chance of being selected. For example, the nearest-neighbour distance distribution function $D(r)$ describes the distribution of the distance from a typical point x in N to the nearest point in $N \setminus \{x\}$, i.e. to the nearest neighbour of x in N. Mathematically, the idea of the typical point is precisely described by means of the Palm distribution theory, as sketched in Section 4.1. In intuitive terms, the Palm distribution probabilities are conditional probabilities of point process events given that a point (the typical point) has been observed at a specific location.

There are two heuristic approaches to the definition of the Palm distribution. The global approach is described in Section 4.1, while the local approach is used here. It is demonstrated in the following example.

Calculation of the nearest neighbour distance distribution function for the homogeneous Poisson process

Consider a conditional nearest-neighbour distance d.f. $D_\varepsilon(r)$, i.e. assume that there is a point of N in $b(o, \varepsilon)$. The conditional probability

$$D_\varepsilon(r) = 1 - \mathbf{P}(N(b(o, r) \setminus b(o, \varepsilon)) = 0 \,|\, N(b(o, \varepsilon)) = 1)$$

is the probability that the distance from a point in the small sphere $b(o, \varepsilon)$ to its nearest neighbour in N is smaller than r, under the condition that there is indeed a point of N in the small sphere. It is well defined for positive ε smaller than r, since

$$\mathbf{P}(N(b(o, \varepsilon)) = 1) = \lambda b_d \varepsilon^d \exp\left(-\lambda b_d \varepsilon^d\right)$$

is positive. Using the definition of conditional probability and property (2) of the homogeneous Poisson process,

$$
\begin{aligned}
D_\varepsilon(r) &= 1 - \frac{\mathbf{P}(N(b(o, r) \setminus b(o, \varepsilon)) = 0)\mathbf{P}(N(b(o, \varepsilon)) = 1)}{\mathbf{P}(N(b(o, \varepsilon)) = 1)} \\
&= 1 - \mathbf{P}(N(b(o, r) \setminus b(o, \varepsilon)) = 0) \\
&= 1 - \exp(-\lambda(\nu(b(o, r)) - \nu(b(o, \varepsilon)))).
\end{aligned}
$$

It is reasonable to regard the nearest-neighbour distance d.f. $D(r)$ as the limit of the above as $\varepsilon \to 0$. Setting $D(r) = \lim_{\varepsilon \to 0} D_\varepsilon(r)$ yields the result

$$D(r) = 1 - \exp(-\lambda \nu(b(o, r)))$$

or

$$D(r) = 1 - \exp\left(-\lambda b_d r^d\right) \qquad \text{for } r \geq 0. \tag{2.5.8}$$

This result can be established in a rigorous way by means of the global approach; see Stoyan et al. (1995).

The right-hand sides of (2.5.8) and (2.3.6) are equal, i.e. the spherical contact d.f. and the nearest-neighbour distance d.f. of the homogeneous Poisson process coincide,

$$D(r) = H_s(r) \qquad \text{for } r \geq 0. \tag{2.5.9}$$

The mean and variance corresponding to $D(r)$ in the case $d = 2$ are thus given by (2.3.7) and (2.3.8).

The distribution functions $D_2(r), D_3(r), \ldots$ of the distances to the 2nd, 3rd, ... nearest neighbours are

$$D_k(r) = 1 - \sum_{j=0}^{k-1} \exp(-\lambda \pi r^2) \frac{(\lambda \pi r^2)^j}{j!} \qquad \text{for } r \geq 0, \qquad (2.5.10)$$

and the corresponding probability density functions are

$$d_k(r) = \frac{2(\lambda \pi r^2)^k}{r(k-1)!} \exp(-\lambda \pi r^2) \qquad \text{for } r \geq 0;$$

see Figure 2.4.

The corresponding jth moments are

$$m_{k,j} = \frac{\Gamma(k + \frac{1}{2} j)}{(k-1)!(\lambda \pi)^{j/2}} \qquad \text{for } j = 1, 2, \ldots,$$

and the position of the mode (maximum of density function) is

$$r_k = \sqrt{\frac{k - \frac{1}{2}}{\lambda \pi}}.$$

Similar to the case of $D(r)$ above, the *reduced second-order moment measure function* or *Ripley's K-function* $K(r)$ can also be calculated explicitly for the homogeneous Poisson process. The term $\lambda K(r)$ is the mean number of points other

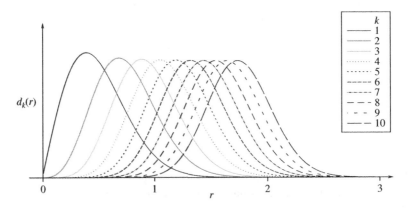

Figure 2.4   Density functions of the distances to the kth nearest neighbours for a homogenous Poisson process of intensity 1, for $k = 1, 2, \ldots, 10$.

than the typical point in a ball of radius r centred at the typical point. In the Poisson process case

$$\lambda K(r) = \mathbf{E}(N(b(o, r))), \qquad (2.5.11)$$

i.e.

$$\lambda K(r) = \lambda b_d r^d$$

and so

$$K(r) = b_d r^d \qquad \text{for } r \geq 0. \qquad (2.5.12)$$

The so-called L-function is obtained by

$$L(r) = \left(\frac{K(r)}{b_d} \right)^{\frac{1}{d}}$$

as

$$L(r) = r \qquad \text{for } r \geq 0, \qquad (2.5.13)$$

and, similarly, the pair correlation function $g(r)$ is given by

$$g(r) = 1 \qquad \text{for } r \geq 0, \qquad (2.5.14)$$

due to the general relation to $K(r)$,

$$g(r) = \frac{\mathrm{d}K(r)}{\mathrm{d}r} \bigg/ d b_d.$$

Using the symbol $\mathbf{P}_o$ for the Palm distribution, formulas (2.5.9) and (2.5.12) can be rewritten as

$\mathbf{P}$(distance from typical point to nearest neighbour in $N \leq r$)

$\quad = \mathbf{P}_o$(distance from o to nearest neighbour in $N \leq r$)

$\quad = \mathbf{P}$(distance from o to nearest neighbour in $N \leq r$)

and

$\mathbf{E}$(number of points in sphere of radius r around typical point,

which is not counted)

$\quad = \mathbf{E}_o(N(b(o, r) \setminus \{o\})) = \mathbf{E}(N(b(o, r))).$

This means that Palm characteristics of homogenous Poisson processes can be calculated as stationary characteristics, if a point in the origin o is added to N. This is the important *Slivnyak–Mecke theorem*, which states that the Palm distribution of a homogeneous Poisson process coincides with that of the point process obtained by adding the origin o to the homogeneous Poisson process. In particular,

$$\mathbf{P}_o(N \in \mathcal{A}) = \mathbf{P}(N \cup \{o\} \in \mathcal{A})$$

and

$$\mathbf{E}_o(N(B)) = \mathbf{E}(N(B)) + \mathbf{1}_B(o).$$

As in Section 1.6, $\mathcal{A}$ denotes a point process event; for example, '$N \in \mathcal{A}$' may mean that the point process N has n points in some set B.

The above theorem will be applied repeatedly in this book. The interested reader may find a formal proof in Stoyan et al. (1995). Heuristically, the theorem says that the probabilistic behaviour of a homogeneous Poisson process is the same whether it is seen from an independent test location or from a randomly chosen point in the process.

A comparison of the formulas for $D(r)$ and $K(r)$ for the Poisson process and the periodic binomial process reveals that these processes differ slightly. This shows again that they do not model the same concept and hence that the periodic binomial process is not an appropriate CSR model.

2.5.3 Summary characteristics of the homogeneous Poisson process

This section provides an overview of the results for the characteristics of the homogeneous Poisson process (see Tables 2.1–2.4). In practice, when a pattern's

Table 2.1 Functional summary characteristics for the homogeneous Poisson process.

Characteristic	Formula	Page
K-function, $K(r)$	$b_d r^d$	77
L-function, $L(r)$	r	77
Pair correlation function, $g(r)$	1	77
Nearest-neighbour distance d.f., $D(r)$	$1 - \exp(-\lambda b_d r^d)$	75
Spherical contact d.f., $H_s(r)$	$1 - \exp(-\lambda b_d r^d)$	68
J-function, $J(r)$	1	91

$\lambda =$ intensity = point density, $r =$ inter-point distance, $b_d = 2$ $(d=1)$, π $(d=2)$ and $\frac{4}{3}\pi$ $(d=3)$.

Table 2.2 Numerical summary characteristics for the homogeneous Poisson process.

Characteristic	Numerical value	Page
Index of dispersion, *ID*	1	195
Aggregation index, *CE*	1	196
Pielou index of randomness, *PI*	1	196
Mean-direction index, $\overline{R}_4$	1.799	93

Table 2.3 Formulas for the Poisson–Voronoi tessellation, planar case.

Characteristic	Mean	Variance
Cell area	λ^{-1}	$0.2802\lambda^{-2}$
Perimeter	$4\sqrt{\lambda}/\lambda$	$0.9455\lambda^{-1}$
Number of vertices	6	1.7808

Table 2.4 Formulas for the Poisson–Voronoi tessellation, spatial case.

Characteristic	Mean	Variance
Cell volume	λ^{-1}	$0.179\lambda^{-2}$
Surface	$5.821\lambda^{-2/3}$	$2.19\lambda^{-4/3}$
Number of vertices	27.071	43.99

properties are assessed with regard to CSR, these formulas are regularly applied. They will also be used as a reference throughout this book when other processes are compared to the homogeneous Poisson process.

Formulas for further tessellation characteristics may be found in Stoyan et al. (1995). The exact distributions of area, volume, perimeter, surface, and edge number are unknown. For volume and area, approximations have been derived which use gamma distributions.

2.6 Estimating the intensity

The most fundamental statistical question for the homogeneous Poisson process concerns the estimation of the intensity λ, which is the only parameter in the model. Several estimation methods are available. In a specific situation, the sampling conditions determine which estimation method may be the most suitable.

Counting method

If all points in the sampling window W can be counted, the number of points $N(W)$ divided by the area or volume $\nu(W)$ of the window is the natural estimator:

$$\hat{\lambda} = \frac{N(W)}{\nu(W)}. \tag{2.6.1}$$

This estimator has some nice statistical properties as it is unbiased and the maximum likelihood estimator. Its variance is

$$\mathbf{var}(\hat{\lambda}) = \frac{\lambda}{\nu(W)}. \tag{2.6.2}$$

Note that $\hat{\lambda}$ is also unbiased for *all* stationary point processes, even for those that are not Poisson processes; see Section 4.2.3.

Distance methods

In applications, it may be practically impossible (e.g. if it is too time-consuming) to count all points. In this case, distance methods can be used. The idea behind these methods is to measure the distances to points from deterministic test locations or from points in the process. Unfortunately, this old idea (see Section 1.3.1) can only be applied with acceptable precision if the distribution of the point process is known. As noted, the distribution is not explicitly known for most point processes, and so the only case where it may be realistically applied is the Poisson process.

Distance methods for the Poisson case are discussed in detail in Diggle (2003) and Krebs (1998). Here, only the case of deterministic test locations is considered, assuming that the distances between the locations are large enough for the nearest-neighbour distances from the test locations to be approximately independent.

Point-quarter method

More information can be obtained from each test location (indexed by j) by determining four nearest-neighbour distances, namely those in the north-east quadrant,

h_{1j}, ..., and in the north-west quadrant, h_{4j}. Under the assumption of a Poisson process, for each of the test locations the nearest-neighbour distance in a quadrant follows the distribution function

$$H_{s/4}(r) = 1 - \exp\left(-\frac{\pi}{4}\lambda r^2\right) \qquad \text{for } r \geq 0.$$

If the quadruplets of nearest-neighbour distances for different locations were independent (which is not exactly true but holds approximately if the inter-test-point distances are large) λ can be estimated by the maximum likelihood method (Cottam et al., 1953; Diggle, 1982):

$$\hat{\lambda} = \frac{4}{\pi \overline{h^2}} \tag{2.6.3}$$

with

$$\overline{h^2} = \sum_{j=1}^{k} \sum_{i=1}^{4} h_{ij}^2 / 4k,$$

where k is the number of test locations.

Confidence intervals

Confidence intervals for λ can be based on (2.6.1) since $\hat{\lambda} \cdot \nu(W) = N(W)$ has a Poisson distribution. For large $N(W)$, Armitage et al. (2001) and Krebs (1998) provide a simple approximate $100(1 - \alpha)\%$ confidence interval for λ, employing the normal approximation and a continuity correction. For example,

$$\left(\frac{z_{\alpha/2}}{2} - \sqrt{N(W)}\right)^2 \leq \lambda\nu(W) \leq \left(\frac{z_{\alpha/2}}{2} + \sqrt{N(W)+1}\right)^2. \tag{2.6.4}$$

Here the $z_{\alpha/2}$ are quantiles of the standard normal distribution:

$$z_{\alpha/2} = 1.65, \ 1.96, \ 2.58, \qquad \text{for } \alpha = 0.10, \ 0.05, \ 0.01.$$

In practice, this confidence interval is a simple tool for determining the window size required for a given accuracy of estimation. If δ is the desired (full) width of the confidence interval and α is the required confidence level, then

$$\delta \cdot \nu(W) \simeq \left(\frac{z_{\alpha/2}}{2} + \sqrt{\lambda\nu(W)+1}\right)^2 - \left(\frac{z_{\alpha/2}}{2} - \sqrt{\lambda\nu(W)}\right)^2$$

yields the approximation

$$\nu(W) \simeq \frac{4z_{\alpha/2}^2 \lambda}{\delta^2}, \tag{2.6.5}$$

where λ itself must be estimated based on data which perhaps have been derived from a pilot study or by using a priori information.

Example 2.1.   Gold particles: intensity estimation
Consider the pattern of gold particles discussed on p. 6 and assume for the moment that it can be modelled by a homogeneous Poisson process. Formula (2.6.4) yields the confidence interval (0.000 754, 0.000 984) for λ for $\alpha = 0.05$. For the same α and $\lambda = 8 \cdot 10^{-4}$ and $\delta = 10^{-4}$, Formula (2.6.5) suggests a window size of 1 229 300. That is to say, a square of size 1100×1100 is sufficient to obtain 95 % confidence with an accuracy of 10^{-4}. If the true point process is not Poisson, this sample size may still be used as an approximation, which is probably a little too large.

Testing homogeneity of point density

Often, the observation window is large and the question may arise whether the point distribution is really homogeneous (i.e. whether the intensity function $\lambda(x)$ is really constant). For this purpose, the sampling window W may be split into two subregions W_1 and W_2. To test the hypothesis that the point densities in W_1 and W_2 are equal, one may consider the observed numbers of points n_1 and n_2 and subregion measures $\nu(W_1)$ and $\nu(W_2)$.

Under the null hypothesis of equality, i.e. the hypothesis that the whole pattern is a realisation a homogeneous Poisson process, the quantity

$$F = \frac{\nu(W_1)(2n_2 + 1)}{\nu(W_2)(2n_1 + 1)},$$

follows approximately the F-distribution with $2n_1 + 1$ and $2n_2 + 1$ degrees of freedom. Note that this assumes that the order of the subscripts 1 and 2 is chosen such that F is greater than one. The equality hypothesis is rejected at a significance level α if

$$F > F_{2n_1+1,2n_2+1;\alpha/2}.$$

Formally this test is correct only if the subregions W_1 and W_2 are chosen a priori, before collecting the data, perhaps to investigate specific possibilities of a trend in the point density. For example, such a trend might relate the point density of tree locations to some measure of nutrient levels or the water contents in the soil. It is not appropriate to apply the test a posteriori to a given point pattern that has unexpected subregions of very high or very low point densities on inspection.

2.7 Testing complete spatial randomness

2.7.1 Introduction

Testing the CSR hypothesis is an important part of exploratory data analysis of point patterns: if the hypothesis is accepted, one can assume that the given point pattern is completely spatially random, which has two main consequences:

(a) There is no need to consider a more complicated model; the simple Poisson process model can be used, with all its consequences.

(b) If no additional information or data on underlying processes is available it is not possible to find indicators of interesting interaction between the points based on the geometry of the observed pattern alone.

Using tests of CSR, Tomppo (1986) showed that 30 % of the permanent inventory plots in Finland of thinned forests in mineral soil areas can be considered CSR. But note with respect to point (b) that there are examples from ecology which show that Poisson-process-like point patterns can result from ecological processes operating on clustered patterns. Note, further, that real patterns are never truly Poisson process patterns; not rejecting the CSR hypothesis only means that the pattern is 'close to' a Poisson process sample.

If the CSR hypothesis is rejected, the more interesting part of point process statistics begins, in particular the search for spatial correlations in the given pattern. Nevertheless, the initial analysis provides valuable information on the direction of the deviation from CSR, as well as on the cause of the deviation to guide further analysis.

A large number of tests of the CSR hypothesis have been developed, and research in this area is still ongoing. Many of the summary characteristics of point processes can be used to construct such tests, and those that yield extremal or simple values in the Poisson case are particularly successful. Experience shows that it is not possible to derive a 'best' test if it is only based on a single criterion, and any test is only capable of assessing particular aspects of CSR behaviour. Which test is the most appropriate in a given situation depends on the nature of the alternative hypothesis envisaged. In practical applications, the choice of the appropriate test also depends on the limitations imposed by sampling methods. Nevertheless, rejection of the CSR hypothesis by any of the tests means final rejection. (Perhaps this last sentence should be reformulated by replacing 'test' by 'standard test', since for true CSR patterns there are always small deviations from a Poisson process behaviour which can be used, after inspection of the given pattern, in a highly specialised test for rejecting CSR.)

In the following, a series of CSR tests will be presented. Note that some of these tests use point process characteristics which have not yet been explained. These will be introduced in Chapter 4. Glance over the text below and then go to Chapter 4 to obtain the necessary information and return to this chapter.

The CSR tests are presented in order of data quality, from the coarsest to the most detailed data type:

- quadrat counts,
- distance measurement,
- measurement of directions to neighbours as angles,
- use of mapped data.

When the point pattern has been mapped exhaustively, such that measurements of the locations of all points in the pattern are available, it is worthwhile to use more sophisticated statistics and tests. These rely on computers and point process statistics software. Two approaches have been devised.

- The traditional approach uses point process statistics software, for example, for the estimation of the L- or J-function.

- The more statistical approach is based on the fact that the point distribution of a Poisson process is uniform. If the window W is a rectangle or a parallelepiped then it can be transformed to the unit square or cube and the given point pattern can be considered a sample from a binomial process or even a periodic binomial process. It is then possible to use discrepancy measures from quasi-Monte Carlo methods in numerical integration techniques (Hua and Wang, 1981; Niederreiter, 1992, Hickernell et al., 2005), which evaluate the discrepancy between the empirical point distribution and the uniform distribution; see Ho and Chiu (2007a). Alternatively, the usual summary characteristics of point processes under the periodic binomial process assumption are estimated with an adapted edge-correction; see Section 2.7.7.

A quite natural approach is the application of goodness-of-fit tests for the uniformity hypothesis. This includes classical goodness-of-fit tests, such as the χ^2-test in Section 2.7.2 and the Kolmogorov–Smirnov test in Section 2.7.7. The authors' favourite test is the L-test based on the L-function, which requires mapped data. For the sake of the exposition, the two-dimensional case is mainly considered here. In most cases a generalisation to the three-dimensional case is straightforward.

General idea of CSR tests

Most CSR tests are constructed as follows. A summary characteristic is estimated for the data and compared with the relevant theoretical summary characteristic for a Poisson process. If there is a large difference between both characteristics, the Poisson null hypothesis is rejected. The tests may be based on either numerical summary characteristics, i.e. a single value, or functional summary characteristics, i.e. a function of distance r.

Numerical summary characteristic M: Let $\hat{M}$ be the estimator of some summary characteristic M and M_P the theoretical value for a Poisson process. Reject the CSR hypothesis if

$$|\hat{M} - M_P| > M_\alpha,$$

where M_α is some critical value. The test is more powerful if it depends on the alternatives 'clustering' and 'regularity': reject the CSR hypothesis if

$$\hat{M} > M_\alpha \qquad \text{if clustering alternative}$$

and

$$\hat{M} < M_{1-\alpha} \qquad \text{if regularity alternative.}$$

An example of such a test is the mean-direction test; see Section 2.7.5. Note that the terms 'clustering' and 'regularity' should not be used naively, since the behaviour of a point process can be different at different spatial scales.

Functional summary characteristic S(r): Let $\hat{S}(r)$ be an estimator of some summary characteristic $S(r)$ and let $S_P(r)$ be its theoretical Poisson counterpart. Reject the CSR hypothesis if

$$\max_{r \le s} |\hat{S}(r) - S_P(r)| > S_\alpha^m \qquad (2.7.1)$$

or if

$$\int_0^s (\hat{S}(r) - S_P(r))^2 \mathrm{d}r > S_\alpha^i. \qquad (2.7.2)$$

The test statistics in (2.7.1) and (2.7.2) are called the *maximum statistic* and *integral statistic,* respectively (Thönnes and Van Lieshout, 1999). In the integral of (2.7.2), $(\cdot)^2$ can be replaced by $|\cdot|$ or $|\cdot|^\gamma$ with some $\gamma > 0$.

The choice of the maximum r-value s is crucial. Advice on the appropriate choice of s is given for the specific tests. Choosing values for s that are too large reduces the power of the tests. This is because r is usually an (inter-point) distance and, for large r, $S(r)$ cannot be estimated from a bounded window of observation or only with large variance.

Finally, the method used for the calculation of $\hat{S}(r)$ has an impact on the result. This clear fact was ignored for a long time but demonstrated by Ho and Chiu (2006) for the L-test.

In some cases, critical values for CSR tests are known. If this is not the case, simulation tests have to be used. A test of this type is discussed on p. 96 in the context of the L-test.

The alternative hypotheses are usually *clustering* or *regularity,* and it is assumed that background knowledge suggests which alternative is most suitable in the given context. Indeed, this assumption is realistic, since some anecdotal evidence or prior knowledge is typically available: for example, in forestry it is known that young trees in natural forests tend to show clustering as the seedlings sprout in proximity to the parent tree, while old trees in cultivated forests tend to be regular. In materials science applications, knowledge of the processes that lead to the structures enables a prediction as to whether the alternative hypothesis should be one of regularity or clustering. Visual inspection of the pattern is often helpful.

The alternative hypothesis has to be chosen prior to data collection. It is not a good idea to estimate a summary characteristic, to establish the alternative hypothesis based on a comparison with the corresponding Poisson process value and to then show that the observed deviation is significant.

There are, of course, other types of deviation from the CSR hypothesis: the point pattern may show clustering and regularity in combination, for example at different spatial scales or in different regions. Also, *inhomogeneity* or non-stationarity is an important case of non-CSR. Often inhomogeneity causes CSR tests to indicate (spurious) clustering.

Example 2.2.   Gold particles: CSR testing
In the following, the pattern of gold particles introduced in Chapter 1 (see Figure 1.3) will serve as a test pattern and all CSR tests will be applied to it. Visual inspection encourages this choice: the pattern looks rather homogeneous, hence the assumption that it is a sample of a stationary and isotropic point process can be accepted.

The following will reveal the results based on the statistical methods for the gold pattern, eventually yielding a clear result. Prior to reading this, readers might now take a moment to write down what they think of this pattern after visual inspection, and why.

Figures 4.14, 4.18 and 4.20 may be consulted which show the empirical nearest-neighbour distance d.f., the L-function and the pair correlation function in comparison to the corresponding Poisson process characteristics. Figure 4.18 shows the estimated L-function for the gold particle pattern and the maximum and minimum envelopes from 99 simulations of binomial processes with 218 points in a 630×400 window. There are big deviations for small r (up to $r = 5.65$) because of the hard-core of the empirical pattern. Note also that $\hat{L}(r)$ is close to r at distances as close as $r = 30$, which may mean that there is some clustering, combined with the hard-core effect.

The pattern of positions of *Phlebocarya* in Figure 1.4 shows such a high degree of clustering that a CSR does not make sense; it cannot be regarded as a sample of a homogeneous Poisson process.

2.7.2   Quadrat counts

The following tests are suitable if the sampling window W has been divided into k different subregions of equal area $\nu(Q)$. In this case, the point pattern has been sampled as counts of numbers of points falling into these defined subregions. In applications, these are often *quadrats* or squares, i.e. the data have been collected on a grid with rectangular grid cells of equal size. Note that mapped data may be transformed into quadrat counts by admitting a grid on the pattern, and the following tests are equally applicable then. However, since this results in a loss of valuable information, the use of the better adapted methods described below (e.g. Section 2.7.7) is recommended.

Of course, the index-of-dispersion test and quadrat count tests in general have the disadvantage that quite different processes may yet yield quadrat counts of a similar distribution. Furthermore, the test depends on the number k of quadrats, which can be chosen freely.

Under the hypothesis of a homogeneous Poisson process, the random number of points counted in these quadrats follows a Poisson distribution of mean $\lambda \cdot \nu(Q)$ and counts in disjoint quadrats are independent. This is convenient, as statistical tests can be based on these distributional properties and it is not necessary to apply simulation-based tests.

First the index-of-dispersion test is introduced, which is the simplest test. It is followed by its refined variant, the Greig-Smith method, which investigates the extent to which the independence properties are valid.

Index-of-dispersion test

The index of dispersion I is defined by

$$I = \frac{(k-1)s^2}{\bar{x}}, \tag{2.7.3}$$

where k is the number of quadrats, $\bar{x}$ is the mean number of points per quadrat, and s^2 is the sample variance of the number of points per quadrat.

This test statistic may seem to have been inspired by the fact that the mean and variance of a Poisson distribution are equal. But this is not the case: it is exactly the same test statistic as that of a χ^2 goodness-of-fit test of the hypothesis that the n points are independently and uniformly distributed in W. Consequently, the index I follows approximately a χ^2-distribution with $k - 1$ degrees of freedom provided that $k > 6$ and $\lambda \cdot \nu(Q) > 1$ (Diggle, 2003).

Hence, if I exceeds $\chi^2_{k-1;\alpha}$ or I is smaller than $\chi^2_{k-1;1-\alpha}$ the test rejects the CSR hypothesis at a significance level of $100\alpha\,\%$.

In the first case, the alternative hypothesis is that the variability in the process is stronger than for the Poisson process, i.e. that there is aggregation. Analogously, in the second case the alternative hypothesis is that there is some regularity in the point pattern as the variability is smaller.

Table 2.5   Results of the index-of-dispersion test for the gold particles.

k	$\chi^2_{0.95}$	I	$\chi^2_{0.05}$
4	0.352	0.46	7.82
16	7.26	13.05	25.0
64	46.6	70.88	82.5
256	218.7	319.83	293.0

Table 2.6   Quadrat counts for the gold particles and $k = 64$.

3	4	6	2	8	4	0	1
5	1	2	4	2	2	5	2
6	1	3	6	2	7	4	2
3	2	2	7	5	4	3	6
4	4	1	3	4	3	0	6
2	6	2	3	3	8	3	1
2	1	2	2	3	2	6	4
4	3	6	7	2	3	2	2

Example 2.3.   Gold particles: index-of-dispersion test

Now the index-of-dispersion test is applied to the gold particle data. In addition to illustrating its use, the test's heavy dependence on the choice of grid can be illustrated well with this example.

Table 2.5 shows the results of the index-of-dispersion test for $k = 4, 16, 64$ and 256. For the smaller k, i.e. the larger quadrats, the CSR hypothesis is not rejected, which might show that on a larger scale the pattern behaves like a Poisson process. However, for $k = 256$ the Poisson hypothesis is clearly rejected – at small distances there are deviations from Poisson behaviour, i.e. clustering.

The quadrat counts for the case of $k = 64$ are shown in Table 2.6. Here the 'quadrats' are rectangles of side lengths 630/8 and 400/8.

Greig-Smith method

This method, introduced in Greig-Smith (1964), is an improvement on the simple quadrat count as it does not consider a single quadrat size but also uses counts grouped by neighbouring quadrats. In this way, clustering at different scales may be detected. The example below illustrates this.

It is assumed that the number of quadrats is $k = 2^q$ for some integer q. The quantities $s_1, s_2, \ldots$ are calculated by

$$s_1^2 = \sum_{(j)} \left(\begin{array}{c} \text{number in the} \\ j\text{th quadrat} \end{array} \right)^2 - \frac{1}{2} \sum_{(k)} \left(\begin{array}{c} \text{number in the } k\text{th} \\ \text{pair of quadrats} \end{array} \right)^2,$$

$$s_2^2 = \sum_{(j)} \left(\begin{array}{c} \text{number in the } j\text{th} \\ \text{pair of quadrats} \end{array} \right)^2 - \frac{1}{2} \sum_{(k)} \left(\begin{array}{c} \text{number in the } k\text{th} \\ \text{quartet of quadrats} \end{array} \right)^2,$$

and so on.

For $q=6$, the quadrats can be numbered in chessboard notation, since $2^q = 64$. The pairs of quadrats are then the pairs $(a1, a2), (a3, a4), \ldots, (b1, b2), (b3, b4), \ldots$. The quartets of quadrats are then $(a1, a2, b1, b2), (a3, a4, b3, b4), \ldots$.

The quantities $s_j^2/2^{q-j}$ are unbiased estimators of the variance of the random number of points in a small quadrat. Thus, the index-of-dispersion formula may now be generalised, the indices

$$I_j = s_j^2 / \bar{x}$$

may be calculated and the χ^2-test applied to each of the indices.

Here $\bar{x}$ is the same as in the index-of-dispersion test. The degrees of freedom for the χ^2-statistics $I_1, I_2, \ldots$ are $2^{q-1}, 2^{q-2}, \ldots$.

The results of the tests should be interpreted as follows. The hypothesis of a homogeneous Poisson process can be accepted if all the indices I_j lie between two-sided critical values of the appropriate χ^2-statistics. Otherwise, the values of the indices indicate the nature of the deviation from the hypothesis. Thus if, for example, $I_3/2^{q-3}$ is significantly greater than one, there is evidence of clustering on the scale of groups of four quadrats, etc.

2.7.3 Distance methods

In some circumstances, the measurements most readily available are distances between pairs of points of the pattern, and between points of the pattern and chosen test locations. For example, this is often the case when point patterns of forest tree locations are being investigated. Distance methods take these limitations into account and work with these measurements alone. However, when the point pattern has been mapped, better tests are available, i.e. those using L- and J-functions.

Equations (2.3.6) and (2.5.8) for the distance d.f.s of the Poisson process form the basis of distance methods as applied to testing the CSR hypothesis. Deviations of the estimators $\hat{D}(r)$ and $\hat{H}_s(r)$ from $1 - \exp(-\pi\lambda r^2)$ indicate deviations from CSR. A further starting point is equation (2.5.9)

$$D(r) = H_s(r), \qquad \text{for } r \geq 0.$$

This equation asserts that the nearest-neighbour distance distribution function $D(r)$ and the spherical contact distribution function $H_s(r)$ coincide for a homogeneous Poisson process. One might hope to obtain empirical distribution functions corresponding to D and H_s based on distance measurements and then test for their equality. For example, Diggle (2003) recommends the test statistic

$$\int\limits_0^t \left(\hat{H}_s(r) - \hat{D}(r) \right)^2 dr$$

for some t, which may play the role of M on p. 85. The critical values M_α can be determined by simulation; only large deviations are relevant.

A particular difficulty arises in the distance approach: in order to obtain nearest-neighbour distances one has to sample the points of the pattern at random. However, this should be done without enumerating all points of the pattern as otherwise distance methods lose their practical advantage. How can the points of the pattern be sampled at random without a bias towards points occurring in particular locations?

Byth and Ripley (1980) give an example of a method that overcomes this problem. A number $2m$ of locations are chosen in the window W by semi-systematic sampling. Half of these are used as test points from which to measure distances to nearest points in the pattern. These distances are denoted by $u_1, \ldots, u_m$. The other half of the locations are used to define subregions within which the point pattern is surveyed exhaustively. A common size for all subregions is chosen, with an expected number of about five points of the pattern falling into each subregion. A point in the pattern is selected at random from each subregion and the nearest-neighbour distance is measured for each of these points. This gives rise to m further distances $v_1, \ldots, v_m$. Under the CSR hypothesis (since then $D(r) = H_s(r)$) these two samples of m numbers each are drawn from the probability distribution given by

$$F(r) = 1 - \exp(-\lambda \pi r^2) \qquad \text{for } r \geq 0,$$

(ignoring edge effects). The samples are approximately independent and independent of each other, if m is not too large. Byth and Ripley (1980) suggest that m should be around 5 % of the total number of points in the window W. As a consequence of the CSR hypothesis the quantities $u_1^2, \ldots, u_m^2, v_1^2, \ldots, v_m^2$ are approximately independent and exponentially distributed with parameter $\lambda \pi$. Thus the statistics

$$h_F = \sum_{i=1}^m u_i^2 \left/ \sum_{i=1}^m v_i^2 \right. \tag{2.7.4}$$

and

$$h_N = \frac{1}{m} \sum_{i=1}^{m} u_i^2 \big/ (u_i^2 + v_i^2) \tag{2.7.5}$$

are approximately F-distributed with $(2m, 2m)$ degrees of freedom (h_F) and normally distributed with mean $1/2$ and variance $1/(12m)$ (h_N), respectively. The measurements u_i, v_i are paired for the formula involving h_N; this pairing may be made at random.

These approximate distributional results allow the construction of statistical tests. Excessively large or small values for either h_F or h_N indicate departures from the CSR hypothesis. Which statistic is appropriate depends on the alternative hypothesis one has in mind. If the alternative involves clustering, h_F is more suitable, whereas h_N is preferred if the alternative assumes regularity in the pattern.

Details of these two tests and comparisons with other tests can be found in Byth and Ripley (1980) and Holgate (1965). For other references to distance methods, see Diggle (2003).

2.7.4 The *J*-test

The J-test uses distances but requires mapped data. It is based on the J-function, which is given by

$$J(r) = \frac{1 - D(r)}{1 - H_s(r)}. \tag{2.7.6}$$

For a Poisson process $J(r) \equiv 1$.

From (2.7.1) and (2.7.2) above, the test statistics are

$$\tau_{\max} = \max_{r \leqslant s} |\hat{J}(r) - 1|$$

and

$$\tau_{\text{int}} = \int_0^s (\hat{J}(r) - 1)^2 \mathrm{d}r.$$

The J-function has the advantage of measuring both the strength and range of interaction and of allowing a simple interpretation; see Section 4.2.7.

This test has been thoroughly investigated by Thönnes and Van Lieshout (1999). They show that its power is strongly influenced by the choice of maximum distance s: for large values of r, $1 - H_s(r)$ is close to zero and thus $J(r)$ can have large fluctuations (which makes the J-test slightly difficult). Therefore, they recommend values of s close to the interaction radius of the point process or, more simply, for which $H_s(r)$ is 'sufficiently below 1'. The authors of the present book recommend applying several values of s and comparing the results.

Critical values for this test are obtained by simulation.

2.7.5   Two index-based tests

The *Clark–Evans test* is based on a numerical summary characteristic, the Clark–Evans index CE; see Section 4.2.4. It compares the mean nearest-neighbour distance of a given pattern with the mean nearest-neighbour distance of a Poisson process of the same intensity as a given process. The estimation of CE is described on p. 198. Mapped data are not required; it suffices to measure nearest-neighbour distances.

For a Poisson process the mean value of CE is 1. Reject the CSR hypothesis if

$$CE > CE_\alpha \qquad \text{for clustering alternative}$$

or

$$CE < CE_{1-\alpha} \qquad \text{for regularity alternative.}$$

Table 2.7 gives some critical values CE_β, for $\beta = \alpha$ or $\beta = 1 - \alpha$, for a square window. These only depend on the number of points n. The values for rectangular windows are slightly smaller for α (larger for $1 - \alpha$) since edge effects are stronger and cause more variability.

Statistical experience shows that this test is not very powerful for regular patterns.

The *mean-direction test* is based on the numerical summary characteristic mean-direction index $\overline{R}_4$; see p. 197. It uses the length of the sum of the unit vectors pointing from the typical point to its four closest neighbours. The estimation of $\overline{R}_k$ is described on p. 199. Mapped data are not required; measurement of the angles for the directions from the points to their neighbours is sufficient.

Table 2.7   Critical values CE_β by number of points n for a square window.

β	n					
	25	50	100	200	400	600
0.99	0.58	0.72	0.80	0.86	0.90	0.92
0.95	0.68	0.77	0.84	0.89	0.93	0.94
0.05	1.12	1.06	1.04	1.03	1.02	1.01
0.01	1.22	1.11	1.08	1.05	1.03	1.03

Table 2.8 Critical values $\overline{R}_{4,\beta}$ as a function of number of points n for a square window.

β	n = 25	50	100	200	400	600
0.99	0.90	1.31	1.50	1.60	1.65	1.68
0.95	1.14	1.43	1.57	1.66	1.69	1.72
0.05	2.40	2.16	2.04	1.94	1.90	1.87
0.01	2.73	2.33	2.11	2.00	1.95	1.91

For a Poisson process the theoretical value of $\overline{R}_4$ is 1.799. This was obtained by simulation from 10^6 quadruples of unit vectors.

The CSR hypothesis is rejected if

$$\overline{R}_4 > \overline{R}_{4,\alpha} \qquad \text{for clustering alternative}$$

or

$$\overline{R}_4 < \overline{R}_{4,1-\alpha} \qquad \text{for regularity alternative.}$$

Again, this test shows only weak reactions against hard-core distances in the pattern.

Table 2.8 gives some critical values $\overline{R}_{4,\beta}$, $\beta = \alpha$ and $\beta = 1 - \alpha$, for a square window, which only depend on the number n of points (Corral-Rivas, 2006). The influence of the shape of the window is small.

2.7.6 Discrepancy tests

If the pattern has been sampled in a rectangular window, it makes sense to test the CSR hypothesis by means of classical methods of statistics, by comparison of the two-dimensional empirical distribution function $U_n(x)$ of the uniform distribution on $\boxed{1}$ with its theoretical counterpart $U(x)$. For this purpose the original points $x_i = (\xi_{i,1}, \xi_{i,2})$ $(0 \leq \xi_{i,1} \leq a,\ 0 \leq \xi_{i,2} \leq b)$ are transformed to the unit square $\boxed{1}$ by $\eta_{i,1} = \xi_{i,1}/a$ and $\eta_{i,2} = \xi_{i,2}/b$. Then

$$U(\eta) = \eta_1 \eta_2$$

and

$$U_n(\eta) = \frac{1}{n} \sum_{i=1}^{n} \mathbf{1}(\eta_i < \eta),$$

for $\eta = (\eta_1, \eta_2)$ and $\eta_i = (\eta_{i,1}, \eta_{i,2})$. A suitable test statistic is the L^2-star discrepancy

$$\omega^2 = \int_0^1 \int_0^1 (U(\eta_1, \eta_2) - U_n(\eta_1, \eta_2))^2 d\eta_1 d\eta_2.$$

This test characteristic ω^2 is 'flawed because its value depends on which corner of W one chooses as the origin of the co-ordinate system' (Zimmerman, 1993). Therefore, Zimmerman decided to choose the average of analogous characteristics for all four corners, which leads to the test characteristic

$$\overline{\omega}^2 = \frac{1}{4n} \sum_{i=1}^n \sum_{j=1}^n (1 - |\eta_{1,i} - \eta_{1,j}|)(1 - |\eta_{2,i} - \eta_{2,j}|)$$

$$- \frac{1}{2} \sum_{i=1}^n \left(\eta_{1,i}^2 - \eta_{1,i} - \frac{1}{2}\right)\left(\eta_{2,i}^2 - \eta_{2,i} - \frac{1}{2}\right) + \frac{n}{9}.$$

A large value of $\overline{\omega}^2$ indicates deviations from uniformity such as clustering or heterogeneity, while small values indicate that the pattern is more regular. Zimmerman (1993) gives the critical values for $n > 20$ shown in Table 2.9.

If $\overline{\omega}^2$ exceeds ω_α (if the alternative is clustering or inhomogeneity) or is smaller than $\omega_{1-\alpha}$ (if the alternative is regularity) the test rejects the CSR hypothesis at significance level $100\alpha\%$. According to Zimmerman (1993), this test is particularly interesting for patterns which might be inhomogeneous. Ho and Chiu (2007a) consider other discrepancies, or measures of uniformity, that are also independent of the choice of origin, and show that their discrepancies are powerful statistics against not only inhomogeneous but also clustered alternatives. However, neither Zimmerman's nor Ho and Chiu's discrepancies are very successful at detecting regularity.

The ideas can be easily extended to the general d-dimensional case.

Table 2.9 Critical values ω_β of the discrepancy test.

β	ω_β
0.99	0.043
0.98	0.049
0.95	0.057
0.05	0.281
0.02	0.342
0.01	0.389

2.7.7 The *L*-test

The *L*-test is based on the fact that the *L*-function of a Poisson process has the simple linear form

$$L(r) = r \quad \text{for } r \geq 0.$$

Deviations of empirical *L*-functions from $L(r) = r$ can be used to test the CSR hypothesis. It is indeed better to use the *L*-function and not the *K*-function, which also has a simple form,

$$K(r) = \pi r^2.$$

The reason is that the square root transformation

$$L(r) = \sqrt{K(r)/\pi} \qquad \text{for } r \geq 0$$

stabilises variance as discussed in Section 4.3.1.

It is then natural to use the test statistics

$$\tau = \max_{r \leq s} |\hat{L}(r) - r| \qquad (2.7.7)$$

with

$$\hat{L}(r) = \sqrt{\hat{K}(r)/\pi},$$

where $\hat{K}(r)$ is some estimator of the *K*-function and s is a suitable maximum distance.

It is advisable to run the *L*-test based on standard point process software, such as the `spatstat` library in R; see Baddeley and Turner (2005, 2006). Good choices for $\hat{K}(r)$ are Ripley's isotropic estimator of the *K*-function (see Section 4.3),

$$\hat{K}(r) = \hat{\kappa}_R(r)/\hat{\lambda}^2,$$

or the stationary estimator

$$\hat{K}(r) = \hat{\kappa}_{st}(r)/\hat{\lambda}_v(r)^2.$$

If τ is large, then the CSR hypothesis is rejected. (The alternative hypothesis is 'no Poisson process' without further specification.) The critical value of τ for the significance level $\alpha = 0.05$ is approximately

$$\tau_{0.05} = 1.45\sqrt{a}/n, \qquad (2.7.8)$$

where $a = \nu(W)$ is the window area and n is the number of observed points (Ripley, 1988, p. 46). This value was obtained by simulations and can be used if nus^3/a^2 is small for a 'wide range of' maximum distances s. Here, $u = U(W)$ is the boundary length of W. For $\alpha = 0.01$ the analogous value is

$$\tau_{0.01} = 1.75\sqrt{a}/n; \qquad (2.7.9)$$

see Chiu (2007). So far statisticians have been unable to determine the exact critical values analytically.

If Ripley's approximation appears to be inappropriate, simulations may be used to provide a *Monte Carlo test*. Here k (e.g. $k = 999$) independent samples from a binomial process with n points are simulated in the window W. For each of the samples the L-function is estimated and the value

$$\tau_i = \max_{r \leq s} |\hat{L}_i(r) - r| \qquad \text{for } i = 1, 2, \ldots, k,$$

is determined. These values and the corresponding value τ for the empirical data are ranked in ascending order. If the rank of τ exceeds $(1 - \alpha)(k + 1)$, the CSR hypothesis is rejected for the error probability α. The position $n(\tau)$ of τ gives an approximation of the p-value; it is around $1 - n(\tau)/(k + 1)$.

A simplified test uses the minimum inter-point distance $\hat{r}_0$ in the pattern (Ripley and Silverman, 1978). If this is too large, the CSR hypothesis may be rejected and some regularity alternative accepted. The rule is: reject CSR if

$$\hat{r}_0 > c\sqrt{a}/n,$$

with $c = 1.38$ for $\alpha = 0.05$ and $c = 1.71$ for $\alpha = 0.01$ for the planar case and $c = 1.20$ for $\alpha = 0.05$ and $c = 1.48$ for $\alpha = 0.01$ for the spatial case.

In the literature, different recommendations have been given for the choice of the maximum distance s of r. Ripley (1979) suggests that s should be $0.25l$ for $n = 25$ and $0.125l$ for $n = 100$ if W is a square of side length l, while Diggle (2003) suggests that s should not be bigger than $0.25l$. Ho and Chiu (2006) demonstrate that values of s in the order of $\hat{r}_0$ or of the mean cluster diameter also perform well. However, the same authors show that if the L-function is estimated by $\hat{\kappa}_{st}(r)$ and $\hat{\lambda}_V^2(r)$ (see Section 4.2.3) the variance-stabilising effect of the use of the square-root transformation $\hat{K} \rightarrow \hat{L}$ in combination with the adapted intensity estimator is so strong that the power of the L-test is nearly independent of s, and so half the diagonal length can be safely recommended.

According to Ho and Chiu (2007b) the L-test can be improved by introducing weight functions $w(r)$, which reduce the importance of very small or large r,

$$\tau = \max_{r \leq s} \{|\hat{L}(r) - r| \cdot w(r)\}.$$

They show that decreasing functions such as $w(r) = 1 - \frac{2r}{l}$ or $w(r) = 10^{-r/l}$ are helpful in the case of processes with short-range correlation, while weight functions such as $w(r) = \frac{r}{l} e^{-r/l}$ seem to be suitable for processes with long-range correlation; l is the side length of the square window as above. They suggest using the empirical pair correlation function $\hat{g}(r)$ to determine whether the range of correlation may be long or short before applying their approach.

2.7.8 Other tests and recommendations

Many other CSR tests have been developed and used; some of them are mentioned here for the reader's convenience.

Instead of the L- or the J-function, $H_s(r)$ and $D(r)$ were used in (2.7.2) as functional summary characteristics; see Diggle (1979), Thönnes and Van Lieshout (1999) and Myles et al. (1995). The latter paper uses the d.f. of the cell area for the Dirichlet tessellation with respect to the points in the sample. Since this function is not known analytically, it was determined by simulation; see also Chiu (2003). Mecke and Stoyan (2005) use the morphological function $n(r)$ (see p. 202) for a CSR test, which seems to have power comparable to that of the L-test. A more sophisticated test is the Q^2-test of Grabarnik and Chiu (2002), which uses the numbers of points with k neighbours within a given distance r. The length distribution of the minimal spanning tree is used in Hoffman and Jain (1983). The basic idea is that the minimal spanning tree tends to be shorter for a cluster process and longer for a regular process, in comparison to the CSR case. Spectral tests are discussed in Mugglestone and Renshaw (2001).

So far, no one has systematically compared all CSR tests. Myles et al. (1995) compare the K- and L-test with the index-of-dispersion test and the tests based on $D(r)$ and Dirichlet cell area distribution and favour the L-test. They use the Thomas cluster process and the RSA process as alternative models; see Section 6.3.2 and 6.4.4. Grabarnik and Chiu (2002) compare tests based on their summary characteristic Q^2, on $D(r), K(r)$, the reduced third moment function and on the variance–area curve. Their results are rather complex and show that there is no uniformly best test.

Recommendations

- If you have mapped data and do not want to apply a simulation test, then use the L-test with (2.7.8) and (2.7.9).

- If you can carry out a simulation test then use the L-test based on the periodic binomial process. Compare the results with that of a classical L-test and a J-test.

- If you have only quadrat counts as data or if you want to work without a computer, for instance in the field, use the index-of-dispersion test.

- If you do not have a map of the points but you can visit each point, then use the CE-test and the mean-direction test.

Example 2.4.   Gold particles: advanced CSR testing
The more advanced tests described in Sections 2.7.4–2.7.7 yield the following results.

J-test. The test was carried out with the maximum statistic and estimates of the *J*-function constructed by means of edge-corrected estimates of $D(r)$ and $H_s(r)$. For the distance s the values $s = 8$, 10 and 20 were used in parallel. For $s = 8$ a *P*-value of 0.058 was obtained by simulation, but for $s = 10$ and 20 the *P*-values were very small (less than 0.0001) and 0.009, respectively. Thus, the CSR hypothesis is not rejected for $s = 8$ but is rejected for the larger values of s. The reader should consider Figures 4.17 and 4.20 in Chapter 4 in order to understand the choice of the values of s.

Mean-direction test. The empirical mean-direction index $\overline{R}_4$ is 1.9095, i.e. between the critical values for $n = 200$ and $\alpha = 0.05$; the CSR hypothesis is not rejected.

Discrepancy test. The test statistic $\overline{\omega^2}$ results in a value of 0.0766 and lies between the limits 0.057 and 0.281. This test does not reject the CSR hypothesis either.

L-test. The test statistic is $\tau = 5.65$. This value results from the minimum inter-point distance of 5.65 in the pattern, which implies $\hat{L}(5.65) = 0$. The critical value $\tau_{0.05}$ is

$$\tau_{0.05} = \frac{\sqrt{650 \cdot 400}}{218} = 3.39,$$

and so the Poisson hypothesis is rejected. The choice of s does not make any difference to the test results for the gold particle pattern. The test based on $\hat{r}_0$ already rejects the CSR hypothesis.

In their original analysis of the gold particle pattern, Glasbey and Roberts (1997) also reject the CSR hypothesis, based on a test using

$$\int_0^s \left(\hat{D}(r) - (1 - \exp(-\lambda \pi r^2)) \right)^2 dr.$$

In summary, the gold particle pattern should not be regarded as a sample from a homogeneous Poisson process. It can perhaps be considered as a point pattern which is globally similar to a Poisson process, but locally there are heavy deviations because of the hard-core distance of $\hat{r}_0 = 5.65$ and because of small-scale clustering.

The behaviour of the tests when applied to the data set in Example 2.4 may support the prejudgemental idea that the more complicated tests are also more powerful than the simple discrepancy and index-based tests. But the reader should note that the authors have experienced cases where the *L*-test confirms CSR and the mean-direction test rejects it. It is always useful to study the nature of a given pattern and to identify a summary characteristic that indicates well any deviations from CSR and then to use a test based on exactly that summary characteristic.

3

Finite point processes

A finite point process is a model for point patterns with a finite number of points. These points form either a single cluster (or a finite number of clusters) or exist within a bounded set, which usually coincides with the observation window. This set, which could be called the 'window of existence', has an influence on the geometry of the point patterns, for example its edge is not an arbitrarily chosen edge but a 'natural' one and may attract or repulse the points. For these types of patterns the classical statistical methods for stationary point processes are not appropriate, and the influence of the window should not be eliminated by edge correction. It may be helpful to consider Figures 3.1, 3.2, 3.9(a) and 3.14 before reading the text.

This chapter presents methods for the analysis of such finite patterns. It starts with methods for the estimation of the intensity function $\lambda(x)$, based on both parametric and non-parametric approaches. Then inhomogeneous Poisson processes, probably the most important models for inhomogeneous and finite point patterns, are considered. Section 3.5 presents a series of functional summary characteristics for finite point processes which are analogues of characteristics commonly used in the context of stationary patterns such as the K-function, the nearest-neighbour distance distribution function and the spherical contact distribution function.

Statistical Analysis and Modelling of Spatial Point Patterns   J. Illian, A. Penttinen, H. Stoyan and D. Stoyan
© 2008 John Wiley & Sons, Ltd

The long concluding Section 3.6 presents the theory and statistical methods for finite Gibbs processes, which are specific Markov point processes. They may be used to model the interactions among the points in an elegant and plausible way. Note that the Gibbs process methodology, originally developed for finite point patterns, has often also been applied to point patterns which are not truly finite. But rather than regarding such a point pattern as a sample from a stationary process, methods for finite processes have been applied because very strong parameter estimation methods have been developed for these, which are based on the maximum likelihood approach.

Those readers who are not interested in finite point processes and Gibbs processes may skip this chapter. Section 3.6 may serve as a useful preparation to facilitate the understanding of the theory of stationary Gibbs processes covered in Section 6.6. Some sections of Chapter 3 assume a basic knowledge of Chapter 4, which, however, is completely independent of this chapter.

3.1   Introduction

Finite point processes are relevant in many applications. All observed point patterns are given in a bounded set and consist of only a finite number of points. Nevertheless, it is often convenient and appropriate to apply models that are based on infinitely many points and to analyse the observed finite point patterns statistically as if they were finite samples of infinite patterns. The corresponding statistical methods have been developed for processes termed 'stationary point processes' and are discussed in full detail in Chapter 4 and 5.

However, the assumption that an observed pattern is a finite sample of an infinite pattern cannot always be made and thus the stationarity approach is not suitable in all situations. Applying statistical methods that were developed for stationary point processes in these situations may lead to unsatisfactory results. Nevertheless, many of the characteristics and methods discussed in this chapter were originally defined for the stationary case, adapted to finite processes.

Many aspects of the theory of finite point processes are mathematically elementary, since these processes can be defined with respect to multivariate densities familiar from classical statistics. A finite point pattern can be regarded as one sample from a multivariate distribution (of $n = N(W)$ points) that defines the finite point process. This approach is distributional and many readers will doubtless be familiar with it. In contrast, stationary point processes are defined in a more abstract way, which is comparable with the theory of time series. Since finite point processes have a simpler structure, these are discussed before stationary point processes in this book.

Consider a finite point process N that consists of $N(W)$ points in a bounded window W. The set W is both the window of observation and the life space or window of existence of the points. These may sometimes 'be aware' of the edge of W and prefer or avoid positions close to it; in other cases they may be attracted by some centre. Furthermore, the number of points $N(W)$ may be random or deterministic. In the random case, p_n denotes the probability $\mathbf{P}(N(W) = n)$. The window of existence is often a random set such as in Example 3.6.

This chapter discusses statistical methods that may be applied if the analysed point pattern is clearly finite in nature and if it is inappropriate to regard it as a sample from a stationary process. The decision to use methods for finite processes is often quite clear, dictated by the nature of the patterns; see Examples 3.1, 3.4 and 3.6 below: the window of existence or the bounded region in which the point cluster lies is given by the scientific problem or the data. In other situations the finite approach is used for pragmatic reasons, when the pattern is not small and the finite approach can be considered as a good approximation to the stationary approach. This may be more convenient for statistical reasons; see Example 3.14. Similarly, for pragmatic reasons a given pattern may be considered 'small' or 'large' and the appropriate methods may then be applied to it; it seems clear that the patterns in Figure 3.14 are small and that in Figure 3.24 is large.

A number of different finite point processes may be considered. The following attempts to provide a classification of the different types of situations that may be modelled with a finite point process.

(a) *Small patterns in well-defined regions.* In some applications, one considers small finite patterns that occur in a well-defined region. This means that the points are observed in a set W which is given naturally or a priori and has not simply been chosen as an observation window. Consider, for example, the locations of trees in a city park surrounded by asphalt, collections of hard spheres in a container, centres of spots on leaves of trees resulting from a fungal disease (see Figure 3.14 on p. 134), centres of pores of metallic foams in bounded bodies (see Figure 3.1), or positions of animals in some fixed region which they do not leave, such as horses in a paddock or fish in an aquarium. Specific point process models which take the special geometry of the window W into account have to be applied in situations like these since it is possible that the objects represented by the points are in some way 'aware' of the geometry of W. This may influence their behaviour and hence the characteristics of the resulting pattern. To properly analyse such a pattern it is usually necessary to work with several independent replicates, i.e. several samples of point patterns in the same W. The number of points in each sample may be random or fixed.

(b) *Single clusters.* Another type of finite pattern are patterns that have typically not been observed in a well-defined region. Thus the observation window has been chosen based on subjective decision criteria around the clump of points, and maybe a rectangle or circle, say. Consider, for example, single large clusters such as the pattern formed by herds of mammals in the wild where each herd is a point, or local geological phenomena such as sinkholes,

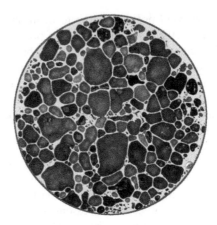

Figure 3.1   Section through a cylinder with pores of a metallic foam. Data courtesy of U. Martin.

volcanoes or basalt bodies in areas of some tens of square kilometres (see Figure 3.2 on p. 111), or locations of seeds or fruits or offspring around the mother tree as in Example 3.4. In situations like these, often only a single sample is given but it is still possible to analyse the pattern appropriately.

(c) *Quasi-homogeneous large patterns.* The two types of spatial patterns discussed in (a) and (b) are often not homogeneous by their very nature. However, there are situations where one may want to analyse spatial patterns with methods for finite patterns even though the pattern may also be considered as a part of a much larger homogeneous pattern and be analysed by the statistical methods for stationary processes. Example 3.14 aims to illustrate this approach.

As in all areas of statistics, both parametric and non-parametric methods may be applied in the statistical analysis of finite point processes. The analysis of patterns of type (a) and (b) often starts off based on estimators of the intensity function $\lambda(x)$ which inform the modelling process. In addition, non-parametric summary characteristics similar to the summary characteristics of the stationary case may be used, such as the K-function and the nearest-neighbour distance d.f. $D(r)$. These may serve as tools in data analysis and for model tests. Finally, a number of different point process models may be fitted to the patterns, where Cox and Gibbs point processes play a major role.

In many cases, a single sample is sufficient for the statistical analysis, namely for i.i.d. clusters, inhomogeneous Poisson processes and Gibbs processes with many points. More than one sample has to be analysed if the random number of points in the whole window is of interest or if an inhomogeneous Cox process is analysed. For a reliable estimation of intensity function trends one sample may be not sufficient either.

The fundamentals of the theory of finite point processes are given in the books by Daley and Vere-Jones (2003) and Van Lieshout (2000). These books are somewhat technical and the text below is addressed to those readers who prefer a less technical explanation or do not have the necessary mathematical background. In addition to the *intensity function* $\lambda(x)$ and *product densities* $\varrho^{(k)}(x_1, \ldots, x_k)$ discussed in Section 1.5, a further family of characteristics is used here, the *location density functions* $f_n(x_1, \ldots, x_n)$.[1] A finite point process in W can be defined by construction using

- a discrete probability distribution $p_n = \mathbf{P}(N(W) = n)$ and

- a family of multivariate density functions $f_n(x_1, \ldots, x_n)$, which are symmetric in their arguments.

'Symmetric in their arguments' means that, independent of the order of the points x_i, the location density functions take on the same value. This requirement is natural because the order of points in a point pattern has no meaning. Mathematically, $f_n(x_1, \ldots, x_n)$ is a density with respect to the Lebesgue measure on W^n.

To gain a better understanding of the location density functions, consider n infinitesimally small spheres $b_1, \ldots, b_n$ of volumes $dx_1, \ldots, dx_n$ centred on the different locations $x_1, \ldots, x_n$. The quantity

$$f_n(x_1, \ldots, x_n)dx_1 \ldots dx_n$$

denotes the probability that the first point is in b_1, the second in b_2 and so on, under the condition that $N(W) = n$, that the point process has exactly n points. On the other hand, the probability of observing the points $x_1, \ldots, x_n$, exactly one in each b_i, is

$$p_n n! f_n(x_1, \ldots, x_n)dx_1 \ldots dx_n.$$

Here the probability p_n appears because the number of points is random, and the factorial is needed because all permutations of the n points yield the same value of the location density function and their number is $n!$. The density function $p_n n! f_n(x_1, \ldots, x_n)$ is the *likelihood* of the finite point process and has an important role in the estimation of the model parameters from the point pattern data $x_1, \ldots, x_n$.

The simulation of such a process follows the construction, provided that the p_n and f_n are tractable: first a random non-negative number n is drawn from the distribution $\{p_n\}$ and then the n point positions are generated using the density $f_n(x_1, \ldots, x_n)$. However, in many important cases the formulas for the p_n and f_n are too complicated and other simulation methods must be employed.

---

[1] In Van Lieshout (2000) f_n is denoted by j_n.

In particular, if the number of points N is random, it is useful to regard point patterns as unordered sets $\{x_1, \ldots, x_n\}$ instead of vectors, which means that only the points matter, not their order.

The samples of a finite point process are finite subsets of W. They are members of the 'configuration space' $\mathcal{N}_{\text{fin}}$. This space consists of the empty set, all singletons $\{x_1\}$ with x_1 in W, all pairs $\{x_1, x_2\}$ with x_1 and x_2 in W, and so on. The set notation $\{x_1, \ldots, x_n\}$ corresponds to this approach.

Sometimes, a finite point process can be defined in $\mathcal{N}_{\text{fin}}$ based on another probability density $p(\{x_1, \ldots, x_n\})$ with respect to the Poisson process of unit intensity. This construction is described in more detail in Section 3.6.3. The location density function f_n then results by conditioning: given that the number of points is n, $N(W) = n$, the location density function is

$$f_n(x_1, \ldots, x_n) = \frac{p(\{x_1, \ldots, x_n\})}{\int_W \cdots \int_W p(\{x_1, \ldots, x_n\}) \mathrm{d}x_1 \cdots \mathrm{d}x_n}.$$

3.2 Distributions of numbers of points

When a pattern of type (a) or (b) is considered and replicated patterns are analysed, the first step in a statistical analysis concerns the random number $N(W)$ of points in the window W. Here standard methods of probability theory for discrete distributions can be used. Johnson et al. (2005) provides a detailed treatment of discrete distributions, but for the reader's convenience those distributions that are of particular importance for the analysis of finite point patterns are briefly covered here.

Note that the suitability of a specific distribution as a model for a given data set may be assessed based on a goodness-of-fit test. The χ^2-test is a standard approach and may be applied here as well.

3.2.1 The binomial distribution

The binomial distribution is closely related to the so-called Bernoulli process. It models a series of n independent trials, each of which has only two possible random outcomes: 'success' or 'failure'. The probability of a success is p for all trials. The number of successes in n trials is denoted by X, and its distribution is given by

$$\mathbf{P}(X = k) = \binom{n}{k} p^k (1 - p)^{n-k} \qquad \text{for } k = 0, 1, \ldots, n. \tag{3.2.1}$$

This is the binomial distribution with parameters n and p. Its mean is

$$\mathbf{E}X = np \tag{3.2.2}$$

and its variance

$$\mathbf{var}X = np(1 - p). \tag{3.2.3}$$

If the notion of 'trial' is suitably interpreted, many situations can be described by this model. Two examples are as follows.

1. Consider a crystal lattice with n sites that can be randomly and independently occupied by foreign atoms. Let the probability of occupying a certain site be p and let it be the same for all sites. Then the total number X of occupied sites has a binomial distribution with parameters n and p.

2. In a large region B, n independent points are uniformly distributed. Let X be the number of points in window W, a subregion of B. This number of points has a binomial distribution with parameters n and p, where

$$p = \frac{\nu(W)}{\nu(B)}.$$

Statistics for the binomial distribution

If n is known, only p has to be estimated. If the outputs $X_1, \ldots, X_m$ from m independent trials are given, p may be estimated by

$$\hat{p} = \sum_{i=1}^{m} \frac{X_i}{mn}. \tag{3.2.4}$$

Also, formulas for confidence intervals for p are known (see Armitage et al., 2001). Furthermore, tests have been developed to test the hypothesis that p has a specific value p_0.

However, if n is unknown the situation is more complicated. The model parameter can be estimated as follows:

$$\hat{n} = \max\left\{ S^2 \frac{\phi^2}{\phi - 1}, X_{\max} \right\}, \tag{3.2.5}$$

where

$$\phi = \begin{cases} \frac{\overline{X}}{S^2} & \text{for } \overline{X} \geq \left(1 + \sqrt{\frac{1}{2}}\right) S^2, \\ \max\left\{ \frac{X_{\max} - \overline{X}}{S^2}, 1 + \sqrt{2} \right\} & \text{otherwise.} \end{cases}$$

Here

$$\overline{X} = \frac{1}{m} \sum_{i=1}^{m} X_i, \qquad S^2 = \frac{1}{m} \sum_{i=1}^{m} (X_i - \overline{X})^2$$

and

$$X_{\max} = \max\{X_1, \ldots, X_m\}.$$

The estimator in (3.2.5) is a stabilised version of the moment method estimator (Olkin et al., 1981).

If n has been estimated based on (3.2.5) or (3.2.6), p can be estimated using (3.2.4), where n is replaced by $\hat{n}$.

The Bayesian approach may be suitably applied here. First an a priori distribution for p is assumed (Carroll and Lombard, 1985; Günel and Chilko, 1989). In particular, a beta distribution with integer parameters a and b may be chosen as a prior distribution, where the density function of p is proportional to $p^a(1-p)^b$. If a uniform improper prior distribution is assumed for n, the marginal posterior distribution of n, the conditional distribution of n, given the sample $X_1, \ldots, X_m$, is proportional to

$$P(\nu) \propto \prod_{i=1}^{m} \binom{\nu}{X_i} \left((m\nu + a + b + 1) \binom{m\nu + a + b}{a + \sum_{i=1}^{m} X_i} \right)^{-1} \qquad \text{for } \nu \geq X_{\max}. \tag{3.2.6}$$

The ν-value $\hat{n}$ yielding the maximum of $P(\nu)$ is the maximum *a posteriori* estimator of the unknown n.

3.2.2   The Poisson distribution

The Poisson distribution is fundamental to the theory of point processes, since the most important model, the Poisson process, is based on it (see Chapter 2).

A random variable X has a Poisson distribution if

$$\mathbf{P}(X = k) = \frac{\mu^k}{k!} e^{-\mu} \qquad \text{for } k = 0, 1, \ldots, \tag{3.2.7}$$

for some $\mu > 0$. The distribution has only one parameter μ, which is both the mean and the variance of X:

$$\mathbf{E}X = \text{var}X = \mu. \tag{3.2.8}$$

The Poisson distribution is often used in applications. This may be partly due to the Poisson limit theorem which states that the binomial distribution can be approximated by the Poisson distribution if the probability of a success is small and n large. The term 'law of rare events' is for this reason sometimes used in this context.

More specifically, consider a sequence of Bernoulli schemes with number of trials n and success probabilities p_n, for $n = 1, 2, \ldots$. Let the success probabilities tend towards zero with $np_n = \mu$ for all n. Then the number of successes X_n in the nth scheme satisfies

$$\lim_{n \to \infty} \mathbf{P}(X_n = k) = \frac{\mu^k}{k!} e^{-\mu} \qquad \text{for } k = 0, 1, \ldots.$$

In the context of spatial point patterns the Poisson distribution may be used as a model for the number of objects in a given set, where the pattern is a sparse pattern of irregularly distributed small objects.

Statistics for the Poisson distribution

Consider a sample of m numbers $X_1, \ldots, X_m$. Then the parameter μ is estimated by

$$\hat{\mu} = \frac{1}{m} \sum_{i=1}^{m} X_i = \overline{X}. \tag{3.2.9}$$

An approximate confidence interval with a confidence level of $1 - \alpha$ can be calculated as

$$\frac{1}{m} \left(\frac{1}{2} z_{\alpha/2} - \sqrt{m\overline{X}} \right)^2 \leq \mu \leq \frac{1}{m} \left(\frac{1}{2} z_{\alpha/2} + \sqrt{m\overline{X} + 1} \right)^2 \tag{3.2.10}$$

for large $m\overline{X}$; see Armitage et al. (2001). Here $z_{\alpha/2}$ is as on p. 81 the $1 - \frac{1}{2}\alpha$ quantile of the normal distribution.

For further statistical methods for the Poisson distribution refer to textbooks on statistics. For example, there are tests of the hypotheses that $\mu = \mu_0$ or $\mu_1 = \mu_2$.

In applications, the binomial and Poisson distribution are often not flexible enough as they require that the variance-to-mean ratio does not exceed 1. More specifically:

$$\frac{\mathbf{var} X}{\mathbf{E} X} = 1 - p < 1 \text{ for binomial} \quad \text{and} \quad \frac{\mathbf{var} X}{\mathbf{E} X} = 1 \text{ for Poisson.}$$

However, in the context of cluster processes overdispersion is often observed, which is indicated by ratios larger than 1.

The following thus discusses two important classes of discrete distributions, which overcome this problem and have many applications in spatial statistics (Pielou, 1977; Cliff and Ord, 1981).

3.2.3 Compound distributions

In compound distributions one or more of the parameters of a given distribution are considered random variables. The following illustrates this for the case of the

Poisson distribution, but similar approaches can be applied to other distributions. (Note that the construction principle underlying these compound distributions is closely related to Cox processes; see Sections 3.4.2 and 6.4.) Now assume that μ is a non-negative random variable and construct the random variable X following the compound distribution in a two-step process. In the first step, the actual value of μ is generated based on some random mechanism following a density function $f(\mu)$. In the second step a Poisson-distributed random variable is generated, with the value μ drawn in the first step as the corresponding parameter.

The probabilities of the corresponding distribution are given by

$$\mathbf{P}(X=k) = \int_0^\infty P_k(\mu)f(\mu)\mathrm{d}\mu, \qquad (3.2.11)$$

where $P_k(\mu) = (\mu^k/k!)e^{-\mu}$ for $k = 0, 1, \ldots$. The mean and variance of this random variable are

$$\mathbf{E}X = m_1 \quad \text{and} \quad \mathbf{var}X = m_1 + m_2 - m_1^2$$

with

$$m_l = \int_0^\infty x^l f(x)\mathrm{d}x \qquad \text{for } l = 1, 2.$$

Consider the special case where $f(\mu)$ is a gamma distribution density,

$$f(\mu) = \frac{\sigma^r}{\Gamma(r)}\mu^{r-1}e^{-\sigma\mu} \qquad \text{for } \mu \geq 0, \ \sigma > 0 \text{ and } r > 0,$$

with shape parameter r and scale parameter $\frac{1}{\sigma}$. Then X has a so-called *negative binomial distribution:*

$$\mathbf{P}(X=k) = \frac{\Gamma(r+k)\sigma^r}{k!\Gamma(r)(1+\sigma)^{r+k}} \qquad \text{for } k = 0, 1, \ldots \qquad (3.2.12)$$

with

$$\mathbf{E}X = \frac{r}{\sigma} \qquad (3.2.13)$$

and

$$\mathbf{var}X = \frac{r(1+\sigma)}{\sigma^2}. \qquad (3.2.14)$$

These equations may be used to estimate the parameters r and σ by the method of moments. The variance-to-mean ratio of a negative binomial distribution is $1 + \frac{1}{\sigma}$.

The negative binomial distribution may also be derived as follows. Again, consider a Bernoulli process as in the case of the binomial distribution. Let the success probability be $(1 + \sigma)^{-1}$. Denote by X_i the random number of successes between the $(i - 1)$th and ith failure, $i = 1, 2, \ldots$. Then the integer-valued random variable X,

$$X = X_1 + X_2 + \cdots + X_r,$$

has a negative binomial distribution with parameters r and σ.

The negative binomial distribution is a very popular distribution in biostatistics and useful for finite point processes. However, there is no stationary spatial point process with negative binomial numbers of points in sample sets; see Diggle (2003), p. 64.

3.2.4 Generalised distributions

Generalised distributions are closely related to patterns with a random number of clusters of random size. Let ν be the number of clusters, where ν is another integer-valued random variable, and let the number of points in cluster i be C_i, which are non-negative i.i.d. integer random variables, independent of ν. The random variable

$$X = \sum_{i=1}^{\nu} C_i$$

is the total number of points and has a generalised distribution. Its mean and variance are

$$EX = E\nu EC_1 \quad \text{and} \quad \text{var} X = E\nu \text{var} C_1 + (EC_1)^2 \text{var} \nu.$$

The formula shows that the variance-to-mean ratio is larger than 1.

The probabilities $p_k = \mathbf{P}(X = k)$ can be calculated on the basis of so-called generating functions

$$H(z) = \sum_{k=0}^{\infty} p_k z^k, \qquad G(z) = \sum_{k=0}^{\infty} \mathbf{P}(\nu = k) z^k, \qquad g(z) = \sum_{k=0}^{\infty} \mathbf{P}(C_1 = k) z^k,$$

since

$$H(z) = G(g(z)) = \sum_{k=0}^{\infty} \mathbf{P}(\nu = k) \left(\sum_{j=0}^{\infty} \mathbf{P}(C_1 = j) z^j \right)^k \qquad \text{for } 0 \leq z \leq 1,$$

where the p_k are coefficients of z^k in the power series expansion.

If ν and the C_i are Poisson-distributed, with parameters μ_ν and μ_c respectively, the resulting generalised distribution is called the *Poisson-Poisson distribution.* Here

$$\mathbf{E}X = \mu_\nu \mu_c, \qquad \mathbf{var}X = \mu_\nu \mu_c (1 + \mu_c),$$
$$G(z) = e^{\mu_\nu (z-1)}, \qquad g(z) = e^{\mu_c (z-1)},$$

thus

$$H(z) = \exp(\mu_\nu e^{\mu_c (z-1)} - 1).$$

Let ν now have a Poisson distribution with parameter μ and let the C_i have a 'logarithmic' distribution:

$$\mathbf{P}(C_1 = k) = -\frac{\alpha^k}{k \ln(1 - \alpha)} \qquad \text{for } 0 < a < 1, \ k = 1, 2, \ldots.$$

Then X has a negative binomial distribution with parameters

$$\sigma = \frac{1 - \alpha}{\alpha} \quad \text{and} \quad r = \frac{\mu}{\ln(1 + 1/\alpha)}.$$

3.3   Intensity functions and their estimation

The intensity function $\lambda(x)$ was introduced in Section 1.5. It describes the mean number of points of a point process N in a set B as

$$\mathbf{E}(N(B)) = \Lambda(B) = \int_B \lambda(x)\mathrm{d}x. \qquad (3.3.1)$$

In a local interpretation, $\lambda(x)$ is proportional to the point density at location x. A classical point process model which is entirely based on the intensity function is the inhomogeneous Poisson process; see Section 3.4.1.

The locally variable point densities of several independent samples from a finite point process follow the deterministic $\lambda(x)$; in particular, regions of high or low point density occur in similar parts of the observation window. Based on a single point pattern or a sample of patterns, the underlying function $\lambda(x)$ can be estimated for all x in W. This results in a two-dimensional smooth function or intensity surface. This procedure is a valuable form of data analysis or regionalisation, which converts the point pattern into a function. Estimation methods for $\lambda(x)$ are described below.

Note that more general models may also be considered where $\lambda(x)$ is random rather than fixed. In these models a random mechanism generates an individual $\lambda(x)$ for every sample in the first step, and in the second step the points are scattered based on $\lambda(x)$. The classical model here is the Cox process, which is discussed in Sections 3.4.2 and 6.4. For Cox processes, the functions $\lambda_i(x)$ that describe the variable point density of sample i may usefully be estimated. However, unlike for the inhomogeneous Poisson process, if k samples are analysed, the functions $\lambda_1(x), \ldots, \lambda_k(x)$ may vary so strongly that the average contains only little information. It is even possible that this mean is constant, e.g. when the Cox process is stationary. The methods described in the subsequent sections can be used to estimate the function $\lambda_i(x)$.

3.3.1   Parametric statistics for the intensity function

In applications, the point pattern(s) often suggest(s) a theoretical form of $\lambda(x)$. In other words, a parametric model can be used to estimate the intensity function and hence the parameters of this model have to be estimated. The simplest and somewhat trivial case of a uniform point pattern, which leads to the binomial process, has already been considered in Chapter 2. If W, n and p are given, no further parameters have to be estimated.

Typical non-trivial examples include point patterns whose intensity function shows a trend (see Figures 3.7(a) and 4.49), which can be determined by regression. Other examples are patterns that consist of single clusters (see Figure 3.2). A simple model in this context is the *i.i.d. cluster model* as discussed in Daley and Vere-Jones

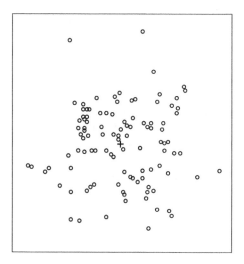

Figure 3.2   Basalt formations in an area of the Swabian Alps. The centre of the point pattern is close to the town Dettingen an der Erms, known for a thermal anomaly with thermal sources. The distance between the two rightmost points is 5.0 km, + is the estimated centre.

(2003), a point process specified by a probability density and having n points. Here a d-dimensional probability density function $f(x)$ is given,

$$f(x) = f(\xi_1, \dots, \xi_d) \qquad \text{for } x = (\xi_1, \dots, \xi_d).$$

The point process is formed by n points that are independent and identically distributed with respect to $f(x)$. The random number of points $N(B)$ in any set B follows a binomial distribution with parameters n and $p = \int_B f(x)\mathrm{d}x$. (With reference to the previous section, note that if the number of points in the cluster is random, $N(B)$ follows a compound binomial distribution.)

The intensity function is

$$\lambda(x) = \mathbf{E}(N(W))f(x) \qquad \text{for } x \in W, \tag{3.3.2}$$

and the location density function is

$$f_n(x_1, \dots, x_n) = f(x_1) \cdot \dots \cdot f(x_n). \tag{3.3.3}$$

The parameters from $f(x)$ can be estimated by classical methods of multivariate statistics, both for one and for k samples.

Example 3.1.   Basalt formations in the Swabian Alps
Figure 3.2 shows the centres of 105 Tertiary basalt formations (volcanic pipes that are likely to be related to small-scale thermal anomalies in the Earth's upper mantle; Goes et al. 1999) in the Swabian Alps. Clearly, the points are randomly distributed (but not uniformly!) and concentrated in the central region. The aim of the following analysis is to fit a simple model to the points in order to gain further understanding of the underlying geological processes. Figure 3.4 on p. 116 shows two non-parametric estimates of the intensity function, derived from kernel estimation (as discussed below). The left-hand figure appears to be too complex, so that it is not clear what type of model could be suitably fitted to it; see the discussion below and in Example 3.2.

A different and much simpler approach is to fit an i.i.d. cluster with a two-dimensional Gaussian density to the pattern:

$$f(\xi_1, \xi_2) = \frac{1}{2\pi\sigma_1\sigma_2\sqrt{1-\varrho^2}}$$

$$\times \exp\left\{ -\frac{1}{2(1-\varrho^2)} \left(\frac{(\xi_1-\mu_1)^2}{\sigma_1^2} - 2\varrho\frac{(\xi_1-\mu_2)(\xi_2-\mu_2)}{\sigma_1\sigma_2} + \frac{(\xi_2-\mu_2)^2}{\sigma_2^2} \right) \right\}.$$

This model is defined on the whole plane $\mathbb{R}^2$.

The Gaussian density function has ellipses as isolines; the equations of these ellipses are

$$\frac{(\xi_1 - \mu_1)^2}{\sigma_1^2} - 2\varrho\frac{(\xi_1 - \mu_1)(\xi_2 - \mu_2)}{\sigma_1\sigma_2} + \frac{(\xi_2 - \mu_2)^2}{\sigma_2^2} = c^2,$$

where c is a constant. The ellipse given by the constant c contains the probability mass $P(c)$ given by

$$P(c) = 1 - \exp\left(-\frac{c^2}{2(1-\varrho^2)}\right).$$

The coordinates are defined with respect to a coordinate system that was chosen a priori and independently of the pattern. The ξ_1-axis is the bottom edge of the figure (west–east direction) and the ξ_2-axis is the left edge (south–north direction).

The parameters may be easily interpreted: μ_1 and μ_2 are the means of the ξ_1- and ξ_2-coordinates respectively, σ_1^2 and σ_2^2 are the corresponding variances, and ϱ is the correlation coefficient of the point coordinates ξ_1 and ξ_2. The standard estimators from classical statistics can be applied, i.e. the sample means $\overline{\xi}_1$, $\overline{\xi}_2$, the sample variances s_1^2, s_2^2 and the empirical correlation coefficient r.

The centre of the ellipses (see above) is located at the point (μ_1, μ_2) and their major semi-axis lies on the line defined by

$$x_2 = \mu_2 + \text{sign}(\varrho)\frac{\sigma_2}{\sigma_1}(x_1 - \mu_1), \qquad \text{sign}(\varrho) = \begin{cases} 1 & \text{for } \varrho > 1, \\ 0 & \text{for } \varrho = 0, \\ -1 & \text{for } \varrho < 0. \end{cases}$$

For $\varrho = 0$ this yields a circle.

For the point pattern in Figure 3.2 the central point $(\overline{\xi}_1, \overline{\xi}_2)$ is shown graphically by +, while $s_1 = 8.5$, $s_2 = 8.3$ and $r = 0.027$. (The unit of measurement is the kilometre, as in Figure 3.2.) These values are used as estimates of the parameters $\mu_1, \ldots, \varrho$.

For the isotropic Gaussian cluster $(\sigma_1 = \sigma_2 = \sigma)$ the intensity function $\lambda(r)$ depends only on the distance from the centre and is given by

$$\lambda(r) = \frac{m}{2\pi\sigma^2}\exp\left(-\frac{r^2}{2\sigma^2}\right) \qquad \text{for } r \geq 0,$$

where m is the mean number of points of N; if the number of points were random, m would be the mean.

Figure 3.3 shows the result of another approach which fits a radial spanning tree to the pattern with respect to the estimated central point. The shape of the graph clearly reflects the single-cluster nature of the pattern and the model seems to be suitable. Thus, the basalt formations seem to result from a unique tectonic process and are grouped around a centre.

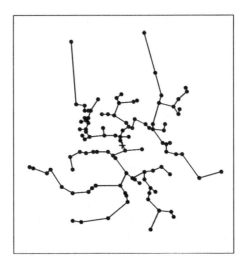

Figure 3.3   The radial spanning tree (see Section 1.8.5) for the point pattern of Figure 3.2. The graph was constructed with respect to the centre $(\bar{\xi}_1, \bar{\xi}_2)$.

Centred processes are a generalisation of the simple i.i.d. cluster model considered above which are strictly defined relative to a central point. In the following, the origin o is chosen as this centre and is assumed not to be a point of the pattern. In this construction it is useful to work with polar coordinates $x = (r, \varphi)$.

Of particular interest is the isotropic case with rotation-invariant point distribution. Then the first-order behaviour is described by an intensity function $\lambda(r)$ yielding the mean number of points in the sphere $b(o, r)$ as

$$\mathbf{E}N(b(o, r)) = \int_0^r db_d r^{d-1} \lambda(r) dr,$$

where $db_d r^{d-1}$ is the surface area of the ball of radius r.

The second-order behaviour can be described by adapted characteristics; see Daley and Vere-Jones (2008, Section 15.3). It suffices to use r_1, r_2 and $\varphi_1 - \varphi_2$ rather than the variables $x_1 = (r_1, \varphi_1)$ and $x_2 = (r_2, \varphi_2)$ of the second-order product density $\varrho^{(2)}(x_1, x_2)$. The one-dimensional point process $\{r_1, r_2, \dots\}$ of ordered distances from o is also of statistical interest; its intensity function is $\mu(r) = db_d r^{d-1} \lambda(r)$.

3.3.2   Non-parametric estimation of the intensity function

The intensity function $\lambda(x)$ is frequently estimated with non-parametric methods that do not assume a specific parametric model for the first-order behaviour. The estimation approach is typically based on kernel methods.

In this context, Diggle (1983) discusses the simple circular (spherical) kernel, which is the most commonly used approach. Here the estimator of $\lambda(x)$ is

$$\hat{\lambda}(x) = \frac{N(b(x, h))}{\nu(b(x, h))} \qquad \text{for } x \in W, \qquad (3.3.4)$$

or in the planar case

$$\hat{\lambda}(x) = \frac{N(b(x, h))}{\pi h^2} \qquad \text{for } x \in W, \qquad (3.3.5)$$

where h is the *bandwidth* of the kernel, which determines the degree of smoothing.

The approach described in (3.3.4) is quite natural: the point density $\lambda(x)$ in the vicinity of x is estimated by the point density in a sphere of radius h centred at x. If the point number is large and h small, one can expect precise estimation results. The estimation can be refined by edge-correction; see Diggle (1985).

Equation (3.3.4) may be generalised and be based on a general kernel function $k(z)$, yielding

$$\hat{\lambda}(x) = \sum_{x_i \in N} k(x - x_i). \qquad (3.3.6)$$

Note that (3.3.4) is obtained by the specific kernel function

$$k(z) = \mathbf{1}_{b(o,h)}(z)/\nu(b(o, h)).$$

The main difficulty with both (3.3.4) and (3.3.5) is the choice of an appropriate bandwidth h. For any kernel function a small value of h may result in an estimated surface $\hat{\lambda}(x)$ that is too spiky, whereas a large h leads to smoother surfaces but may ignore local features of $\lambda(x)$. Diggle (2003, p. 116) discusses the choice of h in the context of a specific model, a stationary Cox process, and recommends a technique which minimises the mean squared error (mse) of the estimator.

In general, however, no simple recipe for the choice of bandwidth exists. Normally, the point patterns themselves contain little information either on the appropriate choice of smoothness required or on the appropriate choice of h. Hence purely data-based methods usually do not produce satisfactory results. Background information on the objects that form the pattern, such as dispersal distances for plants, might inform bandwidth choice. However, in the absence of this the user should simply consider a number of values of h and choose the one that gives the most plausible result in the specific context. An old recipe by Diggle (in his 1983 book, but not in the revised version of 2003) recommends choosing h proportional to $1/\sqrt{n}$ in the planar case. Also, if the pattern is large and 'homogeneous enough' such that the density function $h_s(r)$ of the spherical contact d.f. can be estimated (see Section 4.2.5) then the r value of the (first) maximum of $h_s(r)$ should be determined and used as an upper bound for h. Note that all methods for choosing h based on formulas or recipes should be only regarded as rough guidelines.

Example 3.2.   Non-parametric estimation of $\lambda(x)$ for the basalt formations in the Swabian Alps

Recall the pattern discussed above and shown in Figure 3.2. The intensity of this pattern has now been estimated using the kernel method and is displayed using isolines in Figure 3.4, as discussed above, for two different bandwidths.

The box kernel function

$$k_h(x) = \frac{1}{4h^2}\mathbf{1}_{[-h,h]\times[-h,h]}(x)$$

and Diggle's edge-correction have been applied. Due to the small bandwidth that was used in the figure on the left-hand side, the empirical intensity function shows several maxima, which seems to be inappropriate. Probably the simpler Gauss model in Example 3.1 or the right-hand figure corresponding to the larger bandwidth are more realistic.

Interesting applications of the kernel estimation method in histological research may be found in Muche et al. (2000).

An alternative estimator of the intensity function $\lambda(x)$ based on the Voronoi tessellation, which is particularly useful for clustered patterns, is

$$\hat{\lambda}(x) = \frac{1}{\nu(V(x))} \qquad \text{for } x \in W, \tag{3.3.7}$$

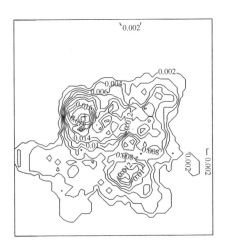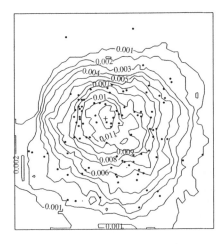

Figure 3.4   Estimated intensity functions for the point pattern in Figure 3.2 shown by isolines, constructed with formula (3.3.5): (left) bandwidth $h = 3.9$ km; (right) $h = 9.0$ km. Perhaps the left-hand figure shows too much detail since the bandwidth is too small. The right-hand figure may be more satisfying. The location of the maximum of the empirical intensity function is quite close to the centre of gravity, shown as $+$.

where $V(x)$ is the Voronoi cell which contains x; see Bernardeau and Van de Weygaert (1996). By construction it is constant in the cells. This estimator is well adapted to strong variation in point density, which is typical of clustered patterns.

3.3.3 Estimating the point density distribution function

Recall that the point density d.f. $G(t)$, introduced in Section 1.5, describes the frequency of the values of the intensity function. Clearly, an estimator $\hat{\lambda}(x)$ of the intensity function $\lambda(x)$ induces an estimator of the point density d.f. $G(t)$ by

$$\hat{G}(t) = \frac{\nu(\hat{W}_t)}{\nu(W)}. \tag{3.3.8}$$

Here $\hat{W}_t$ is defined similarly to W_t in (1.5.9) but $\lambda(x)$ is replaced by its estimate $\hat{\lambda}(x)$, yielding

$$\hat{W}_t = \{x \in W : \hat{\lambda}(x) \le t\}.$$

Example 3.3. Point density d.f. for dislocation density
Figure 3.5 shows a rather irregular point pattern derived from a planar section through a sample of crystalline silicon, and the points show locations where dislocation lines intersect the section plane. There are millions of these points, and $\hat{\lambda}(x)$ varies strongly both within the same sample and between different samples, whereas

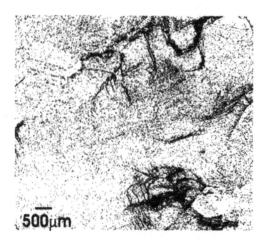

Figure 3.5 A typical sample of a dislocation pattern on a silicon wafer. The darker regions indicate a higher density of dislocation clusters. The maximum density is around $10^7 \, \mathrm{cm}^{-2}$. Data courtesy of M. Rinio.

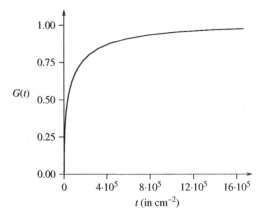

Figure 3.6   The empirical point density d.f. for the pattern in Figure 3.5. A Weibull distribution provides an excellent fit, i.e. $G(t) = 1 - \exp\{-\lambda t^{\alpha}\}$ with parameters $\lambda = 0.007$ and $\alpha = 0.44$.

the point density d.f. (Figure 3.6) is more stable and provides useful summary information which can be used in further modelling; see Ghorbani et al. (2006).

3.4   Inhomogeneous Poisson process and finite Cox process

3.4.1   The inhomogeneous Poisson process

Basic properties

The homogeneous Poisson process may be generalised in a straightforward way by introducing inhomogeneity to yield the *inhomogeneous Poisson process*. This means that the constant intensity λ of the homogeneous Poisson process is replaced by an intensity function $\lambda(x)$ whose value varies with the location x. This is reflected in the properties of the inhomogeneous Poisson process; the fundamental property (1) of the homogeneous Poisson process (see p. 66) is generalised, whereas (2) remains unchanged.

(1) *Poisson distribution of point counts.* The number of points of N in any bounded set B has a Poisson distribution with mean $\int_B \lambda(x) \mathrm{d}x$.

(2) *Independent scattering.* The random numbers of points of N in k disjoint sets are independent random variables, for arbitrary k.

The basic properties (a), (b) and (d) of the homogeneous Poisson process discussed on pp. 67–68 have natural analogues in the inhomogeneous case. In particular, the conditional distribution of the points of N in a bounded set W given that $N(W) = n$ is not uniform, unlike in the homogeneous case. The corresponding location density function of the n points is

$$f_n(x_1, \ldots, x_n) = \prod_{i=1}^{n} \frac{\lambda(x_i)}{\Lambda(W)} \qquad \text{for } x_1, \ldots, x_n \in W, \qquad (3.4.1)$$

with $\Lambda(W) = \int_W \lambda(x)\mathrm{d}x$, i.e. the n points form a sample of n independent points with a probability density function proportional to $\lambda(x)$.

Finally, the product density of the inhomogeneous Poisson process is

$$\varrho^{(n)}(x_1, \ldots, x_n) = \prod_{i=1}^{n} \lambda(x_i). \qquad (3.4.2)$$

Simulation of the inhomogeneous Poisson process

The logic of the algorithm used in the simulation of inhomogeneous Poisson processes applies to a wide range of situations. It is often used in simulations and called the 'rejection method'; see Ripley (1987, p. 60). In this approach, the simulation of an inhomogeneous Poisson process in a bounded set W consists of two steps. First, a sample of the homogeneous Poisson process with intensity $\lambda = \lambda^*$ is generated, with

$$\lambda^* = \max_{x \in W} \lambda(x),$$

using the method described in Section 2.4. Of course, λ^* is assumed to be finite. The number of points in the resulting pattern is much higher than in the final pattern, which is generated in the second step by *independent, location-dependent thinning* with thinning function $p(x) = \lambda(x)/\lambda^*$, as introduced by Lewis and Shedler (1979).

In practice, this means that, based on the thinning function, a decision is made for each point $x_1, \ldots, x_n$ in the sample of the homogeneous Poisson process with intensity λ^* as to whether to 'retain' or 'thin' it. A point x_i is retained with probability $p(x_i) = \lambda(x_i)/\lambda^*$, each point being retained or deleted independent of what happens to any of the other points. Note that any value larger than λ^* may also be used instead of λ^* if it is difficult to determine λ^*. However, this results in a less efficient simulation since more points have to be rejected.

Figure 3.7 shows two simulated samples from an inhomogeneous Poisson process obtained by the method described above. Part (a) shows a linear trend with $\lambda(x, y) = a(x + y)$, and part (b) is a sample with Gaussian scattering which is a particular case of an *isotropic centred Poisson process*. For this type of process the intensity function depends only on the distance from origin o; it is given by

$$\lambda(r) = \frac{md(r)}{db_d r^{d-1}}, \qquad (3.4.3)$$

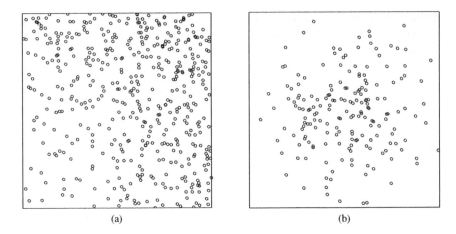

Figure 3.7   Samples from an inhomogeneous Poisson process: (a) linear trend, $\lambda(x, y) = a(x + y)$; (b) Gaussian scattering, $\lambda(r) = c\exp(-dr^2)$, where r is the distance from the origin.

where m is the mean number of points in W, $m = \mathbf{E}(N(W))$, and $d(r)$ is the probability density of the distance of a randomly chosen point from the origin o.

A pattern with a linear trend may occur due to a trend in soil quality in a plant pattern, and a pattern that may be described by the isotropic centred Poisson process is discussed in Example 3.4 below; the pattern in Example 3.1 may also be regarded as a sample from such a process.

Parametric intensity functions

A single cluster at a fixed position x_0 may be modelled with the intensity function

$$\lambda(x) = m\phi(x - x_0).$$

Here, m is the mean number of points and ϕ is a multivariate probability density function. This is an example of an i.i.d. cluster with a Poisson-distributed number of points. In Example 3.4,

$$\phi(x - x_0) = \phi(\|x\|) = p(r).$$

When there are c clusters at fixed positions x_k, the intensity function

$$\lambda(x) = \sum_{k=1}^{c} m_k \phi(x - x_k)$$

may be used. Here m_k is the mean number of points of the kth cluster and x_k its centre. Figure 3.8 shows a $\lambda(x)$ of this form for $c = 3$ and Gaussian $\phi(x)$.

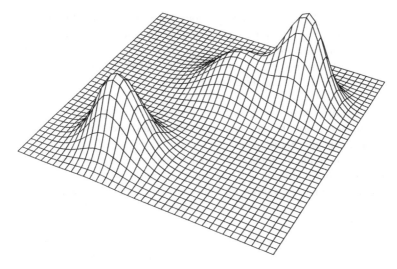

Figure 3.8   Intensity function with three Gaussian clusters.

For a thinned homogeneous process, a suitable intensity function is

$$\lambda(x) = \lambda_o p(x),$$

where $p(x)$ is the retention probability. Lepš and Kindlmann (1987) consider the case

$$p(x) = \min\left\{1, \sum_{k=1}^{l} \exp\left(-a\|x - x_k\|^2\right)\right\}$$

in a forestry application.

Statistics for the inhomogeneous Poisson process

The intensity function $\lambda(x)$ can be estimated using non-parametric methods as described in Section 3.3.

If the form of the intensity function $\lambda(x; \theta)$ with parameter θ is known, the unknown parameter θ can be estimated by the maximum likelihood method, i.e. a parametric method. The likelihood function is given by

$$L(x_1, \ldots, x_n; \theta) = \lambda(x_1; \theta) \cdot \cdots \cdot \lambda(x_n; \theta) \exp\left(-\int_W \lambda(x; \theta)\mathrm{d}x\right). \qquad (3.4.4)$$

Bootstrap methods may be used to assess the quality of estimated intensity functions; see Davison and Hinkley (1997, p. 419), Cowling et al. (1996) and

Section 7.3 below. Either the inhomogeneous Poisson process with intensity function $\hat{\lambda}(x)$ is simulated or n points are sampled at random with replacement from the observed points and then the intensity function is again estimated by means of (3.3.4) or (3.3.6). This is repeated k times (e.g. $k = 1000$) and the resulting intensity function estimates are then analysed statistically.

Example 3.4.   Statistical analysis of fruit dispersal of anemochorous forest trees
Consider an anemochorous tree, i.e. a tree with wind-dispersed fruits or seeds such as ash, lime or maple. Foresters are interested both in estimating the total number of seeds of an individual (mother) tree and in understanding the dispersal pattern. To this end, traps of area a (e.g. $0.25\,\mathrm{m}^2$) are put up around the tree and the random numbers of fruits falling into the traps are counted. Based on such data, the statistical problem may be solved; see Ribbens et al. (1994) and Stoyan and Wagner (2001).

The point process formed by the positions of fruits may be modelled by an isotropic centred Poisson process with intensity function $\lambda(r)$ with origin at the mother tree. Then the mean total number m of fruits is

$$m = 2\pi \int_0^\infty \lambda(r) r \, dr. \tag{3.4.5}$$

Set

$$p(r) = \lambda(r)/m \qquad \text{for } r \geq 0. \tag{3.4.6}$$

The probability density function $d(r)$ of the distance of a randomly chosen fruit position from the mother tree satisfies

$$d(r) = 2\pi r p(r) \qquad \text{for } r \geq 0. \tag{3.4.7}$$

Different types of density functions have been used for $d(r)$, for example Weibull or lognormal distributions. In the latter case, which is considered in the following,

$$d(r) = \frac{1}{\sigma r \sqrt{2\pi}} \exp\left(-\frac{(\ln r - \mu)^2}{2\sigma^2}\right) \qquad \text{for } r \geq 0.$$

For this choice of $d(r)$, the intensity function $\lambda(r)$ depends on the parameters m, μ and σ, i.e. θ is a vector of parameters $\theta = (m, \mu, \sigma)$.

The three parameters are estimated based on distances r_i of traps from the mother tree and the corresponding numbers of fruits n_i in the traps. Table 3.1 shows the data for an example, taken from Wagner (1997).

The likelihood function is

$$L(\theta) = \prod_{i=1}^n \frac{\mu_i^{n_i}}{n_i!} e^{-\mu_i} \qquad \text{with } \mu_i = amp(r_i),$$

Table 3.1   Distances r_i (in metres) and numbers n_i of fruits for 66 traps around an ash tree.

r_i	n_i	r_i	n_i	r_i	n_i	r_i	n_i	r_i	n_i
5	1	5	4	5	5	5	5	15	3
15	3	15	3	15	3	15	7	15	8
15	10	15	14	30	0	30	1	30	1
30	1	30	1	30	2	30	2	30	3
30	3	30	3	30	3	30	4	30	4
30	4	30	5	30	5	50	0	50	0
50	0	50	0	50	1	50	1	50	1
50	1	50	1	50	1	50	1	50	1
50	1	50	2	50	3	50	3	70	0
70	0	70	0	70	0	70	0	70	0
70	1	70	1	70	1	70	1	70	1
70	1	70	1	70	2	70	2	70	3
90	0	90	0	90	0	90	0	90	1
90	1								

where a is the trap area ($= 0.25\,\mathrm{m}^2$), m the mean number of fruits, and $p(r_i)$ is given by (3.4.6) and (3.4.5). For these data numerical methods yield the estimates $\hat{m} = 179\,800$, $\hat{\mu} = 3.93$ and $\hat{\sigma} = 0.94$.

By the way, in the neighbourhood of the maximum the likelihood function turned out to be very flat. For this reason it was important that the numerical maximisation procedure was carried out with high precision.

This approach was generalised to the non-isotropic case in Wagner et al. (2004), while Näther and Wälder (2003) addressed the issue of experimental design, i.e. the appropriate choice of the trap positions.

A successful method for testing the goodness of fit of a fitted model is to use residuals as sketched in Section 4.6.5; see Baddeley et al. (2005) and related papers.

3.4.2   The finite Cox process

Basic properties

Cox processes may be considered in the context of finite as well as of infinite (even stationary) processes. In this section the finite case is discussed; refer to Section 6.4 for a treatment of the stationary case.

Cox processes are a class of very flexible and popular point process models. They are a generalisation of inhomogeneous Poisson processes where the intensity function $\lambda(x)$ is random. Similar to the compound distributions discussed on p. 107, a Cox process can be regarded as the result of a two-stage random mechanism;

for this reason Cox processes are sometimes termed 'doubly stochastic Poisson process'. In the first step, a non-negative intensity function $\lambda(x)$ is generated. Conditional on this, an inhomogeneous Poisson process with intensity function $\lambda(x)$ is constructed in the second step. In order words, given $\lambda(x)$, the point distribution is completely random. This approach is a special case of the hierarchical modelling approach, which is commonly used in model construction in many areas of classical statistics.

Note that it is not possible to distinguish a finite Cox process from an inhomogeneous Poisson process based on a single sample. A number of samples from the same process in the same window may indicate that $\lambda(x)$ is a random variable.

Models for finite Cox processes can be derived by randomising the parameters of the inhomogeneous Poisson process models discussed in Section 3.4.1, i.e. the model parameters m, μ, σ, x_k or a, l, x_k become random variables. A rather simple example of this approach is the mixed Poisson process, where the intensity is a random variable and $\lambda(x)$ is constant for every sample. (The process is made finite by restricting it to the window W.)

Another class of finite Cox processes may be derived by considering a stationary Cox process N_{stat} as in Section 6.4 and by intersection with the window W,

$$N = N_{\text{stat}} \cap W. \tag{3.4.8}$$

In particular, the Neyman–Scott process may serve as N_{stat}; see Provatas et al. (2000) and Waagepetersen (2007). A finite Cox process also results from restricting a stationary Poisson process to a random set W.

The fundamental distributional characteristics of finite Cox processes can be formally expressed as expectations of the respective characteristics for inhomogeneous Poisson processes. For example, the nth-order product density is

$$\varrho^{(n)}(x_1, \ldots, x_n) = \mathbf{E}(\lambda(x_1) \cdots \lambda(x_n)), \tag{3.4.9}$$

and the location density function is

$$f_n(x_1, \ldots, x_n) = \frac{\mathbf{E}\left(\exp(-\Lambda(W)) \prod_{i=1}^{n} \lambda(x_i)\right)}{\mathbf{E}\left(\exp(-\Lambda(W))(\Lambda(W))^n\right)} \tag{3.4.10}$$

with $\Lambda(W) = \int_W \lambda(x)\mathrm{d}x$. The total number of points $N(W)$ of a finite Cox process follows a compound Poisson distribution.

The finite Cox processes can be simulated in a straightforward way, based on the hierarchical nature of the model. In a first step the intensity $\lambda(x)$ is generated (or its parameters such as m, μ and σ) and in a second step the point pattern is simulated given $\lambda(x)$ using the same method as for inhomogeneous Poisson processes.

Statistics for finite Cox process

For the model in (3.4.8) methods for stationary Cox processes can be used, such as the use of pair correlation function or K-function; see Section 6.4.2.

Recall that several samples in the same window W are required to distinguish a Cox process from an inhomogeneous Poisson process. The techniques developed for inhomogeneous Poisson processes can be applied to each sample from a finite Cox process in the same window W. This results in several intensity function estimates $\hat{\lambda}_1(x), \ldots, \hat{\lambda}_k(x)$ or model parameters such as $\hat{m}_1, \ldots, \hat{m}_k$, $\hat{\mu}_1, \ldots, \hat{\mu}_k$ and $\hat{\sigma}_1, \ldots, \hat{\sigma}_k$.

These data can be analysed statistically to understand the first-step variability of the Cox process. Even the simple variability of the number of points per sample may be assessed: in the case of an inhomogeneous Poisson process it follows a Poisson distribution, whereas a finite Cox process is only appropriate if the variability is larger, in particular if the variance-to-mean ratio is larger than 1.

3.5   Summary characteristics for finite point processes

A number of non-parametric summary characteristics can be used to gain an understanding of the spatial distribution of the points of finite point processes. These describe the inter-point distances and the relative positions in the patterns in various ways. Those readers who are familiar with the statistics of stationary point processes will recognise some of these summary characteristics but will also notice differences, in particular the strong influence of the window W. Indeed, it does not make sense to eliminate the influence of the window for finite point processes as is commonly done in the stationary case, and therefore edge-correction methods are applied only in the case (c) of quasi-homogeneous processes as explained on p. 102. If the local point distribution in W is of interest, e.g. the point density in dependence on the distance from the window's boundary, the intensity function should be used (see Figure 3.9).

Those readers who are less familiar with the methods for stationary point processes are referred to Chapter 4, in particular to Sections 4.2.5, 4.2.6 and 4.3.

Recall that the binomial process is the null model for finite point processes similar to the Poisson process in the context of stationary point processes.

Several summary characteristics use nearest-neighbour distances, i.e. analogues of the nearest-neighbour distance d.f. $D(r)$ and of the spherical contact d.f. $H_s(r)$ are considered. However, the definition of these is entirely based on finite characteristics. In addition, an analogue of the K-function may be suitably applied. These summary characteristics can be compared to those for the binomial process. Unfortunately, unlike in the stationary approach, where the Poisson process is the null model, some formulas become a little complicated in the finite case, which uses the binomial process as a null model. Note, finally, that the summary characteristics

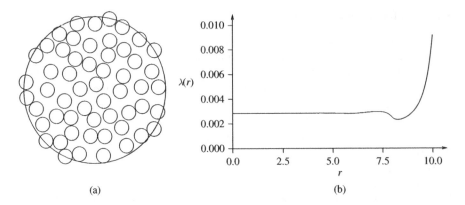

Figure 3.9   (a) A sample of hard discs of radius 1 in a circular window of radius 10. (b) The intensity function $\lambda(r)$ for the disc centres of the model underlying the sample, as a function of distance r from the midpoint of the circle, obtained by simulation of many samples; see p. 132 below for more explanation.

discussed in this section have been chosen such that they can be estimated using existing software for stationary point processes.

Assume throughout this section that m point patterns $N_1, \ldots, N_m$ in the same (or similar) window W are being analysed. If the window and the number of points are not too small, $m = 1$ may be sufficient. The summary characteristics are usually determined for each of these patterns N_l and then regarded as a sample of numerical values or functions.

3.5.1   Nearest-neighbour distances

The finite nearest-neighbour distance d.f. $D_{\text{fin}}(r)$ is defined as

$$D_{\text{fin}}(r) = \mathbf{E}\left(\frac{1}{N(W)} \sum_{i=1}^{N(W)} \mathbf{1}(d(x_i) \leq r) \right) \qquad \text{for } r \geq 0, \qquad (3.5.1)$$

where $d(x_i)$ is the distance of the point x_i to its nearest neighbour. The corresponding estimator is

$$\hat{D}_{\text{fin}}(r) = \frac{1}{m} \sum_{l=1}^{m} \hat{D}_l(r) = \frac{1}{m} \sum_{l=1}^{m} \frac{1}{n_l} \sum_{i=1}^{n_l} \mathbf{1}(d(x_{li}) \leq r) \qquad \text{for } r \geq 0, \qquad (3.5.2)$$

where n_l is the number of points of N_l and $d(x_i)$ is the distance of point x_i in N_l to its nearest neighbour in N_l. The estimator $\hat{D}_{\text{fin}}(r)$ is clearly unbiased. This estimator does not contain an edge-correction factor and was used, for example, in Gignoux et al. (1999) in the context of ecological point patterns.

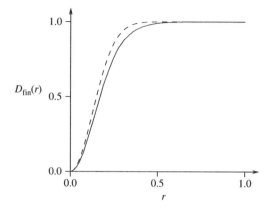

Figure 3.10   Nearest-neighbour distance d.f. $D_{\text{fin}}(r)$ of binomial process in ⊡ with $n = 10$ (solid line) in comparison to $D(r)$ of a homogeneous Poisson process of intensity 10 (dashed line).

Clearly, the nearest-neighbour distance d.f. $D_{\text{fin}}(r)$ for a binomial process as defined in (3.5.1) deviates from $D(r)$ for a homogeneous Poisson process of equal point density, given in (2.5.8). Figure 3.10 shows the two d.f.s for the case of $W = ⊡$ and $n = \lambda = 10$. The differences result from edge effects. If the point process is not constrained to the window W the nearest neighbours of points close to the boundary of ⊡ may actually be outside the window.

Note that the $\hat{D}_l(r)$ in (3.5.2) can easily be calculated using a simple trick, the *large-window trick,* based on existing software for the stationary case: apply the border estimator (see p. 209) to a very large window W_d that has the original window W and the points $x_1, \ldots, x_n$ at its centre, as shown in Figure 3.11. In this way, window effects are eliminated, and numerically the border estimator yields exactly the same result as $\hat{D}_l(r)$.

3.5.2   Dilation function

The dilation function $S(r)$ is in some sense analogous to the spherical contact distribution function $H_s(r)$ (see Section 4.2.5), as defined for stationary processes. It is a very suitable characteristic for describing the mutual positions of the points in N. For a pattern of n points $S(r)$ is formally defined as

$$S(r) = \mathbf{E}\left(\nu\left(\bigcup_{i=1}^{n} b(x_i, r)\right)\right) \qquad \text{for } r \geq 0. \tag{3.5.3}$$

This means that the dilation function reflects the expected area (volume) of the union of discs (spheres) of identical radius r centred at the points x_i. These discs do not overlap for regular patterns even for relatively large values of r. Hence,

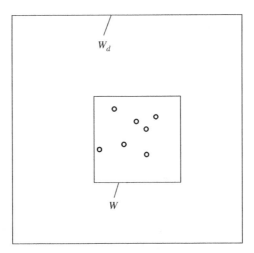

Figure 3.11   The true window W surrounded by a constructed enlarged artificial window W_d. Here, each point has its nearest neighbour in W and the border estimator applied to W_d yields the same result as an estimator which uses the nearest neighbours within W, ignoring potential points outside.

their union covers a relatively large area. Thus, large values of $S(r)$ indicate a regular point distribution. The value of $S(r)$ for those r for which the discs do not overlap is

$$S(r) = nb_d r^d.$$

In contrast, the discs overlap for small values of r in a clustered pattern, and hence $S(r)$ takes on relatively small values.

A natural estimator of $S(r)$ is

$$\hat{S}(r) = \nu \left(\bigcup_{i=1}^{n} b(x_i, r) \right) \qquad \text{for } r \geq 0. \tag{3.5.4}$$

For a sample of m patterns the arithmetic mean can be used to estimate the overall dilation function,

$$\hat{S}(r) = \frac{1}{m} \sum_{l=1}^{m} \hat{S}_l(r) \qquad \text{for } r \geq 0. \tag{3.5.5}$$

The function $\hat{S}(r)$ can be calculated with any program that estimates the spherical contact d.f. $H_s(r)$ as explained in Section 4.2.5, using the large-window trick: determine $\hat{H}_s(r)$ for a very large window W_s which has the original window W at its centre and set

$$\hat{S}(r) = \hat{H}_s(r)\nu(W_s).$$
(3.5.6)

3.5.3 Graph-theoretic statistics

Valuable summary characteristics can be derived from geometric graphs as introduced in Section 1.8.5, if these are constructed for given finite point patterns. Particularly useful are subgraphs or counts of components and corresponding distributions, such as

- the number of edges or components (e.g. triangles containing no other points)
- node degrees (number of edges emanating in nodes or vertices)
- metric characteristics such as edge lengths.

See Example 3.6 for an application of this approach.

3.5.4 Second-order characteristics

Ripley's K-function has proved very versatile as a second-order summary characteristic (see Section 4.3) in the context of stationary point processes, where $\lambda K(r)$ describes the mean number of further points in a disc (sphere) of radius r that is centred at the typical point.

In the context of finite point processes an analogous quantity $K_{\text{fin}}(r)$ can be defined. Consider first a deterministic number of points n in the window W. Here the finite characteristic is constructed such that it can be estimated by the estimators of the classical K-function $K(r)$ for stationary point processes. (For large inhomogeneous patterns the inhomogeneous K-function $K_{\text{inhom}}(r)$ introduced in Baddeley et al., 2000, may be considered – see Section 4.10 below – but this is a rather different approach.) Note that an analogue of the pair correlation function, which is discussed in Section 4.3 and is in some sense a derivative of the K-function, is not considered here and cannot be recommended due to the small sizes of the samples in the finite case.

The finite K-function is defined as

$$K_{\text{fin}}(r) = \mathbf{E}\left(\sum_{i=1}^{n}\sum_{\substack{j=1\\j\neq i}}^{n}\frac{\mathbf{1}(\|x_i - x_j\| \leq r)}{\nu(W_{x_i} \cap W_{x_j})}\right) \Bigg/ \frac{n(n-1)}{\nu(W)^2}.$$
(3.5.7)

The function $K_{\text{fin}}(r)$ has the same structure as the stationary estimator $\hat{\kappa}_{st}(r)\big/\hat{\lambda}^2$ given by (4.3.27). The expectation term in (3.5.7) is equal to

$$\int\limits_{W}\int\limits_{W} \frac{\mathbf{1}(\|x-y\|\leq r)}{\nu(W_x\cap W_y)}\varrho^{(2)}(x,y)\mathrm{d}x\mathrm{d}y, \qquad (3.5.8)$$

where $\varrho^{(2)}(x,y)$ is the second-order product density of N.

Explicit formulas for $K_{\text{fin}}(r)$ for the binomial process are known. They depend on the shape and size of W, but are independent of n (which is larger than 2):

$$K_{\text{fin}}(r)=b_d r^d \qquad \text{for } r\leq r_1 \qquad (3.5.9)$$

and

$$K_{\text{fin}}(r)=2^d\nu(W) \qquad \text{for } r\geq r_2, \qquad (3.5.10)$$

where r_1 is the largest radius such that $b(o,r)\subseteq W\oplus\check{W}=\{x-y:x,y\in W\}$ and r_2 the smallest radius such that $W\oplus\check{W}\subseteq b(o,r)$. Figure 3.12 shows $K_{\text{fin}}(r)$ for circular and quadratic W. Equation (3.5.10) only applies if the window W is central-symmetric, e.g. a disc or rectangle. In other cases $2^d\nu(W)$ has to be replaced by $\nu(W\oplus\check{W})$. If W is a circular window, $W=b(o,R)$, then $r_1=r_2=2R$, and if W is a square with side length a, then $r_1=a$ and $r_2=a\sqrt{2}$. For $r_1\leq r\leq r_2$,

$$K_{\text{fin}}(r)=\nu(W\oplus\check{W}\cap b(o,r)). \qquad (3.5.11)$$

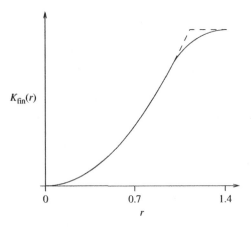

Figure 3.12   The finite K-function $K_{\text{fin}}(r)$ for binomial processes with a quadratic (solid line) and circular (dashed line) window, both of area 1.

For rectangular W the formula for the volume on the right-hand side of (3.5.11) may be found in Appendix B.

If the number $N(W)$ of points is random, $K_{\text{fin}}(r)$ can be defined by replacing n by $N(W)$ in (3.5.7).

An unbiased estimator of $K_{\text{fin}}(r)$ may be obtained by simply considering the right-hand side of (3.5.7) without the expectation **E**:

$$\hat{\kappa}_{\text{fin}}(r) = \frac{1}{m} \sum_{l=1}^{m} \left(\sum_{i=1}^{n_l} \sum_{\substack{j=1 \\ j \neq i}}^{n_l} \frac{\mathbf{1}(\|x_{li} - x_{lj}\| \leq r)}{\nu(W_{x_{li}} \cap W_{x_{lj}})} \middle/ \frac{n_l(n_l - 1)}{\nu(W)^2} \right). \tag{3.5.12}$$

Again, the estimator can be calculated using software for the estimation of Ripley's K-function based on $\hat{\kappa}_{st}$ (see p. 228) and using the estimator (4.3.34) of the squared intensity in Section 4.3. It is ratio-unbiased by construction.

The variance of the estimator in the case of a binomial process is

$$\mathbf{var}\hat{\kappa}_{\text{fin}}(r) = 2 \left(\frac{a}{n} \right)^2 \left(\frac{\pi r^2}{a} + \frac{2ur^3}{3a^2} + 1.34 \frac{n}{a} \cdot \frac{ur^5}{a^2} \right) \qquad \text{for } r \leq r_1, \tag{3.5.13}$$

where $a = \nu_2(W)$ and u is the perimeter of W; see Ripley (1988).

By analogy with the stationary case a finite L-function may be considered, defined as

$$L_{\text{fin}}(r) = \sqrt[d]{\frac{K_{\text{fin}}(r)}{b_d}} \qquad \text{for } r \geq 0. \tag{3.5.14}$$

In order to eliminate the influence of the shape of the window W even more, the *normalised K-function* $K_{\text{nor}}(r)$, defined by

$$K_{\text{nor}}(r) = \frac{K_{\text{fin}}(r)}{K_{\text{fin,b}}(r)} \qquad \text{for } r \geq 0, \tag{3.5.15}$$

may be applied, where $K_{\text{fin}}(r)$ is the finite K-function of the finite point process that is being analysed and $K_{\text{fin,b}}(r)$ the finite K-function of the binomial process in W with the same number of points n.

The corresponding statistical estimator is

$$\widehat{K}_{\text{nor}}(r) = \frac{\hat{\kappa}_{\text{fin}}(r)}{K_{\text{fin,b}}(r)} \qquad \text{for } r \geq 0, \tag{3.5.16}$$

where equations (3.5.9) to (3.5.11) yield $K_{\text{fin,b}}(r)$. Note also that for large r there is no model-independent value of $K_{\text{nor}}(r)$, despite the normalisation.

Another second-order characteristic that may be considered is the *random pair distance d.f.*, the distribution function of the random distance between an arbitrary pair of points of N,

$$P(r) = \mathbf{E}\left(\sum_{i=1}^{n}\sum_{\substack{j=1\\j\neq i}}^{n}\mathbf{1}(\|x_i - x_j\| \leq r)\right)\Big/ (n(n-1)) \qquad \text{for } r \geq 0, \qquad (3.5.17)$$

which can be expressed in terms of the product density as

$$P(r) = \int_W \int_W \mathbf{1}(\|x - y\| \leq r)\varrho^{(2)}(x, y)\mathrm{d}x\mathrm{d}y. \qquad (3.5.18)$$

Hence, this is clearly a second-order characteristic. It depends, of course, on the shape and size of W.

For a binomial process in W the probability density corresponding to $P(r)$ is

$$p(r) = \frac{2\pi^{d/2}r^{d-1}\overline{\gamma}_W(r)}{\Gamma(d/2)\nu(W)^2} \qquad \text{for } r \geq 0, \qquad (3.5.19)$$

where $\overline{\gamma}_W(r)$ is the isotropised set covariance of the window W; see Appendix B. The approximation formula

$$p(r) = \frac{2\pi r}{a} - \frac{2r^2u}{a^2} + o(r^2)$$

holds for small r for arbitrary convex W, where the notation is the same as in (3.5.13).

The function $P(r)$ can be estimated by

$$\hat{P}(r) = \frac{1}{m}\sum_{l=1}^{m}\left(\sum_{i=1}^{n_l}\sum_{\substack{j=1\\j\neq i}}^{n_l}\mathbf{1}(\|x_{li} - x_{lj}\| \leq r)\Big/ (n_l(n_l - 1))\right) \qquad \text{for } r \geq 0; \qquad (3.5.20)$$

see Bartlett (1964). Software for the estimation of the K-function can be modified to yield an estimator of $P(r)$, in a similar way to that discussed above for $D_{\mathrm{fin}}(r)$.

Example 3.5.   Finite planar RSA process in a circular window
Consider a finite point process N consisting of n points in the disc $W = b(o, R)$. The points are the centres of hard discs $b(x_i, \varrho)$ with $x_i \in W$ and $\varrho < R$. These find their

positions by the random sequential adsorption principle (see also Section 6.5.4) where

- the location of first point is chosen following a uniform distribution in W,

- once the kth point has been allocated a position, the $(k+1)$th point finds its position following a uniform distribution among those positions in W which guarantee that there is no overlap with the first k discs.

It is possible that in simulations not all n points can be allocated a position – these configurations are rejected here. This means that the distribution is based on the assumption that all n discs find a position.

Figure 3.9(a) shows a simulated pattern of the planar RSA process with $n = 55$, $R = 10$ and $\varrho = 1$. Clearly, the pattern looks like a sample from an isotropic process (whose distribution is invariant with respect to rotations around o), but the point centre density is not constant in W: it is constant for $r \leq 8$, but has a minimum at $r = 8.2$ and for $r > 8.2$ larger values, as shown in Figure 3.9(b). This clearly results from the fact that it is easier to place discs close to the boundary of W than in its interior, since there are no hard discs outside W which have to be taken into account.

Finally, Figure 3.13 shows the normalised K-function $K_{\text{nor}}(r)$ for the planar RSA process obtained by simulation in comparison to that of the binomial process. It indicates that the process clearly deviates from a binomial process – in particular, the hard-core behaviour of the process is reflected well: $K_{\text{nor}}(r)$ is equal to 0 for $r \leq 2.60$ and still smaller than 0.5 for $r \leq 3.64$. However, the inhomogeneity of the pattern is not apparent in $K_{\text{nor}}(r)$.

Example 3.6.   Black spots on maple leaves
Figure 3.14 shows 10 small point patterns, derived from the centres of black spots of diameter 1–2 cm of the tar spot disease (*Rhytisma acerinum*), a fungal disease

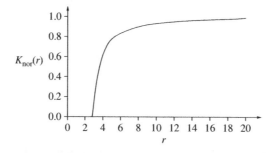

Figure 3.13   Normalised K-function $K_{\text{nor}}(r)$ for the planar RSA process. The value for $r = 20$ is slightly smaller than 1.

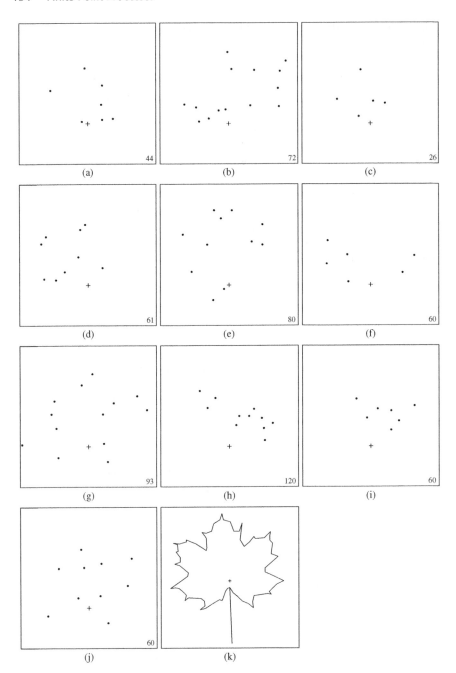

Figure 3.14   (a)–(j) Ten point patterns of tar spots on maple leaves in squares of side length of c. 14 cm. (k) Contour of the largest leaf (h) on a smaller scale. The + marks the origins of local coordinate systems.

found in maple trees, on the leaf surface of Norway maple. The spores of these fungi attack the leaves in spring, when clouds of spores surround the trees.

Note that the leaves are both windows of existence and observation windows, and in this example these are of variable, random size.

In the statistical analysis two issues are considered:

1. Is the spot density higher at the base of the leaf than across the whole leaf? (In other words, the question is whether the leaves still grew after having been infected by spores.)

2. Can the patterns of spot centres be regarded as a part of samples from a homogeneous Poisson process? From the infection mechanism of the leaves one might suspect CSR, i.e. a uniform random distribution of the spots resembling samples from stationary point processes. However, visual inspection suggests that there is a weak tendency towards regularity with a considerable hard-core distance.

The sample mean and variance of the number of points per leaf are $\bar{x} = 9.3$ and $s^2 = 9.12$, i.e. both values are very similar, $\bar{x} \sim s^2$. However, due to the compound nature of the distribution of the number of points, resulting from the random size of the leaves, one would expect $s^2 > \bar{x}$; see the treatment of compound distributions on p. 107, where the variable μ corresponds to variable leaf size. But the sample size $m = 10$ is rather small; if hundreds of leaves were analysed, the true variance might be found to be larger than the true mean.

The spot density across all the leaves is $0.138\,\text{cm}^{-2}$, calculated as $\hat{\lambda} = \frac{n}{a} = \frac{93}{676}$, where a is the total area of all leaves and n the total number of all spots. In addition, the spot density was also estimated in a subwindow close to the leaf base, the rectangle $[-2, 2] \times [0, 3]$ positioned at the origin (which is contained in all 10 leaves). This yields the estimate $0.15\,\text{cm}^{-2}$, which is rather close to $0.138\,\text{cm}^{-2}$. One may thus assume that the spot density is uniform on the leaf surfaces.

Next, the Poisson process hypothesis is assessed. Due to the complicated geometrical shape of the leaves and since the spots appear to be homogeneously distributed across the leaves, the analysis applies characteristics that are based on short interpoint distances:

- the d.f. $D_{\text{fin}}(r)$,
- the d.f. $E_2(r)$ of the edge length of the 2-neighbour graph. The value $k = 2$ was chosen in order to avoid repeating the case $k = 1$, which is closely related to $D_{\text{fin}}(r)$, and to avoid problems with edge effects. These are likely to be substantial for larger values of k in these rather small patterns,
- the d.f. $P(r)$.

Figure 3.15 shows $\hat{D}_{\text{fin}}(r)$ for all 10 leaves in comparison to $D(r)$ for a homogeneous Poisson process. The corresponding intensity was estimated using

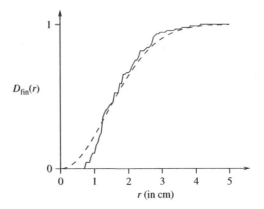

Figure 3.15   The empirical nearest-neighbour distance d.f. $\hat{D}_{\text{fin}}(r)$ for the spots (solid line) in comparison to $D(r)$ for a Poisson process of intensity $\lambda = 0.146$ (dashed line).

the mean nearest-neighbour distance and equation (2.3.7), yielding the value $0.146\,\text{cm}^{-2}$, which is close to the two intensity estimates above.

Considering the minimum inter-point distance of 0.71 cm, one may clearly doubt the Poisson hypothesis. Indeed, based on the hard-core test (see p. 96) the CSR hypothesis has to be rejected: the critical value $r_{0.05}$ is 0.39 cm, which is much smaller than $\hat{r}_0 = 0.71$ cm.

Figure 3.16 shows the empirical edge length d.f. $\hat{E}_2(r)$ for the 2-neighbour graph. The mean length is 2.38 cm, which corresponds to an intensity $\lambda = 0.08\,\text{cm}^{-2}$ for a homogeneous Poisson process. As the finiteness of the spot pattern is apparently well reflected here, an approximation by a homogeneous Poisson process does not make sense.

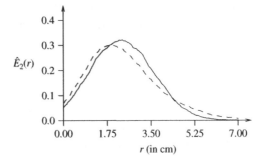

Figure 3.16   The empirical edge length d.f. $E_2(r)$ for the 2-neighbour graph (solid line) and its counterpart for a restricted homogeneous Poisson process on W_{leaf} with $\lambda = 0.15$ (dashed line).

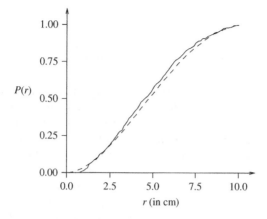

Figure 3.17   Two $P(r)$-functions: empirical (solid line), restricted Poisson process with $\lambda = 0.14$ (dashed line).

The same figure shows the theoretical $E_2(r)$ for a finite Poisson process for a window W_{leaf} of area $71\,\text{cm}^2$ which is shaped like an idealised maple leaf, a symmetric pentagon. Now the intensity is taken as $\lambda = 0.15\,\text{cm}^{-2}$, which reproduces the mean edge length.

The behaviour of $P(r)$ is similar to that of $E_2(r)$: with an intensity λ of $0.13\,\text{cm}^{-2}$ the theoretical function of a restricted Poisson process on W_{leaf} is close to the empirical $P(r)$-function, as shown in Figure 3.17.

In summary, the statistical analysis leads to the conclusion that the spots globally follow a homogeneous Poisson process but that there is some local inhibition. The deviations from CSR are apparently more clearly indicated by $E_2(r)$ than by $P(r)$. In this example a rigorous application of the finite approach suggests that the stationary approach is more suitable. This is in contrast to Example 3.5. Clearly, the edges of the windows of existence play a quite different role in the two examples.

3.6   Finite Gibbs processes

3.6.1   Introduction

The finite Poisson and Cox process models discussed above have many applications in situations where the objects (or points) are mutually independent of each other, i.e. are not interacting: in Poisson processes the points are independently distributed in W and in Cox processes they are conditionally independent given the intensity function. However, in many cases the analysis of a spatial point pattern focuses on the detection and characterisation of interaction among the points. This interaction is often a mutual repulsion leading to patterns of some regularity. However, in some applications the points interact positively, i.e. attract each other, which is reflected in clustered patterns.

The waterstrider data described in Section 1.2.4 are an example of a situation where any reasonable point pattern model should include some form of interaction between the points. Recall that waterstriders are arthropods living on the surface of ponds. They are able to communicate by sending and receiving signals using the water surface as a medium. In addition, ecologists know from experiments that the later larval stages and male adults form their own territories and tend to prevent other males from entering these. Hence, it is not very realistic to assume any kind of independence among the individuals a priori since very close pairs of points are unlikely.

Similar patterns appear in many other biological applications, but also in physics and materials sciences. The jargon in physics describes the interaction among the points in a point pattern as 'forces', which may be repulsive as well as attractive, and the pattern is a result of these forces. In physics process parameters (the so-called pair potential) are usually given that describe these forces and the resulting point process is studied. In a statistical approach, however, point patterns, i.e. samples from a point process, are studied and the corresponding parameters are sought. Since many different assumptions are possible for the forces, different point process models can be formed in this way, leading to the class of *Gibbs processes*. Note that the term 'Gibbs process' derived from physics is preferred in this book, even though the term 'Markov point process' is popular among statisticians. Also note that a 'Markov chain', despite the similarity in name, is a very different concept, a stochastic process in discrete time. Markov chains are used frequently in the simulation of Gibbs processes; see pp. 143ff. below as well as Van Lieshout (2006b).

Gibbs processes are very versatile models, in particular for point patterns with repulsion, but they are mathematically rather complicated. In order to help readers who do not have the requisite mathematical background, this section introduces Gibbs processes step by step. The exposition first discusses the case of processes with a fixed number of points in W, which is the window of existence of the process. Physicists call these types of processes 'canonical ensembles'. In the next step, processes with a random number in W are discussed, termed 'grand canonical ensembles' in physics. As with other parts of this book, a reader new to spatial point processes should first concentrate on the simulation methods (see pp. 143 and 149) and examples and return to the mathematical details later.

A large number of spatial point processes can be very usefully and elegantly defined through a density function. Simple examples are the binomial process, the finite homogeneous and inhomogeneous Poisson process, and the finite Cox process. Hence, despite the difficulties, constructing point processes based on a density function is not necessarily complicated. In the context of Gibbs processes this modern approach turns out to be very powerful and quite intuitive.

In the discussion in the previous sections, finite point processes were constructed based on counting probabilities and intensity functions. In the following, point processes are defined through probability densities.

3.6.2   Gibbs processes with fixed number of points

Consider n points randomly distributed in W, where n is fixed. Assume that their positions are given by a multivariate probability density, the location density function

$$f_n(x_1, \ldots, x_n) \qquad \text{for } x_1, \ldots, x_n \in W, \tag{3.6.1}$$

which does not depend on the order of the points. This multivariate density defines a point process with exactly n points in W.

As a preparation for more complicated constructions below, consider the following interpretation of (3.6.1). Point patterns that follow this distribution may be simulated as follows. Generate a realisation $\{x_1, \ldots, x_n\}$ from a binomial process in W with n points. Accept this realisation with a probability proportional to $f_n(x_1, \ldots, x_n)$. The accepted point patterns exactly follow the density function f_n which is, mathematically speaking, defined with respect to the Lebesgue measure on $(\mathbb{R}^d)^n$.

Recall that the binomial process, or equivalently, the conditional Poisson process with n points, is in this class of point processes with density $f_n(x_1, \ldots, x_n) = 1/\nu(W)^n$. Similarly, the inhomogeneous Poisson process conditional on the number of points n was defined in (3.4.1) in terms of the multivariate density function with

$$f_n(x_1, \ldots, x_n) = \prod_{i=1}^{n} \frac{\lambda(x_i)}{\Lambda(W)} \qquad \text{for } x_1, \ldots, x_n \in W,$$

where $\Lambda(W) = \int_W \lambda(x)\mathrm{d}x$. This process is equivalent to the i.i.d. cluster process generated by the d-dimensional density

$$f(x) = \frac{\lambda(x)}{\Lambda(W)} \qquad \text{for } x \in W,$$

with n independent points. Note that due to the independence among the points the multivariate density in these first two examples is a simple product of n one-dimensional densities, i.e. $f_n(x_1, \ldots, x_n) = f(x_1) \cdots f(x_n)$. It is clearly symmetric in the x_i.

This section focuses on processes which may be used to model patterns exhibiting inter-point interactions, and thus more complicated types of multivariate densities have to be considered.

A very elegant and useful example of such a process is the *Gibbs process with a fixed number of points*. Consider the location density function

$$f_n(x_1, \ldots, x_n) = \exp\left(-\sum_{i=1}^{n-1} \sum_{j=i+1}^{n} \phi(\|x_i - x_j\|) \right) \bigg/ Z_n \tag{3.6.2}$$

for $x_1, \ldots, x_n \in W$. The function $\phi(r)$ in this formula is often called the *pair potential*, a name that originates in physics: it measures the 'potential energy' caused by the interaction among pairs of points (x_i, x_j) as a function of their distance $\|x_i - x_j\|$. It is through the pair potential function that the interaction among the points is expressed.

Clearly, more complicated ways of modelling the interaction of points have been considered theoretically; see Section 3.6.5. However, statistical experience shows that in many cases pairwise interaction is sufficient.

The pair potential attains values in the range $-\infty < \phi(r) \leq \infty$ with the convention that $\exp(-\infty) = 0$. The term Z_n in (3.6.2) is the 'configurational partition function', a normalising factor ensuring that (3.6.2) is really a probability density, i.e. that its integral is 1. In most cases it is extremely difficult to calculate Z_n or even to find a satisfactory approximation.

The double sum over all pairs of different points in the exponent of (3.6.2),

$$U(x_1, \ldots, x_n) = \sum_{i=1}^{n-1} \sum_{j=i+1}^{n} \phi(\|x_i - x_j\|), \qquad (3.6.3)$$

is often called the 'total energy' of the system of points with pairwise interaction defined by the pair potential function $\phi(r)$. The exponential form of the density function (3.6.2) is neither arbitrary nor a convention. It is motivated in physics by the aim of maximising the entropy

$$-\int_W \cdots \int_W f_n(x_1, \ldots, x_n) \ln(f_n(x_1, \ldots, x_n)) dx_1 \cdots dx_n$$

for a fixed expected total energy

$$\int_W \cdots \int_W U(x_1, \ldots, x_n) f_n(x_1, \ldots, x_n) dx_1 \cdots dx_n.$$

Refer to (3.6.2) to understand the influence of n, $\phi(r)$ and W on the density and hence the nature of the interaction structure and configuration of the pattern. Pairs of points with distances r where $\phi(r) > 0$ contribute only little to the exponent in (3.6.2) and thus these distances are less likely to occur. This is reflected in the pattern as 'repulsion' among pairs of points with this distance. Values of $\phi(r) < 0$ have the opposite effect and thus result in 'attraction', such that there are many points with an interpoint distance of r.

If $\phi(r) = 0$ there is no interaction at distance r and if $\phi(r) \equiv 0$ for all $r > 0$, there is no interaction at any distance. In other words, this yields the binomial process, which hence can be regarded as a special case of the finite Gibbs process with fixed n. If there is an r_{max} with $\phi(r) = 0$ for all $r > r_{max}$ then r_{max} is called the *range of interaction;* this is the distance beyond which there is no interaction. (Note that in Chapter 4 the notion of 'range of correlation' is defined, which is a rather different concept. It is denoted by r_{corr} and is usually larger than r_{max}.)

Many different types of pair potential functions may be considered, and the following presents some examples which have been applied in various contexts. The exposition starts with some simple pair potential functions that result in models of considerable practical interest. More flexible models have also been discussed and the readers are encouraged to develop new pair potentials for their applications.

Example 3.7.   Gibbs hard-core process with fixed number of points
This process is defined through the pair potential function

$$\phi(r) = \begin{cases} \infty \text{ for } r \leq r_0, \\ 0 \text{ for } r > r_0, \end{cases} \tag{3.6.4}$$

with a fixed *hard-core* distance r_0. There are no pairs of points that are closer than the distance r_0. This results in a 'regular' point pattern, if n is not small and W not too large. (For small W, large n and r_0 it is possible that the process does not exist.) This model is useful for example in the situation where the points are centres of spherical or circular non-elastic particles of the same size and r_0 is the diameter of these particles, which is the same for all particles.

Example 3.8.   Strauss process with fixed number of points
Following Strauss (1975), define

$$\phi(r) = \begin{cases} \beta \text{ for } r \leq r_1, \\ 0 \text{ for } r > r_1, \end{cases} \tag{3.6.5}$$

with $0 < \beta < \infty$. The parameter r_1 is the same as $r_{\max}$. Here the location density function is

$$f_n(x_1, \ldots, x_n) = \exp(-\beta n_2(r_1))/Z_n \qquad \text{for } x_1, \ldots, x_n \in W,$$

where

$$n_2(r) = \frac{1}{2} \sum_{x_i, x_j \in N}^{\neq} \mathbf{1}(\|x_i - x_j\| \leq r) \qquad \text{for } r > 0$$

denotes the number of pairs of points with an interpoint distance of r or less. Here $n_2(r_1)$ is a sufficient statistic for β, i.e. all information on β in the point pattern is reflected in the number of r_1-close pairs.

Example 3.9.   Overlap model
In the Strauss process all pairs of points with at most a (fixed) distance r_1 have the same strength of interaction. However, this might not be very realistic in applications. In models of competition in plant ecology or forestry, the strength of interaction between two competing individuals might vary with distance relative to the respective areas of influence.

The pair potential function of the Strauss process discussed above has thus been refined by Penttinen (1984). Rather than assuming a constant strength of interaction for inter-point distances less than r_1, he defines the strength as $\nu(b(x_i, \frac{1}{2}r_1) \cap b(x_j, \frac{1}{2}r_1))$, the area of the intersection of two discs (spheres) of radius $\frac{1}{2}r_1$ with centres x_i and x_j. The interaction is higher for smaller distances and attains its maximum when $r \downarrow 0$, it is decreasing when r increases, and for values of $r > r_1$ the points are not interacting at all, as in the Strauss process. This results in the pair potential function

$$\phi(r) = \begin{cases} \theta\left(\frac{r_1^2}{2}\arccos\left(\frac{r}{r_1}\right) - \frac{r}{2}\sqrt{r_1^2 - r^2}\right) & \text{for } 0 < r < r_1, \\ 0 & \text{otherwise,} \end{cases}$$

where θ is a positive model parameter. Møller and Waagepetersen (2004) have generalised this model further by introducing random radii.

Other pair potential models are, for example:

- the *hard-core Strauss* or *square well* (Figure 3.18(a)),

$$\phi(r) = \phi_{r_0, \beta, r_{max}}(r) = \begin{cases} \infty & \text{for } r \leq r_0, \\ \beta & \text{for } r_0 < r \leq r_{max}, \\ 0 & \text{for } r > r_{max}; \end{cases}$$

- the *multiscale* (Penttinen, 1984),

$$\phi(r) = \begin{cases} \infty & \text{for } r \leq r_0, \\ \beta_1 & \text{for } r_0 < r \leq r_1, \\ \beta_2 & \text{for } r_1 < r \leq r_2, \\ \vdots & \\ \beta_k & \text{for } r_{k-1} < r \leq r_{max}, \\ 0 & \text{for } r > r_{max}; \end{cases}$$

- the *very-soft-core* (Ogata and Tanemura, 1984; see also Figure 3.18(b)),

$$\phi(r) = -\ln\left(1 - \exp\left(-\frac{r^2}{\sigma^2}\right)\right) \quad \text{for } 0 < r < \infty;$$

- and the *Lennart–Jones potential* (e.g. Ogata and Tanemura, 1984),

$$\phi(r) = \alpha_1\left(\frac{\sigma}{r}\right)^{n_1} - \alpha_2\left(\frac{\sigma}{r}\right)^{n_2}$$

for $0 < r < \infty$ and $\alpha_1, \sigma > 0$, $\alpha_2 \geq 0$ and $n_1 > n_2$.

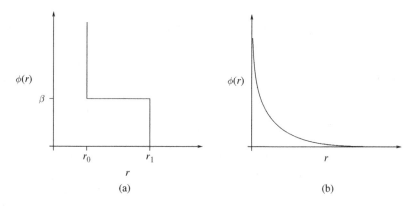

Figure 3.18   (a) Hard-core Strauss (square well) and (b) very soft-core potentials.

Interaction function

As mentioned above, the potential function $\phi(r)$ originates in statistical physics. In this context it is interpreted as a measure of energy: it measures the contribution of a pair of points (x_i, x_j) with $\|x_i - x_j\| = r$ to the energy of the whole system of points. The pairwise interaction between points may also be described in terms of the pairwise *interaction function*

$$h(r) = e^{-\phi(r)}; \tag{3.6.6}$$

see Ripley and Kelly (1977), Ripley (1977), Van Lieshout (2000) and Møller and Waagepetersen (2004).[2] The location density function of the Gibbs process in Equation (3.6.2), for example, can be re-expressed as

$$f_n(x_1, \ldots, x_n) = \prod_{i=1}^{n-1} \prod_{j=i+1}^{n} h(\|x_i - x_j\|) \Big/ Z_n. \tag{3.6.7}$$

The interaction function satisfies $h(r) = 1$ if the members of a pair of points (x_i, x_j) with interpoint distance r do not interact. Pairs of points (x_i, x_j) with distance r exhibit repulsion if $h(r) < 1$ and exhibit clustering if $h(r) > 1$.

Simulation of a Gibbs process with fixed number of points

The simulation of Poisson and Cox processes as discussed above is quite straightforward and follows directly the construction of the models. For the simulation of

---

[2] Note that in the literature the function $h(r)$ has also been denoted by $\phi(r)$.

Gibbs processes, however, one has to resort to more complicated methods. Here an iterative procedure called the Markov chain Monte Carlo method is applied. The idea is to construct a Markov chain (note: not a Markov point process!) the states of which are point configurations in W and the stationary distribution of which is that of the Gibbs process. This method was developed by physicists (Metropolis et al., 1953) and has been applied in many contexts in statistics where it would otherwise be very difficult to simulate directly from a distribution. The algorithm simulates the Markov chain for a long time. Once the chain is in its stationary regime, every state of it is a sample from the stationary distribution of the chain. Møller and Waagepetersen (2004) provide a recent detailed overview of MCMC algorithms for point pattern simulation.

The MCMC approach is often implemented as a simulation algorithm based on the spatial birth-and-death process or, alternatively, as the Metropolis–Hastings algorithm. In the following, the birth-and-death algorithm is described in detail to provide the reader with the necessary knowledge to understand the mechanism and to implement it. Initially a simple but non-trivial example is discussed, followed by a description of the general birth-and-death algorithm.

The approach is iterative in so far as that the simulation commences with an initial point pattern which is modified in a step-by-step fashion by deleting some points ('death') and generating others ('birth'). This procedure is repeated many times such that the algorithm eventually converges and generates patterns that may be considered realisations of a process with the specified density.

Here, the initial configuration is an arbitrary point pattern in W with n points for which the location density function attains a positive value. The properties of the initial configuration are irrelevant since the impact of the initial configuration quickly disappears, often after roughly $10n$ iteration steps (Ripley, 1987).

Example 3.10.   Simulation of a Gibbs hard-core process with n *points*
Consider first the special case of a Gibbs hard-core process with a fixed number of points and with hard-core distance r_0 as discussed in Example 3.7.

The starting configuration may be any n-point pattern in W in which the closest inter-point distance is not less than r_0, e.g. a grid of n points. Assume that the current state of the Markov chain after l steps is $\{x_1, \ldots, x_n\}$. The $(l+1)$th step is as follows:

(i) A point in the set $\{x_1, \ldots, x_n\}$ is deleted at random, where each point in $\{x_1, \ldots, x_n\}$ has the same probability $1/n$ of being deleted. The deleted point is, say, x_k.

(ii) A new point x is drawn from within W with uniform distribution. If

$$\min\{\|x - x_i\| : i = 1, 2, \ldots, n, \ i \neq k\} \geq r_0,$$

then x is chosen to replace the point previously deleted to form a new set of $n + 1$ points $\{x_1, \ldots, x_{k-1}, x, x_{k+1}, \ldots, x_n\}$. Otherwise, if

$$\min\{\|x - x_i\| : i = 1, 2, \ldots, n, \ i \neq k\} < r_0,$$

the point x is rejected and a new point is drawn until the proposed point can be accepted. When n and r_0 are large, it often takes a long time to find such a point.

Note that, although the number of points n is large, this algorithm only simulates from d-dimensional distributions since the proposal x is a point of $\mathbb{R}^d$. Figure 3.19 shows various simulated point patterns obtained by the algorithm.

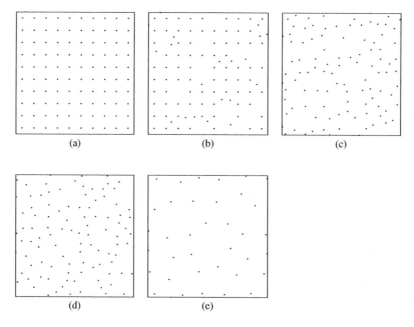

Figure 3.19   Samples simulated from a Gibbs hard-core process with $n = 100$ points in $\boxed{1}$. The hard-core distance is $r_0 = 0.06$. (a) Initial configuration. Configuration after (b) 20 steps, (c) 1000 steps and (d) 2000 steps. The point patterns (c) and (d) are statistically similar and can be regarded as samples from the hard-core process. (e) Another pattern for 2000 steps with $n = 36$ points with $r_0 = 0.15$. Note that it is not possible to generate patterns with 100 points in $\boxed{1}$ with a hard-core distance as large as 0.15. The pattern with 36 points exhibits some boundary effect called 'drift towards the boundary': the point density along the edge of the window is slightly higher than in its interior.

General birth-and-death algorithm

The general algorithm used to generate simulated samples from a Gibbs process with a finite number of points has a similar structure to the algorithm described for the special case of the Gibbs hard-core process above. A randomly chosen point in the pattern is deleted and a new point is generated following the conditional probability density function as above. The simulation starts with an initial point configuration of positive density, which does not 'contradict' the pair potential.

Suppose that after the first l steps the point pattern is $\{x_1, \ldots, x_n\}$.

(i) A point in the set $\{x_1, \ldots, x_n\}$ is deleted at random, where each point in $\{x_1, \ldots, x_n\}$ has the same probability $1/n$ of being deleted. The deleted point is, say, x_k.

(ii) Simulate a new point based on the conditional probability density function

$$f_n(x|x_1, \ldots, x_{k-1}, x_{k+1}, \ldots, x_n) = \frac{f_n(x_1, \ldots, x_{k-1}, x, x_{k+1}, \ldots, x_n)}{\int_W f_n(x_1, \ldots, x_n)\mathrm{d}x_k}.$$

The new configuration $\{x_1, \ldots, x_{k-1}, x, x_{k+1}, \ldots, x_n\}$ is the result after $l+1$ steps.

As above for the Gibbs hard-core process, the x in step (ii) may be simulated by the rejection method: let M be an upper bound of the non-normalised conditional density function

$$\varphi(x) = \exp\left(-\sum_{j=1, j\neq k}^{n} \phi(\|x - x_j\|)\right).$$

A random point x is generated uniformly in W along with an independent uniform random number u on $[0, 1]$. The point x is accepted if $\varphi(x) \geq Mu$. Otherwise a new point x is proposed.

Note that the algorithm for the Gibbs hard-core process follows this approach with $\varphi(x) = 1$ if $\|x - x_j\| > r_0$ for all $j = 1, \ldots, n, j \neq k$, and zero everywhere else.

Example 3.11.   Simulation of a Strauss process with a fixed number of points
Here the non-normalised conditional density function is

$$\varphi(x) = \exp\left(-\beta \sum_{j=1, j\neq k}^{n} \mathbf{1}(0 < \|x - x_j\| \leq r)\right),$$

where the sum in the exponent describes the number of points in the sequence $x_1, \ldots, x_{k-1}, x_{k+1}, \ldots, x_n$ with a distance less than r_0 from x. Assume that $\beta > 0$; then the upper bound M can be assumed to be 1. In the birth step a point, sampled uniformly from within W, is accepted if $\varphi(x) \geq u$ for a u sampled uniformly from

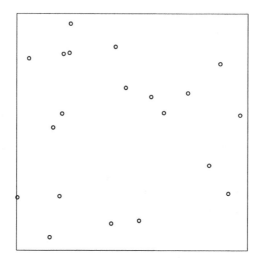

Figure 3.20   A sample simulated from a Strauss process in $\boxed{1}$ with $n = 20$, $r_{\max} = 0.08$ and $\beta = 0.4$.

the interval $[0, 1]$. Figure 3.20 shows an example of a simulation from a Strauss process with a fixed number of points.

3.6.3   Gibbs processes with a random number of points

The above discussion should provide sufficient preparation for the reader to embark on the case of Gibbs processes with a random number of points. These processes are also described by density functions, but now the set-theoretic notation $\mathbf{x}$ is used, where $\mathbf{x}$ is a point configuration in $\mathcal{N}_{\text{fin}}$, with an arbitrary number of points in W; $\{x_1, \ldots, x_n\}$ is the configuration consisting of the points $x_1, \ldots, x_n$.

The distribution of the process N is given by a probability density function $p(\mathbf{x})$ with respect to the distribution of a Poisson process of unit intensity restricted to W, satisfying

$$\int_{\mathcal{N}_{\text{fin}}} p(\mathbf{x}) \Pi(d\mathbf{x}) = 1,$$

as

$$\mathbf{P}(N \in \mathcal{A}) = \int_{\mathcal{A}} p(\mathbf{x}) \Pi(d\mathbf{x}). \tag{3.6.8}$$

The notation here is similar to the notation on pp. 27 and 104, i.e. $\mathcal{A}$ is a subset of $\mathcal{N}_{\text{fin}}$ or a point process event, Π denotes the distribution of the Poisson process restricted to W. The assumption of intensity 1 is only a formality, the same process would be obtained for any other intensity, and only the Z in (3.6.11) would be affected.

The distribution given in (3.6.8) is best understood by initially considering the simulation approach: first a sample $\mathbf{x}$ from the Poisson process is generated as described in Section 2.4. This is accepted as a realisation with probability proportional to $p(\mathbf{x})$. The latter can be realised by von Neumann's rejection sampling method (Ripley, 1987). For this, the maximum M of p has to be determined. A random number u uniformly distributed in $[0, M]$ is generated and compared with the value $p(\mathbf{x})$ for the Poisson sample $\mathbf{x}$. If $u < p(\mathbf{x})$ then $\mathbf{x}$ is accepted; otherwise it is rejected and a new sample is proposed. If, for example, the density function p vanishes for configurations with minimum inter-point distance smaller than r_0, samples of hard-core processes are generated. This simple simulation method has indeed been used in practice, but this book discusses some more efficient methods below.

The probabilities p_n and location density functions f_n are given by

$$p_n = \frac{e^{-\nu(W)}}{n!} \int_W \cdots \int_W p(\{x_1, \ldots x_n\}) \mathrm{d}x_1 \cdots \mathrm{d}x_n \qquad (3.6.9)$$

and

$$f_n(x_1, \ldots, x_n) = \frac{p(\{x_1, \ldots, x_n\})}{\int_W \cdots \int_W p(\{x_1, \ldots, x_n\}) \mathrm{d}x_1 \cdots \mathrm{d}x_n}. \qquad (3.6.10)$$

The Gibbs process with a random number of points is defined through the following density with respect to a Poisson process of unit intensity:

$$p(\{x_1, \ldots, x_n\}) = \exp\left(-\left(\alpha n + \sum_{i=1}^{n-1} \sum_{j=i+1}^{n} \phi(\|x_i - x_j\|)\right)\right) \bigg/ Z \qquad (3.6.11)$$

with $x_1, \ldots, x_n \in W$ for $n = 0, 1, \ldots$. In this equation the term α, the *self-potential* or *chemical activity*, appears, which is an important term in the total energy

$$U(x_1, \ldots, x_n) = \alpha n + \sum_{i=1}^{n-1} \sum_{j=i+1}^{n} \phi(\|x_i - x_j\|). \qquad (3.6.12)$$

Similar to the case of a fixed number of points, the pair potential function $\phi(r)$ describes the interaction, whereas the self-potential together with the pair potential determine the $p_n = \mathbf{P}(N(W) = n)$, i.e. the distribution of the number of points. The same pair potential functions as for Gibbs processes with fixed point number (see p. 142) may be also used for Gibbs processes with a random number of points. The normalising factor Z in (3.6.11) depends on W, α and the pair potential function $\phi(r)$. It is usually analytically intractable.

After all these difficult facts, there is finally a good message: the Papangelou conditional intensity (see p. 28) is

$$\lambda(x|\{x_1, \ldots, x_n\}) = \frac{p(\{x_1, \ldots, x_n, x\})}{p(\{x_1, \ldots, x_n\})}$$

$$= \exp\left(-\alpha - \sum_{i=1}^{n} \phi(\|x - x_i\|)\right) \quad (3.6.13)$$

and fortunately does not contain a normalising factor. Only those points x_i that interact with x, i.e. the points for which $\phi(\|x - x_i\|) \neq 0$, contribute to the sum in the exponent. Often the number of these points is very small compared to the total number of points. This easy formula is of great relevance in the simulation of Gibbs processes.

Simulation of a Gibbs process with a random number of points

The simulation of a Gibbs process with a random number of points is again based on an iterative MCMC procedure, just as for a Gibbs process with fixed number of points discussed above. Hence, again, the simulation starts with an initial configuration associated with a positive density value which is modified during the simulation. At each simulation step, either a point is added or a point is deleted or nothing happens. Several algorithms have been described in the literature that may be used for this purpose. A common feature of all these algorithms is that they are based on ratios of densities. As a result, the intractable normalising factors conveniently cancel out.

A successful MCMC algorithm is the birth-and-death algorithm, a modified version of the algorithm used in the simulation of a Gibbs process with fixed number of points as described on p. 143 above; see also Stoyan et al. (1995, p. 185) or Stoyan and Stoyan (1994, p. 324). Another common MCMC method is the Metropolis–Hastings algorithm (Metropolis et al., 1953; Hastings, 1970), which has been modified for Gibbs process simulation by Geyer and Møller (1994). In what follows, the Metropolis–Hastings algorithm will be described in detail. Further discussion may be found in Møller and Waagepetersen (2004). Geyer and Møller (1994) argue that it is often preferable to the birth-and-death algorithm and may be easier to implement.

Metropolis–Hastings simulation algorithm

The main idea of the Metropolis–Hastings simulation algorithm is that at every step a random proposal is made for a change in the current configuration by birth or death, which is accepted or not, depending on chance. For Gibbs processes it is based on the non-normalised density

$$\varphi(\{x_1, \ldots, x_n\}) = \exp\left(-\left(\alpha n + \sum_{i=1}^{n-1} \sum_{j=i+1}^{n} \phi(\|x_i - x_j\|)\right)\right)$$

with $x_1, \ldots, x_n \in W$ for $n = 0, 1, \ldots$. Proposals are controlled by the following simulation parameters:

- $b(\{x_1, \ldots, x_n\})$, the probability that the 'birth' of a new point will be proposed, while $1 - b(\{x_1, \ldots, x_n\})$ is the probability that the 'death' of a point will be proposed;

- $q_{\text{birth}}(\{x_1, \ldots, x_n\}; x)$, the proposal density function for the location of the new point x;

- $q_{\text{death}}(\{x_1, \ldots, x_n\}; x_k)$, the probability for the proposal to delete the point x_k from the set $\{x_1, \ldots, x_n\}$.

The simulation starts with a point pattern $\{z_1, \ldots, z_m\}$ for which

$$p(\{z_1, \ldots, z_m\}) > 0,$$

i.e. which does not contradict the basic properties of the process. (For example, it is a hard-core pattern if a hard-core process is required.)

Suppose that after l iteration steps the configuration is $\{x_1, \ldots, x_n\}$. At step $l + 1$ initially a decision is made as to whether a new point may be added to the pattern (birth) or removed (death), with probability $b(\{x_1, \ldots, x_n\})$ or $1 - b(\{x_1, \ldots, x_n\})$, respectively.

If step $l + 1$ is a birth, the position of the new point $x \in W$ is proposed from the density function $q_{\text{birth}}(\{x_1, \ldots, x_n\}; x)$ and the proposal is accepted with probability

$$\alpha_{\text{birth}} = \min\{1, \rho_{\text{birth}}\},$$

where

$$\rho_{\text{birth}} = \frac{\varphi(\{x_1, \ldots, x_n, x\})(1 - b(\{x_1, \ldots, x_n, x\}))q_{\text{death}}(\{x_1, \ldots, x_n, x\}; x)}{\varphi(\{x_1, \ldots, x_n\})b(\{x_1, \ldots, x_n\})q_{\text{birth}}(\{x_1, \ldots, x_n\}; x)}$$

is the so-called Metropolis–Hastings birth ratio. With this ratio and the corresponding death ratio the Markov chain is controlled such that its stationary distribution is the desired point process distribution. If the proposal is accepted, the new point is added to the point configuration. If the proposal is not accepted, the configuration does not change in this step.

If step $l+1$ is a death and if $\{x_1, \ldots, x_n\}$ is not empty, a point x_k in $\{x_1, \ldots, x_n\}$ is proposed for deletion with probability $q_{\text{death}}(\{x_1, \ldots, x_n\}; x_k)$. This deletion is accepted with probability

$$\alpha_{\text{death}} = \min\{1, \rho_{\text{death}}\},$$

where

$$\rho_{\text{death}} = \frac{\varphi(\{x_1, \ldots, x_n\} \setminus \{x_k\}) b(\{x_1, \ldots, x_n\} \setminus \{x_k\}) q_{\text{birth}}(\{x_1, \ldots, x_n\} \setminus \{x_k\}; x_k)}{\varphi(\{x_1, \ldots, x_n\})(1 - b(\{x_1, \ldots, x_n\})) q_{\text{death}}(\{x_1, \ldots, x_n\}; x_k)}$$

is the Metropolis–Hastings death ratio. Here $\{x_1, \ldots, x_n\} \setminus \{x_k\}$ denotes the point pattern $\{x_1, \ldots, x_n\}$ without the point x_k, $k \in \{1, \ldots, n\}$. If the proposal is accepted, the point x_k is removed from the configuration. If the proposal is not accepted, the configuration does not change at this step. (If the configuration is empty, the empty configuration does not change at a death step.)

Note that in ρ_{birth}

$$\frac{\varphi(\{x_1, \ldots, x_n, x\})}{\varphi(\{x_1, \ldots, x_n\})} = \lambda(x|\{x_1, \ldots, x_n\}),$$

and in ρ_{death}

$$\frac{\varphi(\{x_1, \ldots, x_n\} \setminus \{x_k\})}{\varphi(\{x_1, \ldots, x_n\})} = \frac{1}{\lambda(x_k|\{x_1, \ldots, x_n\} \setminus \{x_k\})}.$$

In other words, the conditional intensity forms an integral part of the Metropolis–Hastings algorithm.

Simple concrete choices of the parameters of the algorithm are as follows:

$$b(\{x_1, \ldots, x_n\}) \equiv \frac{1}{2},$$

$$q_{\text{birth}}(\{x_1, \ldots, x_n\}; x) = \frac{1}{\nu(W)} \qquad \text{for } x \in W,$$

$$q_{\text{death}}(\{x_1, \ldots, x_n\}; x_k) = \frac{1}{n} \qquad \text{for } x_k \in \{x_1, \ldots, x_n\}.$$

Experience shows that tuning the algorithm increases its efficiency, i.e. it converges earlier to the stationary state and 'mixes' better, and the correlations between subsequent samples decrease faster.

Any MCMC algorithm, and hence both the birth-and-death algorithm and the Metropolis–Hastings algorithm, only generates patterns from the required point process model after a certain number of iterations, after some 'burn-in' period, the length of which is not known a priori. This implies that the user has to observe the simulation process and decide if the algorithm is likely to have converged. Possible indicators are the current number of points n or the current energy $U(\{x_1, \ldots, x_n\})$ given by Equation (3.6.12); see the plot in Figure 3.22. The computing times can be long, in particular for Gibbs processes with a high point density.

Example 3.12.   Simulation of a Strauss process with a random number of points
The Strauss process with parameters $\alpha = -8.0$, $\beta = \exp(0.3) = 1.35$ and $r_1 = 0.08$ is simulated in [1]. The parameters of the Metropolis–Hastings algorithm are chosen as above.

Suppose the configuration at step l is $\{x_1, \ldots, x_n\}$. At first, a decision has to be made as to whether to add or delete a point, either of which happens with probability $\frac{1}{2}$. If a point is to be added to the configuration, the candidate x is chosen uniformly within W. Then the value of the Metropolis–Hastings ratio for a birth is calculated,

$$
p_{\text{birth}} = \frac{1}{n+1} \exp\left(-\alpha - \beta \sum_{i=1}^{n} \mathbf{1}(0 < \|x - x_i\| \le r_1)\right).
$$

The proposal x is accepted with probability

$$
\alpha_{\text{birth}} = \min\{1, p_{\text{birth}}\}.
$$

This means that a random number u is drawn from $[0,1]$ and, if $u \le \alpha_{\text{birth}}$, the proposal x is accepted and the new configuration is $\{x_1, \ldots, x_n, x\}$. If $u > \alpha_{\text{birth}}$, the old configuration $\{x_1, \ldots, x_n\}$ is not changed. (Note that a similar mechanism is used above to decide whether the next step is a potential 'birth' or 'death'.)

If a point is to be deleted, a number k is randomly chosen from $\{1, \ldots, n\}$ and x_k is proposed for removal. The Metropolis–Hastings ratio for a death

$$
p_{\text{death}} = n \exp\left(\alpha + \beta \sum_{i=1, i\ne k}^{n} \mathbf{1}(0 < \|x_k - x_i\| \le r_1)\right)
$$

and $\alpha_{\text{death}} = \min\{1, p_{\text{death}}\}$ are evaluated. The proposal is accepted, i.e. x_k is removed, with probability α_{death} as above. If the proposal is accepted, the new configuration is $\{x_1, \ldots, x_{k-1}, x_{k+1}, x_n\}$, otherwise the configuration does not change.

In this simulation example, the initial configuration is a realisation from a binomial process with 120 points. Figure 3.21 shows the simulation results after 100, 5000 and 10000 iterations, where the random numbers of points were 120, 166 and 166, respectively. (It is only by coincidence that the last two numbers are the same.) Figure 3.22(a) shows the evolution of the number of points $N(W)$ in the 10000 iteration steps.

In this particular case, the burn-in took around 2000 iterations, after which the generated point patterns were considered realisations from the Strauss process. Two point patterns are shown in Figure 3.21 (c) and (d). In Figure 3.21(e) the

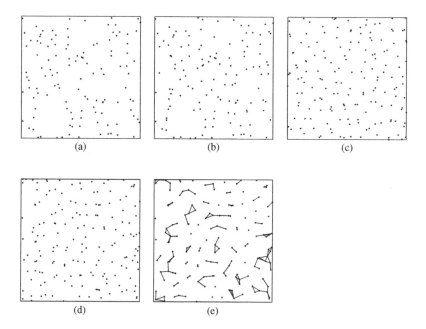

Figure 3.21   Simulated samples from a Strauss process with random number of points generated by the Metropolis–Hastings algorithm on [1]. The parameters of the process and the specific implementation of the algorithm are described in the text. (a) Initial configuration ($n = 120$). Configuration after (b) 100 steps ($n = 120$), (c) 5000 steps ($n = 166$) and (d) 10 000 steps ($n = 166$). The point patterns (c) and (d) are statistically similar and may be considered samples from the Strauss process. A tendency towards regularity can be detected but close pairs are also allowed. (e) The disc graph $G(N, 0.08)$ for the pattern in (d). Points which interact are connected by edges.

interacting points of the pattern (d) are connected by edges forming a disc graph $G(N, 0.08)$.

The number of points in the point patterns at iterations 2001 to 10 000 may be used to inspect the distribution of $N(W)$. Clearly, $N(W)$ is a random variable. If the process's finite nature itself is really of interest, this distribution is investigated. The histogram of this distribution is shown in Figure 3.22(b). In particular, an estimate of $E(N(W))$ seems to be 164.38.

Conditional simulation

A common problem is the conditional simulation of point processes, where the generated patterns have to meet some specific condition. Consider, for example, that a spatial point process N has to be simulated in a window W, given that there

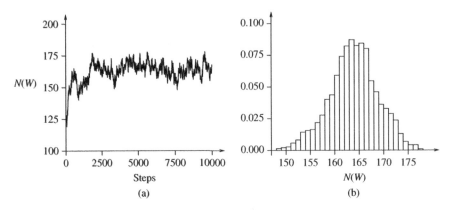

Figure 3.22   (a) A plot of the number of points $N(W)$ at iterations 1 to 10 000 of a run of the Metropolis–Hastings algorithm described in the text. (b) The corresponding histogram of the distribution of the random number of points $N(W)$ derived from the simulations after 2000 steps.

are points of N at locations $y_1, \ldots, y_l$, while other points may have random positions. With the simulation mechanisms discussed above it is straightforward to simulate Gibbs processes with an additional condition like this. The points at locations $y_1, \ldots, y_l$ are simply fixed points in all configurations that are generated by the algorithm and are never changed during the simulation, but contribute to the rates ρ_{birth} and ρ_{death}.

Perfect simulation

An important attempt to solve the problems with the burn-in period is *perfect simulation*, which can be realised by 'coupling from the past' or other techniques; see Kendall and Thönnes (1999), Møller (2001), Møller and Waagepetersen (2004) and Van Lieshout and Stoica (2006). This approach has already been applied successfully in many situations, but not yet in cases of high point density or large interaction radii. When planning long series of simulations one should consult an expert.

3.6.4   Second-order summary characteristics of finite Gibbs processes

Gibbs process may be naturally characterised by their pair potential $\phi(r)$, whereas closed-form expressions for the general summary characteristics discussed in Section 3.5 have often not been found. These depend on $\phi(r)$, the number of points n or chemical activity α, and the window W and can usually only be approximated by simulation. Usually, there is not even a formula for the mean number of points $\mathbf{E}(N(W))$, and it is only by simulation that $\mathbf{E}(N(W))$ may be estimated, as, for example, for the patterns in Figure 3.21. The complex relationship between the pair potential $\phi(r)$ and second-order characteristics has been intensively investigated in the physics literature; see Section 6.6.1.

As in classical statistics, large-sample approximations for finite Gibbs processes with asymptotics for extreme point densities have been considered. This concerns the asymptotic behaviour of sparse data with a decreasing point density and highly dense systems, which tend to be similar to crystals with regular point positions.

For example, the finite K-function $K_{\mathrm{fin}}(r)$ of the hard-core Strauss process has been approximated by sparse data approximation in Saunders et al. (1982); see also Ripley (1988), Stoyan et al. (1995) and Gubner et al. (2000). This is based on the approximate distribution of the number of close pairs in the process, where both the area (volume) of W and n tend to infinity in such a way that $n^d/\nu(W)$ is kept constant. The term 'sparse data' is used as the point density is decreasing in $\mathbb{R}^2$ at a rate of $\frac{1}{n}$ when the point number n increases. The approximation is appropriate for large W and a short range of interaction.

Sparse data approximation for the K-function for the hard-core Strauss process

Recall that a finite hard-core Strauss process $N = \{x_1, \ldots, x_n\}$ in W is a finite Gibbs process defined by the pair potential function

$$\phi(r) = \begin{cases} \infty & \text{for } r \leq r_0, \\ \beta & \text{for } r_0 < r \leq r_{\max}, \\ 0 & \text{for } r > r_{\max}. \end{cases} \qquad (3.6.14)$$

Setting $r_0 = 0$ and $\beta > 0$ yields the classical Strauss process, and setting $r_0 = 0$ and $\beta = 0$ the binomial process. The double sum

$$n_2(r) = \frac{1}{2} \sum_{x_i, x_j \in N}^{\neq} \mathbf{1}(\|x_i - x_j\| \leq r) \qquad \text{for } r > 0, \qquad (3.6.15)$$

is the number of pairs of points with inter-point distance not exceeding r. In general, the distribution of $n_2(r)$ is unknown. However, for sparse point patterns, the following formula may be used as an approximation:

$$\mathbf{E}n_2(r) = \begin{cases} 0 & \text{for } r \leq r_0, \\ c(r^2 - r_0^2)e^{-\beta} & \text{for } r_0 < r \leq r_{\max}, \\ c(r_{\max}^2 - r_0^2)e^{-\beta} + c(r^2 - r_{\max}^2) & \text{for } r_{\max} < r < R_{\max}, \end{cases} \qquad (3.6.16)$$

where $R_{\max}$ is a value beyond the range of interaction and

$$c = \frac{1}{2} \frac{\pi}{\nu(W)} n(n-1).$$

The approximation (3.6.16) ignores edge effects and in this sense does not distinguish between finite and stationary Gibbs processes, which is irrelevant in the

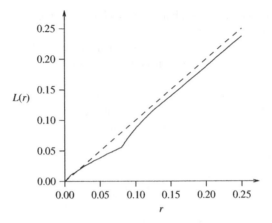

Figure 3.23   The finite L-function of the Strauss process with parameters $\alpha = -8.0$, $\beta = 1.35$ and $r_{max} = 0.08$ as in Example 3.12. In the simulation, 5000 iterations of the Metropolis–Hastings algorithm were calculated and the L-functions were derived as means from 50 samples from every 500th iteration.

large-sample approximation. (Note that stationary Gibbs processes are discussed in Section 6.5.) Equation (3.6.16) yields an approximation for $K_{fin}(r)$ for small r:

$$K_{fin}(r) \approx \frac{2 \cdot \nu(W)}{n(n-1)} E n_2(r) \quad \text{for } r > 0. \tag{3.6.17}$$

Figure 3.23 shows the finite L-function for the Strauss process considered in Figure 3.21. As this process has no hard-core distance, $K_{fin}(r)$ starts to increase at $r = 0$. It has, just as (3.6.17) predicts, a cusp point at $r = 0.08$.

A generalised theory can be found in Jammalamadaka and Penrose (2000).

3.6.5   Further discussion

Inhomogeneous Gibbs processes

The Gibbs processes considered so far may be referred to as being defined in a 'homogeneous environment'. For large windows W the patterns can be regarded as samples from stationary processes, whereas for small W there is some weak influence of the boundary, and for repulsive processes there is some 'drift to boundary'; see Figure 3.19. Sometimes, however, Gibbs processes are desired with stronger forms of non-stationarity, for example with some trend of point density within the window. The homogeneous Poisson process was generalised in Section 3.4.1 for inhomogeneous environments allowing the intensity $\lambda(x)$ to be location-dependent. Similarly, Gibbs processes with interacting points may be generalised to account for inhomogeneous environments. This may be done by no longer assuming that the chemical activity α in (3.6.11) is constant in W but that it depends on location. That is, $\alpha(x)$ reflects how likely it is that a point is located in $x \in W$.

Then the process density function has the form

$$p(\{x_1, \ldots, x_n\}) = \exp\left(-\sum_{i=1}^{n} \alpha(x_i) - \sum_{i=1}^{n-1} \sum_{j=i+1}^{n} \phi(\|x_i - x_j\|)\right) \Big/ Z; \qquad (3.6.18)$$

see Ogata and Tanemura (1986) and Stoyan and Stoyan (1998). Note that the model in (3.6.18) still assumes that the interactions are homogeneous and do not depend on location. Inhomogeneous Gibbs processes with a fixed number of points may be constructed similarly.

The conditional intensity

$$\lambda(x | \{x_1, \ldots, x_n\}) = \exp\left(-\alpha(x) - \sum_{j=1}^{n} \phi(\|x_i - x\|)\right) \qquad (3.6.19)$$

might reflect the situation more clearly; the density at a specific point x depends on the environment through $\alpha(x)$ and on the interaction with other points $x_1, \ldots, x_n \in W$ through the pair potentials.

A statistical question is how to model $\alpha(x)$. If location-dependent covariates $z_1(x), \ldots, z_p(x)$, describing environmental heterogeneity, have been observed, these can be used as explanatory variables for $\alpha(x)$. The function $\alpha(x)$ may also be modelled parametrically.

Area-interaction process

The Strauss process in W with random number of points is defined through the probability density

$$p(\{x_1, \ldots, x_n\}) = \exp\left(-\alpha - \beta n_2(r_1)\right) / Z$$

with respect to the Poisson process of unit intensity. Here $n_2(r_1)$ is, as above, the number of pairs of points of $\{x_1, \ldots, x_n\}$ with an inter-point distance of at most $r_{\max} = r_1$. Usually, the interaction parameter β is assumed to be positive. That is, due to this restriction the model can only model inhibition. Even though the density is also finite for $\beta < 0$ if the number of points is fixed, the realisations are not stable; the points tend to form a single cluster at a random position. If the number of points is not fixed, it increases during the simulation. The conclusion may be that a Gibbs process based on pairwise interactions is not a model for clustered point patterns.

One approach to constructing point processes that are suitable for modelling aggregation is to 'penalise' the effect of a high number of $r_{\max}$-close pairs. Baddeley and Van Lieshout (1995) demonstrate how this penalisation can be done parsimoniously. Define the density

$$p(\{x_1, \ldots, x_n\}) = \exp(-\alpha n - \gamma S(R))\big/ Z, \qquad (3.6.20)$$

where

$$S(R) = \nu \left(\bigcup_{i=1}^{n} b(x_i, R) \cap W \right)$$

is similar to the dilation function introduced in (3.5.4). The model has three parameters, α, γ and R, and describes the area (volume) of the union of discs (spheres) of radius R centred at the points x_i within W. Here again, the normalising factor Z is intractable.

The area-interaction process expresses the interaction in the pattern in terms of the area of the union covered by the 'areas of influence' of the points with radius R. In biological applications, the points may represent locations of plants or birds' nests and the 'area of influence' may be the area in which a plant takes up nutrients from the soil or the areas in which the animals forage. Scarce resources may force the plants or animals to maximally exploit resources in the environment, which translates to maximising $S(R)$, the environment used by the whole population or colony. In terms of the area-interaction process this corresponds to a negative parameter γ and results in a tendency towards regularity in the point pattern. In other situations, facilitation among plants or mutual protection in animals translates to minimising the area $S(R)$, and thus $\gamma > 0$ and the point pattern tends to be clustered. Both cases result in a valid density. The value $\gamma = 0$ yields a Poisson process.

Note that the area-interaction process is not based on simple pairwise interactions but on a much more complex structure of interactions. Nevertheless, the model has only three parameters, making it both parsimonious and interpretable; see Van Lieshout (2000) for more details. It may be generalised to different types of points with different radii; see Kendall et al. (1999). Another generalisation are morphological Gibbs processes, which have turned out to be useful in physical applications; see Brodatzki and Mecke (2002) and Stoyan and Mecke (2005).

Marked Gibbs processes

In many realistic applications the interaction among the objects represented by the points is likely to depend on the distances among the objects but also on the objects' properties such as their sizes. For this purpose, marked Gibbs processes may be defined in a natural way. Now the energies also depend on marks; e.g. the pair potential has the form $\phi(m_1, m_2, r)$, where m_1 and m_2 are the marks of the two points of distance r. Refer to Ogata and Tanemura (1985) and Diggle et al. (2006), who study bivariate patterns, and Møller and Waagepetersen (2007), who use a pair potential for a tree pattern which is influenced by the sizes of trees.

Markov point processes

Gibbs processes as introduced above describe interactions in terms of interactions among pairs of points. This allows inhibition and (sometimes) clustering to be modelled. However, modelling more complicated relationships between the points, e.g. with local alignments of points or connected components, is beyond the scope of this approach. A more general class of processes has been defined which may provide suitable models in these situations.

Kelly and Ripley (1976) and Ripley and Kelly (1977) suggest the general class of *Markov point processes* which exploits the famous Hammersley–Clifford theorem; see Van Lieshout (2000) for a fine presentation. This theorem provides an explicit factorisation of the density of such a process in terms of interactions between the points. A finite Markov point process is defined through a density

$$p(\{x_1, \ldots, x_n\}) = \exp\left(-\sum_C V(C)\right)\bigg/ Z \qquad \text{for } x_1, \ldots, x_n \in W, \qquad (3.6.21)$$

with respect to the Poisson process of unit intensity, where $\{C\}$ is the family of all non-empty subsets of $\{x_1, \ldots, x_n\}$ and V a function which assigns non-negative numbers to these sets. If $V(C) = 0$ for all subsets with more than two points, the pairwise interaction Gibbs process is obtained.

The general form of (3.6.21) allows the generalisation of the pairwise interaction approach to three-point interactions, etc.

Ecological applications

Note that the logic behind Gibbs processes is in many ways akin to an approach in ecology which has been termed 'ecological field theory', as introduced by Wu et al. (1985). Ecological field theory models the growth and survival of an individual plant on the influence of competing plants in terms of their distance and their dimension. In this context a number of competition indices (different from those in Sections 4.2.4 and 5.2.4) together with ecological modelling of growth of species over time have been discussed (see, for example, Kuuluvainen and Pukkala, 1989; Miina and Pukkala, 2002; Kühlmann-Berenzon et al., 2005; Schneider et al., 2006). Clearly, Gibbs processes as such do not model the growth of individuals based on distances. However, in modelling the spatial locations of individuals they may be considered as modelling the survival of individuals in the given locations dependent on distances from competitors (Illian et al., 2008) and competitor properties if marks are included in the model (Møller and Waagepetersen, 2007).

Degenhardt (1999) analyses the evolution of a forest. At various points in time pair potentials are estimated, which clearly show the increasing regularity of the forest as well as the decreasing influence of the tree size marks. She found that the interaction radius in forests can be determined by

$$r_{max} = \frac{a}{\sqrt{\lambda}}$$

with a between 1.2 and 1.5.

3.6.6   Statistical inference for finite Gibbs processes

Statistical inference for finite point processes may be difficult in places, and this is not different for finite Gibbs processes. The simplest part is inference on the number of points $n = N(W)$, which is straightforward if the analysis is based on m, sufficiently many, independent samples, and if the window W is clearly defined by the problem itself, as it is in the maple leaves example (Example 3.6). The situation becomes even easier in cases where it is clear from the start of the analysis whether n should be regarded as random or as fixed. But in both cases the estimation of the density of the Gibbs process is far from being an elementary task.

However, often only a single sample is available. In this case, statistical inference on the point number $N(W)$ is impossible of course; one cannot decide whether n is random or fixed. Thus the main interest focuses on the estimation of the pair potential $\phi(r)$. For this reason, only the case of a fixed number of points is considered in the following. If the point pattern appears to be homogeneous and the window W is part of a larger region with similar point distribution, it seems to be mainly a question of taste whether the analysis will be based on finite Gibbs processes or on stationary Gibbs processes. There are arguments in favour of finite Gibbs processes since the maximum likelihood method can be applied to them, a method which many statisticians believe to be the best estimation method. Indeed, classical statistics provides theoretical results which justify this argument. However, in the context of spatial point processes a proof that maximum likelihood estimators are really superior in general is yet to be given. Fortunately, experience has shown that it is acceptable not to clearly distinguish finite and stationary Gibbs processes with regard to statistical inference. In Diggle et al. (1994) stationary and finite methods were applied in parallel to the same simulated samples and the different pair potential parameter estimation methods showed very similar behaviour. Therefore, this section focuses mainly on examples based on a single-sample case with fixed n and discusses mainly maximum likelihood methods. See Section 6.5 for stationary inference approaches.

Sparse data methods

In most cases statistical inference for Gibbs processes is computationally intensive, as will become obvious in the following. Nevertheless, there are some easier approximate methods, which are particularly suitable for sparse patterns, for large W and n and weak interaction.

Cusp-point method. The first method of this type is the cusp-point method, suggested in Hanisch and Stoyan (1983) and Stoyan and Grabarnik (1991a). It is based on the sparse data approximation discussed in Section 3.6.4 in the context of approximative calculation of the K-function. Consider a finite Gibbs process with pair potential (3.6.14) for a data set of points $\{x_1, \ldots, x_n\}$ in W. This pair potential has two discontinuities, at r_0 and r_{max}, which have to be estimated statistically, along with the interaction parameter β.

The location of the first discontinuity, the hard-core radius r_0, can simply be estimated by the maximum likelihood method, i.e. using the minimum inter-point distance in the data.

One can show that for large windows and sparse data the right and left derivatives of the K-function at r_{max} appear in the relationship

$$\frac{\lim_{r \uparrow r_{max}} K'_{fin}(r)}{\lim_{r \downarrow r_{max}} K'_{fin}(r)} = e^{-\beta}, \tag{3.6.22}$$

which may be exploited in parameter estimation. First the 'cusp point' of the K-function at r_{max} (the range of interaction), corresponding to the discontinuity of $\phi(r)$, is estimated from the nonparametric estimator of $K_{fin}(r)$ obtained with traditional estimators of the K-function. Then β can be estimated from (3.6.22). Admittedly, this approach ignores the finite nature and hence the edge effects.

Note here that the empirical J-function also leads to estimates of r_{max}, by (4.2.51).

Maximum likelihood estimation. The second approach to sparse data approximation is maximum likelihood estimation. The corresponding estimators can usually be applied together with MCMC simulations (see below). With modern computing power this is not a shortcoming in the data analysis since a number of efficient simulation algorithms have been developed for Gibbs processes as described in Section 3.6.2 and well-tested software – the `spatstat` library in R – exists; see Baddeley and Turner (2005, 2006). Note that the simulation approach is less approximate and does not ignore edge effects.

In the case of finite Gibbs processes with fixed n, the n observed points are considered as the data, while the parameter θ (which may also be a vector $\theta = (\theta_1, \ldots, \theta_p)$) of a pair potential $\phi(r)$ of fixed form $\phi_\theta(r)$ has to be estimated. Examples of these functions can be found in Section 3.6.2. Given one sample, the likelihood function is simply the location density function given by (3.6.2), and the log-likelihood function is

$$l(x_1, \ldots, x_n, \theta) = -\ln Z_n(\theta) - \sum_{i=1}^{n-1} \sum_{j=i+1}^{n} \phi_\theta(\|x_i - x_j\|). \tag{3.6.23}$$

Note that the likelihood function also contains the normalising factor Z_n which depends on θ and is thus referred to as $Z_n(\theta)$. The maximum likelihood estimator is the value $\hat{\theta}$ which maximises this function for fixed $x_1, \ldots, x_n$ with respect to θ.

After this rather optimistic start it has to be admitted that the problem is rather complicated due to the normalising factor $Z_n(\theta)$. It depends on θ, n and the window W and is intractable in most cases, as noted in Section 3.6.2. Since $Z_n(\theta)$ is an integral part of the log-likelihood, conventional (numerical) methods cannot be applied when determining the maximum likelihood estimators. In addition, even if it were possible to find maximum likelihood estimators in some way, little could be said about the asymptotic variances of the estimators as no sound theoretical results exist.

Consequently, in maximum likelihood estimation approaches the normalising factor is usually approximated, either by sparse data methods or by Monte Carlo methods.

Ogata and Tanemura (1981) propose a sparse data approximation

$$Z_n(\theta) \approx v(W)^n \left(1 - \frac{a(\theta)}{v(W)}\right)^{n(n-1)/2}, \tag{3.6.24}$$

where

$$a(\theta) = \int_0^\infty (1 - \exp(-\phi_\theta(r)))db_d r^{d-1}dr,$$

in which db_d is the surface area of the unit ball in $\mathbb{R}^d$ (i.e. 2π in $\mathbb{R}^2$). This approximation relies on the so-called cluster or virial expansion originating in statistical physics. The validity of this approximation has been discussed in Ogata and Tanemura (1984), Gates and Westcott (1986) and Mateu and Montes (2001).

For the planar Strauss process in Example 3.8 this yields

$$a(\beta) = (1 - e^{-\beta})\pi r_{\max}^2.$$

If it is assumed that $r_{\max}$ is known, the approximate maximum likelihood estimator of β is

$$\hat{\beta}_{\text{appr}} = -\ln\left(\frac{n_2(r_{\max})(v(W) - \pi r_{\max}^2)}{\pi r_{\max}^2 \left(\frac{1}{2}n(n-1) - n_2(r_{\max})\right)}\right),$$

where

$$n_2(r_{\max}) = \frac{1}{2}\sum_{x_i,x_j \in N}^{\neq} \mathbf{1}(0 < \|x_i - x_j\| \le r_{\max}), \tag{3.6.25}$$

i.e. the number of $r_{\max}$-close pairs in the observed point pattern.

For the planar hard-core Strauss process, assuming that r_0 and $r_{\max}$ are known, this yields

$$a(\beta) = \left(1 - e^{-\beta}\right)\pi\left(r_{\max}^2 - r_0^2\right)$$

and

$$\hat{\beta}_{\text{appr}} = -\ln\left(\frac{n_2(r_{\max})\left(\nu(W) - \pi\left(r_{\max}^2 - r_0^2\right)\right)}{\pi\left(r_{\max}^2 - r_0^2\right)\left(\frac{1}{2}n(n-1) - n_2(r_{\max})\right)}\right).$$

In practice, r_0 and $r_{\max}$ have to be estimated separately, as discussed earlier.

Penttinen (1984) proposes a similar sparse data approximation for the normalising constant of more general planar Gibbs processes assuming a finite range of interaction, i.e. the existence of a constant $r_{\max} > 0$ such that $\phi(r) = 0$ for all $r > r_{\max}$. Then

$$Z_n(\theta) \approx \exp\left(\frac{1}{2}n(n-1)\pi r_{\max}^2 \left(A(\theta) - 1\right)\nu(W)\right) \tag{3.6.26}$$

with

$$A(\theta) = \frac{2}{r_{\max}^2}\int_0^{r_{\max}} r\exp(-\phi_\theta(r))\mathrm{d}r.$$

This approximation yields similar results to (3.6.24). The resulting approximate maximum likelihood estimator for the Strauss process is

$$\hat{\beta}_{\text{appr}} = -\ln\left(\frac{n_2(r_{\max})\nu(W)}{\frac{1}{2}n(n-1)\pi r_{\max}^2}\right),$$

and for the hard-core Strauss process

$$\hat{\beta}_{\text{appr}} = -\ln\left(\frac{n_2(r_{\max})\nu(W)}{\frac{1}{2}n(n-1)\pi(r_{\max}^2 - r_0^2)}\right).$$

Even though the sparse data approximations are quite simple, they can be expected to work well for really sparse data. They yield preliminary approximations for less sparse data at best. The crucial parameter in the planar case is roughly $nr_{\max}/\nu(W)$, where $r_{\max}$ is the range of interaction. If n or $r_{\max}$ increases or $\nu(W)$ decreases, the approximations become increasingly unsuitable. In particular, these approximations do not work for very regular, nearly lattice-like point patterns. A further drawback of the sparse data approximations is that these ignore boundary effects, which might have an effect for small samples.

Monte Carlo maximum likelihood

Monte Carlo approximation of the likelihood. Recall that $Z_n(\theta)$ is intractable in most cases, which implies that the likelihood function is also intractable. In this situation, the idea in Monte Carlo maximum likelihood approaches is to approximate the logarithm of the likelihood function of the finite Gibbs distribution or its derivatives by simulating it for a set of parameter values θ. This implies approximation of the normalising factor. This approach is well known in statistics in general – it is an application of Monte Carlo integration, or equivalently, evaluation of expectations by means of simulation; see Ripley (1987). In what follows the method of Geyer and Thompson (1992) is discussed in detail; see also Geyer (1999) and Møller and Waagepetersen (2004).

To be more specific, Monte Carlo integration involves the following. Let X be a random variable with probability density function $f(x)$ and assume that (independent or dependent) simulations $X_1, \ldots, X_L$ are available. Then the expectation

$$I = \mathbf{E}(h(X); f) = \int h(x) f(x) \mathrm{d}x$$

for some function $h(x)$ is approximated by the mean

$$\frac{1}{L} \sum_{l=1}^{L} h(X_l).$$

Here, $\mathbf{E}(\,\cdot\,; f)$ denotes the expectation with respect to the distribution with density $f(x)$.

In the application considered here, Monte Carlo integration has to yield the log-likelihood function for many parameter values for the optimisation, which is computationally expensive. A clever idea which makes the algorithm more efficient is to approximate the logarithm of the likelihood ratio by

$$\ln\left(\frac{L(x_1, \ldots, x_n, \theta)}{L(x_1, \ldots, x_n, \psi)}\right) = l(x_1, \ldots, x_n, \theta) - l(x_1, \ldots, x_n, \psi),$$

or $l(\theta) - l(\psi)$ for short, where ψ is some fixed parameter value already close to the maximum likelihood solution, found perhaps by one of the sparse data methods. Recall that the normalising factor $Z_n(\theta)$ is of the form

$$Z_n(\theta) = \int_W \cdots \int_W \exp(-U_\theta(\{x_1, \ldots, x_n\})) \mathrm{d}x_1 \cdots \mathrm{d}x_n, \qquad (3.6.27)$$

where

$$U_\theta(\{x_1, \ldots, x_n\}) = \sum_{i=1}^{n-1} \sum_{j=i+1}^{n} \phi_\theta(\|x_i - x_j\|)$$

is the total energy corresponding to the parametrised pair potential $\phi_\theta(r)$. Equation (3.6.27) yields that

$$
\begin{aligned}
\frac{Z_n(\theta)}{Z_n(\psi)} &= \int_W \cdots \int_W \frac{\exp(-U_\theta(\{x_1, \ldots, x_n\}))}{\exp(-U_\psi(\{x_1, \ldots, x_n\}))} \\
&\quad \times Z_n(\psi)^{-1} \exp(-U_\psi(\{x_1, \ldots, x_n\})) dx_1 \cdots dx_n \\
&= \mathbf{E}\left(\exp\left(-\left(U_\theta(\{x_1, \ldots, x_n\}) - U_\psi(\{x_1, \ldots, x_n\})\right)\right); \psi\right),
\end{aligned}
\tag{3.6.28}
$$

where the expectation is with respect to the finite Gibbs distribution with known parameter value ψ.

Generate a sequence of length L (where L is large) of point patterns $\mathbf{x}^{(l)} = \{x_1^{(l)}, \ldots, x_n^{(l)}\}$ for $l = 1, \ldots, L$ in W from the Gibbs distribution with pair potential $\phi_\psi(r)$. Then for parameter values θ 'close' to the fixed value ψ applied in the simulation, the Monte Carlo integration approach yields

$$
\begin{aligned}
&\mathbf{E}\left(\exp\left(-\left(U_\theta(\{x_1, \ldots, x_n\}) - U_\psi(\{x_1, \ldots, x_n\})\right)\right); \psi\right) \\
&\approx \frac{1}{L} \sum_{l=1}^{L} \exp\left(-\left(U_\theta(\{x_1^{(l)}, \ldots, x_n^{(l)}\}) - U_\psi(\{x_1^{(l)}, \ldots, x_n^{(l)}\})\right)\right).
\end{aligned}
$$

This leads to the following approximate log-likelihood ratio with respect to ψ:

$$
\begin{aligned}
l(\theta) - l(\psi) &\approx -\left(U_\theta(\{x_1, \ldots, x_n\}) - U_\psi(\{x_1, \ldots, x_n\})\right) \\
&\quad - \ln\left(\frac{1}{L} \sum_{l=1}^{L} \exp\left(-\left(U_\theta(\{x_1^{(l)}, \ldots, x_n^{(l)}\}) - U_\psi(\{x_1^{(l)}, \ldots, x_n^{(l)}\})\right)\right)\right).
\end{aligned}
\tag{3.6.29}
$$

The maximum likelihood estimator is the value of θ which maximises (3.6.29); it is determined by numerical methods. Note that even though the derivation of the approximation is general, there might be problems in practice when one aims to estimate the log-likelihood function for parameter values θ that are not close to the fixed value ψ that was used in the simulation. Example 3.13 below illustrates this issue for the Strauss process.

Note also that the computation of an expectation using simulations from a different distribution, as above, is known as 'importance sampling', see e.g. Ripley (1987), p. 122.

Example 3.13.   Monte Carlo maximum likelihood for the Strauss process
Refer to Example 3.8 for the potential function. Assume that $r_{\max}$ is known a priori or has been estimated using the cusp-point method or some other methods such that β is the only unknown parameter. The likelihood function is

$$
L(x_1, \ldots, x_n, \beta) = \exp\left(-\beta n_2(r_{\max})\right)/Z_n(\beta) \qquad \text{for } \beta > 0,
$$

with $n_2(r_{max})$ as in (3.6.25). Let $\psi > 0$ be a preliminary parameter value for β chosen a priori or perhaps an estimate of β obtained through a sparse data approximation.

A sequence of point patterns $\{x_1^{(l)}, \ldots, x_n^{(l)}\}$ for $l = 1, 2, \ldots, L$ is simulated from the Strauss process with parameter value ψ and the sufficient statistic $n_2^{(l)}(r_{max})$ is derived from each simulation. Then

$$l(\beta) - l(\psi) \approx -(\beta - \psi)n_2(r_{max}) - \ln\left(\frac{1}{L}\sum_{l=1}^{L}\exp\left(-(\beta - \psi)n_2^{(l)}(r_{max})\right)\right). \quad (3.6.30)$$

Its derivative is

$$\frac{d}{d\beta}l(\beta) \approx -n_2(r_{max}) + \frac{\sum_{l=1}^{L} n_2^{(l)}(r_{max})\exp\left(-(\beta - \psi)n_2^{(l)}(r_{max})\right)}{\sum_{l=1}^{L}\exp\left(-(\beta - \psi)n_2^{(l)}(r_{max})\right)},$$

and the approximate estimation equation can be written as

$$n_2(r_{max}) = \sum_{l=1}^{L} w\left(\beta, \psi, n_2^{(l)}(r_{max})\right) n_2^{(l)}(r_{max}), \quad (3.6.31)$$

where

$$w\left(\beta, \psi, n_2^{(l)}(r_{max})\right) = \frac{\exp\left(-(\beta - \psi)n_2^{(l)}(r_{max})\right)}{\sum_{k=1}^{L}\exp\left(-(\beta - \psi)n_2^{(k)}(r_{max})\right)} \quad \text{for } l = 1, 2, \ldots, L$$

$$(3.6.32)$$

are so-called importance weights. Equation (3.6.31) can be solved by numerical methods.

If β is not close to ψ, it is possible that a few of these weights are very different from zero and the approximation may be poor. When β is close to ψ the weights are evenly distributed around $1/L$, and the approximation is expected to work well.

Penttinen (1984) suggests another Monte Carlo algorithm for the maximum likelihood method. It is based on the Newton–Raphson numerical algorithm where the first and second derivatives of the log-likelihood function are approximated by the Monte Carlo method. Heikkinen and Penttinen (1999) combine the two algorithms. First the stochastic Newton–Raphson algorithm is applied to find a parameter value close to the maximum likelihood estimate. The result is then used as the fixed parameter value ψ in the Geyer–Thompson algorithm as described above.

Profile likelihood method.   In the Strauss process and hard-core Strauss process there are two types of parameters: the 'range' parameters r_0 and r_{max} and the strength-of-interaction parameter β. Baddeley and Turner (2006) call the former parameters 'irregular' and the latter 'regular'. The irregular parameters cause diffi-culties in the maximum likelihood procedure and are estimated directly from the data or derived from prior knowledge. The *profile likelihood method* is a modifica-tion of the maximum likelihood method which can also handle irregular parameters. Consider the Strauss process as an example. The profile likelihood of r_{max} is defined as

$$L(x_1, \ldots, x_n, r_{max}) = \text{argmax}_\beta L(x_1, \ldots, x_n, r_{max}, \beta),$$

and the profile likelihood estimator $\hat{r}_{max}$ maximises the profile likelihood $L(x_1, \ldots, x_n, r_{max})$. In practice, the profile likelihood method may become compu-tationally demanding for Gibbs processes.

Pseudo-likelihood method

A method that was developed early in the history of spatial point process statistics to circumvent the problems with the normalising factor in parameter estimation for finite Gibbs processes is the pseudo-likelihood method (Besag, 1975, 1978), see also Møller and Waagepetersen (2004), p. 171. The observed data $\{x_1, \ldots, x_n\}$ and a parametrised pair potential are linked by the pseudo-likelihood function

$$PL(\theta) = \left(\prod_{i=1}^{n} \lambda_\theta(x_i | \{x_j, j \neq i\}) \right) \exp \left(- \int_W \lambda_\theta(x | \{x_1, \ldots, x_n\}) dx \right). \qquad (3.6.33)$$

The first term is the contribution of the observed points, while the integral part is related to the 'empty space'. Note that (3.6.33) resembles the likelihood function

$$L(\theta) = \left(\prod_{i=1}^{n} \lambda_\theta(x_i) \right) \exp \left(- \int_W \lambda_\theta(u) dx \right)$$

of the inhomogeneous Poisson process in the sense that the intensity function $\lambda(x)$ is replaced by the conditional intensity.

Practical experience shows that *PL* estimation may be a poor method when the interactions in the model are very strong. Another drawback is that in some situations the integral in (3.6.33) can be difficult to calculate even by numerical methods.

Baddeley and Turner (2000) show that when the integral in *PL* (θ) is approx-imated by a finite sum in a clever way this results in an approximate pseudo-likelihood which is formally equivalent to the weighted likelihood of a log-linear model with Poisson responses. (This procedure is referred to as the 'Berman–Turner device'.) Hence standard software developed for generalised linear models may be used to compute the $PL(\theta)$ estimates. However, the asymptotics used to derive standard errors that are provided by the software are not theoretically justified. An advantage of the approach is that information on covariates influencing the point density can easily be taken into account. This may be useful in a preliminary study on the role of potential covariates. However, the maximum likelihood approach is preferable as a method for obtaining the final results.

The Monte Carlo maximum likelihood method solves the parameter estimation problem but no reliable method for variance estimation is known. In practical applications the parametric bootstrap may be used; see Chapter 7.

Baddeley and Turner (2006) present a general estimation procedure including both maximum likelihood and pseudo-likelihood methods for a wide class of finite Gibbs processes. Huang and Ogata (1999) improve the pseudo-likelihood estimation as follows. An MCMC chain is generated based on the pseudo-likelihood estimate as a parameter estimate to estimate the score and Fisher information. These values are used in the Fisher scoring algorithm, yielding an approximation for the maximum likelihood estimate.

Example 3.14.   The positions of 69 Spanish towns
The example originates from the seminal paper by Glass and Tobler (1971) and has been further considered by Ripley (1977, 1988) and Stoyan et al. (1995). Figure 3.24 shows the positions of 69 towns in a square window W of side length 40 miles

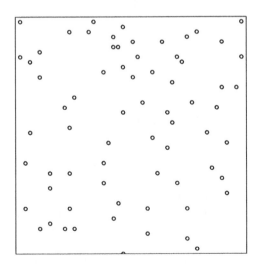

Figure 3.24   Positions of 69 Spanish towns in a 40×40 mile square. Data courtesy of B.D. Ripley.

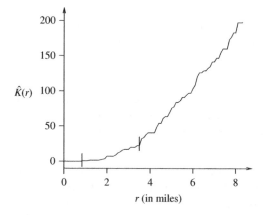

Figure 3.25 The empirical K_{fin}-function for the Spanish towns. The hard-core distance and the chosen cusp point are marked by vertical line segments.

within the Spanish Plateau south-east of Madrid; for details, see Glass and Tobler (1971).

The plot of the K_{fin}-function is shown in Figure 3.25, revealing two properties. There is a minimum inter-point distance of 0.84 miles and inhibition up to 3.5 miles, where a cusp point can be detected. Ripley (1988) suggests the Strauss hard-core point process for this data set based on second-order analysis. In what follows, four methods will be applied for parameter estimation, i.e. for the estimation of r_{max} and β, $\theta = (r_{\text{max}}, \beta)$. In all approaches the value $r_0 = 0.83$ obtained by (4.2.46) is used for the hard-core distance.

(i) *The cusp-point method.* The cusp point is estimated visually as the distance $\hat{r}_{\text{max}} = 3.5$ miles. The relation (3.6.22) is applied by fitting two local regression lines, which yields the estimate $\hat{\beta} = 0.859$. This method uses mainly information of the K_{fin}-function near the cusp point.

(ii) *The mimimum contrast method.* The parameters are determined by minimising the integral

$$\Delta(\beta, r_{\text{max}}) = \int\limits_0^{R_{\text{max}}} \left(\hat{L}_{\text{fin}}(r) - L_{\beta, r_{\text{max}}}(r) \right)^2 \, dr$$

with respect to β and r_{max}; see Section 7.2. Here $\hat{L}_{\text{fin}}(r)$ is an estimate of $L_{\text{fin}}(r)$ computed from data and $L_{\beta, r_{\text{max}}}(r)$ is its theoretical counterpart for the finite Gibbs process with parameters β and r_{max}. In this application, the latter is computed with the approximation (3.6.16), but, as an alternative, simulation can be used. The L_{fin}-function is applied instead of the K_{fin}-function because its estimator has a more stable variance for different values of r. For R_{max} the value 7.5 miles is used. The estimates are $\hat{r}_{\text{max}} = 2.16$ and $\hat{\beta} = 0.81$. The minimum contrast method is based on the information in the $\hat{K}_{\text{fin}}$-summary over the interval $(0, R_{\text{max}})$.

(iii) *The pseudo-likelihood method.* The logarithm of the pseudo-likelihood function (3.6.33) for fixed r_{max} is minimised with respect to β and the value of r_{max} is chosen that maximises the maximum log-pseudo-likelihood. The estimates obtained are $\hat{r}_{\text{max}} = 3.50$ and $\hat{\beta} = 0.67$.

(iv) *Approximate maximum likelihood.* The useful approximate method of Huang and Ogata (1999) starts from the pseudo-likelihood method. Then MCMC simulation is used to improve the initial estimate resulting in an approximation for the maximum likelihood estimator. Again the value $\hat{r}_{\text{max}} = 3.50$, obtained through the pseudo-likelihood method, is used; for β the estimate $\hat{\beta} = 0.76$ is obtained.

Note that in (iii) and (iv) the estimation algorithms provided in `spatstat` (Baddeley and Turner, 2005, 2006) were used. The results for all four methods are shown in Table 3.2.

Ripley (1988) obtained the estimates $\hat{r}_{\text{max}} = 3.5$ and $\hat{\beta}$ between 0.59 and 0.67, while in Stoyan et al. (1995) the estimates were 3.5 and 0.85, respectively. Due to strong inhibition the pseudo-likelihood method underestimates the interaction; hence the maximum likelihood solution is an improvement. The cusp-point method performs well, while the minimum contrast method gives a very small estimate for r_{max}. The minimum contrast method differs from pseudo- and maximum likelihood estimation in the sense that it also includes large interpoint distances (larger than r_{max}).

In Section 7.4 the goodness-of-fit of the hard-core Strauss model for the parameters obtained by the maximum likelihood method is tested, with a positive result.

Note that the pattern of the gold particles cannot be fitted by the hard-core Strauss process. A Gibbs process with estimated parameters $\hat{r}_0 = 5.6$, $\hat{r}_{\text{max}} = 10.0\,\mu\text{m}$

Table 3.2   Estimation results of hard-core Strauss process fits for the Spanish towns data.

Parameter	Estimation method			
	Cusp point	Minimum contrast	Pseudo-likelihood	Maximum likelihood
r_{max}	3.50	2.17	3.50	3.50 (fixed)
β	0.86	0.81	0.67	0.76

and $\hat{\beta}$ between -1.159 and -1.048 shows dense clusters of points, which do not exist in the empirical pattern. The explanation is that attraction in the Gibbs process model with these parameters on the medium scale is too strong. This clearly indicates that Gibbs processes may not be very suitable for clustered patterns even if there is a small hard-core distance.

4

Stationary point processes

This very long chapter forms the core of this book as it deals with the classical non-parametric methods of point process statistics, which assume stationarity or homogeneity and also implicitly that the samples are parts of infinite point patterns. This assumption makes the statistical methodology much simpler and more elegant than in the case of finite patterns. It is realistic in many applications, where the patterns are indeed homogeneous or where homogeneous subsamples in larger inhomogeneous patterns may be chosen with the aim of analysing the local interaction of the points.

The chapter discusses in detail the intensity or point density λ, which is a fundamental first-order characteristic, as well as Ripley's K-function, Besag's L-function and the pair correlation function, which are second-order characteristics. However, these are only some of the characteristics which may be used to analyse point patterns. Other concepts, including indices, nearest-neighbour characteristics and higher-order and topological characteristics, are discussed as well.

The presentation takes a systematic approach and therefore commences with some basic definitions and theoretical concepts which might seem technical but will be relevant throughout the chapter and the book. This includes a discussion of several edge-correction methods, which are necessary to eliminate the influence of the choice of the observation window W.

Statistical Analysis and Modelling of Spatial Point Patterns   J. Illian, A. Penttinen, H. Stoyan and D. Stoyan
© 2008 John Wiley & Sons, Ltd

Furthermore, a number of aspects of high practical relevance are considered such as issues of window choice, replicated patterns, detection of anisotropies as well as outliers and missing values. Finally, modern ideas for the analysis of inhomogeneous patterns are discussed that are based on concepts originally designed for stationary processes.

4.1   Basic definitions and notation

The concept of a stationary point process was introduced and explained in Section 1.6. 'Stationary' in the context of spatial point patterns means 'spatially homogeneous' and does not refer to the temporal behaviour of the data. This property has a strong impact on the statistical analysis.

Statistical methods for stationary point processes consist of those classical approaches that are usually subsumed under the umbrella term 'point process statistics'. This includes well-known concepts such as intensity, K-function and edge-correction, which are directly connected to stationarity. This section presents the relevant theory.

Recall that a point process $N = \{x_1, x_2, \ldots\}$ in $\mathbb{R}^d$ is *stationary* (or *statistically homogeneous*) if N and the translated point process

$$N_x = \{x_1 + x, x_2 + x, \ldots\}$$

have the same distribution for all x, or, in mathematical notation, if

$$N \overset{d}{=} N_x.$$

In other words, both point density and configuration of the process N randomly fluctuate in the same way throughout the whole space.

The reader will soon realise that the statistical methods for stationary point processes are simpler and more elegant than those for finite point processes, which were covered in Chapter 3 and are important examples of non-stationary point processes. Stationarity also facilitates the derivation of formulas for summary characteristics. In the following, this will be demonstrated in detail for the intensity function, void probabilities and second-order characteristics.

Intensity

The mean behaviour of a stationary point process is summarised by a single number, the *intensity* λ. The mean number of points in B satisfies

$$\mathrm{E}(N(B)) = \lambda \cdot \nu(B), \tag{4.1.1}$$

where B is any subset of $\mathbb{R}^d$.[1]

---

[1] Technically, B is a Borel set (as in Section 1.5).

The interpretation of λ is straightforward. Consider, for example, B as a set of unit area (or volume), $\nu(B) = 1$. Then the right-hand side of equation (4.1.1) is simply λ, the mean number of points in N per unit area (volume). Hence, λ is called the *intensity* or *point density*. Note that in the literature symbols based on the Greek letter rho such as ρ, ϱ or ϱ_0 are sometimes used instead.

Equation (4.1.1) can be derived as follows. In general, the mean behaviour of a point process that is not necessarily stationary is described by the intensity measure $\Lambda(B)$, where

$$\Lambda(B) = \mathbf{E}(N(B)) = \text{mean number of points of } N \text{ in } B.$$

If the distribution of N coincides with that of the translated process N_x, then

$$\mathbf{E}(N(B)) = \mathbf{E}(N_x(B)),$$

and since $N_x(B) = N(B - x)$

$$\mathbf{E}(N(B)) = \mathbf{E}(N(B - x)).$$

As x can be chosen arbitrarily, the same is true for $-x$, yielding $\mathbf{E}(N(B)) = \mathbf{E}(N(B_x))$ or

$$\Lambda(B) = \Lambda(B_x) \qquad \text{for all sets } B \text{ and all } x \in \mathbb{R}^d. \tag{4.1.2}$$

A well-known theorem from measure theory then yields that the translation-invariant measure Λ must be a multiple of the Lebesgue measure or area or volume ν, i.e.

$$\Lambda(B) = \lambda \cdot \nu(B). \tag{4.1.3}$$

Campbell theorem

The Campbell theorem is a very useful tool for the calculation of mean values of point process characteristics or statistical estimators. Many of these have the form

$$\sum_{x \in N} f(x),$$

as discussed in Section 1.5. Basically, the Campbell theorem states that the mean of such a sum can be calculated by solving a volume integral.

In the stationary case the Campbell theorem has a very simple form: equation (1.5.10) becomes

$$\mathbf{E}\left(\sum_{x \in N} f(x)\right) = \lambda \int f(x)\,\mathrm{d}x, \tag{4.1.4}$$

i.e. the mean of the sum $f_1(x_1) + f(x_2) + \ldots$ can simply be expressed as λ times a volume integral over $f(x)$. This is true without any specific distributional assumption, i.e. also for non-Poisson processes provided they are stationary.

Example 4.1.   Use of the Campbell theorem. Continuation of Example 1.1

(1) *Seed density*

$$\mathbf{E}S_f = 2\pi\lambda \int_0^\infty m\exp(-cr)r\mathrm{d}r = 2\pi\lambda \cdot \frac{m}{c^2},$$

using $\mathrm{d}x = 2\pi r\mathrm{d}r$, the area differential in polar coordinates.

(2) *Counting birds*

$$\mathbf{E}S_f = 2\pi\lambda \int_0^\infty p(r)r\mathrm{d}r,$$

and for $p(r) = 1 - ar$ (for $r \leq 1/a$) the value

$$\mathbf{E}S_f = 2\pi\lambda \cdot \frac{1}{6a^2}$$

is obtained.

Spherical contact d.f. or empty-space distribution

In the stationary case the probability $\mathbf{P}(N(b(x, r)) = 0)$ that the sphere of radius r centred at location x is empty does not depend on x. Thus the spherical contact d.f. is also independent of x, and it can be defined with respect to $x = o$, yielding a simplification of (1.5.6):

$$H_s(r) = 1 - \mathbf{P}(N(b(o, r)) = 0) \qquad \text{for } r \geq 0. \qquad (4.1.5)$$

Second-order moments

In the stationary case the second-order product density $\varrho(x_1, x_2)$ depends only on the difference $h = x_1 - x_2$, since by stationarity

$$\varrho(x_1, x_2) = \varrho(x_1 + x, x_2 + x) \qquad \text{for all } x, \ x_1 \text{ and } x_2.$$

For convenience, the notation $\varrho(h)$ is typically used, with $h \in \mathbb{R}^d$. If N is also motion-invariant then the product density depends only on the distance r of x_1 and x_2 or the length of h, $r = \|h\|$. Then the notation simplifies further to $\varrho(r)$, where r is a positive number.

Palm distribution*

The Palm distribution is one of the more complicated topics within the theory of point processes. Nevertheless, it is introduced here as it plays a fundamental role in point process statistics. Important point process characteristics such as Ripley's K-function or the nearest-neighbour distance d.f. are of a 'Palm nature'. Readers familiar with the Palm distribution theory will gain a deeper understanding of these characteristics.

Palm characteristics are probabilities or means that refer to individual *points* in a point process. Consider the following two examples:

(a) Let x be a point in N, $x \in N$, and consider $n_x(r) = N(b(x, r)\setminus\{x\})$, the number of points of N in the sphere of radius r centred at x, not counting x itself.

(b) Consider the event $n_x(r) > 0$ that there is at least one point of N within distance r from x or that the distance $d(x)$ from x to its nearest neighbour $h(x)$ is less than r.

Note that $n_x(r)$ with $x \in N$ and $n_y(r)$ for some given fixed deterministic point y, which is not a point in N, are of a different nature: the fact that x is a point in the point process often has an influence on the value of $N_x(r)$. For example, if there is a minimum inter-point distance r_0 in N, then $n_x(r) = 0$ for $r < r_0$ and all x, while $\mathbf{E}\left(n_y(r)\right) = \lambda b_d r^d$ for all r, using (4.1.1). If x is part of a cluster, $\mathbf{E}\left(n_x(r)\right)$ may be larger than $\lambda b_d r^d$.

Clearly, in the stationary case one aims to define Palm distributional characteristics such that these are independent of the particular position of the random point x since all characteristics should be the same throughout space. The usual approach considers a point at the origin o. However, the probability that a stationary point process has a point exactly at o is zero. Therefore, the probability that N has some property provided that it has a point at o is a difficult quantity.

Statistically, a mean related to (a) and a probability related to (b) with $x = o$ may be determined as follows. Consider an observation window W in which N has $N(W) = n$ points. These points $x_1, \ldots, x_n$ are taken in turn and N is shifted such that the relevant point lies at the origin o, i.e. $N - x_i$ is considered for $i = 1, 2, \ldots, n$.

For (a) the numbers $n_i(r)$ of points in $N - x_i$ in $b(o, r)$ may then be determined. Their average yields an estimate of the mean number of points in a sphere of radius r centred at a point process point, where in all cases the point x_i itself is never counted.

For (b) one checks whether the nearest neighbour of o in $N - x_i$ may be found at a distance smaller than r from o. If so, an indicator t_i is assigned the value 1, otherwise 0. Then the sum $\sum_{i=1}^{n} t_i$ is equal to the number of points with a small nearest-neighbour distance, and the value $\sum_{i=1}^{n} t_i/n$ is an estimate of the probability that a point in the point process has its nearest neighbour at a distance less than r.

Note that in this book the theoretical analogues to the mean and the probability discussed above will be denoted by

$$\mathbf{E}_o(N(b(o, r)\backslash\{o\})) \quad \text{and} \quad \mathbf{P}_o(N(b(o, r)\backslash\{o\}) > 0),$$

where the index 'o' indicates the shifting of the patterns towards o. In the literature the mean and probability above are typically denoted by $\lambda K(r)$ and $D(r)$, respectively.

The exact technical definition of the Palm probability $\mathbf{P}_o$ is the following:

$$\lambda v(W)\mathbf{P}_o(N \in \mathcal{A}) = \mathbf{E}\left(\sum_{x \in N \cap W} \mathbf{1}_A(N - x)\right). \tag{4.1.6}$$

Here W is some 'test set' of positive area (volume) $v(W)$, and $N \in \mathcal{A}$ is a general notation for 'point process N has property $\mathcal{A}$'. Clearly, $\mathcal{A}$ has to be a property which makes sense for a point process with a point at o. An example of this is '$N(b(o, r)\backslash\{o\}) = 0$'. The indicator $\mathbf{1}_A(N - x)$ is 1 if the shifted point process $N - x$ has property $\mathcal{A}$ and 0 otherwise.

Analogously,

$$\lambda v(W)\mathbf{E}_o(\mathcal{S}(N)) = \mathbf{E}\left(\sum_{x \in N \cap W} \mathcal{S}(N - x)\right), \tag{4.1.7}$$

where $\mathcal{S}(N)$ is a (real) number assigned to N. An example is $\mathcal{S}(N) = N(b(o, r) \backslash \{o\}) = n_o(r)$, the number of points of N in $b(o, r)$ excluding o.

Equation (4.1.6) is the definition of the Palm distribution as given by Mecke (1967), who showed that the definition of $\mathbf{P}_o(N \in \mathcal{A})$ is independent of the choice of the set W.

Remarks. $\mathbf{P}_o(N \in \mathcal{A})$ is often called the conditional probability that N has property $\mathcal{A}$ given that a point of the process is in o. This condition has probability zero for a stationary point process. Thus it is clear that the conditional probability discussed cannot be defined in the classical way as $\mathbf{P}(A|B)$ for $\mathbf{P}(B) > 0$.

Fortunately, under some conditions $\mathbf{P}_o(N \in \mathcal{A})$ can be obtained as a limit of conditional probabilities with the condition $N(b(o, \varepsilon)) = 1$ for $\varepsilon \to 0$, i.e. that there is a point of N in a small sphere around o; see p. 75 for the Poisson case.

Geometrical relationships between N and its Palm version N_o (the point process having the distribution $\mathbf{P}_o$, with a point at the origin) are studied in Thorisson (2000) and Holroyd and Peres (2005): N and N_o are related by a 'shift coupling', i.e. there is a random point X in N such that the translated process N_{-X} has the same distribution as N_o.

In the literature the term *typical point* is often used and refers to the probability and the mean discussed above. For example,

- $\mathbf{P}_o(N(b(o, r) \setminus \{o\}) > 0)$ is the probability that there is at least one point in a sphere of radius r centred at the typical point;

- $\mathbf{E}_o(N(b(o, r) \setminus \{o\}))$ is the mean number of points in a sphere of radius r centred at the typical point, excluding the typical point itself.

This book also uses this terminology.

Palm distributions can also be constructed with respect to two or more points; see Kallenberg (1983) and Hanisch (1982, 1983).

Campbell–Mecke formula

The Palm mean appears in a refined form of the Campbell theorem, the Campbell–Mecke formula:

$$\mathbf{E}\left(\sum_{x \in N} f(x, N)\right) = \lambda \int \mathbf{E}_o(f(x, N_{-x}))\mathrm{d}x = \lambda \mathbf{E}_o\left(\int (f(x, N_{-x})\mathrm{d}x\right). \qquad (4.1.8)$$

Note that here $f(x, N)$ depends not only on x but also on other points in N. Section 4.2.6 applies the Campbell–Mecke formula in the context of the nearest-neighbour distance distribution function.

4.2 Summary characteristics for stationary point processes

4.2.1 Introduction

As explained in Section 1.7, the aim of summary characteristics is to provide a brief and concise description of point patterns using numbers, functions or diagrams. Summary characteristics for stationary and isotropic point processes are particularly efficient. Well-known examples include the intensity λ (see Section 4.2.3) and the K-function (see Section 4.3).

Prior to the application of summary characteristics that are suitable for stationary processes, the pattern should be investigated at least by visual inspection to verify that the stationarity assumption holds, as recommended in Section 1.6. A reasonably stationary pattern would be one that appears as if it could be continued in space in the same way many times. It is risky, for example, to apply stationary summary characteristics to a pattern which looks like a single cluster in a window, with sparsely distributed points close to the margin. In this case it is not clear if there are further clusters in the pattern outside the window or whether the pattern is more uniform elsewhere.

It is often useful to adapt the observation window W to the point pattern, i.e. to either reduce W by removing empty marginal regions or to use a larger window (which may involve collecting more data). This may lead to irregular, non-rectangular or non-circular windows; see Wiegand and Moloney (2004). The window may be adapted if the original window was chosen based on theoretical considerations, and the choice was not made on the basis of the nature of the statistical problem. Sometimes, inappropriate windows lead to strange behaviour in the summary characteristics, which is not typical of summary characteristics of stationary processes. This is the case in particular for large arguments (i.e. large distances) of functional summary characteristics such as the pair correlation function $g(r)$; see Example 4.17. In this way, the summary characteristics themselves indicate that the window might have to be modified. The issue of window adaptation is discussed in greater detail in Section 4.8.

In the context of stationary point processes, *edge-correction* of statistical estimators of summary characteristics is particularly relevant. With these corrections many estimators are *unbiased* and the influence of the window is reduced.[2] Edge-corrections are necessary since many estimators take information on the neighbourhoods of all points in a pattern into account. If a data set only provides information on the points inside the window, the complete information on the neighbourhood is not available for those points that are close to the window's boundary or edge. The following section shows what can be done to still obtain unbiased estimators.

The text above may appear to imply that the use of summary characteristics for stationary processes was questionable or complicated. However, the following will demonstrate that with sufficient knowledge of potential difficulties this is not the case. Stationary summary characteristics are quite natural, in particular if the focus of the statistical analysis is on short-range fluctuations or local interactions.

The remainder of this chapter is organised as follows: initially the various summary characteristics are introduced in terms of probability theory. This is necessary for an exact and thorough understanding of the nature of the characteristics and their properties, but may appear rather theoretical. This is followed by a discussion of the statistical estimation of the characteristics, which may be more relevant for those readers who are more interested in applications.

Formulas for the summary characteristics of specific point process *models* are presented in Chapter 6. These may be used to identify a suitable model for a specific data set and serve as a basis for parametric statistical approaches.

4.2.2 Edge-correction methods

When working with the statistics of stationary point processes one often faces a difficult problem: the data are given for a bounded observation window W only, but the pattern is (implicitly) assumed to be infinite and the summary characteristic to be estimated is defined independently of W and should not show any traces of

---

[2] See Appendix A for an explanation of the terms 'unbiased' and 'ratio-unbiased'.

W. However, natural estimators of the summary characteristic would need information from outside W, in particular for unbiased or ratio-unbiased estimators. This problem is typical of spatial statistics, but it also appears in a similar way in survival analysis, where information is often 'censored' as only a finite time interval can be considered.

The issue is explained here for two typical point-related summary characteristics. For the case of location-related characteristics refer to p. 189. In later sections of this chapter and in the following chapters it is always stated which edge-correction method is suitable for the given characteristics.

Consider the following two summary characteristics for stationary point processes:

(i) the mean distance $m_D = \mathbf{E}_o(d(o))$ between the typical point and its nearest neighbour;

(ii) the mean number $\lambda K(r) = \mathbf{E}_o(n_o(r)) = \mathbf{E}_o(N(b(o, r) \setminus \{o\}))$ of points in a disc of radius r centred at the typical point (where the latter is not counted).

The 'natural' estimator of $\mathbf{E}_o(d(o))$ is

$$\overline{d} = \frac{1}{n} \sum_{i=1}^{n} d_i, \tag{4.2.1}$$

where $n = N(W)$ is the number of points in the observation window W and the d_i are the distances to the respective nearest neighbours of these points. This estimator is not unbiased, since n is random. However, under certain conditions it is ratio-unbiased (see below).

Unfortunately, in this naive form the estimator $\overline{d}$ cannot be applied if only information on the points in W is available: for points close to the boundary of W, the true distances d_i cannot be accurately determined. Of course, every point in W has a nearest neighbour *in* W, but beyond the border of W (i.e. outside W) there might be closer neighbours including the true nearest neighbour, see Figure 4.1.

The situation is similar for $\mathbf{E}_o(n_o(r))$. The naive natural estimator is

$$\overline{n}(r) = \frac{1}{n} \sum_{i=1}^{n} n_i(r), \tag{4.2.2}$$

where n is as above and $n_i(r) = N(b(x_i, r) \setminus \{x_i\})$ is the number of points of N within distance r of point x_i, again excluding x_i itself. Note that $\overline{n}(r)$ may be rewritten as a double sum

$$\overline{n}(r) = \frac{1}{n} \sum_{i=1}^{n} \sum_{j=1}^{n \neq} \mathbf{1}(\|x_i - x_j\| \leq r) \tag{4.2.3}$$

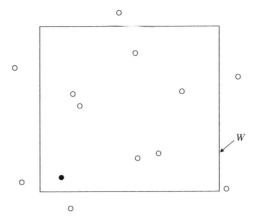

Figure 4.1   Points within a window W and outside. The nearest neighbour of the point marked by a filled circle is outside W. If information on points outside W is not available, it is not possible to correctly determine the nearest-neighbour distance of this point.

since the number of points in the disc $b(x_i, r)$ is equal to the number of points in N with a distance equal to or smaller than r from x_i,

$$n_i(r) = \sum_{j=1}^{n} {}^{\neq} \mathbf{1}(\|x_i - x_j\| \leq r).$$

(The superscript $\neq$ indicates that $j \neq i$ in the sum.) As with $\bar{d}$, it is often impossible to determine $\bar{n}(r)$ if only information on points within W can be used: for points close to the border of W the discs $b(x_i, r)$ do not completely lie within W, in particular for large r. As a consequence, $n_i(r)$ cannot be correctly calculated for all x_i. Or in other words, if only the points within W are given, an incomplete list of pairs (x_i, x_j) as in (4.2.3) has to be used.

Almost all types of estimators of point-related summary characteristics considered in this book correspond to one of the two cases discussed above. For these, two different edge-correction strategies may be used:

- All points x_i in W for which the nearest neighbour cannot be correctly determined are simply excluded; often even further points in W are excluded. The remaining points are weighted to compensate for the resulting loss of information.

- Only those pairs (x_i, x_j) for which x_j is in W are used. In order to compensate for the pairs (x_i, x_j) that are excluded because x_j is outside W, the retained pairs are weighted.

Many edge-correction methods follow these strategies. These are discussed in the following, but for the sake of completeness four other methods of a different character are mentioned as well. In addition, the missing-data approach is also mentioned here (Geyer, 1999) as well as the reconstruction method (see Section 6.7).

Note that edge-effects problems become more aggravated in higher-dimensional spaces. This is easy to understand: consider a square and a cube of side length a and compare its area and volume with that of a square and a cube where the side length is reduced by the same value α in order to exclude points close to the boundary. Clearly $(a - \alpha)^3/a^3 < (a - \alpha)^2/a^2$.

Plus-sampling

Plus-sampling estimators assume that the estimators (4.2.1) and (4.2.2) can be applied. This implies that more information than that contained in W has to be available. For example, this situation may occur in forestry research, when a specific stand is investigated and the forester can determine nearest neighbours outside W for those trees that are close to the border of the stand, if necessary. For example, the estimator $\overline{d}$ of m_D, given by (4.2.1), is ratio-unbiased since $\mathbf{E}n = \mathbf{E}(N(W)) = \lambda \nu(W)$ and

$$\mathbf{E}\left(\sum_{i=1}^{n} d_i\right) = \mathbf{E}\left(\sum_{x \in N} \mathbf{1}_W(x) d(x)\right) = \lambda \cdot \mathbf{E}_o(d(o)) \cdot \nu(W),$$

by the Campbell theorem for stationary marked point processes (see p. 302). Here $d(x)$ denotes the nearest-neighbour distance of point x. Plus sampling may be analogously applied to derive an estimator for $\lambda K(r)$.

No edge-correction

If both the window W and the number of points n are large, edge effects can simply be ignored. This means that in (4.2.1) the d_i are simply the nearest-neighbour distances *in* the window, i.e. the search for the nearest neighbour of point x_i is restricted to the points in W. Analogously, the $n_i(r)$ in (4.2.2) are then the numbers of points in the discs $b(x_i, r)$ intersecting with W,

$$n_i(r) = N(b(x_i, r) \cap W).$$

It is clear that $\overline{d}$ tends to be (a little) too large and $\overline{n}(r)$ (a little) too small. This simple method can be recommended for the estimation of the distributional indices in Section 4.2.4, as shown in Pommerening and Stoyan (2006).

The following two methods, periodic and reflection edge-correction, may be regarded as 'speculative' in as much as they are unrelated to the structure of point processes. They are (rough) attempts to generate large patterns for which plus sampling can be used. The window W is assumed to be a rectangle or parallelepiped,

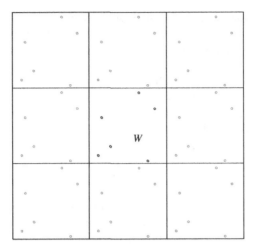

Figure 4.2   A point pattern in a rectangular window W and its periodic continuation.

but attempts have also been made to extend the methods to circular windows; see Windhager (1997). The authors of the present volume do not recommend their use.

Periodic edge-correction

The point pattern in W is enlarged by periodic continuation as shown in Figure 4.2; see also Section 2.2 on the periodic binomial point process. Clearly, this is only possible if W is a rectangle or parallelepiped.

The resulting point pattern can be analysed statistically by means of plus-sampling methods. To do this, the distances have to be redefined. In the planar case this is termed a *torus metric*: if W is the rectangle with side lengths a and b with left lower vertex at the origin, the distance between two points x and $y \in W$ with $x = (\xi_1, \xi_2)$ and $y = (\eta_1, \eta_2)$ is

$$\|x - y\| = \sqrt{(\min\{|\xi_1 - \eta_1|, a - |\xi_1 - \eta_1|\})^2 + (\min\{|\xi_2 - \eta_2|, b - |\xi_2 - \eta_2|\})^2}.$$

Clearly, this method is suitable when the pattern in W is a result of a simulation with 'periodic boundary conditions' or in the context of statistics for binomial point processes on a rectangle or parallelepiped. Otherwise strange and spurious point configurations may appear along the borders, and the method is merely a cheap trick to provide more points.

Reflection edge-correction

The point pattern in W is enlarged by reflection at the borders, as shown in Figure 4.3. Again the resulting large point pattern can be analysed by means

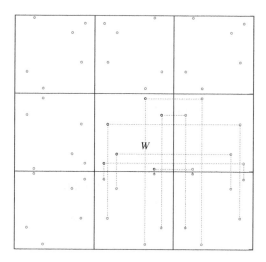

Figure 4.3 A point pattern in a rectangular window W and its continuation by reflection.

of plus-sampling methods. This method is even more problematic than periodic edge-correction as it constructs configurations close to the edges that may differ substantially from the usual configurations in the point process N. It is difficult to predict its statistical properties.

Conditional simulation outside W

This method, which is a missing-data approach, uses simulation to generate a larger sample that includes the investigated pattern given in W, such that plus sampling can be applied. This method may look very laborious as it comprises two difficult steps. First, one has to statistically analyse the given data to obtain a model or, at least, summary characteristics for the reconstruction method described in Section 6.7. Then conditional simulation of the point process has to be carried out in a large region with W at the centre. 'Conditional' means that the data points in W are fixed, while new points are generated outside W which form configurations with the border points in W that are similar to those in the interior of W. This method was introduced in forestry by P. Biber in 1997 (see Pretzsch, 2002); Section 6.7 presents a variant which differs only in the simulation method.

The following now describes the methods mentioned on p. 182.

Minus sampling or border method

Assume that only the neighbours within a distance r are relevant for each point. This is exactly the situation in the context of the estimation of $\mathbf{E}_o(n_o(r))$. It also

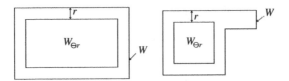

Figure 4.4 Rectangular and irregular window W and corresponding reduced window $W_{\ominus r}$.

applies when $\mathbf{P}_o(d(o) \leq r)$ is estimated, the probability that the nearest-neighbour distance of the typical point is smaller than r.

Here the window W is reduced to the smaller window $W_{\ominus r}$. This is the subset of W the points in which are in the interior of W and have a distance larger than r from the boundary ∂W (see Figure 4.4).

Note that $W_{\ominus r}$ is simplified notation for $W \ominus b(o, r)$, which uses notation from mathematical morphology, where $\ominus$ denotes Minkowski subtraction and $b(o, r)$ is the usual symbol for a disc or sphere of radius r centred at o. The symbol $W \ominus b(o, r)$ denotes the set of all points x in W with $b(x, r) \subseteq W$. If W is a rectangle with side lengths a and b, then $W_{\ominus r}$ is again a rectangle and its side lengths are $a - 2r$ and $b - 2r$. Clearly, $W_{\ominus r}$ is the empty set if $r \geq a$ or $r \geq b$. If W is a disc of radius R, then $W_{\ominus r}$ is a disc of radius $R - r$. See Appendix B for more information on aspects of mathematical morphology relevant to this book.

The statistical estimation of $\mathbf{E}_o(d(o))$ and $\mathbf{E}_o(n_o(r))$ uses only the points in $W_{\ominus r}$ as reference points $x_1, x_2, \ldots$, while all points in W are used to determine the d_i and $n_i(r)$. The minus-sampling estimators are

$$\overline{d}_{\ominus r} = \frac{1}{n_{\ominus r}} \sum_{i=1}^{n_{\ominus r}} d_i \qquad (4.2.4)$$

and

$$n(r) = \frac{1}{n_{\ominus r}} \sum_{i=1}^{n_{\ominus r}} n_i(r), \qquad (4.2.5)$$

where $n_{\ominus r}$ is the number of points of N in $W_{\ominus r}$. Both estimators are ratio-unbiased.

However, much better estimators for $\mathbf{E}_o(d(o))$ and $\mathbf{E}_o(n_o(r))$ have been developed, as shown below. This is not surprising as the edge-correction described above is rather rough and does not refer to individual points. Nevertheless, it is used in some cases, in particular for location-related summary characteristics such as the spherical contact d.f. and the morphological functions; see Section 4.2.5.

Nearest-neighbour edge-correction

The nearest-neighbour edge-correction is a refined version of the border method, an 'individual' edge-correction method which is suitable for summary characteristics related to nearest-neighbour distances (this refers to both first and kth neighbours). For each individual point x in W a decision is made as to whether it may safely be used in the estimation, where 'safely' means that the point's nearest-neighbour distance $d(x)$ is shorter than its distance $e(x)$ to the boundary of W.

A point x included in the estimation is assigned the weight $1/v(W_{\ominus d(x)})$, where $W_{\ominus d(x)} = W \ominus b(o, d(x))$. This weight is large if $d(x)$ is large. This weight choice is plausible as points with a large nearest-neighbour distance $d(x)$ in W are rare.

The nearest-neighbour estimator of $\mathbf{E}_o(d(o))$ is

$$\overline{d}_{nn} = \frac{\sum_{x \in W} \mathbf{1}(d(x) \le e(x)) \cdot d(x) / v(W_{\ominus d(x)})}{\sum_{x \in W} \mathbf{1}(d(x) \le e(x)) / v(W_{\ominus d(x)})}. \tag{4.2.6}$$

The numerator is an unbiased estimator of $\lambda \mathbf{E}_o(d(o))$ and the denominator an adapted unbiased estimator of λ; the simplified notation of p. 26 is used. Despite its complicated form, the calculation is easy: take the points x in W in turn, and check whether $d(x) \le e(x)$! If this is the case use the point x in the estimation and divide by the respective weight, i.e. the area $v(W_{\ominus d(x)})$. If W is a rectangle with sides a and b, then

$$v(W_{\ominus d(x)}) = (a - 2d(x))(b - 2d(x)) \qquad \text{for } a > 2d(x) \text{ and } b > 2d(x),$$

and if W is a disc of radius R, then

$$v(W_{\ominus d(x)}) = \pi(R - d(x))^2 \qquad \text{for } R > d(x).$$

If the inequalities concerning $d(x)$ do not hold then the estimation does not make sense.

The use of the more straightforward estimator which does not use any weights, i.e.

$$\tilde{d}_{nn} = \frac{\sum_{x \in W} \mathbf{1}(d(x) \le e(x)) \cdot d(x)}{\sum_{x \in W} \mathbf{1}(d(x) \le e(x))},$$

cannot be recommended since it is not ratio-unbiased.

Second-order edge-corrections

These are probably the most common edge-correction methods. The estimators of second-order characteristics typically have a structure similar to (4.2.3). That is, they contain double sums of pairs of points in W, and operate on inter-point distances.

For a given point x_1 and inter-point distance r, many partner points x_2 will not be in the window W. In order to construct an unbiased estimator, a large weight is assigned to pairs of points (x_1, x_2) with large inter-point distance and both x_1 and x_2 in W.

Two types of weights or corrections are used:

- *Stationary* or *translational edge-correction,* applicable to all stationary point processes, with

$$\text{weight} = 1/\nu(W_{x_1} \cap W_{x_2}).$$

Here $\nu(W_{x_1} \cap W_{x_2}) = \nu(W \cap W_{x_2-x_1})$ is the area (volume) of the intersection of W_{x_1} and W_{x_2}, where W_x is the translated window $W_x = \{z + x : z \in W\}$, or of W and the translated window $W_{x_2-x_1}$ (see Figure 4.5 and Appendix B).

- *Isotropic* or *rotational edge-correction,* applicable only to stationary and isotropic point processes, with

$$\text{weight} = 1/w(x_1, x_2).$$

In the planar case, the quantity $w(x_1, x_2)$ is the boundary length in the window W of the circle of radius $\|x_1 - x_2\|$ centred at x_1 divided by the circle perimeter length $2\pi\|x_1 - x_2\|$,

$$w(x_1, x_2) = \frac{\nu_1(\partial b(x_1, \|x_1 - x_2\|) \cap W)}{2\pi\|x_1 - x_2\|} \tag{4.2.7}$$

(see Figure 4.6). In the three-dimensional case, surface area is used instead of boundary length.

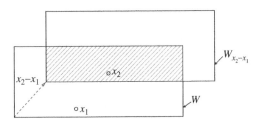

Figure 4.5 The observation window W, the translated window $W_{x_2-x_1}$ and their intersection.

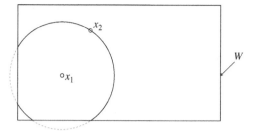

Figure 4.6   The observation window W and the relevant part of the circle with centre x_1 and radius $\|x_2 - x_1\|$.

Edge-corrections for location-related characteristics

The standard edge-correction method for location-related characteristics is minus sampling; see Section 4.2.5 for examples.

4.2.3   The intensity λ

The intensity or point density λ is the fundamental first-order characteristic for stationary point processes, as explained in Section 4.1. For many models formulas for λ are known, but usually the initial analysis applies statistically estimated intensities. Therefore, this section will now discuss a number of estimators of this characteristic in greater detail. The intensity satisfies

$$\mathbf{E}(N(B)) = \lambda \nu(B), \qquad (4.2.8)$$

i.e. the mean number of points of N in any set B is equal to λ multiplied by the area (volume) of B. Since the number of points appears in the first power, λ is called a first-order characteristic.

The intensity also admits a local characterisation. Consider an infinitesimally small disc (sphere) $b(x)$ of area (volume) dx centred at the arbitrary location x. Then the probability $p_1(x)$ that there is a point of N in $b(x)$ is

$$p_1(x) = \lambda dx \qquad (4.2.9)$$

(see p. 28).

A simple stationarity 'test' is closely related to the intensity: determine statistically the intensity function $\lambda(x)$ and verify that a plot of the estimate obtained shows only local irregularities but no general trend.

The *standard estimator* of the intensity is

$$\hat{\lambda} = \frac{N(W)}{\nu(W)}. \qquad (4.2.10)$$

This follows directly from the definition of intensity as it is a quotient of number of points $N(W)$ in W and area (volume) $\nu(W)$. This estimator is unbiased and, if N is ergodic, consistent, i.e. as W increases it converges to the true value λ. This holds independent of the specific distribution of N, whereas the variability of the estimator $\hat{\lambda}$ is of course distribution-dependent.

The variance of $\hat{\lambda}$ is given by

$$\text{var}\hat{\lambda} = \frac{\text{var}N(W)}{\nu(W)^2},\qquad (4.2.11)$$

where $\text{var}N(W)$ can be calculated using (4.3.22) with B replaced by W or by the approximation in (4.3.25). For a Poisson process it is

$$\text{var}\hat{\lambda} = \frac{\lambda}{\nu(W)}.\qquad (4.2.12)$$

This value is a good approximation even in the non-Poisson case. For more regular processes it is an upper bound, for more irregular (clustered) processes a lower bound. Heinrich and Prokešová (2006) discuss a statistical estimator of $\text{var}\hat{\lambda}$, a statistical estimator for the variance of an estimator of λ.

However, the estimator (4.2.10) assumes that it is possible to count all points in W. This is often difficult or even impossible and therefore other estimation methods have to be used.

Fractionator counting

Often the number $n = N(W)$ of points in W is very large; counting them all is too laborious and cannot be carried out automatically. This is true in tree counting in forestry or in cell nucleus counting in biology and medicine. In this case fractionator counting (Gundersen, 1986) may be used, which does not yield the exact value of n but good estimates. One possibility, called *systematic uniform random sampling*, uses m counting fields regularly distributed in W, e.g. in a randomised grid. If the counting fields are of equal area a, then an unbiased estimator of n is

$$\hat{n} = \frac{\nu(W)}{a} \sum_{i=1}^{m} n_i,$$

where the n_i are the counts in the counting fields. If the locations of the counting fields are randomised the following estimator is also unbiased:

$$\hat{n} = \frac{M}{m} \sum_{i=1}^{m} n_i,$$

where W is exhaustively divided into M counting fields (which now can be of different areas); see Gundersen (2002). The above formula is also valid for *independent uniform random sampling*, where the counting fields are sampled randomly and independently with replacement. Clearly, the estimation variance of independent uniform random sampling is larger than that of systematic uniform random sampling. The latter method can even be improved by clever smoothing; see Gundersen (2002) and Gardi et al. (2006).

The following two methods are examples of estimators that are based on distances. It suffices here to measure a small number of distances from test points rather than to count all points.

Voronoi cell weighting

Test locations, for example lattice points, $y_1, \ldots, y_k$, independent of the points in N are placed in the window W. For each y_i, the nearest point in N and the area (volume) a_i of its Voronoi cell are determined. (This implies that the Voronoi tessellation with respect to the points in N as introduced in Section 1.8.4 has to be (at least partly) constructed.) The test locations must be placed in W such that the information necessary for the determination of the a_i is available. Then λ can be estimated without bias using

$$\hat{\lambda}_V = \frac{1}{k} \sum_{i=1}^{k} \frac{1}{a_i}. \tag{4.2.13}$$

Note that the formula does not contain the inverse of the mean of the a_i. This mean would be an unbiased estimator if *all* cells of the tessellation were used.

To prove the unbiasedness of $\hat{\lambda}_V$, it suffices to consider the case $k = 1$ with $y_1 = o$. The points x in N are assigned marks $V(x)$ that are the corresponding Voronoi cells (the sets) shifted to the origin together with their generating points. The resulting mark distribution is denoted by $\mathcal{M}$. The estimator $\hat{\lambda}$ can be rewritten as

$$\hat{\lambda}_V = \sum_{[x; V(x)]} \frac{\mathbf{1}(o \in V(x) + x)}{\nu(V(x))}.$$

The Campbell theorem for marked point processes (see p. 302) yields

$$\mathbf{E}\hat{\lambda}_V = \mathbf{E}\left(\sum_{[x; V(x)]} \frac{\mathbf{1}(o \in V(x) + x)}{\nu(V(x))} \right) = \lambda \int \frac{1}{\nu(V)} \int \mathbf{1}(o \in V + x) \mathrm{d}x \mathcal{M}(\mathrm{d}V).$$

Since $\mathbf{1}(o \in V + x) = \mathbf{1}_V(-x)$, the inner integral is $\nu(V)$ and $\mathbf{E}\hat{\lambda}_V = \lambda$.

Note that tessellations are also useful in intensity estimation, when the patterns of interest are known to be incomplete, i.e. there are missing points. Then the corresponding Voronoi tessellation and the corresponding cell areas (volumes) can be constructed and analysed statistically. If there are no missing points, the mean cell area (volume) is a reasonable estimator of λ. If points are missing some of the cells are extremely large, and then a more robust estimator of mean cell area

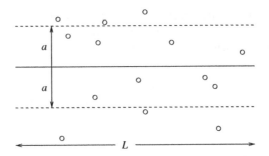

Figure 4.7   Schematic representation of line transect sampling. The observer moves along a transect of length L which is parallel to the strip of width $2a$. The points in the strip are counted.

(volume) than the mean, e.g. the median, can be used to obtain a more realistic estimator of mean cell area and intensity; see Berndt and Stoyan (1997).

Line transect sampling

Place a test line $\mathcal{L}$ of length L in the window and construct a strip $\mathcal{L}_a$ of width $2a$ parallel to $\mathcal{L}$ with $\mathcal{L}$ as central line (Figure 4.7). Count all points which lie inside the strip $\mathcal{L}_a$. Denote the resulting number by $N(\mathcal{L}_a)$. Clearly, (4.2.10) leads to

$$\hat{\lambda} = \frac{N(\mathcal{L}_a)}{2aL},$$

which is an unbiased estimator of λ, since the area of $\mathcal{L}_a$ is $2aL$.

This estimator can be refined if the points have quantitative marks v_i. This is relevant in applications where there is a maximum estimation distance related to the 'visibility' of a point at x_i reflected by v_i. Assume that point x_i is counted if its vertical distance from $\mathcal{L}$ is less than εv_i with some positive scaling factor ε. Then an unbiased intensity estimator is

$$\hat{\lambda}_H = \frac{1}{2L} \sum_{i=1}^{n} \frac{1}{\varepsilon v_i}, \tag{4.2.14}$$

where n is the number of counted points.

To prove the unbiasedness of $\hat{\lambda}_H$, consider marked points $[x; v(x)]$ and denote the mark d.f. by $F_{\mathcal{M}}(m)$ and the distance from x to $\mathcal{L}$ by $\delta(x)$. Then $\hat{\lambda}_H$ can be rewritten as

$$\hat{\lambda}_H = \frac{1}{2L} \sum_{[x; v(x)]} \frac{\mathbf{1}(\delta(x) < \varepsilon v(x))}{\varepsilon v(x)}.$$

The Campbell theorem for marked point processes yields

$$\mathbf{E}\hat{\lambda}_H = \frac{1}{2L}\mathbf{E}\left(\sum_{[x;v(x)]}\frac{\mathbf{1}(\delta(x) < \varepsilon v(x))}{\varepsilon v(x)}\right)$$

$$= \frac{\lambda}{2L}\int\frac{1}{\varepsilon m}\int\mathbf{1}(\delta(x) < \varepsilon m)\mathrm{d}x\mathrm{d}F_{\mathcal{M}}(m).$$

The inner integral is equal to $2\varepsilon mL$ and thus $\mathbf{E}\hat{\lambda}_H = \lambda$ is obtained.

Adapted intensity estimators*

For the sake of completeness a series of further intensity estimators are now presented. These are by no means recommended as estimators of the intensity *per se*, but have an important role in the estimation of other summary characteristics.

Suppose one aims to estimate some summary characteristic C which cannot be estimated without bias. Assume also that it is possible to estimate λC without bias, which leads to a ratio-unbiased estimator of C given by

$$\hat{C} = \frac{\widehat{\lambda C}}{\hat{\lambda}_C}, \qquad (4.2.15)$$

where $\widehat{\lambda C}$ is the unbiased estimator of λC and $\hat{\lambda}_C$ an unbiased intensity estimator.

Clearly, the behaviour of $\hat{C}$ depends on the choice of $\hat{\lambda}_C$. The idea is to choose an intensity estimator 'adapted' to $\widehat{\lambda C}$ in order to ensure that numerator and denominator in (4.2.15) have similar fluctuations such that these partly cancel out through division. It is not a good idea to use a high-precision estimator such as $\hat{\lambda}$ in (4.2.10) if it is not related to $\widehat{\lambda C}$, as then fluctuations might not cancel out. (It can be shown that even the use of the exact λ – if it were known – would not be a good idea; see Stoyan, 2006.)

Many adapted intensity estimators have the form

$$\hat{\lambda}_p = \sum_{x\in N}p(x), \qquad (4.2.16)$$

where $p(x)$ is a function with

$$\int_{\mathbb{R}^d}p(x)\mathrm{d}x = 1. \qquad (4.2.17)$$

It is easy to show that $\hat{\lambda}_p$ is unbiased by means of the Campbell theorem.

Note that the classical intensity estimator $\hat{\lambda}$ is also of the form (4.2.16), with $p(x) = \mathbf{1}_W(x) / v(W)$. In the context of minus sampling with distance r the intensity estimator $\hat{\lambda}(r)$ is used (see Section 4.2.5) which corresponds to

$$p(x) = \mathbf{1}_{W_{\ominus r}}(x) \Big/ v(W_{\ominus r}). \tag{4.2.18}$$

For the estimation of second-order characteristics two further intensity estimators $\hat{\lambda}_V(r)$ and $\hat{\lambda}_S(r)$ are used which correspond to

$$p_V(x, r) = v(W \cap b(x, r)) \Bigg/ \left(db_d \int_0^r t^{d-1} \overline{\gamma}_W(t) dt \right) \tag{4.2.19}$$

and

$$p_S(x, r) = v_{d-1}(W \cap \partial b(x, r)) \big/ (db_d r^{d-1} \overline{\gamma}_W(r)), \tag{4.2.20}$$

both for $d = 2$ or 3 and $r \geq 0$.

The function $p_V(x, r)$ is proportional to the area (volume) of the intersection of the window W and the disc (sphere) $b(x, r)$ of radius r centred at x. The normalising constant in the denominator, which ensures that (4.2.17) holds, contains the isotropised set covariance $\overline{\gamma}_W(r)$ explained in Appendix B.

The function $p_S(x, r)$ is proportional to the boundary length (surface area) $v_{d-1}(W \cap \partial b(x, r))$ of the circular line (sphere surface) $\partial b(x, r)$ of radius r centred at x in the window W. The normalising constant again contains the isotropised set covariance $\overline{\gamma}_W(r)$.

For the cases of a rectangular, parallelepipedal, circular or spherical window W, formulas for $\overline{\gamma}_W(r)$ are given in Appendix B. There are also formulas for the quantities $v(W \cap b(x, r))$ and $v_{d-1}(W \cap \partial b(x, r))$.

All these adapted estimators are worse than the classical $\hat{\lambda}$ in the sense that

$$\mathbf{var}\hat{\lambda}_p \geq \mathbf{var}\hat{\lambda}.$$

However, if the distribution of the point process N is completely known, functions $p(x)$ can be constructed such that

$$\mathbf{var}\hat{\lambda}_p < \mathbf{var}\hat{\lambda};$$

see Mrkvička and Molchanov (2005).

The following intensity estimator has been used in combination with nearest-neighbour edge-correction:

$$\hat{\lambda}_{nn} = \sum_{[x, d(x)]} \frac{\mathbf{1}_{W_{\ominus d(x)}}(x)}{v\left(W_{\ominus d(x)}\right)}, \tag{4.2.21}$$

where $d(x)$ is the nearest-neighbour distance of x. $\hat{\lambda}_{nn}$ is the same as the denominator of (4.2.6), i.e.

$$\hat{\lambda}_{nn} = \sum_{x \in W} \mathbf{1}(d(x) \leq e(x)) \, / \nu \left(W_{\ominus d(x)} \right).$$

This estimator $\hat{\lambda}_{nn}$ is not of the form (4.2.16); it is part of a more general class of estimators of the form

$$\tilde{\lambda} = \sum_{x \in N} p(x, N)$$

with a function $p(x, N)$ which depends not only on x but also on the other points in N. The unbiasedness of $\hat{\lambda}_{nn}$ can be shown in a similar way to that of $\mathcal{D}(r)$ in the context of the estimation of the nearest-neighbour distance d.f. $D(r)$ in Section 4.2.6.

4.2.4   Indices as summary characteristics

Probabilistic definition of some indices

The indices introduced in this section are numerical summary characteristics which describe specific aspects of the distribution of point processes. Some of these were used early in the history of point process statistics and are based on very simple measurement methods.

A good summary index is easy to determine and easy to understand. The first of these properties is probably more interesting and has allowed some of the indices to survive even into the computer age: since it is often very difficult to collect mapped data, point patterns in ecology and forestry are characterised by indices. Two different types of indices are considered: location-related and point-related indices. Location-related indices are determined with reference to deterministic test locations or sampling points, which are chosen independently of the points of the point process, and are usually placed on a lattice. They describe aspects of the point process distribution. In contrast, point-related indices relate to process points and yield information on the typical point, and thus the Palm distribution of the point process.

Location-related indices

Index of dispersion.   A classical example of a location-related index is the ratio of the variance and mean of the number of points, which is known by several different names, e.g. *Clapham's relative variance, Hoel's index of dispersion,* or *Zwicky's index of clumpiness,* (Clapham, 1936; Hoel, 1943; Zwicky, 1953):

$$ID = \frac{\mathbf{var}(N(B))}{\lambda \nu(B)}, \tag{4.2.22}$$

where B is some test set, e.g. a disc, quadrat or cube. (The corresponding location is the centre of the set B.) In other words, ID is the ratio of the variance of counts in sets congruent to B and the corresponding variance of a Poisson process of the same intensity λ. For a clustered process $ID > 1$, while $ID < 1$ for a regular process. Ecologists typically use the terms 'overdispersion' and 'underdispersion' in this context. If ID is considered for the case where B is a disc of radius r, it can be considered as a function of r and represents the variance–mean curve. Since the variance $\mathbf{var}(N(B))$ can be calculated by integration over the product density or the pair correlation function – see (4.3.22) – ID is basically only a by-product of this function. Nevertheless it is a valuable index.

Pielou's index of randomness.   Pielou (1959) introduced the index

$$PI = \pi \lambda \mathbf{E} d^2, \qquad (4.2.23)$$

where d is the random distance from a test point to the nearest point in N. In the CSR case $PI = 1$, while PI is smaller than 1 for regular and larger than 1 for cluster processes.

Other location-related indices used in practice will be discussed in Chapter 5 in the context of marked point processes.

Point-related indices

Formally, point related indices are constructed as follows: for each point x a mark $m(x)$ is constructed and $\mathbf{E}_o(m(o))$, the mean mark of the typical point, is considered. The index is either this value itself or some quantity containing it.

Aggregation or Clark-Evans index.   A classical index of variability based on nearest-neighbour distance marks is the aggregation index CE introduced by the botanists Clark and Evans (1954) in the planar case:

$$CE = 2\sqrt{\lambda} \cdot \mathbf{E}_o(d(o)) = 2\sqrt{\lambda} m_D. \qquad (4.2.24)$$

It may be regarded as the mean of the distance from the typical point to its nearest neighbour divided by the same mean for a Poisson process with the same intensity λ. By (2.3.7) the second mean is $1/(2\sqrt{\lambda})$.

Values of CE greater than 1 indicate that the pattern has a tendency towards regularity, while $CE < 1$ indicates clustering. (The maximum value of CE is 2.1491 for a hexagonal lattice of points.) Section 2.7 discusses how CE may be used in tests of CSR.

Degree of colocalisation.   The proportion of points that have their nearest neighbour within a given distance r' is given by

$$CO(r') = D(r'). \qquad (4.2.25)$$

This index plays an important role in the so-called 'colocalisation analysis'; see Lachmanovich et al. (2003) for details. Since it is the same as the nearest-neighbour distance d.f. $D(r)$ at $r = r'$, it is not discussed here; refer to Section 4.2.6 for more details.

Mean-direction index. This index is based on directions, more specifically on those of the unit vectors pointing from the typical point to its k nearest neighbours ($k \geq 3$). Angles rather than distances have to be measured for each point x. Let $e_1(x), \ldots, e_k(x)$ be these unit vectors. The length of their sum is denoted by $R_k(x)$, i.e.

$$R_k(x) = \|e_1(x) + \ldots + e_k(x)\|. \tag{4.2.26}$$

The corresponding mean $\mathbf{E}_o(R_k(o))$ is the mean-direction index $\overline{R}_k$ introduced in Corral-Rivas (2006). In the planar CSR case its values for $k = 3, \ldots, 6$ are 1.575, 1.799, 2.007, 2.193, respectively. Small values of $\overline{R}_k$ are expected for regular processes since in this case vectors pointing in opposite directions appear frequently in the sum and cancel out. (For a rectangular or hexagonal lattice $\overline{R}_k \equiv 0$ if $k = 4$ or 6, respectively.) On the other hand, $\overline{R}_k$ takes on larger values for cluster processes. Critical values for testing the CSR hypothesis based on $\overline{R}_k$ are given in Section 2.7.5. Angles to nearest neighbours are used also in the uniform angle index introduced in von Gadow et al. (1998).

Degree of hexagonality. In the context of the analysis of highly regular physical point patterns one aim is to characterise deviations of planar point processes from regular hexagonal lattices. For this purpose, Weber et al. (1995) introduced the degree of hexagonality $\Psi(r)$. Here, a slightly modified version is presented which will be denoted by Ψ_6. Similar to the approach used to derive $\overline{R}_k$, the unit vectors $e_1, \ldots, e_6$ of the first six nearest neighbours of the typical point are considered.

Choose the direction of e_1 as the reference direction and denote the direction angles of $e_2, \ldots, e_6$ by $\alpha_2, \ldots, \alpha_6$. The index is then obtained by calculating the number H defined as

$$H = \left| 1 + \sum_{j=2}^{6} \exp(i6\alpha_j) \right|,$$

where $i = \sqrt{-1}$ and the angles are given in radians. For those unfamiliar with complex numbers, H may be rewritten as

$$H = \sqrt{\left(\sum_{j=2}^{6} \sin 6\alpha_j \right)^2 + \left(1 + \sum_{j=2}^{6} \cos 6\alpha_j \right)^2}.$$

If the α_j are integer multiples of $60°$ then $H = 6$. The mean of H is the degree of hexagonality,

$$\Psi_6 = \mathbf{E}_o(H)/6. \tag{4.2.27}$$

In the case of a hexagonal lattice $\Psi_6 = 1$, while for a Poisson process $\Psi_6 = 2.193/6 = 0.366$. Values close to 1 may also occur for cluster processes. In the three-dimensional case an index denoted by Q_6 replaces H_6; see Lochmann et al. (2006a). It uses spherical harmonics.

The three indices CE, $\overline{R}_k$ and Ψ_6 are scale invariant.

Statistical estimators of the indices

The location-related indices are estimated using a lattice of test points. These points can be chosen such that edge effects are avoided, i.e. implicit minus sampling is used by choosing test points far enough from the boundary of the window W.

The estimation of point-related indices is not more complicated. For *in situ* measurement, plus sampling is the natural form of edge-correction, while for mapped data, where no information from outside the window is available, 'no edge-correction' is recommended; see Pommerening and Stoyan (2006).

Plus-sampling estimators. Each point in W is visited and its nearest neighbour(s) and the corresponding distances or angles are determined. If the nearest neighbour is outside W, it is still used for measurement. In most applications this is feasible, since the nearest neighbour is typically only a few steps away from the window's edge. For example:

- Clark–Evans index

$$\widehat{CE} = 2\sqrt{\hat{\lambda}} \cdot \overline{d} \tag{4.2.28}$$

with the classical intensity estimator

$$\hat{\lambda} = \frac{n}{\nu(W)} \quad \text{and} \quad \overline{d} = \frac{1}{n}\sum_{i=1}^{n} d_i,$$

where n is the number of points in W and d_i the nearest neighbour distance of the ith point.

- Degree of colocalisation

$$\widehat{CO}(r') = \frac{1}{n}\sum_{i=1}^{n} \mathbf{1}(d_i < r'). \tag{4.2.29}$$

- Mean-direction index

$$\widehat{\overline{R}}_k = \frac{1}{n} \sum_{i=1}^{n} R_i.$$

Again, n is the number of points in W and R_i is the length of the sum of the corresponding k unit vectors to the nearest neighbours of x_i.

Note that R_i can be determined as follows. First, go to point x_i and determine its nearest neighbour $z_1(x_i)$ and choose the direction $x_i \to z_1(x_i)$ as the reference direction. Then determine the 2nd, ..., kth neighbour of x_i and the corresponding directions and order these with respect to the directions (clockwise). In the next step measure the angles $\alpha_{i2}, \ldots, \alpha_{ik}$ with respect to the reference direction (see Figure 4.8). Then

$$R_i^2 = \left(\sum_{j=2}^{k} \sin \alpha_{ij} \right)^2 + \left(1 + \sum_{j=2}^{k} \cos \alpha_{ij} \right)^2$$

and the square root yields R_i.

No edge-correction. The same estimators as above are used but the nearest neighbours are simply nearest neighbours *within* the observation window W.

4.2.5 Empty-space statistics and other morphological summaries

Spherical contact d.f.

This section presents the first functional summary characteristics. The most important of these is related to the probability that a disc or sphere of radius r does not

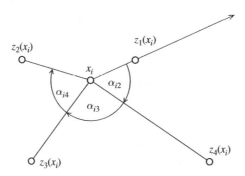

Figure 4.8 Definition of the angles α_{ij} for the determination of R_i for the mean-direction index.

contain a point of N, as a function of r. Due to the stationarity assumption, the centre of the disc or sphere can be any point that is independent of N. Thus the origin o can also be chosen as a centre. In order to obtain a distribution function, the complement of the void probability is taken, which yields the *spherical contact d.f.*

$$H_s(r) = 1 - \mathbf{P}(N(b(o, r)) = 0) \qquad \text{for } r \geq 0. \qquad (4.2.30)$$

This makes sense since the spherical contact d.f. $H_s(r)$ can also be regarded as the distribution of the distance from an 'arbitrary' test location to its nearest neighbour in the point process N.

By definition, $H_s(r)$ is a location-related summary characteristic. In some sense it is an analogue to the nearest-neighbour distance d.f. $D(r)$, which is point-related. Both distributions are compared at the end of Section 4.2.6.

The name 'spherical contact' or 'first contact' d.f. relates to the fact that $H_s(r)$ describes the distribution of $\|x_{\min}\|$, i.e. the smallest radius necessary for a sphere (disc) centred at the origin o to touch or *contact* a point in N.

Note that a different 'geometric shape' may also be used instead of a sphere, for example a cube (square), leading to another contact d.f., the cubic (square). These non-spherical contact d.f.s may be usefully applied to anisotropic point processes.

Morphological functions

The morphological functions $a(r)$, $l(r)$ and $n(r)$ have only been considered very recently in the point process statistics literature. For this reason, and also because they are powerful, they are discussed here in particular detail, for the planar case.

Note first that the spherical contact d.f. can also be defined in a different way, which facilitates generalisation and statistical estimation. In addition to N, consider the random set X_r defined as

$$X_r = N \oplus b(o, r),$$

i.e. the set formed by the union of all discs of radius r centred at the points of N, which was introduced in Section 1.8.2; see Figure 1.11 on p. 43. If N is stationary, the set X_r is stationary as well and it makes sense to consider its area fraction $A_A(r)$. $A_A(r)$ is the fraction of the whole plane which is covered by X_r. Clearly $A_A(0) = 0$ and $A_A(\infty) = 1$. By random set theory

$$A_A(r) = \mathbf{P}(o \in X_r), \qquad (4.2.31)$$

i.e. this fraction is equal to the probability that the origin lies in the set X_r. Since $\mathbf{P}(o \in X_r) = \mathbf{P}(\|x_{\min}\| \leq r)$, it is

$$H_s(r) = A_A(r) \qquad \text{for } r \geq 0, \qquad (4.2.32)$$

which can be considered an alternative definition of $H_s(r)$. However, in order to characterise X_r other geometrical measures than the area may also be considered. For instance, the boundary length and Euler number of X_r may be used as well, as described below.

Denote by $L_A(r)$ the boundary length of X_r per unit area. As a result of overlapping discs, $L_A(r)$ will be smaller than $\lambda \cdot 2\pi r$, in particular for larger r. Clearly, this function $L_A(r)$ also describes aspects of the distribution of X_r and of N. If the point process N is regular, the function $L_A(r)$ is close to $\lambda \cdot 2\pi r$ for small r since there are no overlappings of the discs of radius r centred at process points. For increasing r, $L_A(r)$ takes a maximum and then decreases with increasing r, since more and more disc boundaries vanish due to overlapping. In cluster processes the discs overlap already at small distances r.

The function $L_A(r)$ may be used in two ways. First, the derivative of $A_A(r)$ (regarded as a function of r) is equal to $L_A(r)$,

$$A'_A(r) = L_A(r);$$

see Hansen et al. (1999) for a proof. This means that $L_A(r)$ is equal to the probability density function $h_s(r)$ of $H_s(r)$,

$$H'_s(r) = h_s(r) = L_A(r). \tag{4.2.33}$$

Second, $L_A(r)$ is related to a triplet of morphological functions. The first of these is $a(r)$, defined as

$$a(r) = A_A(r) \big/ (\lambda \pi r^2) \qquad \text{for } r \geq 0. \tag{4.2.34}$$

This is $A_A(r)$ normalised by the area fraction for non-overlapping discs. The corresponding normalisation for $L_A(r)$ yields the function $l(r)$:

$$l(r) = L_A(r)/(2\lambda \pi r). \tag{4.2.35}$$

This is the ratio of the true specific boundary length $L_A(r)$ to the specific boundary length of the disc system leading to X_r ignoring the overlappings of discs.

The third function may be defined in terms of the Euler or connectivity number. Note that the Euler number $\chi(A)$ of a planar set A is defined (in simplified form) as the number of components of the set minus the number of its holes; see Figure 4.9 for an illustration.

For any stationary random set X the *specific Euler number* N_A is defined as

$$N_A = \lim_{K \uparrow \mathbb{R}^2} \frac{\mathbf{E}\chi(X \cap K)}{\nu(K)}, \tag{4.2.36}$$

where $K \uparrow \mathbb{R}^2$ is related to a sequence of growing sets K, for example squares with increasing side length going to ∞. This specific Euler number is used for X_r and

Figure 4.9   A set A consisting of five components with three holes. Its Euler number is 2.

denoted by $N_A(r)$. As the specific Euler number in the case of non-overlapping discs is equal to the intensity λ, it makes sense to normalise $N_A(r)$ to obtain the morphological function

$$n(r) = N_A(r)/\lambda \qquad \text{for } r \geq 0. \tag{4.2.37}$$

It is interesting that there is no simple relationship between $N_A(r)$ and $L_A(r)$ or $A_A(r)$; more specifically, it is not the case that $L'_A(r) = 2\pi N_A(r)$ as some readers might perhaps have expected; see the discussion in Last and Schassberger (2001).

By construction, the morphological functions $a(r)$, $l(r)$ and $n(r)$ correspond to fundamental geometrical characteristics – area, length and Euler number. Their analogues in the three-dimensional case are volume, surface area, integral of mean curvature and Euler number and, in the general d-dimensional case, the Minkowski functionals or intrinsic volumes; see Stoyan et al. (1995).

All of these functions have the value 1 for $r = 0$, but their behaviour differs for larger r. While $a(r)$ and $l(r)$ decrease monotonically and remain between 0 and 1, $n(r)$ is not necessarily monotonic and can even take on negative values, as explained below.

In order to usefully apply morphological functions as summary characteristics in practical applications it is necessary to have information on their behaviour for some point process models which may be used as a reference. For *Poisson processes* the following hold:

$$n_p(r) = (1 - x)e^{-x},$$
$$l_p(r) = e^{-x},$$
$$a_p(r) = (1 - e^{-x})/x$$

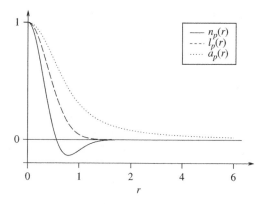

Figure 4.10   The three morphological functions $n(r)$ (solid line), $l(r)$ (dashed line) and $a(r)$ (dotted line) for a Poisson process of intensity $\lambda = 1$.

with $x = \lambda \pi r^2$. Figure 4.10 shows these functions for $\lambda = 1$; $n(r)$ is negative for $r \geq 1/\sqrt{\pi} = 0.564$.

If the points of N are arranged in a square *lattice* with spacings δ, then all morphological functions are constant for $0 \leq r \leq \delta/2$ and vanish for $r > \sqrt{2}\delta/2$. Between $\delta/2$ and $\sqrt{2}\delta/2$, $a(r)$ and $l(r)$ are decreasing, while $n(r)$ takes the value -1. For a *hard-core point process* the behaviour is similar. For small and medium r, all morphological functions are close to 1 and larger than their Poisson process analogues. When, for larger r, the discs touch, the functions decrease rapidly, in particular $n(r)$. The functions can then be smaller than their Poisson process counterparts and approach them from below for increasing r. Negative values appear since, for large r, X_r is totally connected with a large number of holes, which contribute negatively to the Euler number χ.

The behaviour is different for *cluster processes:* the discs overlap already at small r and thus all morphological functions tend to be smaller than their Poisson process counterparts for these r. The further behaviour depends on the spatial arrangement of the cluster centres: if these are distributed regularly and there is large empty space between the clusters, the function $n(r)$ may be constant in some r-interval (while $a(r)$ and $l(r)$ continue to decrease), since enlarging the size of clusters of overlapping discs does not immediately lead to overlapping of these clusters. It is advisable to consider all three functions in parallel, as may be seen from the above explanation.

The factor 1/2 plays an important role in the interpretation of the morphological functions. In the case of a regular point process the downward jump of $n(r)$ takes place at $r \approx \delta/2$, where δ is the lattice spacing (or the mean nearest-neighbour distance). Similarly, the interval of r-values where $n(r)$ is constant ends around $r \approx \Delta/2$ for a cluster process with regularly distributed cluster centres, where Δ is the mean inter-cluster distance.

Statistical estimation of $H_s(r)$ and the morphological functions

The main issue in the estimation of $H_s(r)$ and the morphological functions concerns the determination of the area, boundary length and Euler number of X_r. Some authors apply the point-count method: a lattice of test points is admitted on the window and the number of lattice points in $X_r = N \oplus b(o, r)$ is counted; see Diggle (2003, p. 21). However, better estimation approaches are available: due to the simple structure of X_r as a union of overlapping discs areas, the boundary length and Euler number can be determined exactly; see Edelsbrunner (1995) and Brodatzki and Mecke (2002). In the following, $\nu(B)$, $L(B)$ and $\chi(B)$ denote the area, boundary length and Euler number of a set B, respectively.

As above, edge-correction has to be applied to derive unbiased estimators. In the given case, minus sampling is the most appropriate method. This means that the reduced window $W_{\ominus r}$ is used to estimate the morphological functions for the value r. Furthermore, the adapted intensity estimator $\hat{\lambda}(r)$ is applied to reduce estimation variance,

$$\hat{\lambda}(r) = N(W_{\ominus r}) / \nu(W_{\ominus r}).$$

This leads to the following estimators:

$$\hat{H}_s(r) = \nu(X_r \cap W_{\ominus r}) / \nu(W_{\ominus r}), \tag{4.2.38}$$

$$\hat{a}(r) = \nu(X_r \cap W_{\ominus r}) / (\pi r^2 N(W_{\ominus r})), \tag{4.2.39}$$

$$\hat{l}(r) = l(X_r \cap W_{\ominus r}) / (2\pi r N(W_{\ominus r})), \tag{4.2.40}$$

$$\hat{n}(r) = \chi(X_r \cap W_{\ominus r}) / N(W_{\ominus r}). \tag{4.2.41}$$

Estimation software for the morphological functions is available from `http://www.maths.jyu.fi/~penttine/ppstatistics`.

The probability density function $h_s(r)$ of the spherical contact distribution can be estimated by

$$\hat{h}_s(r) = l(X_r \cap W_{\ominus r}) / \nu(W_{\ominus r}), \tag{4.2.42}$$

and it is not necessary to use a kernel estimator. The estimators $\hat{H}_s(r)$ and $\hat{h}_s(r)$ are unbiased, while the other estimators are ratio-unbiased. $\hat{H}_s(r)$ and $\hat{h}_s(r)$ are discussed in more detail in Chiu and Stoyan (1998).

Example 4.2.   Gold particles: spherical contact d.f.

Figure 4.11 shows the estimated spherical contact d.f. for the point pattern formed by the gold particles. For comparison, the spherical contact d.f. of a Poisson process of the same intensity is shown as well. The graph for the gold particle pattern is below that for a Poisson process, i.e. the distances from test locations to the points tend to be larger than for a Poisson process, indicating clustering in the pattern.

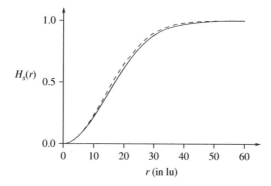

Figure 4.11   Estimate of $H_s(r)$ (solid line) for the pattern of gold particles in comparison to the corresponding function of a Poisson process (dashed line) of equal intensity.

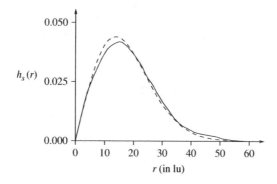

Figure 4.12   Estimate of $\hat{h}_s(r)$ (solid line) for the pattern of gold particles in comparison to the corresponding function of a Poisson process (dashed line).

However, the difference is small in the scale used in Figure 4.11. The estimate $\hat{h}_s(r)$ of the density function shown in Figure 4.12 is a little more informative. The mode of the distances is at $r = 18\,\text{lu}$ ($r = 15\,\text{lu}$ for the Poisson case) and the tail of the H_s-distribution is heavier than in the Poisson case at values larger than $r = 40\,\text{lu}$. A comparison with the nearest-neighbour characteristics $\hat{D}(r)$ and $\hat{d}(r)$ presented in Section 4.2.6 below shows interesting differences and leads to a deeper understanding of the distribution of the gold particles.

Finally, Figure 4.13 shows an estimate of the Euler function $n(r)$ together with $n(r)$ for a Poisson process of equal intensity. It is clearly smaller than its Poisson process equivalent for small r (between 5 and 10) and is a better indicator of the clustering in the gold pattern than $\hat{h}_s(r)$. Only second-order methods are more informative (see p. 221).

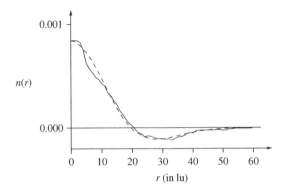

Figure 4.13   Estimate of $n(r)$ (solid line), in comparison to its Poisson process counterpart (dashed line).

Further applications of the morphological functions in classical point process statistics are presented in Mecke and Stoyan (2005). Many interesting applications in physics, for the three-dimensional case, can be found in papers by K. Mecke; see, for example, Mecke (1998, 2000) and Mecke et al. (1994).

4.2.6   The nearest-neighbour distance distribution function

The nearest-neighbour distance d.f. $D(r)$ is the d.f. of the random distance from the typical point to its nearest neighbour. It has already been discussed in Section 4.1 in the context of the Palm distribution. In the notation of that section,

$$D(r) = \mathbf{P}_o(N(b(o, r) \setminus \{o\}) > 0) \qquad \text{for } r \geq 0, \tag{4.2.43}$$

since the distance from the typical point to its nearest neighbour is smaller than r if and only if there is at least one point in the sphere of radius r centred at the typical point.

The nearest-neighbour distance d.f. $D(r)$ can also be defined based on constructed marks $d(x)$ for the points x of the process: $d(x)$ is the distance from x to its nearest neighbour, which is denoted as $z_1(x)$. The resulting marked point process inherits the stationarity property from the original point process, and the corresponding mark d.f. is precisely $D(r)$. This formulation corresponds to the statistical estimation of $D(r)$ with constructed marks. In this notation,

$$D(r) = \mathbf{P}_o(0 < d(o) \leq r). \tag{4.2.44}$$

The nearest-neighbour distance d.f. is a classical tool in point process statistics. Very often beginners start with $D(r)$, probably because it is so easy to understand and 'so natural'. However, $D(r)$ is a summary characteristic which is not very useful in many cases. It is 'short-sighted' as it considers only the nearest neighbour, and is not suitable for describing any behaviour at large distances. In the case of a cluster process it usually only describes aspects of the geometry of the clusters. An extreme example is a point process consisting of isolated pairs of points with distance Δ. Here, all nearest-neighbour distances are constant and equal to Δ. The same nearest-neigbour distances appear in a regular lattice with grid cells of side length Δ. Hence the point process and a lattice cannot be distinguished based on $D(r)$. (Of course, considering the intensity in addition to $D(r)$ may already provide some insight into the type of the pattern and help distinguishing these two cases.)

Usually, the corresponding probability density function $d(r)$,

$$d(r) = D'(r),$$

is used in exploratory analysis, while in statistical tests $D(r)$ is used. The function $D(r)$ has found some application in statistical analyses of building material in the context of mechanical investigations; see Stroeven and Stroeven (2001) and Hubalková and Stoyan (2003).

The mean m_D corresponding to the d.f. $D(r)$ is also a valuable summary characteristic.

This book uses $D(r)$ mainly in combination with other characteristics, more specifically in the J-function (see Section 4.2.7) and in statistical tests, when model parameters have been estimated based on second-order characteristics, with the aim of testing the goodness of fit with a different summary characteristic. However, Example 4.3 below shows that $D(r)$ and $d(r)$ are also valuable in data analysis.

In this context, it is also helpful to use the *nearest-neighbour pair hazard rate* $\delta(r)$. The classical hazard rate used in survival analysis may be reinterpreted in the current context as the probability that the nearest neighbour of the typical point may be found at a distance between r and $r + dr$ conditional on the nearest-neighbour distance being larger than r. The nearest-neighbour pair hazard rate is the classical hazard rate $\frac{d(r)}{1-D(r)}$ corresponding to $D(r)$, suitably normalised. It is thus defined as

$$\delta(r) = \frac{d(r)}{1 - D(r)} \bigg/ (2\pi r \lambda). \tag{4.2.45}$$

Many point processes have a positive *minimum inter-point* or *hard-core distance* r_0, such that

$$D(r) \begin{cases} = 0 & \text{for } r < r_0, \\ > 0 & \text{for } r \geq r_0. \end{cases}$$

For example, when the centres of a random system of identical hard spheres of diameter d are considered, then $r_0 \geq d$. The maximum likelihood estimator $\hat{r}_0$ of r_0

is simply the minimum inter-point distance in the window W. Sometimes $\hat{r}_0$ can be 'improved' by

$$\hat{r}_0 = \frac{n}{n+1} \cdot \text{minimum inter-point distance,} \qquad (4.2.46)$$

where $n = N(W)$ is the number of points in the sample.

Differences between $D(r)$ and $H_s(r)$

$D(r)$ and $H_s(r)$ both describe distances from points to points in the process N. For $D(r)$ the reference point is the typical point in N, whereas for $H_s(r)$ it is a test location, the origin, which is not a point in N. For the three main types of point processes the following can be said:

- *Poisson process*. Both functions coincide:

$$D(r) = H_s(r) \qquad \text{for } r \geq 0.$$

- *Regular process*. The inter-point distances tend to be larger than distances from test locations to process points; as a consequence,

$$D(r) \leq H_s(r) \qquad \text{for } r \geq 0.$$

- *Cluster process*. The inter-point distances are mainly distances between points in the same clusters. Thus, the distances represented by $D(r)$ will be short, while the distance from o to the nearest cluster can be large. Thus

$$D(r) \geq H_s(r) \qquad \text{for } r \geq 0.$$

The d.f. $H_s(r)$ always has a density function, which is not necessarily true for $D(r)$; estimates of $H_s(r)$ are represented by smooth curves if (4.2.38) is used and $\nu(X_r \cap W_{\ominus r})$ and $\nu(W_{\ominus r})$ are determined exactly, but estimates of $D(r)$ usually have discontinuities.

For a Poisson process, $D(r)$ is a Weibull d.f. with parameters $b = \lambda b_d$ and $p = d$. This has led foresters to assume that in their applications $D(r)$ is always a Weibull d.f., but that the shape parameter of this distribution may be different from $p = d = 2$; see von Gadow et al. (2003).

Estimation of D(r)

The estimation of $D(r)$ is a classical example of the use of edge-corrections, and various estimators have been proposed. Here, only those estimators are considered which are conceptionally simple and which are not based on heuristic assumptions. Hence, the estimator in Floresroux and Stein (1996) is excluded.

Border estimator. The border estimator or minus-sampling $D_b(r)$ was introduced by Ripley (1977). It is defined as

$$\hat{D}_b(r) = \sum_{[x;d(x)]} \mathbf{1}_{W_{\ominus r}}(x)\mathbf{1}(0 < d(x) \le r) \Big/ N(W_{\ominus r}) \qquad \text{for } r \ge 0. \qquad (4.2.47)$$

The idea behind the estimator $\hat{D}_b(r)$ is quite simple: when estimating $D(r)$, only those points x in W are considered which have a distance larger than r from the window's boundary. For these points it is thus clear that there are no points closer than r outside the window and the nearest-neighbour distance $d(x)$ can be determined within W.

These points lie in $W_{\ominus r}$. Since $N(W_{\ominus r})$ is the number of those points, (4.2.47) is simply the corresponding ratio estimator. Because numerator and denominator are random, $\hat{D}_b(r)$ is not unbiased. However, if N is ergodic, $\hat{D}_b(r)$ is asymptotically unbiased.

Unfortunately, the estimator $\hat{D}_b(r)$ has two disadvantages. It is not necessarily monotonically increasing in r, and it can exceed 1 in value. (Usually this happens only for small samples.)

There is a better estimator which can be easily understood if $\hat{D}_b(r)$ is rewritten as follows. The quantity

$$\mathcal{D}_b(r) = \sum_{[x;d(x)]} \mathbf{1}_{W_{\ominus r}}(x)\mathbf{1}(0 < d(x) \le r)/\nu(W_{\ominus r}) \qquad \text{for } r \ge 0,$$

is clearly an unbiased estimator of $\lambda D(r)$, which can easily be shown by the Campbell theorem. Division by the adapted intensity estimator corresponding to (4.2.18),

$$\hat{\lambda}(r) = N(W_{\ominus r})\Big/ \nu(W_{\ominus r}),$$

yields $\hat{D}_b(r)$. The use of $\hat{\lambda}(r)$ reduces the estimation variance (while the use of the classical λ results in intolerable biases and large mean squared errors), but it leads to the non-monotonicity mentioned above.

Nearest-neighbour estimator. Hanisch (1984) suggested an unbiased estimator which outperforms the border estimator. It uses the nearest-neighbour edge-correction discussed in Section 4.2.2, which leads to the estimator

$$\hat{D}_n(r) = \mathcal{D}_n(r)\Big/ \hat{\lambda}_{nn} \qquad \text{for } 0 \le r \le R, \qquad (4.2.48)$$

where

$$\mathcal{D}_n(r) = \sum_{[x;d(x)]} \mathbf{1}_{W_{\ominus d(x)}}(x)\mathbf{1}(0 < d(x) \le r)/\nu(W_{\ominus d(x)})$$

with $\hat{\lambda}_{nn}$ in (4.2.21) and

$$R = \sup\{r > 0 : \nu(W_{\ominus r}) > 0\}.$$

The indicator $\mathbf{1}_{W_{\ominus d(x)}}(x)$ can be rewritten as $\mathbf{1}(d(x) < e(x))$, where $e(x)$ is, as before, the distance of x to the boundary of W.

To prove the unbiasedness of $\hat{\mathcal{D}}_n(r)$, the mean $\mathbf{E}\left(\hat{\mathcal{D}}_n(r)\right)$ is calculated by means of the Campbell–Mecke formula (4.1.8). Here, the function f has the form

$$f(x, N) = \mathbf{1}_{W_{\ominus d(x)}}(x)\mathbf{1}(0 < d(x) \le r)\big/\nu(W_{\ominus d(x)});$$

clearly, $d(x)$ depends not only on x but also on its neighbours in N. When N is replaced by N_{-x}, $d(x)$ has to be replaced by $d(o)$ since the translation moves x to o:

$$f(x, N_{-x}) = \mathbf{1}_{W_{\ominus d(o)}}(x)\mathbf{1}(0 < d(o) \le r)/\nu(W_{\ominus d(o)}).$$

Finally,

$$\int \mathbf{1}_{W_{\ominus d(o)}}(x)\mathrm{d}x = \nu(W_{\ominus d(o)})$$

implies

$$\mathbf{E}\hat{\mathcal{D}}_n(r) = \lambda \mathbf{E}_o(\mathbf{1}(0 < d(x) \le r)) = \lambda \mathbf{P}_o(0 < d(x) \le r) = \lambda D(r).$$

Hence, $\hat{D}_n(r)$ is derived by simply using precisely those points x which have their nearest neighbour $z_1(x)$ both within W and at a distance of less than r to x.

Using simulation, Stoyan (2006) compared the mean squared errors of the above two estimators of $D(r)$ and other estimators of $D(r)$ such as the Kaplan–Meier estimator introduced by Baddeley and Gill (1997) for Poisson, cluster and hard-core processes and found that Hanisch's estimator has the smallest mse. Furthermore, estimation without edge-correction produces comparable results for small r.

Readers who are less familiar with the notation of stochastic geometry should note that in the special case of a rectangular window W of side lengths a and b, the area $\nu(W_{\ominus r})$ is equal to $(a - 2r)(b - 2r)$; see Appendix B. For all windows, the indicator $\mathbf{1}_{W_{\ominus d(x)}}(x)$ can be rewritten as $\mathbf{1}(d(x) \le e(x))$, where $e(x)$ is the distance of x from the window's boundary.

Analogous estimators can be constructed for the d.f. $D_k(r)$ of the distance to the kth nearest neighbour for $k = 2, 3, \ldots$; here edge-correction ensures that the kth neighbour can be determined properly.

The mean nearest-neighbour distance m_D can be estimated by

$$\hat{m}_D = \sum_{[x;d(x)]} \mathbf{1}_{W_{\ominus d(x)}}(x) \cdot d(x)\big/\nu(W_{\ominus d(x)}) \Big/ \hat{\lambda}_{nn}.$$

The estimation of the density function $d(r)$ is based on the same principle as that of $D(r)$. The use of the following kernel estimator is recommended:

$$\hat{d}_n(r) = \hat{\vartheta}_n(r) \Big/ \hat{\lambda}_{nn} \qquad \text{for } 0 \le r \le R, \tag{4.2.49}$$

where

$$\hat{\vartheta}_n(r) = \sum_{[x;d(x)]} k(d(x)) \mathbf{1}_{W_{\ominus d(x)}}(x) / \nu(W_{\ominus d(x)}),$$

in which $k(z)$ is the Epanechnikov kernel explained in Appendix A, and $\hat{\lambda}_{nn}$ and R are as above.

The hazard rate can be estimated using the above estimators for $d(r)$, $D(r)$ and λ.

Example 4.3.   Gold particles: nearest-neighbour distance d.f.
Figure 4.14 shows the nearest-neighbour distance characteristics of the point pattern of gold particles; the mean nearest-neighbour distance is estimated as 15.2 lu. The estimated d.f. shows strong deviations from the d.f. for a Poisson process: there is (i) a hard-core distance of 5.66 lu and (ii) some form of clustering, resulting in a large number of nearest neighbours at distances around 10 lu. This situation can be shown even more clearly by the estimate of the density function: there is a large peak at $r = 10$ lu and, perhaps, two shoulders at 22 lu and 30 lu. Nearly all nearest-neighbour distances are smaller than 40 lu, since nearest-neighbour distance characteristics are blind for larger distances. (There is a single point with nearest-neighbour distance larger than 50 lu, namely with 51.9 lu.)

It is interesting to compare the density estimates $\hat{d}(r)$ and $\hat{h}_s(r)$; see Figure 4.15. The estimate for the spherical contact distribution has its mode at $r = 15$ lu, while $\hat{d}(r)$ has its mode at $r = 10$ lu; and the tail of $\hat{h}_s(r)$ is heavier. Hence, the clustering

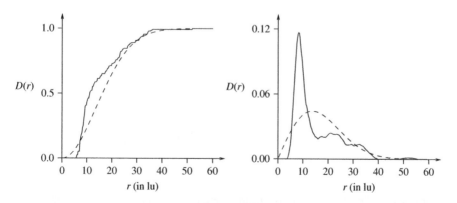

Figure 4.14   Estimates of $D(r)$ (left, solid line) and $d(r)$ (right, solid line) for the gold particle pattern, compared to their Poisson process analogues (dashed lines).

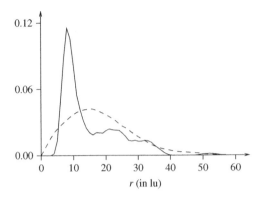

Figure 4.15   Comparison of the density estimates $\hat{d}(r)$ (solid line) and $\hat{h}_s(r)$ (dashed line) for the pattern of gold particles.

in the pattern is clearly shown, and the nearest-neighbour distances are 'smaller' than the distances from test locations.

Example 4.4.   Phlebocarya *pattern: nearest-neighbour distance d.f.*
    Figure 4.16 shows the empirical nearest-neighbour distance p.d.f. $\hat{d}(r)$ for the bandwidths $h = 0.5$ m and 2.0 m. The mean nearest-neighbour distance is 0.66 m. Clearly, there is a big influence of the bandwidth, especially here where the data were collected on a 10×10 cm grid. There are two modes, at $r = 1$ m and 3.7 m. These modes might reflect the fact that *Phlebocarya* typically grows in larger clusters consisting of several smaller clumps.

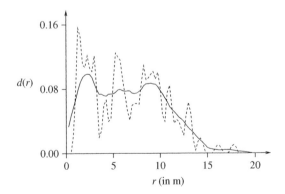

Figure 4.16   Two estimates of the nearest-neighbour distance d.f. $d(r)$ for the *Phlebocarya* pattern with bandwidths 0.5 m (dashed line) and 2.0 m (solid line). The latter seems to be more appropriate given that the data were collected on a grid.

4.2.7 The *J*-function

The *J*-function, introduced by Van Lieshout and Baddeley (1996), is a valuable tool for detecting deviations from a Poisson process and for characterising the interaction of points in terms of its type, strength and range. It is a clever combination of $D(r)$ and $H_s(r)$, defined as

$$J(r) = \frac{1 - D(r)}{1 - H_s(r)} \qquad \text{for } r \geq 0, \text{ with } H_s(r) < 1. \qquad (4.2.50)$$

For a Poisson process, $J(r) \equiv 1$, but this is not unique; non-Poisson processes have been identified that have the same *J*-function (see Bedford and Berg, 1997). From the discussion of $D(r)$ and $H_s(r)$ in Section 4.2.6, it is clear that $J(r) \geq 1$ for regular processes and $J(r) \leq 1$ for cluster processes. For Gibbs point processes with finite radius of interaction r_{max},

$$J(r) = 1 \qquad \text{for } r \geq r_{max}. \qquad (4.2.51)$$

The estimation of $J(r)$ is not trivial. First of all, the denominator $1 - H_s(r)$ is small for large r, and large fluctuations of $\hat{J}(r)$ are to be expected for these r. Also, the nature of numerator and denominator as point- and location-related summary characteristics is different, which results in estimators of a different nature. As a consequence, it is difficult to construct an estimator in which the fluctuations in numerator and denominator cancel out. Probably the choice of the estimator is not so important for small and moderate r.

Paulo et al. (2002) give a good example of an application of the *J*-function, in which the competition of cork oaks in Portuguese forests is considered.

Example 4.5.   Gold particles: J-function

Figure 4.17 shows the *J*-function for the pattern of gold particles. It is greater than 1 for $r \leq 7 \, \text{lu}$ and less than 1 at distances beyond 7. The graph indicates

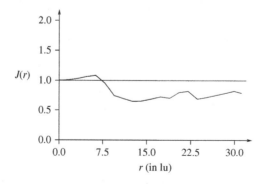

Figure 4.17   Estimate of the *J*-function for the pattern of gold particles. Data courtesy of M.N.M. van Lieshout.

clustering of the particles combined with short-range repulsion. Since the graph is almost horizontal for r between 10 lu and 20 lu, it seems that the interaction between particles is restricted to distances around 10 lu.

4.3   Second-order characteristics

4.3.1   The three functions: K, L and g

Introduction

Ripley's K-function, Besag's L-function and the pair correlation function $g(r)$ have often been regarded as the functional summary characteristics that are the most important tools for the analysis of point patterns. They are often called 'second-order characteristics'. The reason why this name is used is not important at this stage, but will be explained in due course. These functions are believed to be more powerful than the other summary characteristics considered in this chapter, and are therefore treated in a separate section. Physicists in particular follow a 'second-order dogma' and believe that second-order characteristics, especially the pair correlation function, offer the best way to present statistically the distributional information contained in point patterns and all the information that is necessary for describing the correlations of point locations.

Clearly, the distributional indices in Section 4.2.4 yield only aggregated information about the relationship of the typical point to its close neighbourhood but ignore long-range spatial correlations in point patterns. Furthermore, statistical experience suggests that the functional summary characteristics such as $D(r)$, $H_s(r)$ and the morphological functions are 'short-sighted' as they describe the distances to nearest neighbours of reference points, but say little or nothing about the points beyond the nearest neighbour.

The authors of this book also share the opinion that second-order characteristics play a central role in the analysis of point patterns. However, they also recommend the use of the other summary characteristics of Sections 4.2 and 4.4, since these have their own specific strengths and often yield information that cannot be derived from second-order characteristics. This is demonstrated well by Example 4.19 on p. 272. Furthermore, the other summary characteristics can be used to support an analysis based on second-order characteristics, e.g. for testing models that have been found using second-order methods.

In the following, the three functions are described from an applications standpoint. The corresponding theoretical background is presented in Section 4.3.2, including an explanation of why these functions are called second-order characteristics. The statistical estimation is discussed in detail in Section 4.3.3.

Ripley's K-function

The idea behind the K-function is the average number of other points found within the distance r from the typical point. The following provides a motivation for the construction of the K-function.

Let $\lambda K(r)$ denote the mean number of points in a disc (sphere) of radius r centred at the typical point (which is not counted). Using plus sampling, it can be estimated as follows. Let n be the number of points in the window W, and let $n_i(r) = N(b(x_i, r) \setminus \{x_i\})$ be the number of points of N within distance r from point x_i, excluding x_i itself. Then

$$\bar{n}(r) = \frac{1}{n} \sum_{i=1}^{n} n_i(r)$$

is an estimator of $\lambda K(r)$. Using the notation of Section 4.1 yields

$$\lambda K(r) = \mathbf{E}_o\left(N(b(o, r) \setminus \{o\})\right),$$

i.e. $\lambda K(r)$ is precisely the Palm mean discussed above.

In order to separate out the global point density given by λ and local point density fluctuations, the mean $\lambda K(r)$ is divided by λ; this is why it was denoted as $\lambda K(r)$ above. The resulting $K(r)$ is the popular summary characteristic, which is thus defined as

$$K(r) = \mathbf{E}_o(N(b(o, r) \setminus \{o\}))/\lambda \qquad \text{for } r \geq 0. \tag{4.3.1}$$

Clearly, $K(r)$ depends on the radius r: with increasing r the number of points in $b(o, r)$ increases and, as a result, $K(r)$ increases as well. In general, one can expect $K(r)$ to be proportional to r^d; only for small r interesting deviations are observed.

The function has a simple form in the *Poisson process case;* recall from Section 2.5.2 that

$$K(r) = \pi r^2 \qquad \text{for } r \geq 0 \tag{4.3.2}$$

in the planar case and

$$K(r) = b_d r^d \qquad \text{for } r \geq 0 \tag{4.3.3}$$

in the general d-dimensional case.

The shape of $K(r)$ relative to that of the Poisson process provides valuable information on the point process distribution. Typical non-Poisson cases are cluster processes (processes with aggregation) and regular processes (processes with repulsion). For these, the following behaviour of $K(r)$ can be expected:

- *Cluster process:*

$$K(r) > \pi r^2 \quad \text{or} \quad K(r) > b_d r^d.$$

In this case the typical point is part of a cluster and, as a result of this, has some very near neighbours. Thus, the local point density around the typical point is larger than λ and

$$\mathbf{E}_o(N(b(o, r)\setminus\{o\})) > \mathbf{E}(N(b(o, r))) = \lambda b_d r^d.$$

- *Regular process:*

$$K(r) < \pi r^2 \quad \text{or} \quad K(r) < b_d r^d.$$

Now the typical point is isolated, i.e. there is a certain distance between the typical point and its nearest neighbours. Thus, the local point density around the typical point is smaller than λ and

$$\mathbf{E}_o(N(b(o, r)\setminus\{o\})) < \mathbf{E}N(b(o, r)) = \lambda b_d r^d.$$

In practice, point patterns often exhibit a combination of these two extreme cases, e.g. short-range regularity and long-range clustering. In general, the interpretation of $K(r)$ is complicated by the fact that it is a cumulative characteristic: increased numbers of inter-point distances r within an interval $[r_1, r_2]$ lead to large values of $K(r)$ not only for r within the interval $r_1 \leq r \leq r_2$ but also for $r > r_2$, indicating spurious clustering at distances larger than r_2. A similar situation arises for rare distances r where the K-function indicates (spurious) regularity. In such a situation it may be useful to consider the difference $\lambda(K(r) - K(r_2))$, that is, the mean number of points in a ring or shell of radii r and r_2.

Note that there is a simple relationship between $K(r)$ and the d.f.s $D_k(r)$ of the distances to the kth nearest neighbours introduced on p. 210:

$$\lambda K(r) = \sum_{k=1}^{\infty} D_k(r) \qquad \text{for } r \geq 0. \tag{4.3.4}$$

Sometimes the points for statistical analysis are given as nodes of a graph, as discussed in Section 1.8.5. It makes sense to define a graph-adapted K-function via shells around the typical point, where $K(k)$ is the mean number of points in the kth shell, where the kth shell is given by all points connected with the typical point by k or fewer edges. Note that the graph $G(N; r)$ introduced in Section 1.8.5 can be interpreted as a graph counterpart of $K(r)$.

The *L*-function

Modern point process statistics rarely uses $K(r)$, but rather its variant, the *L-function* as introduced by Besag (1977). In the planar case the *L*-function is defined as

$$L(r) = \sqrt{\frac{K(r)}{\pi}} \qquad \text{for } r \geq 0, \tag{4.3.5}$$

and in the general d-dimensional case as

$$L(r) = \sqrt[d]{\frac{K(r)}{b_d}} \qquad \text{for } r \geq 0. \tag{4.3.6}$$

Clearly, $L(r)$ represents the same information as $K(r)$. However, it has both graphical and statistical advantages:

- $K(r)$ is proportional to r^2 in the planar case, i.e. $K(r) \propto r^2$, and $K(r) \sim r^d$ in $\mathbb{R}^d$ but $L(r)$ is always proportional to r,

$$L(r) \propto r,$$

  and in the Poisson case

$$L(r) = r \qquad \text{for } r \geq 0.$$

  Thus in practice, when a pattern is assessed for complete spatial randomness, the graph of its L-function is compared to a line, whereas that of the K-function is compared to a parabolic curve. The second-order behaviour of a point process can be visualised and interpreted more easily based on the L-function than based on the K-function.

- Statistical experience shows that the fluctuations of estimated K-functions increase with increasing r. The root transformation stabilises these fluctuations (both means and variances) and can even make them independent of the distance r.

In the literature, the L-function is sometimes defined as

$$L^*(r) = \sqrt{\frac{K(r)}{\pi}} - r,$$

which leads to

$$L^*(r) = 0 \qquad \text{for } r \geq 0$$

in the Poisson case. In general, the graph of $L^*(r)$ tends to be horizontal, which has some graphical advantages as deviations from a horizontal line are easier to detect than deviations from a diagonal line. However, this book prefers the classical definition in order to emphasise the cumulative nature of the L-function due to an increase in the number of points with increasing r. A better way of obtaining horizontal graphs that characterise the second-order behaviour of point processes is to use the pair correlation function (see below).

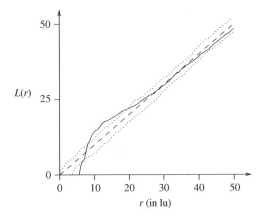

Figure 4.18   The empirical L-function (solid line) for the pattern of gold particles, minimum and maximum envelopes from 99 simulations of a Poisson process of intensity 0.000 865 in a 630×400 rectangle (dotted lines) and the theoretical L-function of a Poisson process (dashed line). The curves indicate micro-scale repulsion and meso-scale clustering.

Example 4.6.   Gold particles: L*-function*

Figure 4.18 shows an estimate of the L-function for the gold particles. This function deviates clearly from the line $L(r) = r$ for the Poisson case: for $r \leq 8$ lu, $\hat{L}(r) \leq r$, and for 8 lu $< r \leq 33$ lu, $\hat{L}(r) > r$, while for larger r there are irregular fluctuations around the line $L(r) = r$. Further, $\hat{L}(r) = 0$ for $r \leq r_0$ since the smallest inter-point distance in the pattern is $r_0 = 5.66$ lu.

The vague initial impression from visual inspection of the point pattern can be confirmed by the graph of $\hat{L}(r)$: at very short distances there is some tendency to regularity (reflected in the positive minimum inter-point or hard-core distance $r_0 = 5.66$ lu). Also, there is strong clustering indicated by a large number of neigbours at distances slightly larger than r_0 resulting in $L(r) > r$. Because of the cumulative nature of the L-function it is difficult to specify the exact range of clustering. The fact that $L(r) \approx r$ for $r \geq 30$ lu shows that there are a smaller number of neighbours at distances smaller than $r = 30$ lu. The discussion of the pair correlation function in Example 4.8 on p. 221 will clarify this point.

The pair correlation function

The authors of this book recommend the pair correlation function $g(r)$ as the best, most informative second-order summary characteristic, even though its statistical estimation is relatively complicated. While it contains the same statistical information as the K- or L-function, it offers the information in a way that is easier to understand, in particular for beginners. Its relationship to $K(r)$ is similar to that of a probability density function $f(x)$ to its corresponding distribution function $F(x)$, with $f(x)$ being the derivative of $F(x)$, i.e.

$$F(x) = \int_{-\infty}^{x} f(t)dt \quad \text{or } f(x) = F'(x).$$

The pair correlation function $g(r)$ is proportional to the derivative of $K(r)$ with respect to r, i.e. in the planar case,

$$g(r) = \frac{K'(r)}{2\pi r} \qquad \text{for } r \geq 0, \tag{4.3.7}$$

and in the general d-dimensional case,

$$g(r) = \frac{K'(r)}{db_d r^{d-1}} \qquad \text{for } r \geq 0. \tag{4.3.8}$$

The following is a heuristic explanation of $g(r)$. First, recall from (4.2.9) that the probability of a point of N being in the infinitesimally small disc (sphere) $b(x)$ of area (volume) dx centred at x is λdx. Consider now a second point y at distance r from x and consider the probability $p_2(x, y)$ that there is a point both in $b(x)$ and in the small sphere $b(y)$ of area (volume) dy and centred at y. This probability can be expressed by the second-order product density $\varrho(x, y)$ (see Section 1.5) as

$$p_2(x, y) = \varrho(x, y)dxdy. \tag{4.3.9}$$

In the isotropic case, $p_2(x, y)$ and $\varrho(x, y)$ depend only on the distance r of x and y, and hence the notation can be simplified to $p_2(r)$ and $\varrho(r)$. Now,

$$p_2(r) = g(r) \cdot \lambda dx \cdot \lambda dy,$$

where $g(r)$ is the pair correlation function, which satisfies

$$g(r) = \varrho(r)/\lambda^2 \qquad \text{for } r \geq 0. \tag{4.3.10}$$

If the point distribution is completely random,

$$p_2(r) = \lambda dx \cdot \lambda dy$$

by the multiplication theorem of probability theory, and thus $g(r) \equiv 1$. In general $\lambda dx \cdot \lambda dy$ has to be multiplied by a correction factor to yield $p_2(r)$ which depends on r, and this factor is precisely $g(r)$.

For large r the function $g(r)$ always takes the value 1, since the events 'there is a point of N in $b(x)$' and 'there is a point of N in $b(y)$' are independent for large r:

$$\lim_{r \to \infty} g(r) = 1. \tag{4.3.11}$$

To be more precise, in order to satisfy (4.3.11) the point process must have some distributional property, the so-called mixing property explained in Section 1.6.4.

If there is a finite distance r_{corr} with

$$g(r) = 1 \qquad \text{for } r \geq r_{\text{corr}}, \tag{4.3.12}$$

then r_{corr} is called *range of correlation*. This means that there are no correlations between point positions at larger distances.

Another interpretation of $g(r)$ is of a conditional and predictive nature. Assume that the typical point of N is at o. What is the probability $\pi_2(r)$ of another point being in an infinitesimally small disc (sphere) of area (volume) dx whose centre has a distance of r from o? This probability is

$$\pi_2(r) = \lambda g(r).$$

By analogy with (4.2.9), $\lambda g(r)$ is sometimes called the *Palm intensity function*. In the literature it has also been termed the O-ring statistic; see Wiegand and Moloney (2004).

It is clear that in the Poisson process or CSR case

$$g(r) = 1 \qquad \text{for } r \geq 0,$$

i.e. the pair correlation function is constant and equal to one. This reflects the fact that the location of any point is entirely independent of the locations of the other points. Again, in typical non-Poisson cases a characteristic behaviour of $g(r)$ may be found:

- *For the cluster process,*

$$g(r) \geq 1,$$

and $g(r)$ can take large values, in particular for small r, and is decreasing as r increases further.

- *For the regular process,*

$$g(r) \leq 1 \qquad \text{for small } r,$$

and many patterns have a hard-core distance r_0 for which

$$g(r) = 0 \quad \text{for } r \le r_0.$$

For larger distances r, $g(r)$ can exceed 1 and $g(r)$ can have interesting shapes, which will be discussed in more detail in Section 4.3.4.

Example 4.7. Phlebocarya *pattern: pair correlation function*
Figure 4.19 shows an estimate of the pair correlation function for the pattern of *Phlebocarya* plants. This curve describes the aggregation in the pattern very well. As the analysis by means of nearest-neighbour distances has already shown, there are two cluster sizes: there are small clusters of a diameter around 80 cm and perhaps larger clusters, perhaps with a diameter of roughly 2 m. Note that for this pattern the CSR hypothesis is clearly rejected by the L-test.

Example 4.8. *Gold particles: pair correlation function*
Figure 4.20 shows an estimate of the pair correlation function for the pattern of gold particles. This curve describes the spatial structure of the pattern very well: the hard-core distance r_0 ($=$ minimum inter-point distance) is 5.66 lu, which is also a very frequent nearest-neighbour distance ($r_1 = r_0$) and the gap distance $r_2 = 20$lu, which corresponds to the empty space between the first and second neighbours of the typical point, are clearly visible. The

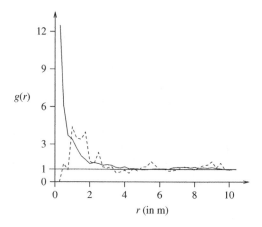

Figure 4.19 Empirical pair correlation functions for the pattern of *Phlebocarya* positions, obtained with bandwidths $h = 0.5$ (dashed line) and 2.0 m (solid line). The large values for small r obtained with $h = 2$ m indicate strong clustering, while the values larger than 1 for r around 8 m may indicate larger clusters. For the smaller h the lattice nature of the pattern becomes apparent. The use of adapted bandwidths makes sense here.

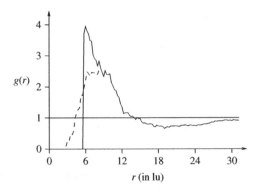

Figure 4.20   The empirical pair correlation function of the pattern of gold particles, obtained with the estimator (4.3.38) and bandwidths $h = 3$ lu for $r \leq 20$ lu and 6 lu for $r > 20$ lu and improved with the reflection method. The dashed line shows the result without this correction. A comparison with Figure 4.18 reveals the advantages of using $g(r)$ as opposed to $L(r)$ as an instructive summary characteristic.

estimated pair correlation function has the typical form of that of a cluster process with local regularity, i.e. repulsion between the points. The minimum at $r_2 = 20$ lu indicates perhaps some further regularity, in particular for the distribution of cluster centres. Of course, the small irregularities of the curve are considered statistical artifacts that do not provide any further useful distributional information.

The fact that $\hat{g}(r) \geq 1$ does not hold for all r shows that the pattern is not a pure cluster process, but exhibits a combination of clustering and regularity at different scales. Clearly, there are many points with distances between $r = 6$ lu and 15 lu, while the values of $\hat{g}(r)$ smaller than 1 for r between 15 lu and 30 lu were reflected in $L(r) \approx r$ for $r \geq 30$ lu as described in Example 4.6.

Note that a statistical trick had to be applied to construct Figure 4.20. The estimator that was used is based on a kernel function. This smoothes the pair correlation function but the smoothing also generates positive values for r smaller than r_0. The reflection method explained in Example 4.9 on p. 237 and Appendix A was applied to construct the function depicted here. The other curve in Figure 4.20 is the estimate without reflection.

In some areas of science, modifications of the pair correlation function $g(r)$ have been used. Astronomers, for example, use what they call the 'correlation function',

$$\xi(r) = g(r) - 1.$$

Some authors call $\varrho(x, y)/\lambda^2$ the pair correlation function and $g(r)$ the radial distribution function; see Torquato (2002, p. 63).

Pair correlation functions can sometimes have poles at $r = 0$, i.e. $g(r) \propto r^{-\alpha}$ for small r, but the order α of these can only be smaller than $d - 1$. For example, astronomers know that $\alpha = 1.8$ for the point pattern of galaxies. In these cases it is reasonable to use the function

$$\eta(r) = (g(r) - 1)r^\alpha.$$

4.3.2 Theoretical foundations of second-order characteristics*

This subsection presents the theory of second-order characteristics, including an explanation as to why K, L and g are called second-order summary characteristics.

The starting point is the *second-order factorial moment measure* $\alpha^{(2)}$. As explained in Section 1.5, $\alpha^{(2)}(B_1 \times B_2)$ is the mean number of pairs of points (x_1, x_2) with $x_1 \neq x_2$ and $x_1 \in B_1$ and $x_2 \in B_2$. (Note that the symbols x_1 and x_2 are used as a convenient notation. They denote any points, which could also be simply referred to as x and y.) If $B_1 \cap B_2 = \emptyset$,

$$\alpha^{(2)}(B_1 \times B_2) = \mathbf{E}(N(B_1)N(B_2))$$

and if $B_1 = B_2 = B$,

$$\alpha^{(2)}(B \times B) = \mathbf{E}(N(B)(N(B) - 1)) = \mathbf{E}(N(B)^2) - \mathbf{E}(N(B)). \tag{4.3.13}$$

The last equation yields the following expression for the variance of the number of points in B:

$$\mathbf{var}(N(B)) = \alpha^{(2)}(B \times B) + \mathbf{E}(N(B)) - (\mathbf{E}(N(B)))^2$$

or, using (4.1.1),

$$\mathbf{var}(N(B)) = \alpha^{(2)}(B \times B) + \lambda\nu(B) - (\lambda\nu(B))^2. \tag{4.3.14}$$

Similarly, the covariances $\mathbf{cov}(N(B_1), N(B_2))$ for arbitrary B_1 and B_2 can also be expressed in terms of $\alpha^{(2)}$ and λ. Hence, it is clear that λ and $\alpha^{(2)}$ completely describe the second-order behaviour of the stationary point process N. Thus it is reasonable to derive simpler expressions for $\alpha^{(2)}(B_1 \times B_2)$, which use the second-order product density ϱ and the so-called reduced second-order moment measure $\mathcal{K}$. Note that

$$\alpha^{(2)}(B_1 \times B_2) = \int\limits_{B_1} \int\limits_{B_2 - x} \varrho(h)\mathrm{d}h\mathrm{d}x \tag{4.3.15}$$

and

$$\alpha^{(2)}(B_1 \times B_2) = \lambda^2 \int\limits_{B_1} \mathcal{K}(B_2 - x)\mathrm{d}x. \tag{4.3.16}$$

In (4.3.15) ϱ, which is in general a function $\varrho(x_1, x_2)$ of two variables x_1 and x_2, appears as a function of $h = x_1 - x_2$ since the value of this function depends only on the difference $x_1 - x_2$ in the stationary case.

Note that in the following the simpler symbols $\varrho(h)$ and $\varrho(r)$ (for the stationary and isotropic case, respectively) are often used.

The symbol $\mathcal{K}$ in (4.3.16) denotes the *reduced second-order moment measure* defined by

$$\lambda \mathcal{K}(B) = \mathbf{E}_o(N(B \setminus \{o\})). \qquad (4.3.17)$$

This means that $\lambda \mathcal{K}(B)$ is equal to the mean number of points of N in the set B conditional on the typical point of N being at o. If $o \in B$ then o is not counted. (The adjective 'reduced' does not indicate the subtraction of $\{o\}$ but the reduction of the number of sets from two [in $\alpha^{(2)}(B_1 \times B_2)$] to one [in $\mathcal{K}(B)$].)

Equations (4.3.15) and (4.3.16) yield

$$\lambda^2 \mathcal{K}(B) = \int_B \varrho(h) \mathrm{d}h. \qquad (4.3.18)$$

It is not difficult to prove formulas (4.3.15) and (4.3.16). Assuming that $\alpha^{(2)}$ has a density ϱ with respect to the Lebesgue measure, called second-order product density, yields

$$\alpha^{(2)}(B_1 \times B_2) = \int_{B_1} \int_{B_2} \varrho(x_1, x_2) \mathrm{d}x_1 \mathrm{d}x_2$$

$$= \int_{B_1} \int \mathbf{1}_{B_2}(x_1) \varrho(x_1, x_2) \mathrm{d}x_1 \mathrm{d}x_2.$$

Introducing the new variables $h = x_1 - x_2$ and $x = x_2$ results in

$$\int_{B_1} \int \mathbf{1}_{B_2}(x + h) \varrho(h) \mathrm{d}h \mathrm{d}x,$$

which is the same as the right-hand side of (4.3.15) since

$$\mathbf{1}_{B_2}(x + h) = \mathbf{1}_{B_2 - x}(h).$$

One may also write

$$\alpha^{(2)}(B_1 \times B_2) = \mathbf{E}\left(\sum_{x_1, x_2 \in N}^{\neq} \mathbf{1}_{B_1}(x_1) \mathbf{1}_{B_2}(x_2) \right)$$

$$= \mathbf{E}\left(\sum_{x \in N} \mathbf{1}_{B_1}(x) N(B_2 \setminus \{x\}) \right).$$

The last mean can be re-expressed by means of the Campbell–Mecke formula as

$$\lambda \int \mathbf{1}_{B_1}(x)\mathbf{E}_o(N((B_2 - x)\setminus\{o\}))dx,$$

which yields (4.3.16) by (4.3.17).

Generalising the Campbell theorem, the calculation of means such as

$$s_f = \mathbf{E}\left(\sum_{x_1, x_2 \in N}^{\neq} f(x_1, x_2)\right)$$

can be reduced to integrals with respect to ϱ or $\mathcal{K}$:

$$s_f = \int \int f(x, x+h)\varrho(h)dhdx \tag{4.3.19}$$

and

$$s_f = \lambda^2 \int \int f(x, x+h)\mathcal{K}(dh)dx. \tag{4.3.20}$$

This yields for the variance of the number of points in B:

$$\mathbf{var}(N(B)) = \lambda^2 \int \gamma_B(h)\mathcal{K}(dh) + \lambda\nu(B) - (\lambda\nu(B))^2 \tag{4.3.21}$$

or

$$\mathbf{var}(N(B)) = \int \gamma_B(h)\varrho(h)dh + \lambda\nu(B) - (\lambda\nu(B))^2. \tag{4.3.22}$$

Here $\gamma_B(h) = \nu(B \cap B_h)$ is the set covariance of B, where $\gamma_B(h)$ is the area (volume) of the intersection of B and its translate B_h (see Appendix B).

By (4.3.17), $\mathcal{K}$ is closely related to Ripley's K-function:

$$K(r) = \mathcal{K}(b(o, r)) \qquad \text{for } r \geq 0. \tag{4.3.23}$$

This equation shows that $K(r)$ can also be used in the general stationary case. However, ideally $K(r)$ is only applied to stationary *and isotropic* point processes where λ and $K(r)$ determine completely the second-order behaviour of the point process.

If $K(r)$ has a derivative $K'(r)$ for all r, then the expression for $\alpha^{(2)}(B \times B)$ simplifies to

$$\alpha^{(2)}(B \times B) = \lambda^2 \int_0^\infty \overline{\gamma}_B(r)K'(r)dr,$$

where $\overline{\gamma}_B(r)$ is the isotropised set covariance of B, the rotational average of $\gamma_B(h)$; see Appendix B. Formula (4.3.8) leads to the final variance formula for the isotropic case,

$$\mathbf{var}(N(B)) = \lambda^2 db_d \int_0^\infty \overline{\gamma}_B(r) g(r) r^{d-1} dr + \lambda \nu(B) - (\lambda \nu(B))^2. \qquad (4.3.24)$$

This formula is valuable as it separates out the geometry of B (given by $\overline{\gamma}_B(r)$) and the point process variability (given by the pair correlation function $g(r)$). For large B, an approximation for $\overline{\gamma}_B(r)$ can be used which yields

$$\frac{\mathbf{var}(N(B))}{\nu(B)} \simeq \lambda + db_d \lambda^2 \int_0^\infty (g(r) - 1) r^{d-1} dr. \qquad (4.3.25)$$

In the planar case this simplifies to

$$\frac{\mathbf{var}(N(B))}{\nu(B)} \simeq \lambda + 2\pi\lambda^2 \int_0^\infty (g(r) - 1) r \, dr. \qquad (4.3.26)$$

Historical remark. Like many other statistical concepts, most second-order summary characteristics of point processes were originally introduced by physicists. The idea of the pair correlation function appeared first in the context of X-ray scattering experiments of von Laue around 1900. Physicists were not very much interested in the statistical aspects as they were typically dealing with rather large samples, but had gained experience in the interpretation of pair correlation functions. The example on p. 242 is in the spirit of these applications.

The K-function was considered much later, first in Bartlett (1964) and finally in the modern form in Ripley (1977), who considered small samples and an edge-corrected estimator. Since then, the K-function has become the main tool of point process statistics and has also been used in situations where the pair correlation function would have been more appropriate. The best pair correlation function estimators have been derived in astronomy, a very important field of applied point process statistics; see Hamilton (1993) and Landy and Szalay (1993).

Uniqueness of the second-order characteristics

In classical statistics it is well known that mean and variance do not uniquely determine the distribution of a random variable. Similarly, in point process statistics, different point processes can have the same intensity λ and K-function. Baddeley and Silverman (1984) identify a planar point process which has the same K-function as the Poisson process, i.e. $K(r) = \pi r^2$. Diggle (1983) and Tscheschel and Stoyan (2006) present families of Neyman–Scott processes with the same λ and $g(r)$;

see p. 411 below. And clearly, two point processes N_1 and N_2 can have the same K, L and g, but different intensities. As a simple example consider N_1 and N_2 where N_2 is a p-thinning of N_1; see Section 6.2, p. 366.

An issue that still has not been resolved completely is the characterisation of pair correlation functions: which functions can be pair correlation functions? One necessary condition is

$$g(r) \geq 0 \qquad \text{for } r \geq 0,$$

which is trivial, and another, more difficult, is

$$S(k) \geq 0 \qquad \text{for } k \geq 0,$$

where $S(k)$ is the function given by

$$S(k) = 1 + \lambda \int_0^\infty f_d(kr)(g(r) - 1)r^{d-1}\mathrm{d}r \geq 0;$$

see Sakai et al. (2002). Physicists call $S(k)$ the 'structure factor'. For $d = 2$, it is $f_2(z) = 2\pi J_0(z)$ where $J_0(z)$ is a Bessel function (of the first kind, order 0); for $d = 3$, $f_3(z) = 2\pi \frac{\sin z}{z}$.

Another issue is that of the existence of a point process with specific λ and $g(r)$; see Uche et al. (2005). It is easy to find examples of λ and $g(r)$ for which no point process exists. It is clear that the existence of point processes for λ and $g(r)$ implies the existence for λ' and $g(r)$ if $\lambda' \leq \lambda$.

The following paragraph is addressed to readers from the fields of traditional mathematical spatial statistics and biostatistics. In these fields, K (or L) has now become fashionable and popular; people have learnt to estimate K and L (here the discussion is restricted to K) and can interpret these. Physicists and material scientists have other traditions; they have never seriously worked with K and have always used the pair correlation function g. This might be due to the fact that in physics sample sizes are usually large. Now, this book recommends working with g rather than with K even if not exclusively as it does recommend the use of the K-function for very small samples and in the context of goodness-of-fit tests; see Sections 2.7 and 7.4. This is by analogy with goodness-of-fit tests in classical statistics, where these are usually *not* based on density functions. However, this book does indeed recommend the use of g for exploratory analysis despite the fact that the estimation of the pair correlation function is more delicate and complicated than that of K due to the serious issues of bandwidth choice and estimation for small r.

Note that the problems with unusual window shapes (i.e. non-rectangular, polygonal, ...) have now been solved. One solution has been implemented in `spatstat` in R, astronomers have developed their own methods (see Hamilton, 1993), and Fourier methods may also be used (see the remark on p. 233 below).

At first sight, it seems as if parameter estimation, via the minimum contrast method, is often based on K; see Diggle (2003) and Section 7.2 below. But in the modern literature contrast functions are also used which contain the derivative of K, which is nearly equivalent to the use of g by (4.3.7). The question which of the two functions, g or K, yields the better minimum contrast estimates remains unanswered and requires further research; the authors' experience indicates that it is the pair correlation function $g(r)$.

4.3.3   Estimators of the second-order characteristics

This subsection introduces and characterises several estimators and advises on the appropriate choice of estimators for the second-order charcteristic of interest. The estimation starts with that of $\lambda^2 \mathcal{K}$ and ϱ. The corresponding estimators are then combined with intensity estimators, to finally yield estimators of K, L and g.

Estimation of $\lambda^2 \mathcal{K}$ (anisotropic case)

In the general case of a process that is not necessarily isotropic $\lambda^2 \mathcal{K}(B)$ is estimated by $\hat{\kappa}_{st}(B)$, where

$$\hat{\kappa}_{st}(B) = \sum_{x_1, x_2 \in W}^{\neq} \frac{\mathbf{1}_B(x_2 - x_1)}{\nu(W_{x_1} \cap W_{x_2})} \tag{4.3.27}$$

is defined for any bounded set B such that $\nu(W \cap W_z)$ is positive for all z in B. If $B = b(o, r)$ and W is a rectangle or parallelepiped, r must be smaller than the shortest edge length of W. This r-value is denoted as r_{st}.

To prove the unbiasedness of $\hat{\kappa}_{st}(B)$, observe that $\hat{\kappa}_{st}(B)$ has the general form

$$\sum_{x_1, x_2 \in N}^{\neq} f(x_1, x_2)$$

with

$$f(x_1, x_2) = \mathbf{1}_W(x_1) \mathbf{1}_W(x_2) \frac{\mathbf{1}_B(x_2 - x_1)}{\nu(W \cap W_{x_2 - x_1})}$$

and $\nu(W \cap W_{x_2 - x_1}) = \nu(W_{x_1} \cap W_{x_2})$. Consequently, (4.3.20) can be applied, which yields

$$\mathbf{E}\hat{\kappa}_{st}(B) = \lambda^2 \int \int \mathbf{1}_W(x) \mathbf{1}_W(x + h) \frac{\mathbf{1}_B(h)}{\nu(W \cap W_h)} dx \mathcal{K}(dh)$$

$$= \lambda^2 \int \frac{\mathbf{1}_B(h)}{\nu(W \cap W_h)} \int \mathbf{1}_W(x) \mathbf{1}_W(x + h) dx \mathcal{K}(dh)$$

$$= \lambda^2 \int \mathbf{1}_B(h) \mathcal{K}(dh) = \lambda^2 \mathcal{K}(B).$$

In technical terms the unbiasedness of $\hat{\kappa}_{st}(B)$ is obtained by a general statistical technique called Horvitz–Thompson weighting (see Baddeley, 1999). The idea is to weight the pairs of points (x_1, x_2) which contribute to the double sum: 'rare' pairs are assigned large and 'frequent' ones only small weights in order to obtain an unbiased estimator. Since the pattern is sampled within the bounded window W large inter-point distances are relatively rare, and the weights $1/\nu(W \cap W_h)$ are indeed large for large h.

Note that it is not difficult to determine the term

$$\nu(W \cap W_{x_1 - x_2}) = \nu(W_{x_1} \cap W_{x_2})$$

if W is a rectangle, parallelepiped, disc or sphere. The relevant formulas are given in Appendix B.

Estimation of $\lambda^2 K$ (isotropic case)

Here $\kappa_i(r)$ is an unbiased estimator of $\lambda^2 K(r)$,

$$\hat{\kappa}_i(r) = \sum_{x_1, x_2 \in W} \frac{\mathbf{1}(0 < \|x_1 - x_2\| \leq r) w(x_1, x_2)}{\nu(W^{(\|x_1 - x_2\|)})} \tag{4.3.28}$$

for $0 \leq r < r_i$, where

$$r_i = \sup\left\{r : \nu\left(W^{(r)}\right) > 0\right\} \quad \text{and} \quad W^{(r)} = \{x \in W : \partial(b(x, r)) \cap W \neq \emptyset\},$$

and, for $d = 2$, $w(x_1, x_2) = 2\pi/\alpha_{x_1 x_2}$ where $\alpha_{x_1 x_2}$ is the sum of all angles of the arcs in W of a circle with centre x_1 and radius $\|x_1 - x_2\|$. If $\alpha_{x_1 x_2} = 0$ then $w(x_1, x_2) = 0$. In the general d-dimensional case $w(x_1, x_2)$ is the ratio {surface area of the sphere $\partial b(x_1 \|x_1 - x_2\|)$ in W} / $\{db_d \|x_1 - x_2\|^{d-1}\}$. If W is a rectangle or parallelepiped then r_i is the length of the diagonal. Formulas for $w(x_1, x_2)$ for windows W of various shapes are given in Appendix B.

The proof of unbiasedness of $\hat{\kappa}_i(r)$ is similar to that of $\hat{\kappa}_{st}(B)$, but is based on polar coordinates.

With $B = b(o, r)$ the estimator $\hat{\kappa}_{st}(B)$ can also be used for the estimation of $\lambda^2 K(r)$, but the range of r-values is reduced since $r_{st} < r_i$. Of course, $\hat{\kappa}_i(r)$ has a smaller estimation variance than $\hat{\kappa}_{st}(b(o, r))$ for a really isotropic point process. However, $\hat{\kappa}_i(r)$ is sensitive to deviations from the isotropy assumption, and thus it may be preferable to always use $\hat{\kappa}_{st}(b(o, r))$ instead of $\hat{\kappa}_i(r)$. All examples in this book are computed by means of $\hat{\kappa}_{st}$.

Both estimators tend to be sensitive to clusters close to the boundary of W yielding spurious large values, which may overestimate the degree of clustering in the pattern.

Doguwa and Upton (1989) discuss an estimator of $\lambda K(r)$ in the spirit of $n_i(r)$ on p. 215 above with weights proportional to the area of $b(x_i, r) \cap W$, which is unbiased in the Poisson process case.

Estimation of $\varrho(r)$

A useful general kernel estimator of $\hat{\varrho}(r)$ is

$$\hat{\varrho}_{st}(r) = \sum_{x_1,x_2 \in W}^{\neq} \frac{k(\|x_1 - x_2\| - r)}{db_d r^{d-1} \nu(W_{x_1} \cap W_{x_2})} \qquad \text{for } 0 \le r \le r_{st}, \qquad (4.3.29)$$

which is analoguous to $\hat{\kappa}_{st}(b(o, r))$.

In the isotropic case the analogue to Ripley's estimator $\hat{\kappa}_i(r)$ is

$$\hat{\varrho}_i(r) = \frac{1}{db_d r^{d-1}} \sum_{x_1,x_2 \in W} k(r - \|x_1 - x_2\|)w(x_1, x_2) \qquad \text{for } 0 \le r \le r_i. \qquad (4.3.30)$$

Ohser and Mücklich (2000, p. 279) present a version of $\hat{\varrho}_{st}(r)$ adapted to the isotropic case,

$$\hat{\varrho}_0(r) = \frac{1}{db_d r^{d-1} \overline{\gamma}_W(r)} \sum_{x_1,x_2 \in W}^{\neq} k(r - \|x_1 - x_2\|) \qquad \text{for } 0 \le r \le r_i. \qquad (4.3.31)$$

Here, the term $\nu(W_{x_1} \cap W_{x_2})$ is replaced by the isotropised set covariance $\overline{\gamma}_W(r)$. This is computationally much simpler than (4.3.30).

All product density estimators use a kernel function k, i.e. a non-negative function $k(z)$ that satisfies $\int_{-\infty}^{\infty} k(z)dz = 1$; see Appendix A. There are several kernel functions to choose from, but the recommended choice for $k(z)$ is the simple *box kernel*,

$$k(z) = \begin{cases} \dfrac{1}{2h} & \text{for } -h \le z \le h, \\ 0 & \text{otherwise.} \end{cases}$$

After (4.3.42) below, an explanation is given as to why this simple kernel (usually used by physicists and astronomers) should be used for product density estimation rather than the Epanechnikov kernel, which was recommended in older publications. The parameter h, termed the *bandwidth*, plays a very important role; its choice is crucial to the quality of the product density estimators and heavily influences their variance. It is also discussed below.

As in the proof of the unbiasedness of $\hat{\kappa}_{st}(B)$ one can show that, for the box kernel,

$$\mathbf{E}\hat{\varrho}_{st}(r) = \frac{1}{2h} \int_{-\min\{h,r\}}^{h} \varrho(r+s)ds; \qquad (4.3.32)$$

thus, as $h \to 0$, $\mathbf{E}\hat{\varrho}_{st}(r) \to \varrho(r)$. For small r $(r < h)$ there is clearly some bias. For example, the mean is $\frac{h+r}{2h} \lambda^2$ for a Poisson process, instead of λ^2. There are two ways to handle the case of $r < h$: one can either apply the reflection method explained in Example 4.9 and Appendix A or, following Guan (2007), change $\hat{\varrho}_{st}(r)$ at two points:

- replace r in the denominator of (4.3.29) by $\|x_i - x_j\|$;
- multiply the estimator by the factor

$$f_h(r) = \begin{cases} \dfrac{2h}{h+r} & \text{for } r \leq h, \\ 1 & \text{otherwise.} \end{cases} \tag{4.3.33}$$

Note that this is only recommended for $r < h$.

Estimation of $K(r)$

Estimators of $K(r)$ are obtained by dividing the estimators of $\lambda^2 K(r)$ described above by estimators of λ^2. Ripley (1976) originally used the standard estimator $\hat{\lambda}$ and simply squared it. Later some authors used

$$\widehat{\lambda^2} = n(n-1)/\nu(W)^2, \tag{4.3.34}$$

where n is the number of points in the window W. This is reasonable since $\widehat{\lambda^2}$ is unbiased for λ^2 in the Poisson case. Thus in this case the resulting estimator is ratio-unbiased. However, $(\hat{\lambda})^2$ is not unbiased for λ^2 and hence Ripley's estimator is not ratio-unbiased. Heinrich (1988, 1991) show its asymptotic normality and constructed a corresponding CSR test.

Both estimators, $\widehat{\lambda^2}$ and $(\hat{\lambda})^2$, are not adapted to $\hat{\kappa}_{st}(b(o, r))$ and $\hat{\kappa}_i(r)$. Their precision may be much higher than that of $\hat{\kappa}_{st}(b(o, r))$ and $\hat{\kappa}_i(r)$, in particular for large r, and so there is little chance that their fluctuations cancel out.

Thus, there is some potential to improve the estimation. Indeed, Stoyan and Stoyan (2000) introduce the estimator $\hat{\lambda}_V(r)$ of λ, which is adapted and indeed improves the quality of estimation of $K(r)$. This intensity estimator was discussed in Section 4.2.3. The estimators of $K(r)$ are

$$\hat{K}_{st}(r) = \frac{\hat{\kappa}_{st}(b(o, r))}{\hat{\lambda}_V(r)^2} \qquad \text{for } 0 \leq r \leq r_{st} \tag{4.3.35}$$

and

$$\hat{K}_i(r) = \frac{\hat{\kappa}_i(r)}{\hat{\lambda}_V(r)^2} \qquad \text{for } 0 \leq r \leq r_i. \tag{4.3.36}$$

Whether different intensity estimators should be used for 'i' and 'st', replacing the general $\hat{\lambda}_V(r)^2$, is still an open question. Note that the estimation of $K(r)$ would not be improved if the exact value of λ were known. Even then the adapted intensity estimator yields a smaller mse.

Estimation of $L(r)$

So far, $L(r)$ has been always estimated based on estimators of $K(r)$, via

$$\hat{L}(r) = \sqrt{\hat{K}(r)/b_d}, \tag{4.3.37}$$

where $\hat{K}(r)$ is one of the estimators of $K(r)$ above. It is not known whether there is an efficient way to estimate $L(r)$ directly.

Estimation of $g(r)$

This book strongly recommends the use of kernel estimators for the product density $\hat{\varrho}(r)$ which are then divided by the squared adapted intensity estimator $\hat{\lambda}_S(r)$ to yield an estimate of $g(r)$. Thus, the estimation of the intensity also plays an important role in the estimation of the pair correlation function. Statisticians used $\hat{\lambda}$ (as in (4.2.10)) squared or $\widehat{\lambda^2}$ (as in (4.3.34)). The astronomers Hamilton (1993) and Landy and Szalay (1993) revealed that it is better to use the estimator $\hat{\lambda}_S(r)$ introduced in Section 4.2.3. This leads to the estimators

$$\hat{g}_{st}(r) = \frac{\hat{\varrho}_{st}(r)}{\hat{\lambda}_S(r)^2} \qquad \text{for } 0 \le r \le r_{st} \tag{4.3.38}$$

and

$$\hat{g}_i(r) = \frac{\hat{\varrho}_i(r)}{\hat{\lambda}_S(r)^2} \qquad \text{for } 0 \le r \le r_i \tag{4.3.39}$$

with $\hat{\varrho}_{st}(r)$ and $\hat{\varrho}_i(r)$ or $\hat{\varrho}_o(r)$ as defined in (4.3.29) and (4.3.30) or (4.3.31), respectively.

The intensity estimators $\hat{\lambda}_V(r)$ and $\hat{\lambda}_S(r)$ are adapted to the estimation of $K(r)$ and $g(r)$. The sphere $b(x, r)$ and the sphere surface $\partial b(x, r)$ are relevant for every point x of N in W. The weighting considers the volume and surface content of $b(x, r)$ and $\partial b(x, r)$ in W and thus the importance of x: if x is close to the boundary it is assigned a small weight since it contributes only little to $\hat{\kappa}(r)$ and $\hat{g}(r)$, respectively. The main effect of the use of $\hat{\lambda}_V(r)$ and $\hat{\lambda}_S(s)$ is that the mse is smaller than for $\hat{\lambda}$.

This book does not recommend methods exploiting the relationship (4.3.7) between $K(r)$ and $g(r)$, which involves the estimation of $K(r)$ and subsequent numerical differentiation of $K(r)$. Kernels can be handled better than smoothers (such as splines) of $K(r)$. Furthermore, the analogy between the estimation of $g(r)$ and the estimation of probability density functions provides an argument in favour of kernel estimators as these are the preferred estimators of probability density functions.

For large samples with a low degree of clustering, edge-correction loses its importance and numerical problems become more important. In these cases the Fourier approach is recommended; see Ohser and Schladitz (2008), Szapudi et al. (2005) and Vio et al. (2007). The idea is to transform the point process into a shot-noise random field (by transforming the original image into a pixel image), to estimate the covariance function $k(r)$ of this field and to use the relationship between $k(r)$ and $g(r)$ given by (6.9.4). Strong numerical methods such as the fast Fourier transform make this approach efficient, even for irregular windows W. The first step is quite natural when the data result from image analysis, e.g. in materials science studies. The smoothing effect of the shot-noise transformation is comparable to that resulting from kernel estimation.

Biases and mean squared errors

In applications, one strives to choose both the right size of the observation window W and an appropriate bandwidth h. This decision is typically based on practical limitations as well as on the performance of the estimators. Thus, it is important to be able to obtain approximations of the bias and the mean squared error or the estimation variance.

Theoretical calculations (see Hamilton, 1993; Landy and Szalay, 1993; Ripley, 1988) and simulations have shown that the biases of the estimators introduced above are negligible under normal conditions. For this reason, this section focuses on mse and estimation variance.

An approximate analytical expression for the variance of $\hat{K}_i(r)$ with $\hat{\lambda}^2$ in the planar case was given by Ripley (1988):

$$\sigma_K^2(r) = \frac{2}{\lambda^2} \left(\frac{\pi r^2}{\nu(W)} + 0.96 \frac{U(W)}{\nu(W)^2} r^3 + 0.13\lambda \frac{U(W)}{\nu(W)^2} r^5 \right), \tag{4.3.40}$$

where $U(W)$ is the perimeter length of W. Note, however, that it was derived for the case of a Poisson process.

Simulations by Stoyan and Stoyan (unpublished) have shown that (4.3.40) is very precise and close to the mse of $\hat{K}_{st}(r)$ with $\hat{\lambda}_V(r)$. This leads to the recommendation to use this estimator, i.e. $\hat{K}_{st}(r)$ as in (4.3.35), since it is robust against deviations from isotropy and shows the same precision as Ripley's estimator.

Little is known of the statistical properties of the estimators of the L-function. Simulations for Poisson processes have shown that, as expected, $L(r)$ estimated via $\hat{\kappa}_{st}(b(o, r))$ combined with $\hat{\lambda}_V(r)$ yields an mse which is practically independent of r for small and medium sized r (see Figure 4.21).

Hamilton (1993) and Landy and Szalay (1993) show that, for the Poisson case, $\hat{g}_i(r)$ (with $\hat{\lambda}_S(r)$) has a small bias and estimation variance, but they do not present simple formulas for the calculation of these quantities that may be used in practical

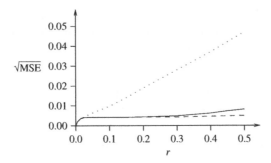

Figure 4.21   Square root of mean squared error for several estimators of $L(r)$ for a planar Poisson process of intensity $\lambda = 100$ for $W = \boxed{1}$: (dashed line) estimator (4.3.35); (solid line) estimator (4.3.36) with standard estimator $\hat{\lambda}$; (dotted line) estimator (4.3.36) with known intensity $\lambda = 100$.

applications of point process statistics. The formula

$$\sigma_g^2(r) = \frac{g(r) \int_{-h}^{h} k^2(z) \mathrm{d}z}{\frac{1}{2} d b_d r^{d-1} \overline{\gamma}_W(r) \lambda^2} \qquad \text{for } 0 \leq r \leq r_{st} \tag{4.3.41}$$

serves this purpose for the variance of $\hat{g}_{st}(r)$ in the general d-dimensional case. In the planar case this simplifies to

$$\sigma_g^2(r) = \frac{g(r) \int_{-h}^{h} k^2(z) \mathrm{d}z}{\pi r \overline{\gamma}_W(r) \lambda^2} \qquad \text{for } 0 \leq r \leq r_{st}. \tag{4.3.42}$$

The term $\int_{-h}^{h} k^2(z) \mathrm{d}z$ with the integral over the squared kernel function is $0.6/h$ for the Epanechnikov kernel and $0.5/h$ for the box kernel (which is the minimum over all kernel functions). It is mainly for this reason that this book recommends the box kernel for pair correlation function estimation rather than the Epanechnikov kernel.

Equation (4.3.41) was derived for general stationary and isotropic point processes using a Poisson approximation based on the idea of rare events. It was recommended in Stoyan et al. (1993) for the estimator $\hat{g}_{st}(r)$ with the classical intensity estimator $\hat{\lambda}$. However, later simulations showed that it is an excellent approximation for the mse of $\hat{g}_{st}(r)$ with $\hat{\lambda}_S(r)$. Figure 4.22 shows $\sigma_g^2(r)$ and the mse of $\hat{g}_{st}(r)$ for a Poisson process and confirms this statement.

The mse and the estimation variance have frequently been found to behave similarly to what is depicted in Figure 4.22: the mse is (nearly) independent of r for medium distances r, but larger for small and large r (not shown here). The variance $\sigma_g^2(r)$ as given by (4.3.41) and (4.3.42) has very large values for $r \to 0$ and for large r (since $\overline{\gamma}_W(r) \to 0$ for large r). This is not surprising as here the number of contributing pairs of points (x_1, x_2) with an inter-point distance close to

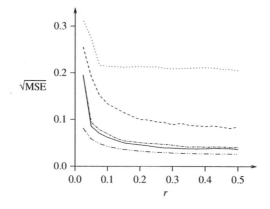

Figure 4.22   Comparison of square roots of mse of $\hat{g}_{st}(r)$ obtained by simulation for a planar Poisson process of intensity $\lambda = 100$ in [1]: (solid line) box kernel, $h = 0.05$; (dash-dotted line) Epanechnikov kernel, $h = 0.05$; (dash-dot-dotted line) $\sigma_g(r)$ as in (4.3.42), $h = 0.10$; (dashed line) box kernel, $h = 0.01$; (dotted line) box kernel, $h = 0.05$, known λ used.

r is small for large r, but their individual contributions are large as a result of the Horvitz–Thompson weighting mentioned on p. 229.

For small r a technical problem may occur: the estimators of $\varrho(r)$ contain the term $1/r$ (planar case) or $1/r^{d-1}$ (d-dimensional case) in the denominator. Thus, for very small r, if the numerator is positive, $\hat{g}_{st}(r)$ or $\hat{g}_i(r)$ will become very large, even though small values of $g(r)$ would have been correct. This problem occurs in particular when a large bandwidth is used as pairs of points with inter-point distances (much) larger than r can contribute. In general this effect appears for values of r smaller than h, see p. 230.

Formula (4.3.42) can be used to tackle the following statistical problems:

1. What size of window ensures a specified precision of estimation of the pair correlation function?

2. What is the appropriate bandwidth h?

The first question is addressed in Section 4.8.2 below in the context of other window issues, while the second is discussed here.

Choice of bandwidth h

Since the choice of h 'is an art' (Martínez et al., 2005) the following section attempts to offer detailed advice to the beginner. Large values of h produce smooth estimates of $g(r)$, but important details may be lost. Conversely, small values of h yield noisy estimates with probably spurious and meaningless spikes. Since there is no strict mathematical theory for the optimal choice of h, the following advice is offered:

- Estimate $g(r)$ for several values of h and compare the results. This may yield an appropriate and interpretable curve for $\hat{g}(r)$. In some cases, two different values of h might be found in this way, one of which is suitable for small and the other one for large r. Such *adapted* bandwidths are highly recommended.

- Start the search for h with a value of the order of

$$h \approx 0.1 / \sqrt{\lambda} \qquad (4.3.43)$$

for the planar case, and

$$h \approx 0.05 / \sqrt[3]{\lambda} \qquad (4.3.44)$$

for the spatial case. The one-dimensional case is discussed in Vio et al. (2007). The relation (4.3.43) has been derived from practical experience with point patterns of 50 to 300 points. Large bandwidths may also be useful since $\sigma_g^2(r)$ as given by (4.3.41) decreases with increasing h for fixed r; this is true in particular for processes that are 'almost' Poisson processes with a slowly varying pair correlation function, whereas smaller details, e.g. connected with hard-core distances, may be smoothed away (see Figure 4.24). Guan (2007) recommends values of h smaller than (4.3.43) for cluster and hard-core processes.

The following uses the above formulas to identify the value of $\sigma_g^2(r)$ that results from $h = 0.1 / \sqrt{\lambda}$ for a planar Poisson process with intensity $\lambda = 80$ for $W = \boxed{1}$. The numerator of (4.3.42) is equal to $0.5/h$ if the box kernel is used. The denominator is

$$\pi r \left(1 - \frac{4}{\pi} r + \frac{r^2}{\pi} \right) \lambda^2 \qquad \text{for } r \leq 1.$$

For $r = 0.5$, this leads to

$$\sigma_g^2(r) = 0.01 \quad \text{or} \quad \sigma_g(r) = 0.1.$$

Thus Equation (4.3.42) yields a standard deviation of 0.1 for a theoretical value of 1 in this example, which may be acceptable.

Consider now the three-dimensional case and try to find an a in

$$h = a / \sqrt[3]{\lambda},$$

which yields the same value as above, i.e. $\sigma_g(0.5) = 0.1$, for $\boxed{1}$, $r = 0.5$ and $\lambda = 80$. Now the numerator of (4.3.42) is $0.5/(a\sqrt[3]{\lambda})$ and the denominator

$$2\pi r^2 \left(1 - 1.5r + \frac{2}{\pi}r^2 - \frac{1}{4\pi}r^3 \right) 80^2$$

and $a = 0.0536$, i.e. a value as suggested in (4.3.44).

Example 4.9. Choosing bandwidths
In the following, two planar point process samples of 200 points in $\boxed{1}$ are considered. These are the results of (conditional) simulation (with a fixed number of 200 points) and the aim is to re-estimate the pair correlation functions. Thus we are in the comfortable situation of knowing the theoretical pair correlation functions and can observe the influence of the bandwidth h and make comparisons. The simulations are based on spatial point process models which are described in greater detail in Chapter 6.

The first model is the Matérn cluster process (see Section 6.3.2) with parameters $\lambda = 200$, $R = 0.05$ and $\mu = 10$. (The range of correlation r_{corr} is $2R = 0.1$.) Figure 6.3 on p. 379 shows a simulated sample. The theoretical pair correlation function $g(r)$ is given by (6.3.6), in particular $g(0) = 7.37$ and $g(r) = 1$ for $r \geq 0.1$. Formula (4.3.43) suggests the value $h = 0.007$ for the bandwidth. Figure 4.23 shows the result: $\hat{g}(0.001) = 22.0$ and $\hat{g}(r) \approx 1$ for $r \geq 0.135$ i.e. $\hat{r}_{corr} = 0.135$.

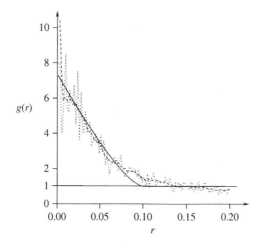

Figure 4.23 The theoretical pair correlation function $g(r)$ of a Matérn cluster process as described in the text (solid line) and its estimates $\hat{g}(r)$ with $h = 0.007$ (dashed line) and 0.001 (dotted line). The irregular fluctuations of the estimates for r larger than 0.1 do not provide any interpretable information.

If information on the process had not been available one might now have concluded that the pattern exhibits clustering with cluster diameters of about 0.135 and that the fluctuations of $\hat{g}(r)$ for large r represent only statistical noise. Thus the estimation is repeated for the smaller r with $h_0 = 0.001$ and the expected effect occurs: $\hat{g}(0)$ is now smaller, 8.0, but still larger than the theoretical value. The value r_{corr} also appears to be 0.135 for the smaller r, i.e. still different from the theoretical value of 0.1. (This kind of deviation is typical of samples from cluster processes that are as small as this considered here.) Again, adapted bandwidths turn out to yield good results.

The second model is the Matérn hard-core process (see Section 6.5.2) with parameters $\lambda = 200$ and $r_0 = 0.039$. Figure 6.7 on p. 392 shows a simulated sample. The pair correlation function $g(r)$ can be derived from (6.5.1) and (6.5.2) and satisfies

$$g(r) = 0 \qquad \text{for } 0 \le r \le r_0$$

and

$$g(r) = 1 \qquad \text{for } r \ge 2r_0,$$

while $g(r)$ is a little larger than 1 for $r_0 \le r \le 2r_0$. Formula (4.3.43) suggests $h = 0.007$. Figure 4.24 shows the result of the experiment: the simulated pair correlation function does not look like that of a typical hard-core process. It rather resembles the pair correlation function of a process that could be called a soft-core process, since it has values significantly smaller than 1 for $r \le 0.044$, and random fluctuations around 1 occur for $r \ge 0.07$. These random fluctuations are not meaningful and should be ignored. If a suitable adapted bandwidth approach were used these would disappear. However, even without the process information

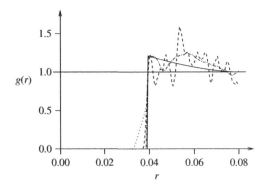

Figure 4.24   The theoretical pair correlation function $g(r)$ of a Matérn hard-core process as described in the text (solid line) and its estimate $\hat{g}(r)$ with bandwidth $h = 0.002$ (dashed line). The dotted line is $\hat{g}(r)$ for $r < 0.039$. The result of the reflection method for $h = 0.007$ is shown as a dot-dashed line.

the minimum inter-point distance would have been estimated to be close to 0.039 and one would have concluded that the values of $\hat{g}(r)$ for $r < 0.039$ are statistical artefacts. Using a smaller bandwidth, for example $h = 0.002$, clearly improves the result for r around 0.04.

The estimate can be improved by the *reflection method* as introduced in Stoyan and Stoyan (1992) for point process statistics, an approach that is commonly used in the context of density estimation (Silverman, 1986); see Appendix A. 'Reflect' the pair correlation function mass below $\hat{r}_0 = 0.039$ on r-values larger than $\hat{r}_0$ by setting:

$$\hat{g}(r) \leftarrow 0 \qquad \text{for } r < \hat{r}_0,$$

$$\hat{g}(r) \leftarrow \hat{g}(r) + \hat{g}(\hat{r}_0 - (r - \hat{r}_0)) \qquad \text{for } r \geq \hat{r}_0.$$

Figure 4.24 shows the resulting further improvement. A suitable approach would be to combine the curves with bandwidths $h = 0.007$ and $h = 0.002$ for $r \geq 0.045$ and $r \leq 0.045$ respectively, and set $\hat{g}(r) = 1$ for $r \geq 0.06$.

Note, in conclusion, that issues such as choosing the bandwidth h are common in classical statistics such as in density estimation with kernel functions and even appear in the context of histogram construction, where bin size or class interval length heavily influence the result.

4.3.4 Interpretation of pair correlation functions

The interpretation of statistically estimated pair correlation functions requires experience, skill and imagination, and the beginner usually needs assistance.

Figure 4.25 shows a schematic description of the shapes of pair correlation functions for the three main classes of point processes: Poisson, cluster (aggregative) and regular (repulsive) processes. In many cases, the pair correlation function helps to identify an appropriate model class for given data or may be applied to verify a priori assumptions. However, point processes are often more complex; regularity and clustering might occur simultaneously but at different spatial scales, and it is useful to be able to detect this type of behaviour.

In the non-trivial cases of clustering (aggregation) and regularity (repulsion) one may consider specific characteristic distances r_0, r_1, ... to quantify the variability of the point distribution as described by the pair correlation function. This section concludes by providing examples outlining this process of interpreting pair correlation functions by referring to a few characteristic distances.

For a *cluster process* with little structure inside the clusters (e.g. Neyman–Scott processes; see Section 6.3.2) the pair correlation function has the form given in Figure 4.23. The most interesting feature here is the range of correlation r_{corr}. It describes the size of the clusters. Indeed, if it is finite (and not infinite as in the case of a Thomas process; see Section 6.3.2) and the clusters are circular (spherical), r_{corr} is the cluster diameter.

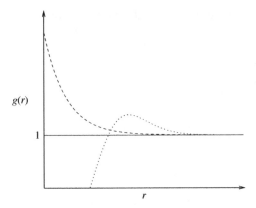

Figure 4.25   Schematic shapes of pair correlation functions for a Poisson process (solid line), a cluster process (dashed line) and a regular process (dotted line).

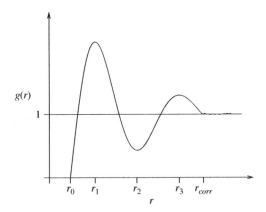

Figure 4.26   Typical form of the pair correlation function of a point process with repulsion. The r_i are explained in the text.

More interesting distances, r_0, r_1, ..., r_{corr}, may be considered for *processes with repulsion* (see Figure 4.26). These are listed below:

$r_0 = $ minimum inter-point distance, hard-core distance;

$r_1 = $ distance at which $g(r)$ has its first maximum, range of most frequent short inter-point distance, distance from typical point to near neighbours;

$r_2 = $ distance at which $g(r)$ has its first minimum after r_1 with $g(r_1) \leq 1$, distance from typical point to regions with a small number of points beyond the nearest neighbours;

$r_3 = $ distance at which $g(r)$ attains its second maximum, range of most frequent longer inter-point distance, distance from typical point to regions with further neighbours.

Many processes only show the first maximum. These processes are informally termed *soft-core processes,* if $0 \leq r_0 < r_1$. The existence of a minimum after r_1 is already an indicator of short-range order in the point process. By contrast, Figure 4.27 shows a pair correlation function for a pattern with long-range order.

The distances r_1 and r_2 and the values $g(r_1)$ and $g(r_2)$ can be used to calculate numerical parameters for the characterisation of the *degree of short-range disorder,* which are useful for the comparison of patterns of different degrees of disorder. Examples are

$$M = \frac{g(r_1) - g(r_2)}{r_2 - r_1} \tag{4.3.45}$$

(see Stoyan and Schnabel, 1990) and

$$O = \frac{g(r_1)}{r_1 - r_0} \tag{4.3.46}$$

(see Hubalková and Stoyan, 2003). Small values of M and O indicate a high degree of short-range disorder.

The beginner may find it helpful to consider a number of different examples of graphs of pair correlation functions and the corresponding interpretation. Here is a list of figures in this book that provide further examples:

- cluster processes: 4.19, 5.6 $(g_{22}(r))$, 6.22, 6.25;
- soft-core processes: 4.52, 5.5 $(g_{11}(r), g_{22}(r))$, 5.6 $(g_{11}(r))$, 6.11 (solid line);
- hard-core processes: 4.24, 4.27, 4.28, 5.20(a), 6.11 (dashed line).

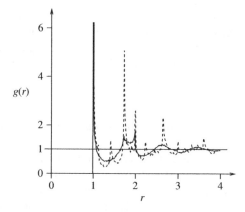

Figure 4.27   Pair correlation functions of random packings of identical hard spheres with diameter 1 and volume fraction $V_V = 0.637$ (solid line), $V_V = 0.668$ (dashed line). With kind permission of Springer Science and Business Media.

Finally, some pair correlation functions for systems of hard spheres are shown and discussed, which represent extreme cases of repulsive processes.

Figure 4.27 shows the estimated pair correlation functions of two three-dimensional systems of random packings of identical hard (non-interpenetrable) spheres (diameter = 1) of volume fractions 0.637 and 0.668. Packings of this type play an important role in physics and materials science, but the centres of cells in biological tissue also follow this or similar models. Each of these packings consists of 10 000 spheres and was obtained by the force-biased algorithm described in Section 6.5.5. The case with 0.637 corresponds to the classical random dense packing of spheres, which is usually assumed to be stationary and isotropic. The case with 0.668 shows tendencies to crystallisation (which is complete for 0.74), i.e. local anisotropies. The pair correlation function for 0.637 has a very typical form which is frequently observed in nature. However, a formula or analytical approach to describe this form has not yet been found.

Since the spheres are hard, there is a hard-core distance r_0, which is of course $r_0 = 1$. Thus here $r_0 = r_1$: the location of the first maximum coincides with the hard-core distance. The nature of this maximum is difficult: since many spheres are in direct contact (as a 'packing' is being discussed), $g(r)$ has a Dirac-measure component at $r = 1$ (the reduced second-moment measure $\mathcal{K}$ gives positive mass to the surface of the unit sphere). Thus the graph presents only a smoothed version of the true behaviour at $r = 1$. Furthermore, there are many spheres close together but not in direct contact. Thus it may be (but this is theoretically not quite clear) that the pair correlation function has a pole at $r = 1$ for $r \downarrow 1$. Moving on, the next feature can be found at the location r_2 of the first minimum. It appears at a characteristic distance of r_2, namely, $\sqrt{3}$. All other minima and maxima also appear at characteristic distances in all well-simulated random dense sphere packings.

The curve for 0.668 is similar and reflects the reduced degree of disorder in the sharper minima and maxima. If the pair correlation function had been determined for a crystalline packing of hard spheres (of so-called fcc or hcp type), one would have observed only peaks (which correspond to Dirac-measure/function components) at the characteristic inter-point distances in the crystal lattices.

Example 4.10.   Application of the pair correlation function in the analysis of the structure of metallic glasses

Metallic glasses are amorphous materials resulting from fast cooling of melted metallic alloys. An example is $Pd_{40}Cu_{30}Ni_{10}P_{20}$, a material which consists of 40 % palladium, etc. Mattern et al. (2003) report on structural investigations of this metallic glass at different temperatures. The physical measurement method used is high-temperature X-ray synchrotron diffraction. This yields the so-called structure factor $S(k)$, which leads to the pair correlation function $g(r)$ of the point process of atom centres by Fourier transform.

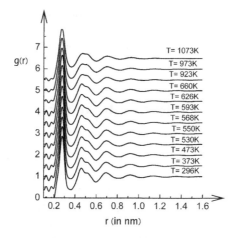

r (in nm)

Figure 4.28   Atomic pair correlation function $g(r)$ of $Pd_{40}Cu_{30}Ni_{10}P_{20}$ at different temperatures (maximum positions R_i for $T = 296\,K$: $R_1 = 0.277$, $R_2 = 0.455$, $R_3 = 0.520$, $R_4 = 0.692$, $R_5 = 0.921$, $R_6 = 1.136$, $R_7 = 1.355\,nm$). Note that the R_i are positions of maxima, in contrast to the notation on p. 240, and that $r_1 = R_1$. Data courtesy of N. Mattern.

Figure 4.28 shows 12 pair correlation functions for different temperatures. Mattern et al. (2003) discuss the curves as follows:

The position of the first maximum in $g(r)$ of $Pd_{40}Cu_{30}Ni_{10}P_{20}$, $r_1 = 0.277\,nm$, corresponds to the atomic diameter of palladium. The estimated $g(r)$ curves represent in the four-component alloy the weighted sum of the 10 partial functions, the $g_{ij}(r)$ as in Section 5.3.2 below for $i, j = 1, \ldots, 4$ with $1 = Pd, \ldots, 4 = P$. The peaks of the pair correlation functions of $Pd_{40}Cu_{30}Ni_{10}P_{20}$ at elevated temperature become broader with increasing temperature. The nearest-neighbour number N_1, which is obtained by integration from 0.20 to 0.35 nm over the first maximum in $g(r)$, is constant within the error limits. The value of $N_1 = 14.2$ is calculated for the glass at room temperature, and is 13.8 for the liquid at $T = 973\,K$. The split second maximum in $g(r)$ is also present in the melt with a reduced height of the first component. . . .

The broadening of the first maximum is asymmetric with increasing atomic pair fractions at the larger distance site. In the supercooled liquid state, additional changes occur in $g(r)$ with temperature. The asymmetry becomes much more extended, and the height of first component of the second maximum at r_2 decreases more distinctively. . . .

The distances of the second split maximum at r_2 and r_3 stay nearly constant in the whole temperature range. The distances of the higher coordination shells $r_4 - r_7$ increase with a thermal expansion coefficient comparable to the macroscopic dilatometer measurements.

Another thorough discussion of pair correlation functions and the structural information they give can be found in Shepilov et al. (2007), also in the context of materials science. Interesting medical applications are discussed in Mattfeldt (2005).

4.4 Higher-order and topological characteristics

4.4.1 Introduction

In most analyses of non-marked stationary point processes, the summary characteristics discussed above are sufficient for characterisation, parameter estimation or model testing. The nature of second-order characteristics on the one hand and nearest-neighbour and morphological characteristics on the other hand is so different that results obtained with summary characteristics from one class can very well be checked by means of summary characteristics from the other class. Nevertheless, this section presents further characteristics, which have specific applications and scope beyond the possibilities of the characteristics that have been considered so far. The main aim pursued with these additional characteristics is to find finer structural differences among samples which look rather similar at first sight.

The characteristics considered here comprise two groups of approaches. The exposition starts with third-order characteristics as a natural follow-on from second-order characteristics. These characteristics can reveal distributional differences in point patterns more easily, more clearly and in a qualitatively different way.

The second group of characteristics is based on Voronoi tessellations as introduced in Section 1.8.4, constructed with respect to the points of the pattern. The corresponding cells and even more the corresponding dual Delaunay simplices may be used to draw valuable conclusions on the point distribution, in particular on fine formations within local clustering or regularity such as points forming lines.

These properties can also be revealed using topological characteristics of the secondary structures as introduced in Section 1.8. As an example the random set X_r is considered which consists of the union of discs (spheres) of (fixed) radius r centred at the points of N (see Sections 1.8.2 and 4.2.5).

4.4.2 Third-order characteristics

Product densities

Theoretically, bearing in mind the success of second-order product densities and pair correlation functions, the most natural third-order characteristic of a point process N is its third-order product density $\varrho^{(3)}(x_1, x_2, x_3)$. However this implies dealing with functions of three d-dimensional variables, which are difficult to interpret and estimate.

In the case of stationarity the description simplifies slightly, similar to ϱ, such that $\varrho^{(3)}$ depends only on two d-dimensional vectors k_2 and k_3, defined as $k_2 = x_2 - x_1$

and $k_3 = x_3 - x_1$. In the planar isotropic case the function still depends on three real variables, the lengths of k_2 and k_3 and the angle between them. As a result $\varrho^{(3)}$ is usually not applied in point process statistics because it is considered too complex. Aggregated or integrated characteristics are used instead, see e.g. Stillinger et al. (2000).

Number of *r*-close triplets

Schladitz and Baddeley (2000) introduced the summary characteristic $T(r)$ given by

$$T(r) = \frac{1}{2\lambda^2} \mathbf{E}_o \left(\sum_{x_1, x_2 \in N \cap b(o,r)}^{\neq} \mathbf{1}(\|x_1 - x_2\| \le r) \right) \qquad \text{for } r \ge 0 \qquad (4.4.1)$$

and investigated its properties. It considers triplets of points with inter-point distances smaller than r consisting of pairs of points along with the typical point. The Palm mean $\mathbf{E}_o(\cdot)$ can be interpreted as the mean number of pairs of points of an inter-point distance larger than 0 and smaller than r in the sphere (disc) $b(o, r)$ provided that the typical point of N is at the origin o, while $x_1 \neq o$ and $x_2 \neq o$.

By means of the Campbell–Mecke formula one can show that $T(r)$ is indeed a third-order characteristic having the form of a mean similar to the mean in (1.5.16) with $k = 3$.

For a Poisson process, $T(r)$ can be calculated analytically as

$$T(r) = \frac{\pi}{2} \left(\pi - \frac{3}{4}\sqrt{3} \right) r^4 \qquad \text{for } r \ge 0$$

in the planar case, and

$$T(r) = \frac{5}{12}\pi^2 r^6 \qquad \text{for } r \ge 0$$

in the three-dimensional case An unbiased estimator of $\lambda^3 T(r)$, which resembles $\kappa_{st}(b(o, r))$ in (4.3.27) and was given in Schladitz and Baddeley (2000), is

$$\hat{\tau}_{st}(r) = \frac{1}{2} \sum_{x_1, x_2, x_3 \in W}^{\neq} \mathbf{1}(\|x_1 - x_2\| \le r)\mathbf{1}(\|x_1 - x_3\| \le r)$$

$$\times \mathbf{1}(\|x_2 - x_3\| \le r) \frac{\mathbf{1}(x_1, x_2, x_3)}{\nu(W_{x_1} \cap W_{x_2} \cap W_{x_3})}, \qquad (4.4.2)$$

where $\mathbf{1}(x_1, x_2, x_3) = 1$ if $W_{x_1} \cap W_{x_2} \cap W_{x_3} \neq \emptyset$, summing over all triples of different points $(x_1 \neq x_2, x_2 \neq x_3, x_1 \neq x_3)$. The admissible r-values are those for which $\mathbf{1}(x_1, x_2, x_3) = 1$ for all x_1, x_2 and x_3 with $\mathbf{1}(\|x_1 - x_2\| \le r)$, $\mathbf{1}(\|x_1 - x_3\| \le r)$ and $\mathbf{1}(\|x_2 - x_3\| \le r)$, $r \le r_\tau$. Note that $W_{x_i} = W + x_i = \{y : y = w + x_i \text{ with } w \in W\}$. The condition $r \le r_\tau$ is satisfied for a rectangular/parallelepipedal W if r is smaller than the side lengths.

Then $T(r)$ can be estimated by

$$\hat{T}(r) = \hat{\tau}_{st}(r) \Big/ \left(\frac{n(n-1)(n-2)}{\nu(W)^3} \right) \qquad \text{for } 0 \le r \le r_{\tau}, \qquad (4.4.3)$$

where $n = N(W)$ is the number of points of N in W. Clearly, the denominator in (4.4.3) is an estimator of λ^3. Just like $L(r)$, $T(r)$ is monotonically increasing and takes larger values with increasing variability.

Schladitz and Baddeley (2000) report that $T(r)$ is highly suitable for detecting clustering and Lochmann et al. (2006a) show that $T(r)$ is a powerful characteristic for indicating fine structural differences in regular patterns, particularly the sphere centres in random dense sphere packings. The following example shows that $T(r)$ may also be used to detect slight deviations of irregular patterns from the Poisson process with the same intensity.

Example 4.11.   Comparing a tree pattern with a Poisson process
This example discusses the pattern shown in Figure 5.4, ignoring the marks. It was discussed in this way in Mecke and Stoyan (2005) and the CSR hypothesis was rejected; the p-value for the L-test is between 0.01 and 0.05.

Figure 4.29 shows the corresponding $\hat{T}(r)$ compared to $T(r)$ for a Poisson process. One may safely regard the pattern as non-CSR as $T(r)$ is partly outside the envelopes obtained by 99 simulations of binomial processes. This example suggests the idea of constructing a CSR test based on $T(r)$.

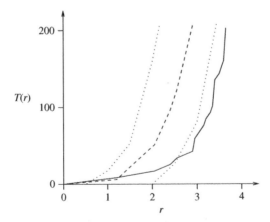

Figure 4.29 $\hat{T}(r)$ for the tree pattern of Figure 5.4 (solid line) compared to the theoretical $T(r)$ for a Poisson process (dashed line) and envelopes from 99 binomial process simulations (dotted lines).

4.4.3 Delaunay tessellation characteristics

Tessellation methods have turned out to be very successful in the statistical characterisation of point processes. This is true both for highly clustered patterns such as systems of galaxies (see Weygaert, 1994; Weygaert and Schaap, 2008) and for patterns of a high degree of regularity. An important application concerning the latter case are systems of centres of spheres in random dense packings of identical hard spheres. For this case, Medvedev and Naberukhin (1987), Naberukhin et al. (1991) and Medvedev (2000) develop the *tessellation method*, which, in essence, characterises a point pattern based on the shapes of the simplices within a tessellation.

This method initially finds the Voronoi tessellation for the point pattern of interest. Then the corresponding Delaunay tessellation is constructed, i.e. a system of simplices (tetrahedra; triangles in the planar case) with vertices at the points of the pattern including each one of the vertices of the Voronoi tessellation. The shapes of these simplices are the main concern here and, in particular, their deviation from regular tetrahedra and regular quartoctahedra. (A regular quartoctahedron is a tetrahedron which forms a regular octahedron together with three other congruent quartoctahedra. It has five edges of equal length and a further edge which is $\sqrt{2}$ times as long. These tetrahedra appear in so-called fcc and hcp lattice systems.)

Suitable shape characteristics for single simplices in the three-dimensional case are Medvedev's shape parameters

$$T = \sum_{i<j}(l_i - l_j)^2 \Big/ \left(15\bar{l}^2\right) \tag{4.4.4}$$

and

$$Q = \left(\sum_{i<j}^{(m)}(l_i - l_j)^2 + \sum_i^{(m)}\left(l_i - l_m/\sqrt{2}\right)^2\right) \Big/ \left(15\bar{l}^2\right), \tag{4.4.5}$$

where the l_i and l_j are the lengths of the edges of the simplex, m is the index of the longest edge and $\bar{l}$ the mean edge length. The '(m)' on the summation symbols means that the summation is only over those indices that are different from m. Finally, 15 is the number of pairs of edges in the simplex.

The parameter T characterises deviations from regular tetrahedra, for which $T = 0$, and Q deviations from regular quartoctahedra, for which $Q = 0$. In this way, the tetrahedra are assigned numerical values which can be used as constructed marks for the vertices of the Voronoi tessellation. As it turns out, this vertice-related approach may be used to find more subtle structural differences than methods which mark the original points with characteristics of the corresponding Voronoi cells, as for example in Stillinger et al. (2000). Indeed, the shape of Voronoi cells is influenced by many points (around 40 in the three-dimensional case as discussed here) and it is clearly difficult to describe their shape since the numbers of faces and edges vary.

Delaunay simplices are determined by only four points and are thus very sensitive to small local fluctuations.

The resulting marked point process can be analysed statistically; in particular, the mark distribution may be estimated and the neighbourhoods of tetrahedra with extreme shapes may be considered; see Lochmann et al. (2006a, 2006b). Anikeenko et al. (2006) extend this approach using methods from statistical shape analysis.

Example 4.12.   Characterisation of two sphere packings
Two random packings of identical spheres, with volume fractions 0.637 and 0.668, were analysed using Delaunay simplices. While the first system can be considered a random dense packing, the second is slightly more regular and shows the beginnings of crystallisation. This is well expressed by T and Q, as demonstrated by the corresponding mark distributions shown in Figure 4.30. In the 0.668 system the indices are more similar to the values corresponding to octahedra.

4.4.4   The connectivity function

The connectivity function $c(r)$ may be considered an example of a topological characteristic which is related to the Euler function $n(r)$ introduced in Section 4.2.5. Like the Euler function, it is based on the set X_R that was used in Section 1.8.2 as an example of a random set associated with a point process N as follows: a disc (sphere) of radius R is assigned to each point x in N; the union of all these discs (spheres) is X_R. (Note that, in order to avoid confusing notation, the radius is denoted here by R, while r, the argument of $c(r)$, denotes distance.)

If the radius R is not too large (below the so-called percolation threshold; see Stoyan et al., 1995, p. 74), the set X_R consists of bounded disjoint components.

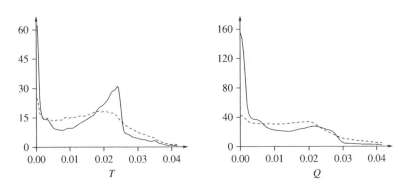

Figure 4.30   Distributions of the T- and Q-marks for the 0.637 (dashed lines) and 0.668 (solid lines) packings. With kind permission of Springer Science and Business Media.

A 'component' is defined as a subset A of X_R such that there is a curve that lies completely within A and connects x_1 and x_2 for any two points x_1 and x_2 in A.

The connectivity function $c(r)$ is defined by

$$c(r) = \mathbf{P}_{or}(o \text{ and } \mathbf{r} \text{ are in the same component of } X_R), \qquad (4.4.6)$$

for $r \geq 0$. Here $\mathbf{P}_{or}(\cdot)$ denotes a two-point Palm probability, namely a probability provided that o is the typical point of N and $\mathbf{r}$ another point of N with $\|\mathbf{r}\| = r$.

A ratio-unbiased estimator of $c(r)$ is

$$\hat{c}(r) = \frac{\hat{\varrho}(o \sim \mathbf{r})}{\hat{\varrho}_{st}(r)}. \qquad (4.4.7)$$

Here $\hat{\varrho}_{st}(r)$ is the estimator of the second-order product density (4.3.29), while $\hat{\varrho}(o \sim \mathbf{r})$ is the following quantity in the planar case:

$$\hat{\varrho}(o \sim \mathbf{r}) = \frac{1}{2\pi r} \sum_{x_1, x_2 \in W}^{\neq} \frac{k(r - \|x_1 - x_2\|)}{\nu(W_{x_1} \cap W_{x_2})} \mathbf{1}(x_1 \sim x_2). \qquad (4.4.8)$$

The indicator $\mathbf{1}(x_1 \sim x_2)$ is equal to one if x_1 and x_2 are in the same component of X_R and zero otherwise. It is not a trivial matter to determine this indicator but the Hoshen–Kopelman algorithm may be used for this purpose. Since X_R is closely related to the sphere graph $G(N, 2R)$, methods from graph theory may also be used; see Babalievski (1998).

The window W and the number of points in N should be large since there are edge effects which cannot be corrected for by the term $\nu(W_{x_1} \cap W_{x_2})$, i.e. those resulting from the construction of X_R using only information from the interior of the window W.

Example 4.13. Gold particles: connectivity function
Figure 4.31 shows the empirical connectivity function $c(r)$ for the pattern of gold particles. The radius R was chosen as 7.5 lu, half the mean nearest-neighbour distance $\hat{m}_D$ of the point pattern. The function has the value 1 for small r since pairs of points of such distances are always in the same component. For larger r, $c(r)$ decreases more slowly than $c(r)$ for a Poisson process of equal intensity, providing an alternative indication of the short-range order of the pattern. However, the clustering of the pattern is not reflected by $c(r)$. Note that for systems of packed hard discs (with mutual contacts), $c(r) = 1$ for all r if R is larger than the disc radii.

Similar connectivity relationships may be studied in forestry or ecology, where the points may be tree positions and the discs of radius R and X_R correspond to crowns of trees and the canopy of the forest, respectively. In this case, $x_1 \sim x_2$ means that the crowns of tree i and j are in contact.

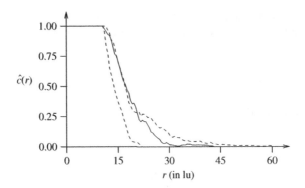

Figure 4.31   The connectivity function $\hat{c}(r)$ for the pattern of gold particles (solid line) in comparison with envelopes from 99 simulations of a Poisson process with the same intensity (dashed lines).

4.5 Orientation analysis for stationary point processes

4.5.1 Introduction

Anisotropy is a relevant issue in the analysis of point patterns since the spatial behaviour in a pattern often varies with direction. This type of behaviour is often closely related to and a result of the process that has led to the formation of a pattern and determines properties of the structure represented by the points, for example in applications in materials science. Figures 4.32 and 4.33 show typical anisotropic point patterns.

Figure 4.32 shows the centres of 573 carbide (Fe_3C) particles in rolled steel in planar section. Here, the anisotropy is the result of the cold rolling process, which leads to the formation of bands of particles in the direction of rolling. In other samples of this material the band structure is often even more pronounced; the sample discussed here was selected in order to make the statistical analysis more interesting. Wiencek (2000) and Wiencek and Satora (1999) systematically studied patterns of this type and analysed them by point process and stereological methods, assuming that the carbide particles are spherical.

As a further example, Figure 4.33 shows the results from a biochemical study. Lachmanovich et al. (2003) studied the distribution of proteins on the surface of cells. The figure shows the distribution of the positions of two mutants of influenza haemagglutinin protein that do or do not associate into mutual oligomers via binding to adaptor complexes. The two protein types differ in their fluorescence intensity and appear in electron microscopic images in green and red.

Visual inspection clearly indicates an anisotropy in the point pattern, mainly in the north-east direction. (Not all patterns of this type show such a clear anisotropy.) Below, an orientation analysis of the pattern is carried out ignoring the marks, i.e.

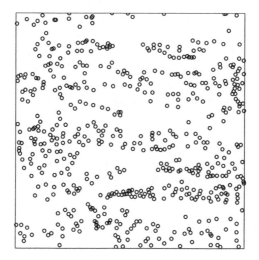

Figure 4.32   A planar section of a sample of rolled steel with 573 centres of carbide particles in a $100 \times 100\,\mu$m square. Data courtesy of K. Wiencek.

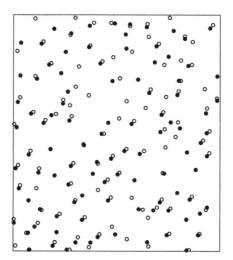

Figure 4.33   Positions of proteins on the surface of a cell in a $107 \times 119\,\mu$m rectangle. The colours discussed in Lachmanovich et al. (2003) are shown as • (green) and ∘ (red). Data courtesy of A. Weiss.

the interaction between the proteins, which was the main objective in Lachmanovich et al. (2003) who used the methods of Section 5.2.5 for qualitative marks.

The following section presents *non-parametric methods* for the statistical analysis of anisotropy in stationary point processes, as introduced in Ohser and Stoyan (1981), Hanisch and Stoyan (1984), Penttinen and Stoyan (1989) and Stoyan and Beneš (1991). This means that anisotropy is not considered in the most general case, but only for 'tame' (stationary) patterns similar to those presented in Figures 4.32 and 4.33. These patterns are regarded here as pieces of larger, translation-invariant structures.

It might seem rather strange that anisotropic processes play only a minor role in this book and in current point process statistics in general. One might perhaps expect that stationary point patterns are typically anisotropic, and isotropic patterns are exceptional cases. However, statisticians prefer to sample in situations where isotropy can be reasonably assumed:

- plants in regions of homogeneous growth conditions,

- cells in the interior of tissues,

- material structures far away from specimen boundaries.

The reason for this is that studies of these 'isotropic' samples typically aim to investigate small-scale relationships and local interactions between the points (trees, plants, cells, particles) rather than a trend or more global irregularities.

4.5.2   Nearest-neighbour orientation distribution

The methods introduced here explore the distribution of the orientation of the undirected line connecting the typical point with its nearest neighbour. Essentially, constructed marks are considered here, and the following describes the estimation of the mark distribution. The discussion focuses on the planar case, where the angle β with respect to the horizontal direction, i.e. the direction of the x-axis, is considered and measured in radians. Thus $0 \leq \beta \leq \pi$, which corresponds to $0° \leq \beta \leq 180°$. Because β is an orientation, values of β close to 0 and π are considered similar. Therefore, a distribution function such as

$$\mathbf{P}_o(\beta \leq t) \qquad \text{for } 0 \leq t \leq \pi$$

is not very natural. Thus, this book prefers to consider the corresponding probability density function $\vartheta(t)$, which satisfies

$$\mathbf{P}_o(\beta \leq t) = \int\limits_0^t \vartheta(u)\mathrm{d}u$$

and

$$\mathbf{P}_o(t_1 \le \beta \le t_2) = \int_{t_1}^{t_2} \vartheta(u)\mathrm{d}u. \tag{4.5.1}$$

The quantity $\mathbf{P}_o(t_1 \le \beta \le t_2)$ is the proportion of the connecting lines with angles of orientation between t_1 and t_2. In the isotropic case $\vartheta(t)$ is constant,

$$\vartheta(t) \equiv \frac{1}{\pi}. \tag{4.5.2}$$

Figure 4.34 shows the nearest-neighbour orientation density functions ϑ for the patterns in Figures 4.32 and 4.33. The 'north-east' anisotropy of the protein pattern is clearly expressed by the curve with maximum at $t = 45°$. Perhaps surprisingly, there is no indication of a similar anisotropy for the carbide pattern. Apparently, in this pattern anisotropy is a global property which cannot be sufficiently visualised by the nearest-neighbour orientations, which are rather 'short-sighted'. By the way, $\overline{R}_4$ also fails to indicate global anisotropy: the empirical values are 1.81 for the carbides and 1.80 for the proteins. The figure also shows $\hat{\vartheta}(t)$ for the oak and beech pattern of Figure 5.4. It may be regarded as an isotropic pattern, but there is some mysterious preference for orientations around $1.92\,\mathrm{rad} = 110°$.

Statistical estimation of $\vartheta(t)$

The density function $\vartheta(t)$ is estimated by a kernel estimator, and ratio-unbiasedness is obtained by nearest-neighbour edge-correction:

$$\hat{\vartheta}(t) = \sum_{i=1}^{n} \frac{\mathbf{1}(d_i \le e_i)k(t - \beta_i)}{\nu(W_{\ominus d_i})} \bigg/ \hat{\lambda}_{nn}. \tag{4.5.3}$$

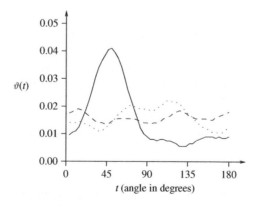

Figure 4.34   Nearest-neighbour orientation density functions for proteins (solid line), carbides (dashed line), and trees (dotted line). The bandwidth in each case is $h = 0.35\,\mathrm{rad} = 20°$.

The estimator is analogous to the estimators in Section 4.2.6, where e_i, d_i and $v(W_{\ominus d_i})$ are explained. Here, β_i is the orientation of the line from the ith point to its nearest neighbour and k is the Epanechnikov kernel function with bandwidth h. Subtraction of angles is done modulo $180°$, i.e. if $t - \beta_i < 0°$ the value $180° - t + \beta_i$ is used.

Section 5.4.2 discusses the application of constructed orientation marks for the detection of inner orientations in isotropic patterns.

4.5.3   Second-order orientation analysis

The second-order orientation analysis is based on the reduced second-order moment measure $\mathcal{K}$. This might sound rather theoretical, but the idea of the main estimator is quite natural. Consider the sample of the orientations of all lines which connect pairs of points of an inter-point distance between r_1 and r_2 and determine the corresponding orientation distribution. The corresponding probability density is denoted by $\vartheta_{r_1 r_2}(t)$.

Before looking at the theoretical derivation and the estimator below, consider Figure 4.35 by way of an example. This shows the estimated density functions $\vartheta_{r_1 r_2}(t)$ for the carbides, proteins and trees considered above. In all three cases the values of r_1 and r_2 were found by experimentation, with the aim of finding a way to detect anisotropies as clearly as possible:

	r_1	r_2
carbides	$0\,\mu m$	$50\,\mu m$
proteins	$0.5\,\mu m$	$7.5\,\mu m$
trees	$1.0\,m$	$5.0\,m$

For the carbides, the second-order characteristic $\hat{\vartheta}_{r_1 r_2}(t)$ detects the anisotropy, with the maximum of $\hat{\vartheta}_{0,0.05}(t)$ around $t = 0$ and π. For the proteins, the result is similar to that of nearest-neighbour orientation, and for the oak and beech pattern the strange anisotropy found on p. 253 is again evident, again with some preference for directions around $1.92 = 110°$.

The estimator $\hat{\vartheta}_{r_1 r_2}(t)$ can be derived theoretically as follows. The reduced second-order moment measure $\mathcal{K}$ can be described by the function $K(r, \alpha)$,

$$K(r, \alpha) = \mathcal{K}(S(r, \alpha)),$$

where $S(r, \alpha)$ is the sector of radius r, centred at the origin and given by the angle α with respect to the x-axis (see Figure 4.36). For fixed r the ratio $K(r, \alpha)/K(r, \pi)$ is a distribution function that is similar to $\mathbf{P}_o(\beta \leq t)$ above. Again, rather than the distribution function, the corresponding density function $\vartheta_r(t)$ is considered. Such

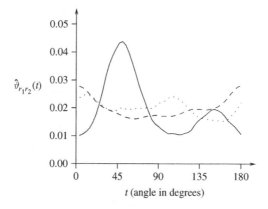

Figure 4.35   Second-order orientation density functions $\hat{\vartheta}_{r_1 r_2}(t)$ for the proteins (solid line), carbides (dashed line), and trees (dotted line). The bandwidth in each case is $0.35\,\text{rad} = 20°$.

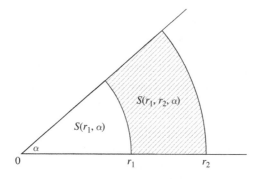

Figure 4.36   The sector $S(r_1, \alpha)$ and the sector ring $S(r_1, r_2, \alpha)$.

a density function was used for the carbide pattern, with $r_1 = 0$. By the way, the case of the four quadrant sectors, where the first (north-east) is the same as $S(r, 90°)$, is applied in Haase (2001) in an ecological context.

If the case with $r_1 > 0$ is of interest, as for the proteins and trees, the description may be refined by introducing two distances r_1 and r_2 $(0 \le r_1 < r_2)$ and considering the sector ring $S(r_1, r_2, \alpha)$ given by

$$S(r_1, r_2, \alpha) = S(r_2, \alpha) \setminus S(r_1, \alpha).$$

The corresponding analogue to $\vartheta_r(x)$ is denoted by $\vartheta_{r_1 r_2}(x)$. This density function describes the distribution of the random orientations of the lines connecting the typical point and other points at distances between r_1 and r_2.

The corresponding edge-corrected unbiased estimator is

$$\hat{\vartheta}_{r_1 r_2}(t) = \sum_{x_1, x_2 \in W}^{\neq} \frac{\mathbf{1}(r_1 \leq \|x_1 - x_2\| \leq r_2) k(t - \beta_{x_1 x_2})}{\nu(W_{x_1} \cap W_{x_2})}. \tag{4.5.4}$$

This estimator is related to (4.3.29); k is the Epanechnikov kernel function and $\beta_{x_1 x_2}$ the orientation angle of the line through the points x_1 and x_2.

Note that the estimator $\hat{\vartheta}_{r_1 r_2}(t)$ (and $\hat{\vartheta}(t)$ in (4.5.3)) deviate from correct density estimators by a constant factor of proportionality, which is irrelevant for directional analysis.

4.6   Outliers, gaps and residuals

4.6.1   Introduction

An important step in the statistical analysis of point patterns is the search for unusual points or unusual point configurations. This includes the issue of *outliers,* i.e. points appearing at locations where they are not expected according to the construction principles of the pattern. Similarly, it is possible that there are unusual *gaps* or *missing points* in the pattern, i.e. areas where, according to the general structure of the pattern, points would have been expected. Such outliers and missing values have to be detected. Both issues are addressed in the following two subsections using simple data-analytic methods, based on constructed marks.

In a more formal but also more powerful approach, both problems may be treated using point process models, more specifically Gibbs processes. Based on these processes it is possible to predict point positions or gaps given the point positions in the neighbourhood. By analogy with classical regression theory (which, in some sense, also predicts points) the term *residual* is used when differences between prediction and reality are considered.

Note that this section assumes that the reader is familiar with the general ideas of the theory of marked point processes; in particular, constructed marks (Section 5.1.3) are relevant in Section 4.6.2, and Gibbs processes (Section 6.6) in Section 4.6.4.

4.6.2   Simple outlier detection

The basic statistical idea for outlier detection is quite simple: assign numerical or functional marks to all points in the pattern, analyse these marks statistically and regard points with extreme marks as outliers. Examples of constructed marks for point x are: $d(x)$, the distance from x to its nearest neighbour; $v(x)$, the volume or area of the Voronoi cell with generating point x; and $E(x)$, the energy needed to add the point x to the pattern $N \setminus \{x\}$ if N is a Gibbs process. Functions may also serve as constructed marks: $K_x(r)$, the *individual K-function* for point x, given by

$$K_x(r) = N(b(x, r)) - 1 \qquad \text{for } r \geq 0 \text{ and } x \in N; \tag{4.6.1}$$

and $L_x(r)$, the *individual L-function* for point x, given by

$$L_x(r) = \sqrt{\frac{K_x(r)}{b_d}} \qquad \text{for } r \geq 0 \text{ and } x \in N. \qquad (4.6.2)$$

The corresponding variability and extremality analysis can be based on bundles of K_x- or L_x-functions, for all points in the pattern. These ideas go back to Getis and Franklin (1987), Doguwa (1989), Wartenberg (1990) and Stoyan and Grabarnik (1991b) and serve as 'local indicators of spatial association' (LISA); see Cressie and Collins (2001) and Dale et al. (2002).

4.6.3   Simple gap detection

The determination of gaps is more difficult than the detection of outliers. Tessellations and graphs may be successfully applied in this context.

A natural tessellation-related procedure, which can be carried out with existing software, is the procedure introduced by Medvedev (2000) in a different context. Construct the Voronoi tessellation for the point pattern and take its vertices as centres of discs (spheres). Determine the largest radius such that the corresponding disc (sphere) only touches points in the pattern for each centre. Discs (spheres) with very large radii may be interpreted as having resulted from gap areas perhaps containing missing points.

A procedure based on graphs uses k-neighbour graphs as introduced in Section 1.8.5. The edges of these graphs connect points that are close together and do not cross larger regions without points, thus indicating gaps. In this way polygonal (polyhedral) gap regions are constructed. The appropriate order of k may be determined by experimentation.

Example 4.14.   Gaps in a forest of Sitka spruce
Figure 4.37 shows a stand of 294 Sitka spruce trees (*Picea sitchensis*) from Clocaenog (Wales) in November 2003 in a 101×100 m rectangle. The pattern looks homogeneous, but there are some gaps and the question is whether or not these can be explained by normal fluctuations of tree density. It is known that in the stand there are small roads that are used to transport the timber and that some trees have fallen in a storm.

Figure 4.38 shows the 4-neighbour graph which indicates some large gaps, in particular that around the point (25, 65). This observation led to an *in situ* investigation, which did indeed find the remains of two fallen trees around this point.

4.6.4   Model-based outliers

This section assumes that an estimator $\hat{\lambda}(u|N)$ of the conditional intensity is available. This is usually a parametric estimator, based on a Gibbs process model as

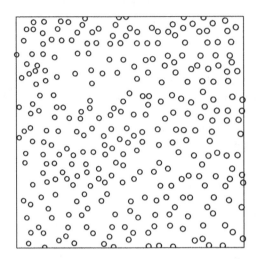

Figure 4.37   The positions of 294 Sitka spruce trees in the Clocaenog 4 stand, which is a Sitka spruce research plot, part of a larger forestry experiment at Clocaenog Forest in North Wales (UK). Data courtesy of University of Wales, Bangor (Arne Pommerening), and Forest Research (Forestry Commission). The authors are grateful to both organisations for permission to use these data.

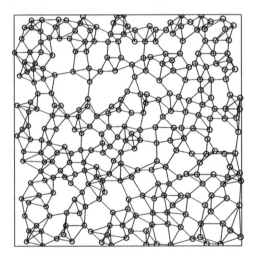

Figure 4.38   The 4-neighbour graph for the stand of Sitka spruce trees indicates some large gaps, one of which was caused by a past storm.

described in Sections 3.6 and 6.6, which is characterised by a parameter θ. The corresponding theoretical conditional intensity depends on θ and is denoted by $\lambda_\theta(u|N)$. The estimator has the form

$$\hat{\lambda}(u|N) = \lambda_{\hat{\theta}}(u|N),$$

where the estimator $\hat{\theta}$ has been plugged in.

The estimator $\lambda_{\hat{\theta}}(u|N)$ can be used to construct marks suitable for outlier detection. Two possibilities are:

$$m_1(x) = \lambda_{\hat{\theta}}(x|N\setminus\{x\}) \tag{4.6.3}$$

and

$$m_2(x) = 1/\left(\lambda_{\hat{\theta}}(x|N\setminus\{x\})\right). \tag{4.6.4}$$

Note that, as explained in p. 28, $\lambda(x|N\setminus\{x\})$ is proportional to the probability that the point process N has a point at x given all other points in N. Thus, large marks $m_1(x)$ confirm existing points, while small marks indicate uncertain, suspicious points. The meaning of 'large' and 'small' can be determined with reference to the corresponding mark distribution.

Mark $m_2(x)$ may be preferable since the fact that for a stationary point process the mean mark μ_2 corresponding to $m_2(x)$ is equal to 1 facilitates comparison; see Stoyan and Grabarnik (1991b) and Baddeley et al. (2005). The mark $m_2(x)$ can be interpreted as an 'exponential energy mark'; see p. 400 below.

4.6.5 Residuals

Two different types of residuals can be considered: residuals for the points of the process (point-related) and for all locations u in the window W (location-related).

Baddeley et al. (2005) introduce a series of residuals of the latter type. Here, only the simplest of these is covered, the so-called *raw residuals*; refer to Baddeley et al. (2005, 2007) for details on the other residuals.

Raw residuals describe the difference between the observed local point density and the expected point density using the conditional intensity $\lambda(x|N)$. The first term is discrete since it is based directly on the points of N, while the second term is continuous and given by an integral over $\lambda(x|N)$. For visualisation purposes it is better to smooth the first term, which leads to the function

$$s(x) = \int_W \lambda(u)k(x-u)\mathrm{d}u - \mathbf{E}\left(\int_W \lambda(u|N)k(x-u)\mathrm{d}u\right) \qquad \text{for } x \in W,$$

where $\lambda(u)$ is the intensity function and $k(z)$ some kernel function, e.g. a rotation-symmetric Gauss kernel (a probability density). The theoretical value of $s(x)$ is close to 0, but its statistical analogue $\hat{s}(x)$ is a valuable indicator of unusual point configurations:

$$\hat{s}(x) = \sum_{y \in W} k(x - y) - \int_W k(x - y)\lambda_{\hat{\theta}}(y|N)\mathrm{d}y \qquad \text{for } x \in W. \tag{4.6.5}$$

Positive values of $\hat{s}(x)$ suggest that the model, which is accounted for by $\lambda_{\hat{\theta}}(y|N)$, underestimates the local intensity and negative values indicate overestimation.

One may also consider point-related residuals. Similar to classical residuals in regression theory, these characterise the difference between the observed points and points expected according to the model. The residual r_i of point x_i is defined as

$$r_i = \hat{x}_i - x_i,$$

where $\hat{x}_i$ is constructed as follows. Consider the disc (sphere) of radius ϱ centred at x_i. Determine the location u in $b(x_i, \varrho)$ where the conditional intensity $\lambda(u|N \setminus \{x\})$ has its maximum; this is the most likely position of a point of N in $b(x_i, \varrho)$, given all points of N other than x_i. If this definition does not determine $\hat{x}_i$ uniquely, the point closest to x_i is chosen. The radius ϱ is a control parameter, which could be called the 'residual radius'. If the fitted model is appropriate, the length of the residual r_i is expected to be small, i.e. either $\hat{x}_i = x_i$ or $\hat{x}_i$ is close to x_i. Clearly, the residuals depend on the residual radius ϱ, the choice of which is often rather crucial. These residuals make sense for non-trivial soft-core processes with or without a hard core.

4.7 Replicated patterns

4.7.1 Introduction

Frequently, in particular in medical, histological or materials science studies, not only a single sample of a point process is available, but several patterns are considered at the same time with the aim of eventually analysing them all together to obtain information on some spatial behaviour that is reflected in them. Examples of this are series of micrographs of tissue sections or of planar sections through steel samples taken at different positions in some organ or block at large distances, or in organs of different animals or different blocks. Hence, the patterns can be considered independent samples. The aim is to determine general summary characteristics by aggregating the statistical results for the single samples. This section presents methods for performing this aggregation.

Assume that m observation windows $W_1, \ldots, W_m$ are given, which may be congruent, but which might also differ in shape and size. The patterns in the W_i are regarded as samples of i.i.d. stationary and isotropic point processes. The aim is to find estimates of their joint summary characteristics such as λ, K, g, D and H_s.

The usual approach is to estimate these characteristics separately for each of the windows W_i and to then aggregate the estimates. If the windows are congruent, simply forming *arithmetic means* is a good strategy. Those readers who do not wish to consider this issue any further might want to stop reading here and apply this simple method. For the interested reader the rest of this section presents refined methods, which also include the case of windows with different size and shape, and some theoretical justification.

4.7.2 Aggregation recipes

In the case of non-congruent windows no unique aggregation method that may be applied to all characteristics is available. The appropriate method depends on the specific summary characteristic and edge-correction method. This subsection details rules for the aggregation of some important summary characteristics.

Intensity

Let $\hat{\lambda}_1, \ldots, \hat{\lambda}_m$ be the estimators of λ for the windows $W_1, \ldots, W_m$ obtained by (4.2.10). Then the natural aggregated estimator is

$$\hat{\lambda} = \sum_{i=1}^{m} \hat{\lambda}_i \frac{\nu_i}{\nu}, \tag{4.7.1}$$

where $\nu_i = \nu(W_i)$ is the area (volume) of W_i and

$$\nu = \sum_{i=1}^{m} \nu_i. \tag{4.7.2}$$

Nearest-neighbour distance d.f. $D(r)$

Let $\hat{D}_1(r), \ldots, \hat{D}_m(r)$ be estimators of $D(r)$ for the windows $W_1, \ldots, W_m$ obtained by the border estimator (4.2.47) or the nearest-neighbour estimator (4.2.48). Then the aggregated estimator is

$$\hat{D}(r) = \sum_{i=1}^{m} \hat{D}_i(r) \frac{n_{r,i}}{n_r} \tag{4.7.3}$$

with $n_{r,i} = N(W_{i \ominus r})$, i.e. $n_{r,i}$ is the number of points in the reduced i-th window, and $n_r = \sum_{i=1}^{m} n_{r,i}$.

Spherical contact d.f. $H_s(r)$

Let $\hat{H}_{s,1}(r), \ldots, \hat{H}_{s,m}(r)$ be estimators of $H_s(r)$ for the windows $W_1, \ldots, W_m$ obtained by (4.2.38). Then the aggregated estimator is

$$\hat{H}_s(r) = \sum_{i=1}^{m} \hat{H}_{s,i}(r) \frac{\nu_i}{\nu}, \tag{4.7.4}$$

where $\nu_i = \nu(W_{i \ominus r})$ for $i = 1, \ldots, m$ and ν is defined by (4.7.2).

Ripley's K-function

Let $\hat{K}(r), \ldots, \hat{K}_m(r)$ be estimators of $K(r)$ for the windows $W_1, \ldots, W_m$ obtained by (4.3.35) and (4.3.36). Then the aggregated estimator is

$$\hat{K}(r) = \sum_{i=1}^{m} \hat{K}_i(r) \frac{n_i}{n}, \tag{4.7.5}$$

where $n_i = N(W_i)$ and $n = \sum_{i=1}^{m} n_i$.

Note that Baddeley et al. (1993) recommend the use of other weights, in particular $n_i^2 / \sum_{i=1}^{m} n_i^2$.

Pair correlation function $g(r)$

Let $\hat{g}_1(r), \ldots, \hat{g}_m(r)$ be estimators of $g(r)$ for the windows $W_1, \ldots, W_m$ obtained by (4.3.38) or (4.3.39). Then the aggregated estimator is

$$g(r) = \sum_{i=1}^{m} g_i(r) \frac{\gamma_i(r)}{\gamma(r)}, \tag{4.7.6}$$

with $\gamma_i(r) = \overline{\gamma}_{W_i}(r)$ for $i = 1, \ldots, m$ and $\gamma(r) = \sum_{i=1}^{m} \gamma_i(r)$. Here $\overline{\gamma}_{W_i}(r)$ is the isotropised set covariance of W_i.

Theoretical justification of the aggregation formulas

One theoretical approach to the aggregation formulas above is based on the following consideration: all windows W_i are regarded as subsets of $\mathbb{R}^2$ (or $\mathbb{R}^3$), and their set-theoretic *union*

$$W = \bigcup_{i=1}^{m} W_i$$

is regarded as the sampling window for the entire analysis ('between images within cases'). The distances between the W_i are assumed to be very large, such that the statistical results in the W_i may be considered mutually independent. Under this assumption, the general formulas in Section 4.2 and 4.3 are used to derive estimators for the union window W yielding (4.7.1), (4.7.3) and (4.7.4). This results in the following:

Intensity   For the union window W,

$$n = N(W) = \sum_{i=1}^{m} n_i = \sum_{i=1}^{m} N(W_i)$$

and

$$\hat{\lambda} = \frac{N(W)}{\nu(W)} = \frac{\sum_{i=1}^{m} n_i}{\sum_{i=1}^{m} \nu_i} = \sum_{i=1}^{m} \hat{\lambda}_i \frac{\nu_i}{\sum_{i=1}^{m} \nu_i}.$$

Nearest-neighbour distance d.f. $D(r)$ The border estimator in (4.2.47) becomes, for the union window,

$$\hat{D}(r) = \sum_{i=1}^{m} \sum_{[x; d(x)]} \mathbf{1}_{W_{i \ominus r}}(x) \mathbf{1}(0 < d(x) \le r) \Bigg/ \sum_{i=1}^{m} N(W_{i \ominus r}),$$

using

$$\left(\bigcup_{i=1}^{m} W_i \right) \ominus b(o, r) = \bigcup_{i=1}^{m} W_i \ominus b(o, r),$$

since for disjoint W_i,

$$N \left(\bigcup_{i=1}^{m} W_{i \ominus r} \right) = \sum_{i=1}^{m} N(W_{i \ominus r}),$$

yielding (4.7.3).

It is difficult to study analytically the behaviour of the nearest-neighbour estimator (4.2.48). Simulations reported in Stoyan (2006) and further simulations performed by Tscheschel (unpublished) confirm that (4.7.3) may also be applied in this case; the pooling method does not seem to have a substantial influence on the mse.

Spherical contact d.f. $H_s(r)$ Equation (4.2.38) yields, for the union estimator,

$$\hat{H}_s(r) = \frac{\nu \left(X_r \cap \bigcup_{i=1}^{m} W_{i \ominus r} \right)}{\nu \left(\bigcup_{i=1}^{m} W_{i \ominus r} \right)} = \frac{\sum_{i=1}^{m} \nu(X_r \cap W_{i \ominus r})}{\sum_{i=1}^{m} \nu(W_{i \ominus r})},$$

which leads directly to (4.7.4).

Ripley's K-function   Diggle (2003, p. 123) recommends (4.7.5) based on simulations in Diggle et al. (1991). Simulations run by the present authors have also shown that (4.7.5) is suitable.

Pair correlation function   For the case of Ohser's estimator of the second-order product density $\varrho(r)$ calculations similar to those for $\hat{H}_s(r)$ can be done, exploiting the fact that here

$$\overline{\gamma}_W(r) = \sum_{i=1}^{m} \overline{\gamma}_{W_i}(r).$$

This leads to the weights $\gamma_i(r)/\gamma(r)$ as in (4.7.6); see also Ohser and Mücklich (2000). This argument probably also applies to the other pair correlation function estimators. It is important to use the local intensity estimators $\hat{\lambda}_i$ rather than $\hat{\lambda}$.

4.8   Choosing appropriate observation windows

4.8.1   General ideas

This section discusses an issue which is highly relevant in applications since the choice of the appropriate size of the observation window W has, of course, a fundamental role in point process statistics. In general, the rule 'the larger the window the better the statistical results' clearly holds. However, in practical applications only a limited amount of financial resources for data collection and time are available and not every potential window may be accessible. Nevertheless, to arrive at any useful conclusion in an analysis of a stationary spatial point process, the window size should be chosen large enough to ensure that all essential information is contained in the window. For instance, observing a pattern consisting of only two clusters makes little sense in the stationary process approach. Windows that are too small may result in misleading information. In some cases the measurement conditions determine the window's shape, i.e. windows have to be chosen which deviate from the standard rectangular and circular shapes or their three-dimensional analogues. This is the case in ecology (see Lancaster and Downes, 2004; Wiegand and Moloney, 2004) and astronomy (see Hamilton, 1993), where, for example, polygonal and complicated three-dimensional windows are used.

If the shape of the window can be chosen freely and there is no restriction on resources for collecting data, circular or square windows are recommended in the planar case. Data collected within long rectangular windows are less informative: they enable the observation of (some) very long inter-point distances, which might be of interest in specific applications, but there are serious problems with edge effects.

If there are restrictions on the choice of W, the issue of identifying 'representative' windows arises. In this context 'representative' means 'large enough to satisfy predefined precision requirements'. This section reduces the issue of determining a window to that of determining its area or volume $\nu(W)$ and discusses it in some detail below.

Finally, another tricky problem may arise, which is particularly relevant in geological and ecological studies. Patterns investigated in these fields are commonly not stationary at all (in truth, they are samples of finite point processes). However, it still makes sense to apply methods for stationary processes if the analysis mainly focuses on short-range interactions. As an example, consider the pattern formed by the locations of mammals in a meadow. Assume the aim of a study is to gain knowledge about the short-range interaction among the individuals, assuming that this interaction would be the same in a larger, stationary pattern. (A similar situation arises in a system of volcanos in some region, which is finite and controlled by some interesting geological interaction.) In these cases, it is useful to adapt the window's size and shape to the pattern in some appropriate way, in order to avoid a potentially large empty boundary zone around the pattern. An example of window adaption is discussed towards the end of this section.

4.8.2   Representative windows

This section discusses the issue of choosing an appropriate window size by ensuring that predefined precision requirements are satisfied or that the window is large enough to ensure that the summary characteristics of interest achieve their asymptotic values in the window. The terminology refers to the three-dimensional case but the methodology, of course, applies equally to two-dimensional data. In this context the term *representative volume element* (RVE) for three-dimensional data is commonly used for the representative window; in the 2D case the word 'volume' may simply be replaced by 'area' and thus *representative area elements* would be considered.

It is well known that the RVE depends both on the variability of the point process of interest and on the summary characteristic considered (see Freudenthal, 1950, for the case of applications in materials science). If the same summary statistic is considered, larger RVEs are necessary when cluster point processes are analysed, as opposed to regular processes were smaller RVEs are sufficient. On the other hand, for a precise estimation of the pair correlation function a larger RVE is necessary than for intensity estimation, say.

In classical statistics, sample size calculations require some prior knowledge of the nature of the data that are analysed, such as their variation. In the context of spatial point processes this is similar. It is impossible to determine the RVE without any a priori knowledge of the distribution of the point process investigated. A straightforward approach to acquiring a priori knowledge is a *pilot study* consisting of a preliminary statistical analysis of a small window, or a small number of windows if a series of windows has to be analysed. The expectation is that the pilot study yields a useful yet rough estimate of the intensity λ and fundamental information on the point process type, i.e. whether it is a regular or a clustered pattern. Based on this, the methods sketched below can be used to obtain rough estimates of RVEs corresponding to the intensity λ and the pair correlation function $g(r)$. For the other summary characteristics similar methods can be used.

Alternative approaches use (a) statistical experiments or (b) reconstruction simulations for the determination of RVEs. Method (a) assumes that it is possible to generate a series of nested windows of increasing size. The summary characteristics of interest are estimated for these in order to identify the minimum window size that can be considered representative, such that any further increase in the window size will essentially not change the statistical results further. Method (b) generates point patterns larger than the original pattern by the reconstruction simulation method described in Section 6.7, assuming that the pattern given in the single original observation window shows the typical process behaviour. Subsequently, method (a) can be applied to the simulated patterns.

Determination of RVE for intensity λ

The aim is to determine a window size $v(W)$ which ensures that the estimation variance $\mathbf{var}(\hat{\lambda})$ of the standard intensity estimator $\hat{\lambda}$ is smaller than a predefined fixed value. In the isotropic case this variance is given by (4.2.11) and (4.3.24) as

$$\mathbf{var}(\hat{\lambda}) = \left(db_d \lambda^2 \int_0^\infty \overline{\gamma}_W(r)g(r)r^{d-1}dr + \lambda v(W) - (\lambda v(W))^2 \right) \Big/ (v(W))^2. \quad (4.8.1)$$

It depends on the intensity λ, the pair correlation function $g(r)$, the volume (area) of window $v(W)$ and the window's isotropised set covariance $\overline{\gamma}_W(r)$, i.e. in some way on the window's shape. However, in most applications $g(r)$ is unknown, therefore it is not very realistic to predict $\mathbf{var}(\hat{\lambda})$ in this way. Also, the integral may be too complicated (but see the approximations given on p. 226). A simpler approach is to determine an upper bound on $\mathbf{var}(\hat{\lambda})$, i.e. to accept a window W that is slightly too large. If the point process is more regular than a Poisson process, the Poisson process variance $\mathbf{var}(\hat{\lambda}_P)$ can be used as such a bound. Equation (4.2.12) yields

$$\mathbf{var}(\hat{\lambda}_P) = \frac{\lambda}{v(W)}. \quad (4.8.2)$$

This approximation requires only a priori knowledge of λ.

Example 4.15.   Gold particles: intensity RVE
Consider again the point pattern of gold particles and regard the sample as coming from a pilot study. In the window with an area of $v(W) = 252\,000\,\text{lu}^2$ the estimated λ is $0.000\,865\,\text{lu}^{-2}$. It is difficult to classify this pattern: there is a positive hard-core distance, but at a larger spatial scale the points form clusters. The pair correlation function in Figure 4.20 is already close to one for $r \geq 30\,\text{lu}$, while the maximum at $r = 10\,\text{lu}$ may be compensated for by the zero values for $r \leq 5\,\text{lu}$ and the smaller values for $12\,\text{lu} \leq r \leq 30\,\text{lu}$ in the integral in (4.8.1). Hence it may be an acceptable approximation to use the Poisson process formula (4.8.2).

Assume that a relative error of 10 % in intensity estimation is deemed acceptable, i.e.

$$\frac{\sqrt{\mathbf{var}(\hat{\lambda}_P)}}{E\hat{\lambda}} \leq 0.1.$$

Applying (4.8.2) for $\mathbf{var}(\hat{\lambda}_P)$ leads to the approximation

$$\sqrt{v(W)} \geq \frac{1}{0.1 \cdot \lambda},$$

and the estimate $\hat{\lambda} = 0.000\,865$ yields

$$\nu(W) \geq 115\,606\,\text{lu}^2.$$

Hence, in order to satisfy the intended target precision of the intensity estimation a window with an area of $115\,606\,\text{lu}^2$ would have been sufficient, i.e. a square window with side length $340\,\text{lu}$, less than half the size of the original 680×400 window, which was, of course, used to obtain more than just only λ.

Determination of RVE for the pair correlation function $g(r)$

Here, the aim is to determine a window size which ensures that the estimation variance of the estimator $\hat{g}_S(r)$ (equation (4.3.41)) is smaller than a predefined target value. For the box kernel estimator with bandwidth h, the variance is approximately given by

$$\mathbf{var}(\hat{g}_S(r)) = \frac{g(r)}{db_d r^{d-1}\overline{\gamma}_W(r)h\lambda^2}; \qquad (4.8.3)$$

see p. 234. The isotropised set covariance of W, $\overline{\gamma}_W(r)$, accounts for window size and shape.

Since the variance depends on r, several values of r can be used to determine the RVE. The authors recommend choosing an r close to r_{corr}, the range of correlation, where $g(r) = 1$ for $r \geq r_{\text{corr}}$.

Example 4.16.   Gold particles: pair correlation function RVE
Consider again the pattern of gold particles. Based on the recommendations on p. 236, the bandwidth h is chosen to be $3\,\text{lu}$. The value $60\,\text{lu}$ is identified as r_{corr}.
Assume a target standard deviation of $\sqrt{\mathbf{var}(\hat{g}_S(r))} = 0.1$ and aim to determine the side length a of a square window W. Then

$$\overline{\gamma}_W(r) = a^2 \left(1 - \frac{4}{\pi}\frac{r}{a} + \frac{1}{\pi}\left(\frac{r}{a}\right)^2 \right) \qquad \text{for } r \leq a.$$

By (4.8.3) with $d = 2$ and $r = 60\,\text{lu}$, this yields the value $a = 282\,\text{lu}$, i.e. the RVE is a square of side length $282\,\text{lu}$. Note that a is clearly larger than r_{corr}.

Finding the appropriate RVE is also relevant in the context of marked point processes. This concerns, for example, the minimum area which guarantees that with high probability all (discrete) marks are observed in W. A simple solution to this problem is given in Pfeifer et al. (1996) with an ecological interpretation, where the marks represent species.

Adapting windows to data

As mentioned above, in practical applications the window often has to be adapted to the data. The following example uses simulated data to explain the procedure and to show that it is not a good idea to use a 'naive stationarity approach', i.e. to not adapt the window to the pattern.

Example 4.17.   Circular cluster

Consider the pattern in Figure 4.39. In a practical application, the points may perhaps be animals and the whole pattern a herd. Assume that one wishes to analyse the pattern using methods for stationary point processes in order to obtain information on the short-range repulsion or attraction between the points (or animals). Technically, Figure 4.39 shows a simulated pattern of 78 points which form a circular sample of a Matérn hard-core process with $r_0 = 0.04$ as introduced in Section 6.5.2.

Using the 'naive stationarity approach', one might now simply take the data as they are as points in a square window and use standard software to estimate the pair correlation function. The result is displayed in Figure 4.40. Based on the recommendation in Section 4.3.4 this would result in the following interpretation: there is a hard-core distance of $r_0 = 0.04$ and some short-range regularity with clustering at a larger scale. This is not a very satisfying result as the estimate of the pair correlation function apparently exhibits strange behaviour. It seems not to reach the limit of 1 for large r, but $\hat{g}(r)$ even approaches values larger than 1 for $r > 0.5$.

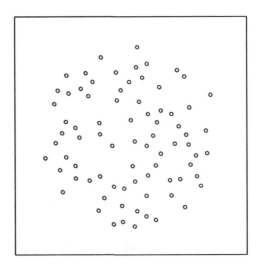

Figure 4.39   A cluster of 78 points in $\boxed{1}$. The statistical analysis of the pattern as a sample from a stationary point process in a square window produces misleading results.

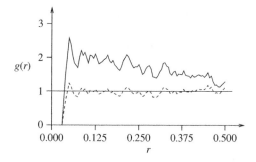

Figure 4.40 Pair correlation functions for the points in Figure 4.39. The first function (solid line) which was obtained for the square window, looks like a pair correlation function of a combination of a hard-core and a cluster process; the second (dashed line), which corresponds to an adapted circular window, looks like the pair correlation function of a hard-core process.

Clearly, both the spurious clustering information and the behaviour of the pair correlation function for large r are a result of a bad choice of window. If one is really interested in information on short-range behaviour, it is recommended to use a smaller, adapted window. The pattern in Figure 4.39 suggests the use of a circular subwindow. For simplicity the disc of radius 0.4 that was used in the simulation is chosen as W in this example. The pair correlation function estimate for the circular window is much better: it does not indicate clustering and tends to 1 for large r.

Note, by the way, that the nearest-neighbour distance d.f. behaves differently and is not affected as much by an inappropriate window choice, such that it is not necessary to adapt the window size. Compare this with the analysis of centres of herds of herbivores in Kenya in Stein and Georgiadis (2006).

One large window or several smaller ones?

Another question is relevant in the context of this section: is it better to use one (or a small number of) large window(s) or some (many) smaller windows, with the same total area (volume) in both cases? Note that the answer to this question still assumes that the point process considered is stationary and isotropic and that the observations in different windows are independent. When a large natural pattern is sampled with strong fluctuations and possible deviations from stationarity it may be preferable a priori to use some smaller windows in order to balance out these fluctuations.

Of course, the point process distribution itself has an impact on the choice of the number and size of the windows. For instance, if it exhibits extreme fluctuations (such as large clusters) large windows are *necessary*. The following example may indicate a tendency in a simple case.

Example 4.18.   Estimation of standard deviations for $\hat{K}_i(r)$ and $\hat{g}_{st}(r)$ for a Poisson process

Consider a Poisson process with intensity $\lambda = 100$ and compare the estimated standard deviations of $\hat{K}_i(r)$ and $\hat{g}_{st}(r)$ (using a bandwidth of $h = 0.01$) for $r = 0.25$ for three cases:

- two unit squares,
- one square with side length $\sqrt{2}$,
- one rectangle with side lengths 1 and 2.

Equations (4.3.40) and (4.3.42) may be used to calculate the estimated variances $\sigma_K^2(0.25)$ and $\sigma_g^2(0.25)$. In the case of two unit squares the resulting value is divided by 2, corresponding to the averaging of the estimates.

The results are presented in Table 4.1. The theoretical values of $K(0.25)$ and $g(0.25)$ are $K(0.25) = 0.196$ and $g(0.25) = 1$. In all cases the $\sqrt{2} \times \sqrt{2}$ square is the best solution, probably since problems with edge effects are less pronounced here. It is typical that the accuracy of estimation of $K(r)$ is better than that of $g(r)$.

Table 4.1   Estimated standard deviations $\sigma_K(0.25)$ and $\sigma_g(0.25)$ for Example 4.18.

Window(s)	$\sigma_K(0.25)$	$\sigma_g(0.25)$
two 1×1	0.00554	0.06735
$\sqrt{2} \times \sqrt{2}$	0.00524	0.06368
1×2	0.00529	0.06424

4.9   Multivariate analysis of series of point patterns

In applications of point process statistics, one sometimes wishes to analyse several point patterns of a similar nature or origin simultaneously. These patterns are usually given in the same or in congruent windows. This yields a situation with an objective similar to applications of multivariate statistics in classical statistics. That is, a classification of the patterns is sought which identifies groups (or clusters[3]) of similar patterns. In addition, one may relate the grouping structure to characteristics of the objects that form the patterns.

---

[3] Note that in this section, in accordance with conventions in classical statistics the term 'cluster' exclusively refers to a group of similar patterns whereas what has been called a 'clustered' pattern elsewhere in the book will be referred to as 'aggregated' and a point cluster as a 'clump' in order to avoid confusion in terminology.

As an illustration, think of an example in materials science, where a number of point patterns derived from different materials are analysed. One may then wish to form groups of similar point patterns with the intention of revealing that the patterns originate from similar (potentially unknown) material structure groups. This may lead to further research. In this situation each sample has been collected in a separate window and all windows are congruent.

In ecological applications, in particular in the context of studies of species diversity in ecosystems, the point patterns formed by the locations of several species have often been collected in the same window. This includes data sets detailing the positions of individuals from different plant species in the same observation area. In this situation the correlation analysis for multivariate patterns may of course be applied to study the associations between individuals from different species (see Section 5.3).

However, in this section the primary interest is in classifying the species by their spatial behaviour rather than considering the inter-species interactions. That is to say, the aim is to find groups of 'similar' point patterns, where 'similar' refers to similarity in the variability of the point distribution.

For example, one might find that one group consists of patterns that are CSR (or close to), another one of patterns with aggregation at small distances, and yet another one of patterns with aggregation at large distances, etc. Further inspection of these groups may then lead to conclusions that these patterns have a similar interaction structure, probably as a result of similar ecological processes, and may facilitate or deepen further ecological research.

There are two approaches to solving the problem:

(a) *Classical multivariate analysis based on numerical characteristics.* This approach is based on a number of suitable numerical characteristics (such as those discussed in Section 4.2.4) and applies standard multivariate methods, such as cluster analysis and principal component analysis, to these characteristics. To be more specific, one calculates k numerical summary characteristics $c_1, \ldots, c_k$ for each pattern. In other words, the patterns are regarded as *objects* or observation units and the summary characteristics c_i as *variables*. Classical methods of multivariate statistics may then be applied to these variables; see Manly (2004) and Mardia et al. (1989) for a general treatment of multivariate methods.

(b) *Functional data analysis methods based on functional summary characteristics.* This approach may be applied to functional summary characteristics $S(r)$ as introduced in Sections 4.2–4.4. Since these summary statistics are functions rather than individual values, classical multivariate methods cannot be applied to them. However, specific statistical methods that operate on functional data have been developed outside spatial statistics. These methods may be suitably adapted to be applied to summary function characteristics of point patterns. Ramsay and Silverman (2002, 2005) provide a general introduction to functional data analysis.

Technically, approach (a) is by far the easier of the two. It is described below in detail for an ecological example. Readers who are familiar with classical multivariate statistical methods will have no difficulties in understanding the text. The functional data analysis approach is methodologically more complicated and requires a thorough understanding of functional data analysis methodology. It is beyond the scope of this book to provide an introduction to this; readers are referred to Illian (2006) and Illian et al. (2006) for more detail. There, a systematic approach uses a feasibility study to show that the method is indeed capable of detecting (known) groups if based exclusively on second-order characteristics. Groups of simulated point patterns with different spatial behaviour were simulated based on point process models that describe different types of spatial behaviour. The method classifies the patterns as expected. A further study assesses the suitability of the method in the context of (realistically) noisy data, i.e. the influence of inaccurately recorded locations and misclassified points on the performance of the method. The results show that the pair correlation function $g(r)$ is the recommended functional summary characteristic for this approach as the classification results based on the pair correlation function were better than those based on the L-function. However, this approach can only be successfully applied with sufficient experience and detailed knowledge of functional data analysis methods. These require that the estimated pair correlation functions are transformed into functional data based on an appropriate smoothing method where the degree of smoothing has to be determined carefully. For instance, since empirical pair correlation functions can be rather irregular, cubic B-splines were applied to yield a smooth representation of the estimated functions.

Example 4.19.   Multivariate analysis of 31 point patterns in a plant community of Western Australia
The data set is a multi-type point pattern formed by a natural plant community in the heathlands of Western Australia described in detail in Armstrong (1991), Illian (2006) and Illian et al. (2006). The point locations were recorded on a fine grid with 10 cm by 10 cm cells in a 22×22 m plot. While 67 species were originally observed, only the most frequent 31 species are considered here. Species no. 57 (*Phlebocarya filifolia*) with 207 points has already been discussed in Section 1.2.3. (In order to facilitate comparison with the papers mentioned above, the original species numbers are used here; see Armstrong, 1991, for a list of the species names.)

The first step in the data analysis based on classic statistical methods for multivariate data (which cannot be presented here due to limitations of space) consists of a visual inspection of the point patterns and the corresponding empirical pair correlation functions. All patterns are clustered, with degrees of clustering ranging from 'close to CSR' to 'stronger clustering'. Four of the patterns (6, 23, 42, 61) do not look homogeneous. This might be due to the fact that the 22×22 m window is perhaps too small for these species. Consequently, the statistical methods for stationary point processes interpret them as patterns with large clusters. Furthermore, 11 patterns have pair correlation functions that indicate that these may be considered as 'good cluster processes', such as no. 57 (see Example 4.7): specifically, nos. 13, 15, 18, 19, 20, 32, 34, 38, 49, 54 and 57. These may be expected to form a group of similar patterns in an analysis.

In the next step, the following numerical summary characteristics were used: n, the number of points in the 22×22 m window W; CE, the Clark–Evans index; $\overline{R}_4$, the mean-direction index; $\mu = \hat{m}_D / \tilde{m}_D$, the ratio of mean nearest-neighbour distance and median nearest-neighbour distance; $r^{(1)}$, the minimum r with $\hat{g}(r) = 1$; and $\hat{g}(2)$, the estimate of $g(r)$ for $r = 2$ m.

The Clark–Evans index CE was chosen as a scale-invariant measure of aggregation (see (4.2.24)), and $\overline{R}_4$ as an alternative scale-invariant measure of aggregation, (see p. 197). These two indices both measure clustering but highlight different nuances, such that it makes sense to include both of them in an analysis. Typically, large values of CE correspond to small values of $\overline{R}_4$, and vice versa. They also provide information that is different from that derived from second-order summary statistics.

The characteristic μ is also scale-invariant and has large values for very irregular patterns. It provides detailed information on the nearest-neighbour behaviour in a pattern, in particular the skewness of $D(r)$.

The characteristics $r^{(1)}$ and $\hat{g}(2)$ were determined using the estimator $\hat{g}_{st}(r)$ given by (4.3.38) with a bandwidth of $h = 1$ m and were chosen in an attempt to describe the shape of the pair correlation function by two numerical values. Due to the high degree of irregularity in the patterns and the high variability of their pair correlation functions the finer measures considered on p. 241 cannot be applied here.

Table 4.2 contains the values of the six summary characteristics for the 31 plant species. Before considering the results of the multivariate analysis note, the following three interesting features in the table:

- CE and $\overline{R}_4$ behave contrarily; large values of CE correspond to small values of $\overline{R}_4$ and vice versa, as expected, where theoretically $CE = 1.0$ corresponds to $\overline{R}_4 = 1.8$.

- 'Good cluster patterns' have large $r^{(1)}$.

- There are some patterns (e.g. species 3, 33 and 47) with $CE \approx 1.0$, $\overline{R}_4 \approx 1.8$, which are clearly close to CSR.

The initial multivariate analysis was based on all six variables, including n. This did not yield informative and clear results, as n varies a lot between the species and thus has a large influence on the results, but is mainly a scale parameter. However, as in the context of indices, scale-invariant results are preferable as the analysis is mainly focused on the difference in spatial behaviour, independent of the total number of points. Therefore, the analysis was repeated without n, in order to produce appropriate and interpretable results.

Table 4.2   Numerical summary characteristics for 31 plant patterns.

Species no.	n	CE	$\overline{R}_4$	μ	$r^{(1)}$	$\hat{g}(2)$
3	977	1.00	1.89	1.11	0.2	1.00
5	689	0.86	1.93	1.13	0.3	1.17
6	91	0.78	2.07	1.23	0.4	1.74
8	26	0.85	2.23	0.99	3.9	0.97
12	26	0.85	2.09	1.07	2.0	0.79
13	176	0.80	1.93	1.29	5.5	0.95
14	30	0.79	2.62	2.48	8.8	2.40
15	108	0.85	1.97	0.97	5.2	1.24
18	266	0.88	2.02	1.09	8.1	1.20
19	61	0.65	2.28	1.50	6.4	1.43
20	28	0.94	2.27	0.95	11.8	1.38
23	167	0.43	2.48	1.14	16.3	1.04
25	207	0.94	2.03	1.18	0.7	0.95
26	65	0.87	2.06	0.94	0.4	1.08
32	96	1.01	1.95	1.12	3.4	0.92
33	148	0.97	1.81	1.02	1.7	1.05
34	134	1.00	2.13	1.11	4.6	1.14
36	96	1.00	2.20	1.23	1.2	1.14
37	69	1.01	1.93	1.10	1.0	1.07
38	124	0.79	2.28	1.25	12.7	1.30
42	154	0.84	2.01	1.22	0.7	1.63
45	61	0.94	2.07	1.16	1.0	1.07
47	657	1.00	1.85	1.06	0.2	1.02
48	251	0.89	1.95	1.12	0.6	1.32
49	304	1.00	2.03	1.04	0.7	1.05
50	299	0.79	2.06	1.20	24.4	1.29
51	377	0.81	2.08	1.14	0.4	0.99
54	79	0.86	2.19	1.13	4.2	0.99
57	207	0.87	2.03	1.06	3.2	1.08
61	171	0.98	1.99	1.13	19.5	1.45
64	27	1.01	2.18	1.18	2.5	0.59

Figure 4.41 shows the result of a cluster analysis based on Ward's method in the form of a dendrogram. This method is an agglomerative clustering algorithm that tries to minimise the increase in total within-cluster error at each agglomeration step (for Ward's methods and other alternative hierarchical clustering algorithms see, for example, Everitt et al., 2001). Four main clusters can be identified from the plot:

(1) 23, 50, 61;

(2) 3, 5, 6, 12, 25, 26, 33, 36, 37, 38, 42, 45, 47, 48, 49, 64;

(3) 8, 13, 15, 19, 32, 34, 54, 57;

(4) 14, 18, 20, 38.

A detailed inspection of the four clusters yields the following, where initially only the geometry of the patterns is considered. Cluster 1 contains a strange mixtures of different patterns. The patterns have small CE values and large $\overline{R}_4$ values but are not the only patterns in the data set that have this property (such as 6 or 14). However, as opposed to these, they have very large $r^{(1)}$ values, the three largest in the data set. This indicates that the patterns in this cluster are aggregated with relatively large and clear clumps.

Cluster 2 is the largest cluster and consists of a mixture of patterns with different structures, including the patterns close to CSR. They have mainly medium-sized values of CE, $\overline{R}_4$ and μ, and medium to small values of $r^{(1)}$. Apparently, this cluster mainly contains weakly aggregated patterns.

Finally, clusters 3 and 4 both consist of aggregated patterns, among them nearly all the 'good cluster patterns' mentioned above. They all have large CE and $\overline{R}_4$ values. In addition, the patterns in cluster 4 have relatively large $r^{(1)}$ values. Furthermore, the dendrogram reveals that of the four clusters clusters 3 and 4 are the most closely related. Apparently, they mainly differ in their $r^{(1)}$ values, i.e. in the size of their clumps. Note that patterns of high intensity may be found in similar frequencies in all clusters, indicating that the analysis has definitely been scale-invariant, as intended.

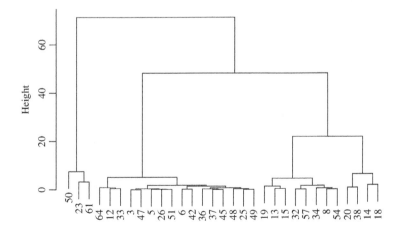

Figure 4.41   Dendrogram for the Western Australian data set from a hierarchical agglomerative cluster analysis (Ward's method).

A thorough biological interpretation of the results of the cluster analysis requires expert knowledge on the properties of each of the species. A detailed description of these properties is of course beyond the scope of this book. However, note that the species in the data set grow in an area that is highly susceptible to reoccurring natural fires. The species have adapted to this by developing specific fire regeneration strategies. These strategies fall into two main types. Plants either store their seeds over long time periods only to release them on the onset of a fire and die in the fire ('seeder plants') or survive the fire under ground and grow back after the fire ('resprouter plants'). Plants of the latter type develop very slowly but are often hundreds of years old. It is interesting to note that clusters 1 and 3 exclusively contain resprouting species, whereas clusters 2 and 4 are mixtures of resprouting and seeder plants. This may be the result of growth habits specific to these types of species. Thus the character of the point pattern formed by the different species is closely related to the species' specific adaption to the environment.

A further analysis that may be applied to the data in Table 4.2 is principal component analysis. It identifies linear combinations ('principal components', PCs) of the numerical summary statistics that explain the largest amount of variation between the different patterns. The principal component analysis (without using the variable n) reveals here that the first two principal components explain 74.3 % of the variance.

Table 4.3 lists the loadings of the five summary characteristics on the first two principal components, i.e. the contribution of each of the summary characteristics to the PCs, and Figure 4.42 shows a biplot of the results of the principal component analysis.

The summary characteristic μ is most strongly associated with the first principal component, followed by $\overline{R}_4$, $r^{(1)}$ and CE, where μ, $\overline{R}_4$, $r^{(1)}$ are negatively associated with it and CE positively. However, none of the characteristics is so strongly associated with the first PC that it dominates the PC so that one could interpret the component primarily in terms of this characteristic. In an interpretation of the results of the scores of the different patterns on the first PC all variables have

Table 4.3   Loadings of the five summary characteristics on the first two principal components.

Summary characteristic	1st PC	2nd PC
CE	0.411	0.480
$\overline{R}_4$	−0.485	0.497
μ	−0.512	−0.102
$r^{(1)}$	−0.463	0.439
$\hat{g}(2)$	−0.344	−0.565

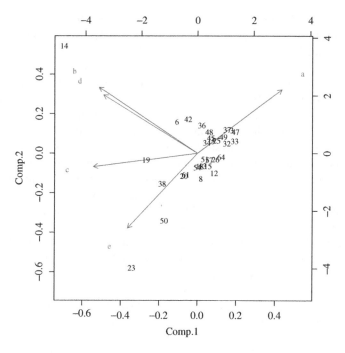

Figure 4.42   Biplot of the principal component analysis of the 31 patterns on the first two principal components.

to be taken into account. This is certainly a result of the fact that the summary characteristics have been chosen to be as informative as possible and to explain aspects of the spatial behaviour that are as relevant as possible to the specific data set. A different choice of characteristics is very likely to have resulted in different scores for the different species on the PCs in the principal component analysis.

It is clear that patterns that have large negative values on the first PC are somewhat aggregated (such as pattern number 23 and 19), while patterns with high positive values on the first PC exhibit as behaviour close to CSR (such as 33 and 47).

The summary characteristic $\hat{g}(2)$ is most strongly (negatively) associated with the second principal component, followed by $\overline{R}_4$, CE and $r^{(1)}$, which are all positively associated with the second PC. Whereas these statistics are relatively similar in their loadings, the loading for μ (which was most strongly associated with the first PC) is rather low and may be ignored in the interpretation. Again, the loadings for $\overline{R}_4$, CE and $r^{(1)}$ on the second principal component show similarly strong associations with this PC. Hence, patterns with high positive scores on the second PC show high values in these characteristics and low values in $\hat{g}(2)$. It is difficult to find a joint interpretation of $\overline{R}_4$, CE and $r^{(1)}$, together with aspects of $\hat{g}(2)$.

The pattern of species 14 stands out as having the lowest value on the first and the highest value on the second PC. A closer inspection of the pattern itself reveals that it is a highly inhomogeneous and clustered pattern where the clusters are all located close to the edge of the observation window. This might indicate that the window is actually too small for this pattern. The inhomogeneity is likely to have exaggerated the clustering and thus led to the extreme values.

When considering the scores of the patterns on the first two PCs in Figure 4.42 as such, little structure can be identified among them, i.e. it is difficult to distinguish separate groups of patterns on the basis of this and hence it is probably not useful to force a structure onto it by applying a clustering algorithm.

For comparison with the analysis described so far, Figure 4.43 shows the dendrogram as obtained in Illian et al. (2006) based on functional data analysis methodology using the second-order characteristic $g(r)$. A comparison of this figure with Figure 4.41 shows that the two methods did not yield the same groupings. However, some similarities may be detected. Most of the species in the second group in Figure 4.41 were classified into groups 1 and 2 in Figure 4.43, and most of the species in the third group in Figure 4.41 were classified into groups 3 and 4 in Figure 4.43. Apparently, there is a strong correspondence between these groups.

Note that the differences between the results from the analysis based on classical multivariate statistical methodology and the results obtained in Illian et al. (2006) are due to fundamental differences in the two statistical approaches and the different information used, as well as showing the complicated character of the distribution of the plants considered.

The functional data analysis method yielded a description of the most distinctive features in the spatial behaviour within the community, i.e. presence or absence of clustering at close distances. The second most distinctive feature was the presence or absence of clustering at larger distances. This may indicate that multi-species

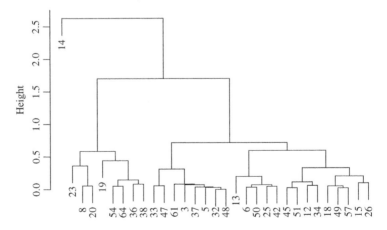

Figure 4.43   Dendrogram for the same point patterns which led to Figure 4.41, but now resulting from functional data analysis; see Figure 8 in Illian et al. (2006). With kind permission of Springer Science and Business Media.

coexistence is facilitated through a strong variation in strength of intra-specific attraction in an environment as poor in nutrients as the heathlands of Western Australia.

The grouping of the species into groups of similar spatial behaviour does not coincide with other classifications such as regeneration method, genus or growth habit. This phenomenon might be due to the fact that the system studied is a very ancient community where the variation in spatial behaviour has been adapted across families to enable coexistence over millions of years.

4.10   Summary characteristics for the non-stationary case

In many applications the point patterns cannot be considered stationary, and hence methods which assume stationarity are not suitable. Therefore, modern point process statistics has developed approaches which may be used to analyse inhomogeneous point patterns. Given replicated patterns, i.e. data sets that consist of several samples of (small) inhomogeneous patterns in the same (small) window, the methods for finite point processes covered in Chapter 3 can be applied. This section discusses methods that may be applied to a single sample in a large window W. These methods are suitable in a situation where one wishes to study short-range interaction among the points in the presence of larger-scale inhomogeneity.

First, there are two simple alternative approaches, which can be employed by means of the methods for stationary processes. In one of these approaches the main focus is on the large-scale behaviour such that only long-range point density fluctuations (which might be perhaps correlated with smoothly variable covariates) are relevant and individual points are of little interest. A useful approach in this context is the application of geostatistical methods rather than point process statistics, applied to a regionalised (i.e. smoothed) point pattern data as briefly discussed in Section 6.11. The other approach focuses on short-range interactions that may vary within the observation window W. For this purpose, a number of (nearly) homogeneous subplots W_1, W_2, ... of W are considered; see Franklin et al. (1985) and Pélissier and Goreaud (2001). Methods for stationary patterns may then be applied to each of these. The resulting estimated summary characteristics will tend to vary among the subpatterns (e.g. controlled by point density) and may be considered in relation to relevant covariates if these are relatively constant within the W_i. This approach may in some sense resemble research plots of foresters, which often are located in 'quasi-homogeneous' regions of forests.

However, this section does not follow either of these approaches but discusses point process methods applied to the whole window W in order to study short-range interaction among the points in the presence of larger-scale inhomogeneity. For this purpose, summary characteristics for inhomogeneous patterns are introduced which are of a similar nature to the stationary summary characteristics and are not location-dependent. The underlying theory is based on Baddeley et al. (2000), Hahn et al. (2003) and Prokešová et al. (2006) and uses intensity reweighting and local scaling. Note that in the non-stationary case there is no typical point and that there is a family of Palm distributions $\mathbf{P}_x$ rather than one Palm distribution $\mathbf{P}_o$, where $\mathbf{P}_x$ describes the distribution of N under the condition that there is a point of N at the deterministic location x.

Towards the end of this section, examples illustrate the application of the methods. Since the underlying mechanisms that generated the pattern are usually unknown in applications and errors resulting from inappropriate statistical methods are difficult to detect, two simulated patterns are used to provide examples where the true underlying mechanisms that generated the patterns are known. This will help the reader to gain some understanding of the potential of these methods. The practical relevance of the approach is illustrated with the real-life Example 4.22.

A first approach to the analysis of inhomogeneous point patterns is the following.

4.10.1 Formal application of stationary characteristics and estimators

The classical summary characteristics that were developed for stationary patterns may still be determined even if the pattern is a sample from an inhomogeneous process. This may yield interesting information about the degree of inhomogeneity and about short-range behaviour. This subsection discusses briefly the results of this approach, which requires experience and a careful interpretation of the results.

Application of the intensity estimator in (4.2.10) leads to an estimate of the quantity

$$\overline{\lambda}_W = \int_W \lambda(x)\mathrm{d}x/\nu(W), \qquad (4.10.1)$$

which may be called the *mean point density*. It depends on the window W and may serve as a benchmark value for $\lambda(x)$ in W, to distinguish between regions of high and of low point density.

Formal application of estimators of the distance characteristics $D(r)$ and $H_s(r)$ will lead to complex results, such as bi- or multimodal distributions. For example, regions of high point density may lead to a mode due to a large number of small nearest neighbours. Second-order characteristics $g(r)$ and $L(r)$ estimated in this way resemble those for cluster processes indicating (spurious) large clusters; refer to Examples 4.20–4.22 at the end of this section. As these examples show, the application of methods for stationary patterns to inhomogeneous data yields results that have to be interpreted carefully and are useful only to experienced statisticians.

Nevertheless, there are examples of summary characteristics where the application of stationary methods to inhomogeneous data does not cause great problems. This is the case, for example, for mark correlation functions such as $k_f(r)$ and $p_{ij}(r)$ as explained in Section 5.3. These quantities are estimated by ratio estimators, i.e. as quotients of other summary characteristics (namely $\varrho(r)$, $\varrho_f(r)$ and $\varrho_{ij}(r)$), and statistical experience shows that errors resulting from inhomogeneity cancel out. In fact, Capobianco and Renshaw (1998) and Lancaster (2006) showed that these estimators are robust and may be applied even without edge-correction, and they may also safely be used in the non-stationary case. The estimation of the nearest-neighbour distance d.f.s $D_{ij}(r)$ may also make sense.

The other two approaches require that enough points have been collected so that it seems reasonable to estimate the intensity function $\lambda(x)$. As discussed in Baddeley et al. (2000), smooth estimates of $\lambda(x)$ should be used; otherwise large- and meso-scale variation cannot be distinguished from small-scale variation resulting from point interaction. If the kernel estimators introduced in Section 3.3.2 are used, the bandwidth for $\hat{\lambda}(x)$ should be chosen larger than that for the summary characteristics such as $g(r)$. Parametric models for the intensity function are recommended, which may be fitted using standard software for generalised linear models (Berman and Turner, 1992) as implemented in the `spatstat` library in R; see Baddeley and Turner (2000, 2005, 2006). In the examples below $\lambda(x)$ is piecewise constant or linear. Diggle et al. (2007) provide an example where $\lambda(x)$ is estimated based on a covariate using additional information, the elevation in a tropical rainforest.

Both approaches assume that the local inhomogeneity due to short-range inter-actions is of a uniform nature across the observation window. They cannot be applied to patterns which are, for example, regular in some subareas and clustered for others. Experience shows that they produce better results for regular than for cluster patterns.

4.10.2 Intensity reweighting

This approach is based on the idea of replacing the classical (constant) intensity estimator by variable intensity function estimators in the estimation of stationary summary characteristics. This means that the estimation is adapted to the variable point density. The results are global, averaged estimates which should be used and interpreted jointly with the intensity function.

The method was introduced in Baddeley et al. (2000) for both $K(r)$ and $g(r)$. The following focuses on the pair correlation function $g(r)$.

Consider first the function

$$g(x, y) = \frac{\varrho(x, y)}{\lambda(x)\lambda(y)} \qquad \text{for } x, y \in \mathbb{R}^d, \tag{4.10.2}$$

where $\varrho(x, y)$ is the second-order product density of N. (If $\lambda(x)\lambda(y) = 0$, then set $g(x, y) = 0$.) In the case of an inhomogeneous Poisson process, $g(x, y)$ is constant and equal to 1, an indication of the usefulness of the function.

Assume now that $g(x, y)$ is independent of the locations x and y and depends only on the distance $r = \|x - y\|$ between x and y. Denote the corresponding 'master' function by $g_{\text{inhom}}(r)$. Together with the intensity function $\lambda(x)$, it completely describes the second-order behaviour of a point pattern. Processes for which such a $g_{\text{inhom}}(r)$ exists are called *second-order intensity-reweighted stationary,* following Baddeley et al. (2000). In addition to the inhomogeneous Poisson processes, all non-stationary point processes which result from a $p(x)$-thinning of a stationary point process (see Section 6.2.1 for explanation) have this property. If the pair correlation function of the original stationary process is $g_b(r)$, then $g_{\text{inhom}}(r) = g_b(r)$.

However, the assumption of second-order intensity-reweighted stationarity is rather restrictive. Many processes are not in this class, including hard-core processes with different hard-core distances that vary within the observation window (see Example 4.22 for an illustration) and processes in which the range of correlation varies.

A simple test of second-order intensity-reweighted stationarity is as follows. Select several subwindows $W_1, \ldots, W_k$ of W in which the point distribution appears homogeneous. Estimate the pair correlation function for each of these. For a second-order intensity-reweighted stationary process the estimates $\hat{g}_1(r), \ldots, \hat{g}_k(r)$ should be similar.

The master function $g_{\text{inhom}}(r)$ can be estimated by a simple modification of the estimators that are used for stationary patterns, (4.3.38) or (4.3.29), given by

$$\hat{g}_{\text{inhom}}(r) = \sum_{x_1, x_2 \in W}^{\neq} \frac{k(\|x_1 - x_2\| - r)}{db_d r^{-1} \nu(W_{x_1} \cap W_{x_2}) \lambda(x_1) \lambda(x_2)}. \tag{4.10.3}$$

The values $\lambda(x_1)$ and $\lambda(x_2)$ cause problems if kernel estimators are used for the intensity function $\lambda(x)$. This is because x_1 and x_2 are data points, which leads to overestimation; see Baddeley et al. (2000) for a detailed discussion.

The estimator $\hat{g}_{\text{inhom}}(r)$ should be used carefully. If the process underlying the observed pattern is really second-order intensity-reweighted stationary, $\hat{g}_{\text{inhom}}(r)$ does indeed estimate $g_{\text{inhom}}(r)$. If not, then a master function $g_{\text{inhom}}(r)$ does not exist and the estimate is not much better than that resulting from the application of a stationary estimator, as illustrated in Example 4.21.

An inhomogeneous K-function can be estimated by analogy with the stationary estimator in (4.10.3), where $k(\|x_1 - x_2\| - r)$ is replaced by $\mathbf{1}(\|x_1 - x_2\| \leq r)$. However, the problems resulting from inappropriately applying a reweighted summary statistic may even be aggravated when the K-function rather than the pair-correlation function is used, due to the cumulative nature of the K-function.

4.10.3   Local rescaling

The aim of this approach, explained here for the planar case, is to find global summary characteristics which are adapted to variable point density by a mechanism that rescales distances relative to local point density. This may be of particular value for a point pattern where the hard-core distance varies with point density. Hahn et al. (2003) introduce a class of so-called locally scaled point process models that have this property. In the particular case of a hard-core pattern it is expected that a successful application of the local rescaling approach yields a master summary characteristic with a unique hard-core distance.

More specifically, the approach of defining master summary characteristics for inhomogeneous patterns rescales the distance according to location. This means that it defines a 'metric' which varies within W. It has small values ('distances become shorter') in regions of high point density and large values ('distance become longer') in regions of high point density. The local point densities are considered relative to the mean point density $\overline{\lambda}_W$. The local metric around any point x is characterised by the definition of distance in the neighbourhood of x: the distance r in the case of a homogeneous pattern of intensity $\overline{\lambda}_W$ corresponds to the distance $\delta_x(r)$ in the neighbourhood of x for the inhomogeneous pattern. In the planar case, $\delta_x(r)$ is chosen such that

$$\overline{\lambda}_W \pi r^2 = \int\limits_{b(x,\delta_x(r))} \lambda(u)\mathrm{d}u, \qquad (4.10.4)$$

i.e. that the mean number of points in a disc of radius r under homogeneity equals the mean number of points in a disc of radius $\delta_x(r)$ centred at x. If $\lambda(x)$ is only weakly variable and may thus be considered constant within a circle of radius r around x, the right-hand side of (4.10.4) simplifies to $\lambda(x)\pi\delta_x(r)^2$ and yields

$$\delta_x(r) = r\sqrt{\frac{\overline{\lambda}_W}{\lambda(x)}}.$$

Nearest-neighbour distances

A local analogue of $D(r)$, denoted by $D_x(r)$, can be defined for every x in W. Here, the nearest-neighbour distance d.f. is not independent of the deterministic observation location x. $D_x(r)$ is formally defined as

$$D_x(r) = \mathbf{P}_x(d(x) \le r) \qquad \text{for } r \ge 0,$$

where $d(x)$ is the nearest-neighbour distance of the point at x. The probability, as before for $D(r)$, is of a Palm distribution nature, i.e. it is a conditional probability given that there is a point of N at x.

The definition of the analogous quantity $H_{s,x}(r)$ is more straightforward. Here

$$H_{s,x}(r) = 1 - N(b(x,r)) = 0) \quad \text{for } r \ge 0,$$

i.e. it is not necessary to consider a conditional probability for the definition.

In contrast, in the rescaling approach it is assumed that there are master nearest-neighbour distance d.f.s $D^*(r)$ and $H_s^*(r)$ such that

$$D_x(r) = D^* \left(r \sqrt{\frac{\lambda(x)}{\overline{\lambda}_W}} \right) \tag{4.10.5}$$

and

$$H_{s,x}(r) = H_s^* \left(r \sqrt{\frac{\lambda(x)}{\overline{\lambda}_W}} \right) \tag{4.10.6}$$

for all $r \geq 0$. There is no guarantee that these functions do indeed exist, even for inhomogeneous Poisson processes, but this may be the case in an approximate sense. Thus this method provides an elegant way of analysing the nearest-neighbour distances in inhomogeneous point processes.

The function $D^*(r)$ is estimated in the same way as $D(r)$ (see Section 4.2.6), but the original distances $d(x_i)$ are multiplied by the factor

$$\sqrt[d]{\frac{\lambda(x_i)}{\overline{\lambda}_W}},$$

which is larger than 1 if $\lambda(x_i) > \overline{\lambda}_W$ and smaller than 1 otherwise.

Pair correlation function

The master pair correlation function $g^*(r)$ is defined similarly based on local rescaling of interpoint distances. The original distance $\|x_1 - x_2\|$ between two points x_1 and x_2 is multiplied by the factor

$$\frac{1}{2} \left(\sqrt[d]{\frac{\lambda(x_1)}{\overline{\lambda}_W}} + \sqrt[d]{\frac{\lambda(x_2)}{\overline{\lambda}_W}} \right).$$

The factor is larger than 1 if $\lambda(x_1) > \overline{\lambda}_W$ and $\lambda(x_2) > \overline{\lambda}_W$ and smaller if the converse inequalities hold. In the estimator $\hat{g}_{\text{inhom}}(r)$ in (4.10.3), the term $\|x_1 - x_2\|$ is simply replaced by

$$\frac{1}{2} \left(\sqrt[d]{\frac{\lambda(x_1)}{\overline{\lambda}_W}} + \sqrt[d]{\frac{\lambda(x_2)}{\overline{\lambda}_W}} \right) \|x_1 - x_2\|.$$

Prokešová et al. (2006) present a similar estimator for the K-function.

The following examples illustrate the application of the techniques in the statistical analysis of inhomogeneous patterns. The first two consider synthetic patterns constructed to illustrate the potential of these methods.

Example 4.20.   Stationary estimates for an inhomogeneous Poisson process
Consider a point pattern that consists of 400 points in a 400×200 window; 200 points are uniformly distributed in the left 100×200 rectangle and 200 points are uniformly distributed in the right 300×200 rectangle, as shown in Figure 4.44. This pattern may be regarded as a sample from an inhomogeneous Poisson process with $\lambda(x) = 0.01$ for locations x on the left-hand side and $\lambda(x) = 0.0033$ for locations x on the right-hand side of the window.

Admittedly, this pattern is rather contrived and unlikely to be observed in reality, but in the ecological literature similar patterns have been used for demonstration purposes; see Pélissier and Goreaud (2001). If a pattern like this were to be analysed in an application the window would probably be divided into two subwindows which would be analysed separately. However, it is used here as an extreme example to illustrate the usefulness of applying appropriate summary characteristics to inhomogeneous patterns. If stationary summary characteristics were applied to this pattern the summary characteristic would indicate (spurious) clustering with a large 100×200 cluster. Intensity reweighting, however, should detect the Poisson process nature of the pattern and not indicate clustering.

Figure 4.45 shows the estimated nearest-neighbour distance and spherical contact probability density functions $d(r)$ and $h_s(r)$ using estimators for stationary patterns. The first function is bimodal due to the two different point densities in the pattern; due to statistical fluctuation the maxima are not at $r = 4$ and $r = 6.9$ as expected. The second indicates large distances between test points and points in the process since the low intensity on the left-hand side has a strong influence. Figure 4.46 shows a formal estimate of the pair correlation function $g(r)$, again using an estimator for stationary patterns. At $r = 0$ its value is 1.33 and the function decreases from there, taking the value 1 at $r = 100$.

These results are basically not interpretable on their own, without a visual inspection of the pattern (or knowledge of the underlying model). The distance of $r = 100$ corresponds to the size of the subwindow of high point intensity, i.e. the subpattern with higher point density may be (mis)interpreted as a large cluster.

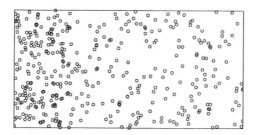

Figure 4.44   A simulated inhomogeneous Poisson process with high intensity on the left-hand side and low intensity on right-hand side.

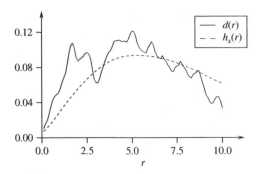

Figure 4.45   Estimated densities $\hat{d}(r)$(solid line) and $\hat{h}_s(r)$ (dashed line) for the inhomogeneous data obtained by misuse of stationary estimators.

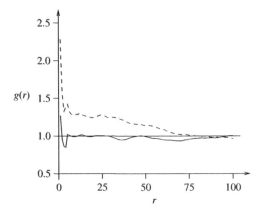

Figure 4.46   Estimated pair correlation function obtained by application of the stationary estimator to inhomogeneous data (dashed line); the values larger than 1 for small r result from the high local point density in the left-hand 100×200 subwindow. $\hat{g}_{\text{inhom}}(r)$ for the same data (solid line). It fluctuates irregularly around 1. For both estimates the bandwidth was $h = 4$ for $r \leq 4$ and $h = 8$ otherwise.

As expected, the intensity-reweighted estimate $\hat{g}_{\text{inhom}}(r)$ is close to 1, as shown in Figure 4.46. (The fluctuations of $\hat{g}_{\text{inhom}}(r)$ for small r are statistical artefacts resulting from the kernel estimation method and the small number of points.) Hence (4.10.3) has been successfully applied to the inhomogeneous pattern. The master $\hat{g}_{\text{inhom}}(r)$ looks like the pair correlation function for a homogeneous Poisson process and does not (spuriously) indicate that the pattern has been derived from a cluster process.

Figure 4.47 shows the estimated probability density function corresponding to the master $D^*(r)$ for the inhomogeneous data. As expected, it is similar to the

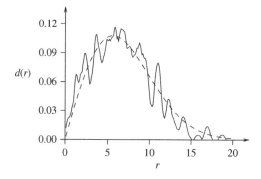

Figure 4.47   Estimated probability density function $d^*(r)$ corresponding to $D^*(r)$ for the inhomogeneous data (solid line), compared to $d(r)$ for a Poisson process of intensity $\overline{\lambda}_W$ (dashed line).

corresponding density function for the Poisson process with intensity $\overline{\lambda}_W$. The root transform in (4.10.5) was successful.

Example 4.21.   Estimates of master second-order characteristics for an inhomogeneous hard-core pattern
This example illustrates the usefulness of local rescaling based on a simulated point pattern which is known not to be second-order intensity-reweighted stationary. The pattern was generated in the same window as the points in Example 4.20. It consists of 200 simulated points from a Matérn hard-core process with intensity 0.01 with hard-core distance $r_0 = 5$ in the left-hand rectangle and 200 points simulated in the same way but with intensity 0.0033 and $r_0 = 9$ in the right-hand rectangle. Since the hard-core distance is different in areas with different densities, this pattern is definitely not a sample from a second-order intensity-reweighted stationary point process. Therefore, analysing this pattern by applying intensity reweighting for short distances is not a good idea. However, the local rescaling approach may yield master summary characteristics corresponding to a unified hard-core distance. This should be around 7 as

$$9\sqrt{\frac{0.0033}{0.005}} = 7.3 \quad \text{and} \quad 5\sqrt{\frac{0.01}{0.005}} = 7.1, \qquad \text{with } \overline{\lambda}_W = 0.005.$$

Figure 4.48 shows a formal estimate of the pair correlation function $g(r)$ using the stationary estimator. As expected, the smaller hard-core distance 5 in the left-hand rectangle dominates, while the right hard-core distance 9 leaves only a thin trace. The estimate has values that are above 1 for $r \geq 10$ because the stationary estimator interprets the point pattern in the left-hand rectangle as a large cluster. $\hat{g}_{\text{inhom}}(r)$ for intensity reweighting shows clearly two hard-core distances and has values around

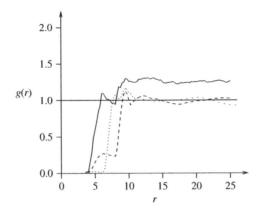

Figure 4.48   Result of pair correlation function estimation obtained by application of the stationary estimator for the inhomogeneous hard-core process data (solid line). Two estimates of master pair correlation functions are also shown, one based on intensity reweighting (dashed line) and the other based on local rescaling (dotted line). The bandwidth was $h = 1$ for $r \leq 12$ and $h = 4$ for larger r.

1 for $r > 10$, as expected for this pattern. Given its limitations, it was successful, but local rescaling yields a better result.

As an improvement of the simple stationary estimate, Figure 4.48 also shows the estimated master function $g^*(r)$ resulting from local rescaling. For small r, it looks like the pair correlation function of a point process with a hard-core distance around 7. Obviously, the local rescaling was successful. The pair correlation function is now quite similar to that of a Matérn hard-core process with r_0 between 5 and 9.

Example 4.22.   A gradient pattern of bronze particles
Figure 4.49 shows a cross-section through a bronze sinter filter, which was also analysed in Hahn et al. (1999). The filter consists of almost spherical bronze

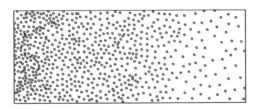

Figure 4.49   A cross-section through a bronze sinter filter. The points are centres of circular section profiles. The length of the longer side of the window is 18 mm. Data courtesy of R. Bernhardt and H. Wendrock.

particles with diameters (not shown) that decrease along the x-axis marking the filtering direction, $0 \le x \le l = 18$ mm. Since the particles are densely packed, the number of particles per unit volume increases as the diameters decrease. This is also observable on sections parallel to the directions of inhomogeneity: the centres of the particle section profiles form an inhomogeneous point pattern.

This point pattern seems to be a good example of a pattern that the methods discussed in this section may be applied to since the character of the distribution of the points is the same throughout the pattern: it is a packing of hard spheres in which only the sphere radii vary. Although the sample is only a planar section, some statistical analysis is useful; the information that local scaling yields plausible results may also be valid in the spatial case, i.e. for the three-dimensional filter.

The estimated intensity function of the pattern is shown in Figure 4.50. Since a gradient structure is given, it suffices to consider $\lambda(x)$, the intensity function in the direction of x-axis. This function is used in the further analysis of the pattern.

Visual inspection already indicates that the inter-point distances behave quite differently in the left- and right-hand parts of the sample. Thus, the idea that the gradient results from an independent thinning operation must be rejected; the physical process of packing contradicts this idea as well. The estimated pair correlation functions for the left 5 mm and for the right 5 mm shown in Figure 4.51 are therefore quite different, and intensity-reweigthing is clearly not suitable for this data set. See also $\hat{g}_{inhom}(r)$ in the same figure, which deviates from both estimates.

Local rescaling yields a plausible master pair correlation function, the solid line in Figure 4.52.

The following describes a rather natural way to derive a master function for cases with gradient structures like those depicted in Figure 4.49. The idea is simply to

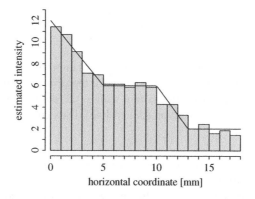

Figure 4.50   The estimated intensity function $\hat{\lambda}(x)$ for the pattern in Figure 4.49. Here x varies along the longer side of the rectangular 18×7 mm window. Data courtesy of U. Hahn.

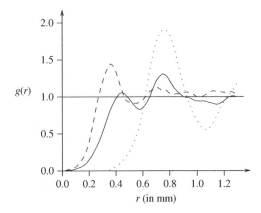

Figure 4.51   Three empirical pair correlation functions for the bronze pattern. Two of these were obtained using the estimator for the stationary case: $\hat{g}(r)$ for the left 5 mm region of Figure 4.49 (dashed line) and for the right 5 mm region (dotted line). $\hat{g}_{\text{inhom}}(r)$ for the whole window is also shown (solid line).

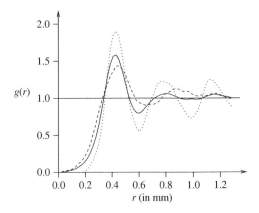

Figure 4.52   Three empirical master pair correlation functions for the bronze pattern: estimates obtained via local rescaling for the whole pattern (solid line), estimates assuming stationarity for the pattern with $x \leq 5$ (dashed line) and an estimate obtained by vertical homogenisation (dotted line). Data courtesy of U. Hahn.

transform the data along the gradient axis, to apply *vertical homogenisation* as in Fleischer et al. (2006), where the gradient was in vertical direction; in Figure 4.49 the gradient is in horizontal direction. In areas where the estimated intensity function $\hat{\lambda}(x)$ is high, the vertical distances are dilated and in areas where $\hat{\lambda}(x)$ is low the distances are compressed such that the point density is $\overline{\lambda}_W$ in the entire window. Figure 4.53 shows the result of this transformation for the pattern in Figure 4.49

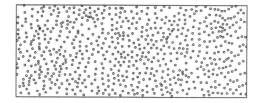

Figure 4.53   The pattern of Figure 4.49 after horizontal homogenisation.

using the intensity function in Figure 4.50. The pattern looks globally isotropic, even though it has only been transformed in horizontal direction. Figure 4.52 shows the corresponding estimated pair correlation function, which is quite similar to the result of local scaling.

In other cases and many applications the patterns have to be transformed more carefully and a theory of homogenisation by transformation has been developed; see Jensen and Nielsen (2000), Nielsen (2000) and Prokešová et al. (2006).

The formula for vertical homogenisation is as follows. Set

$$\Lambda(x) = \int_0^x \hat{\lambda}(t)\mathrm{d}t \qquad \text{for } 0 \leq x \leq l.$$

For the $\hat{\lambda}(x)$ in Figure 4.50 the formula is, for example for $0 \leq x \leq 5$,

$$\Lambda(x) = 12x - 0.6x^2,$$

as obtained by integration. The pattern is transformed horizontally as

$$x \rightarrow \frac{\Lambda(x)}{\Lambda(l)} l \qquad \text{for } 0 \leq x \leq l.$$

This method is used in Fleischer et al. (2006) where the root distribution in pure stands of beeches and spruces was analysed and modelled. Vertical sections yield planar gradient point patterns, where the points correspond to cross-sections of roots of diameter between 2 mm and 5 mm. In this specific application

$$\hat{\lambda}(x) = a\exp(-ax),$$

i.e. the root density decreases exponentially.

5

Stationary marked point processes

The previous chapters have treated patterns of unmarked points. This chapter now discusses marked patterns, the analysis and modelling of which is complex, yet challenging and interesting. Marked point pattern analysis also studies ensembles of objects scattered in space, but the objects are characterised not only by their positions but also by marks, i.e. additional data on each individual object, which may be either quantitative (continuous) or qualitative (discrete or categorial).

This chapter concentrates on the stationary case and presents a definition of stationarity which also includes the marks. It discusses in detail the fundamental first-order characteristics, intensity and mark distribution, and a large number of second-order characteristics. The latter depend on the character of the marks, i.e. on whether the marks are quantitative or qualitative. In the same way, the issues and ideas addressed in analyses of marked point patterns are different for patterns with qualitative and quantitative marks. Hence, several different approaches have to be discussed, and the choice depends on what kind of statistical information contained in a marked point pattern

Statistical Analysis and Modelling of Spatial Point Patterns   J. Illian, A. Penttinen, H. Stoyan and D. Stoyan
© 2008 John Wiley & Sons, Ltd

is of interest in a particular application. For instance, the analysis of patterns with qualitative marks may study the relative positions of the different types of points, i.e. whether there is within-type aggregation or repulsion as well as between-type aggregation or repulsion. In contrast, in the context of quantitative marks the analysis assesses whether the marks vary continuously and to what extent interaction among the points has an influence on the marks. A general fundamental question concerns independence among the marks.

The structure of this chapter is similar to that of Chapter 4. The presentation commences with some basic definitions and theoretical concepts which might appear technical at first but which will turn out to be relevant throughout this chapter and the rest of the book.

5.1   Basic definitions and notation

5.1.1   Introduction

Marked point process statistics is a key method in spatial statistics as it analyses data consisting of observations of variables given at irregularly distributed points. These processes are models for random point patterns where marks that describe properties of the objects represented by the points are attached to the points. In other words, a marked point process M is a sequence of random marked points, $M = \{[x_n; m(x_n)]\}$, where $m(x_n)$ is the mark of the point x_n. A number of practical examples of these processes are discussed in detail throughout this chapter.

Note that marked point pattern data appear to be in some sense similar to geostatistical data, which also consist of both information on locations and associated Z-values. However, the aim of geostatistics is to estimate spatially continuous phenomena (regionalised variables) based on discrete measurements at points chosen for this purpose. In point process statistics, however, the points represent not the locations that have been chosen for measurement purposes but the locations of the objects that are analysed. Hence the analysis of geostatistical data and marked point process data pursues entirely different aims and applies different methods; the misuse of geostatistical methods in point processes may produce incorrect results (see p. 344).

The points and marks in a marked pattern are often correlated. Consider, for example, data from a plant community where the points are plant locations and the marks plant size characteristics. In areas of high point density the marks may tend to be smaller than in areas with low point density resulting from stronger competition for limited resources. In contrast, the absolute values of velocities of galaxies are high in regions of high galaxy density. A marked point process model with [point; mark] = [galaxy centre; velocity] may take this correlation into account. Furthermore, in biological point patterns qualitative marks may characterise different species. Inter-species cooperation may lead to attraction among species, while inter-species competition may lead to repulsion.

5.1.2 Marks and their properties

The marks in a marked point processes may be either *quantitative* (continuous) or *qualitative* (categorial, discrete). Quantitative marks are real-valued and describe, for example, the size or extent of the objects represented by the points or any other physical property, such as tree height, particle diameter, or galaxy velocity. Qualitative marks are discrete or integer-valued categorial variables describing different types of points, such as plant species or shape types. If only two types of points are considered, coded as 1 and 2, say, the point process is called *bivariate,* otherwise it is *multivariate.* Clearly, quantitative marks can be reduced to qualitative marks by binning the marks into discrete classes such as 'small', 'medium' and 'large'. Also qualitative marks can be aggregated, e.g. 'deciduos' or 'coniferous' trees may be considered instead of tree species.

A marked point process may be considered as consisting of several sub-point processes (in each of which all points have the same qualitative marks) with interesting correlations and structures. More complex marks describe, for example, the shape of particles or crowns of trees if the points are particle centres or tree locations, respectively. However, these marks, represented by high-dimensional vectors, are beyond the scope of this book.

The aim of the analysis of bivariate point patterns is often to reveal relationships between points of type 1 and 2. In an attempt to systematically characterise such relationships one could classify them along two axes, dependence and relative degree of dominance (see Table 5.1).

In many point patterns the two types of points have a similar 'relative degree of dominance', i.e. neither type dominates the other; this is often described by words such as 'equality' or 'symmetry'. In forestry, these might for example be trees of similar sizes and similar age but from different species, or trees either damaged or not damaged by wind, frost, insects or disease. As another example, tissue cells of comparable but different function might not dominate each other.

In bivariate patterns with equal degree of dominance the two subpatterns N_1 and N_2 can either be (a) independent (irrespective of any within-type interactions) or (b) dependent, where dependence can either mean attraction or repulsion among the point types.

Table 5.1 Classification of relations between the points of bivariate point processes.

		Dependence	
Relative degree of dominance	equal	independent (a)	dependent (b)
	different	separation (c)	functional/ controlled (d)

In other cases ('asymmetry') one type has a dominating role while the points of the other type are suppressed or even controlled by points of the first type or appear only once the first-class points have taken up their locations. The two subpatterns have different degrees of dominance, and the relationship between the two point patterns of points of different degree of dominance may vary. They may be (c) simply separated as a result of the domination, or there might also be (d) some form of control or even of a functional relationship, such as between parent and daughter points in cluster processes or adults and seedlings.

All in all, four typical cases corresponding to four different combinations of (equal/different) degree of dominance and dependence can be distinguished, as summarised in Table 5.1. Some of the examples below correspond to one of these cases. This is indicated by the coding letter used in this table.

The description in Table 5.1 does certainly simplify reality. In many applications patterns cannot be grouped into either of the four types as easily and in such a clear-cut way.

Note that quantitative marks lead to different types of conclusions and do not imply dichotomous classifications.

5.1.3 Marking models

Marking models describe how the marks in a pattern might have been 'formed' given the points ('marking a posteriori') or how the marked points have been 'generated' ('marking a priori'). Only once a suitable model has been identified for a marked point pattern can it be properly interpreted and simulated. A number of models and modelling approaches have been discussed in the literature. This section provides a short introduction to three different types of marking models which are discussed in detail in Section 6.8.

The *independently marked* or *randomly labelled* point process is the simplest model for a stationary marked point process. In these patterns the point positions may be regarded as given a priori but the characteristics of the objects in these locations are determined independently at random, based on some probability distribution or frequencies of different types of points. In other words, the marks are i.i.d. random variables.

More formally, $M = \{[x_n; m(x_n)]\}$ is constructed as follows. Consider a stationary point process $N = \{x_n\}$ and an independent sequence of i.i.d. random variables $\{m_n\}$. Both are combined to yield $\{[x_n; m(x_n)]\}$ with $m(x_n) = m_n$. In the construction the numbering is arbitrary. This type of process may be simulated by first simulating N and then generating the marks based on the mark distribution in an arbitrary order.

In many applications, this simple process with independent marks is a realistic model. It is used in the very useful random set model called the 'Boolean model' (see Stoyan et al., 1995) and has often been observed for the marks of trees in forests, in particular in managed forests, where typically trees with extreme properties caused by competition are removed such that the remaining trees show only small 'independent' fluctuations. In other applications, such as in natural ecosystems, the processes are more complicated, but independently marked processes may serve as useful and important null models.

Another model whose components have a high degree of independence is the *random superposition model.* This model is relevant for those point processes with qualitative marks for which it is justified to separately consider the subprocesses N_i consisting of points with mark i. These processes are assumed to be independent and the whole point process results from a superposition or set-theoretic union of these N_i. In other words, the points in each of the subprocesses are distributed according to the interaction structure among points of the same type, but points of different types are 'ignored'. In an ecological context this is termed 'population independence' (Goreaud and Pélissier, 2003).

Correlated marks are obtained by what is called 'geostatistical marking', i.e. from the *random field model,* which was probably first used by Mase (1996). Its construction is based on a (non-marked) point process $N = \{x_n\}$ and a stationary random field $\{Z(x)\}$, which is independent of N. The points of the marked point process M are the x_n of N, while the marks are derived from the random field:

$$m(x_n) = Z(x_n). \tag{5.1.1}$$

Here the spatial correlations in the random field are reflected in M. If $\{Z(x)\}$ has positive correlations then points that are close together tend to have similar marks.

The random field model is appropriate in many applications and can be a good approximation in other applications. However, patterns will often deviate from the model at short inter-point distances since it does not model interaction among points, i.e. situations where the mark of a point is dependent on the existence of other points in close vicinity and their marks. Hence, the random field model is rather unsuitable as a model of a pattern resulting from biological competition.

Even more complicated marking models have been discussed in the literature. One of these is the model with *intensity-weighted marks,* a model in which point density and mark sizes are closely coupled – both are controlled by a basic random field, as explained in Section 6.8.

A very important type of marks, already mentioned on p. 196, are *constructed marks.* Constructed marks reflect the geometry of the point configuration of the neighbourhood of the points. Simple examples are $d(x)$, the distance from the point $x \in N$ to its nearest neighbour $z_1(x)$ in N, and $n_r(x) = N(b(x, r)) - 1$, the number of further points within distance r from x. The behaviour of constructed marks may be compared to that of natural marks and thus reveals information on a suitable marking model. For example, consider constructed marks that describe the intensity of the pattern around the points, such as $d(x)$ and $n_r(x)$. In a pattern where high local point density corresponds to large (natural) marks, constructed marks and natural marks are positively correlated, whereas in a pattern where high local point density corresponds to small (natural) marks, they are negatively correlated.

A more realistic approach to constructing marks may be to use a statistical mark construction model, i.e. to model the marks by means of a (linear) function of the location plus random error. Let $c(x)$ be a mark constructed as discussed above. The mark $m(x)$ of point x may then be given by

$$m(x) = c(x) + \varepsilon(x),\tag{5.1.2}$$

where $\varepsilon(x)$ is a random fluctuation. For instance, the constructed mark defined by

$$m(x) = a + bd(x) + \varepsilon(x),$$

where a and b are parameters and $d(x)$ is as above, follows this pattern. This yields marked patterns with small marks in areas of high point density if $b > 0$. A stimulation effect operating in the opposite direction results from $b < 0$, where in areas of high point density the marks tend to be large.

Example 5.1.   Gold particles: mark models
Consider again the gold particle example introduced in Section 1.2.2 but now take the marks of the gold particles into account, i.e. include the particles' diameters in the analysis. Here, two simple models based on the constructed marks $d(x)$ and $n_r(x)$ are considered. Figure 5.1 shows scatterplots for the natural marks $m(x_n)$, i.e. the diameters, as well as the two neighbourhood-related constructed marks $d(x_n)$ and $n_r(x_n)$, based on the distance $r = 30\,\mu\mathrm{m}$. Since there is no apparent structure in the plot one cannot assume that there is a relationship between the natural and constructed marks. Apparently, the relationship between the pattern and its marks is more complicated such that a different marking model would have to be considered to describe this relationship, see p. 469.

Other, more sophisticated constructed marks are the areas or volumes of Dirichlet or Voronoi cells around the points or exponential energy marks as used in the

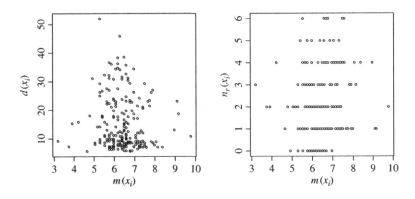

Figure 5.1   Scatterplots of natural and constructed marks for the gold particles: (left) $m(x)$ and $d(x)$; (right) $m(x)$ and $n_r(x)$ for $r = 30\,\mu\mathrm{m}$.

residual analysis in Section 4.6.4. Furthermore, many of the indices discussed in Section 4.2.4 are related to suitable constructed marks.

Another form of close coupling of point positions and marks is realised by marked Gibbs processes; see the references on p. 158. Finally, ecological and forestry models may be mentioned which describe the evolution of plant communities and forests controlled by growth and competition, in which point density decreases over time, and marks, which describe plant size, increase; see, for example, Adler (1996), Berger and Hildenbrand (2000), Bugmann (2001), Canham et al. (2003), Chave (1999), Comas and Mateu (2007), Dubé et al. (2001), Pacala et al. (1993), Pretzsch (2002), Pretzsch et al. (2002), Renshaw et al. (2007), Särkkä and Renshaw (2006) and Uriarte et al. (2004).

5.1.4 Stationarity

The stationarity or homogeneity assumption considerably simplified the statistics of non-marked point processes (see Chapter 4), and this is similar for marked point patterns. Of course, stationarity of a marked point process $M = \{[x_n; m(x_n)]\}$ automatically implies that the point process N of the points x_n without the marks is stationary as defined in Sections 1.6 and 4.1.

However, the formal definition of a stationary marked point process also involves the marks. It is analogous to the definition in the non-marked case, i.e. it considers the process M along with a translated process, denoted by M_x and defined as follows. If $M = \{[x_1; m(x_1)], [x_2, m(x_2)], \ldots\}$ then

$$M_x = \{[x_1 + x; m(x_1)], [x_2 + x; m(x_2)], \ldots\}. \tag{5.1.3}$$

This means that in the translated marked point process the marks stay the same but the points are translated.

This yields the following definition: a marked point process M is stationary if and only if

$$M \overset{d}{=} M_x \qquad \text{for all } x \in \mathbb{R}^d, \tag{5.1.4}$$

i.e. if the marked point process M and the marked point process translated by x have the same distribution. The definition of *isotropy* is analogous.

This invariance property has important consequences, which will become clear below, for example in (5.1.9). The property also determines which marks may be suitably used in a marked point pattern analysis. Quantities which are closely related to the objects that are represented by the points (such as size, diameter, height) are reasonable marks, as are constructed marks such as the distance of a point in the process to its nearest neighbour. However, the methods described in this book may not be applied to patterns with marks such as the distance from a point in the process to a fixed origin o of the coordinate system, since translations change these marks.

5.1.5 First-order characteristics

As indicated above, a larger number of first- as well as higher-order summary characteristics have to be considered for marked point patterns than for non-marked patterns. This is because the characteristics describe both the points and the marks. In addition, different characteristics are used for qualitative and quantitative marks.

In the stationary case two different types of *first-order characteristics* are used, one that concerns the points, i.e. the intensity λ, and another one that describes the marks, i.e. the mark probabilities p_i for qualitative marks or the mark d.f. $F_{\mathcal{M}}(m)$ for quantitative marks.

The *intensity* λ is the same as the intensity introduced in Section 4.2.3 for the non-marked case: the mean number of points per unit area or volume, or, in other words, the intensity of the point process N which results from M by removing the marks.

With qualitative marks, the *mark probability* p_i may be interpreted as the probability that the typical marked point has mark i. If the random variable $\mathcal{M}$ describes the mark of the typical point, then

$$p_i = \mathbf{P}(\mathcal{M} = i) \qquad \text{for } i = 1, 2, \ldots .$$

The quantity p_i is simply the relative intensity

$$p_i = \lambda_i / \lambda ,$$

where λ_i is the intensity of the sub-point process N_i of points with mark i. The mean number of points of M in the set B with mark i is equal to

$$\mathbf{E}(M(B \times \{i\})) = \lambda p_i \nu(B). \tag{5.1.5}$$

With quantitative marks, the *mark distribution function* $F_{\mathcal{M}}(m)$ describes the distribution of the marks irrespective of the point positions. A more technical mathematical definition of $F_{\mathcal{M}}(m)$ is given below. However, it may be easier to understand the notion of a mark distribution function by initially considering the following two heuristic interpretations.

(a) *Frequentist interpretation.* One heuristic explanation may consider the empirical analogue of the mark d.f. This means considering a (large) observation window W and collecting the marks $m(x_1), \ldots, m(x_n)$ for the points $x_1, \ldots, x_n$ in W. As in classical statistics, the corresponding empirical d.f. $\hat{F}_n$ is

$$\hat{F}_n(m) = \frac{1}{n} \sum_{i=1}^{n} \mathbf{1}(m(x_i) \le m), \tag{5.1.6}$$

i.e. $\hat{F}_n(m)$ is the relative frequency of points with marks smaller than m. The statistical estimation of mark d.f.s follows this pattern (see Section 5.2.3).

(b) *Typical-point interpretation.* Assume that it is possible to select a marked point from the infinitely many marked points in M such that each of these has the same chance of being selected. The corresponding mark is a random variable, which is denoted here by $\mathcal{M}$. Then

$$F_{\mathcal{M}}(m) = \mathbf{P}(\mathcal{M} \le m) \,. \tag{5.1.7}$$

This means that the mark d.f. is the classical d.f. of $\mathcal{M}$, i.e. $F_{\mathcal{M}}(m)$ is the probability that $\mathcal{M}$ is smaller than m.

The following provides a rigorous mathematical derivation of the mark d.f. defined in (5.1.9) below. Consider for an arbitrary set B in $\mathbb{R}^d$ the mean number of points of M in B with marks smaller than m. Denote this mean by

$$\mathbf{E}(M(B \times (-\infty, m])) \,.$$

Due to the translation invariance of marks

$$\mathbf{E}(M(B \times (-\infty, m])) = \mathbf{E}(M(B_x \times (-\infty, m])) \qquad \text{for all } B \text{ and all } x \in \mathbb{R}^d \,.$$

Consequently, similar to the explanation of the intensity in Chapter 4 on p. 175, the quantity

$$\mathbf{E}(M(\cdot \times (-\infty, m]))$$

is a multiple of the area or volume (or the Lebesgue measure), i.e.

$$\mathbf{E}(M(B \times (-\infty, m])) = \lambda_m \nu(B) \,,$$

with a non-negative factor λ_m depending on m. Choosing $B = \boxed{1}$ clearly shows that the above may be interpreted as the intensity of the sub-point process of M of points with marks smaller than m. Note that its complement, i.e. the same expression for marks larger than m, is sometimes called an m-at-risk point process.

If $m = \infty$, the marks do not need to be considered since all marks are smaller than ∞ and

$$M(B \times (-\infty, \infty)) = N(B) \,,$$

which yields $\mathbf{E}(M(B \times (-\infty, \infty))) = \lambda \nu(B)$. Hence $\lambda_\infty = \lambda$, whereas of course

$$\lambda_{-\infty} = 0 \,,$$

since there cannot be a point with mark smaller than $-\infty$. Finally, $m_1 \le m_2$ implies $\lambda_{m_1} \le \lambda_{m_2}$.

Due to these properties, the ratio λ_m/λ is a distribution function which depends on the variable m and is termed the mark d.f.,

$$F_{\mathcal{M}}(m) = \lambda_m/\lambda \quad \text{for} \ -\infty < m < \infty. \tag{5.1.8}$$

The mark d.f. satisfies

$$\mathbf{E}(M(B \times (-\infty, m])) = \lambda F_{\mathcal{M}}(m)\nu(B). \tag{5.1.9}$$

Often a *mark probability density function* $f_{\mathcal{M}}(m)$ exists with

$$F_{\mathcal{M}}(m) = \int_{-\infty}^{m} f_{\mathcal{M}}(x)\mathrm{d}x \quad \text{or} \quad f_{\mathcal{M}}(m) = F'_{\mathcal{M}}(m).$$

The *mean mark* μ is the expectation of $\mathcal{M}$ and is given by

$$\mu = \int_{-\infty}^{\infty} m f_{\mathcal{M}}(m)\mathrm{d}m = \int_{-\infty}^{\infty} m\mathrm{d}F_{\mathcal{M}}(m). \tag{5.1.10}$$

The *mark variance* σ_μ^2 is the variance of $\mathcal{M}$ and is given by

$$\sigma_\mu^2 = \int_{-\infty}^{\infty} (m - \mu)^2 f_{\mathcal{M}}(m)\mathrm{d}m = \int_{-\infty}^{\infty} (m - \mu)^2 \mathrm{d}F_{\mathcal{M}}(m). \tag{5.1.11}$$

The intensity function $\lambda(x, m)$ (introduced on p. 35) of a stationary marked point process with mark p.d.f. $f_{\mathcal{M}}(m)$ has the simple form

$$\lambda(x, m) = \lambda f_{\mathcal{M}}(m). \tag{5.1.12}$$

Note here that for an independently marked point process the mark d.f. coincides with the d.f. $F(m)$ of the members of the sequence $\{m_n\}$ mentioned on p. 296,

$$F_{\mathcal{M}}(m) = F(m). \tag{5.1.13}$$

A similar theory exists for qualitative marks.

Campbell theorem

In the stationary case the following simple Campbell theorem holds for marked point processes:

$$\mathbf{E}\left(\sum_{[x; m(x)] \in M} f(x, m(x))\right) = \lambda \int_{\mathbb{R}^d} \int_{-\infty}^{\infty} f(x, m)\mathrm{d}F_{\mathcal{M}}(m)\mathrm{d}x, \tag{5.1.14}$$

where the right-hand side simplifies to

$$\lambda \int_{\mathbb{R}^d} \int_{-\infty}^{\infty} f(x, m) f_{\mathcal{M}}(m) \, dm \, dx \quad \text{or} \quad \lambda \int_{-\infty}^{\infty} \int_{\mathbb{R}^d} f(x, m) \, dx f_{\mathcal{M}}(m) \, dm$$

if a mark p.d.f. exists, and for qualitative marks,

$$\lambda \int_{\mathbb{R}^d} \sum_{(i)} p_i f(x, i) \, dx \quad \text{or} \quad \lambda \sum_{(i)} p_i \int_{\mathbb{R}^d} f(x, i) \, dx .$$

Example 5.2. Two applications of the Campbell theorem
(1) *Seed density.* Consider again the seed dispersal example discussed in Example 1.2(1) on p. 34. The aim is now to calculate the mean of the seed density S_f at $y = o$. Equation (5.1.14) yields

$$\mathbf{E}S_f = \lambda \int_{\mathbb{R}^2} \int_0^{\infty} m \, d(\|x\|) f_{\mathcal{M}}(m) \, dm \, dx$$

$$= 2\pi\lambda\mu \int_0^{\infty} r d(r) \, dr ,$$

where $f_{\mathcal{M}}(m)$ is the mark p.d.f. and μ the corresponding mean.

The choice of $y = o$ is natural in the stationary case since the mean value is the same for all observation points and hence any point may be chosen as an observation point. Note that the mean seed density is constant while the seed density fluctuates randomly across the space.

(2) *Counting birds.* Consider now the birds in Example 1.2(b), which are marked by their sound levels. The mean number of birds heard at position $y = o$, denoted by S_f, is

$$\mathbf{E}S_f = \lambda \int_0^{\infty} \int_{\mathbb{R}^2} \left(1 - \frac{a\|x\|}{m}\right)_+ f_{\mathcal{M}}(m) \, dx \, dm$$

$$= 2\pi\lambda \int_0^{\infty} \int_0^{m/a} \left(1 - \frac{ar}{m}\right) f_{\mathcal{M}}(m) \, dr \, dm$$

$$= 2\pi\lambda \int_0^{\infty} \frac{m}{2a} f_{\mathcal{M}}(m) \, dm = \pi\lambda\mu/a ,$$

where μ is the mean mark. The formula

$$\mathbf{E}S_f = \pi\lambda\mu/a \tag{5.1.15}$$

can be used to estimate λ if a and μ (as 'bird parameters', characterising the distance at which the bird is audible) are known: if n is the number of birds heard at some position, then

$$\hat{\lambda} = n\frac{a}{\pi\mu} \qquad (5.1.16)$$

is an unbiased estimator of λ.

5.1.6 Mark-sum measure

By way of illustration, consider the marked point pattern M of a forest, with x denoting tree location, $m(x)$ the cross-sectional area of the trunk of the tree at x, and $S(B)$ the sum of all cross-sectional areas in a set B. In other words, the aim is to consider $S(B)$, the mark-sum measure of a marked point process M, i.e. the sum of the marks of all points x in the set B.

The mark-sum measure is defined as

$$S(B) = \sum_{[x;m]\in M} m\mathbf{1}_B(x) \qquad (5.1.17)$$

for arbitrary subsets B of $\mathbb{R}^d$. If the marks are all positive, S is a random measure. If the marked point process M is stationary, S is also stationary, i.e.

$$S(B) \stackrel{d}{=} S(B_x) \qquad \text{for all } x \in \mathbb{R}^d \text{ and all subsets } B.$$

Then S has also an intensity, the *mark-sum intensity*, which is denoted by λ_S,

$$\lambda_S = \mathbf{E}\left(S\left(\boxed{1}\right)\right), \qquad (5.1.18)$$

i.e. λ_S is the mean mark-sum per unit area or volume.

Note that in forestry the intensity λ_S based on cross-section area marks is commonly used rather than the intensity λ. Whereas λ_S is called the 'basal area', λ is, clumsily, referred to as 'mean number of stems per hectare'.

By the Campbell theorem (5.1.14) with $f(x, m) = m\mathbf{1}_{\boxed{1}}(x)$, λ_S turns out to be simply

$$\lambda_S = \lambda\mu. \qquad (5.1.19)$$

5.1.7 Palm distribution*

Similar to the approach in Section 4.1, Palm distributions can also be defined for marked point processes. (Note that this section assumes that the reader is familiar with the treatment of Palm distributions in Section 4.1.)

Palm distributions $\mathbf{P}_L$ and means $\mathbf{E}_L$ with respect to mark sets L may be considered, where $L \subset \mathbb{R}$, for example $L = (-\infty, m]$ or $L = \{i\}$. They yield the conditional distribution of M given that there is a point of M with mark in L at the origin o. By analogy with (4.1.6) and (4.1.7), $\mathbf{P}_L$ and $\mathbf{E}_L$ can be defined by

$$\lambda \nu(W) \mathbf{P}_L(M \in \mathcal{A}) = \mathbf{E} \left(\sum_{[x; m(x)] \in M} \mathbf{1}_W(x) \mathbf{1}_L(m(x)) \mathbf{1}_{\mathcal{A}}(M - x) \right), \qquad (5.1.20)$$

where $M \in \mathcal{A}$ means that the marked point process M has property $\mathcal{A}$, and

$$\lambda \nu(W) \mathbf{E}_L(\mathcal{S}(M)) = \mathbf{E} \left(\sum_{[x; m(x)] \in M} \mathbf{1}_W(x) \mathbf{1}_L(m(x)) \mathcal{S}(M - x) \right) \qquad (5.1.21)$$

for any function $\mathcal{S}(M)$ which assigns a real number to the marked point process M. Consider the following two specific cases:

- Let $L = (-\infty, \infty)$ and $\mathcal{A}$ be a process property that is not based on the marks (such as $\mathcal{A} = $ 'M has three points in some set'). Then

$$\mathbf{P}_L(M \in \mathcal{A}) = \mathbf{P}_o(N \in \mathcal{A}),$$

  where, as above, N is the point process M without the marks.

- $\mathcal{A} = $ 'there is a point of M at the origin o with mark $m(o) \le m$'. Then

$$\mathbf{P}_{\mathbb{R}}(M \in \mathcal{A}) = F_{\mathcal{M}}(m).$$

  Hence it is clear that the mark d.f. $F_{\mathcal{M}}(m)$ may be regarded as a characteristic related to the Palm distribution.

The distribution $\mathbf{P}_L$ may be refined based on $\mathbf{P}_{(o,m)}$, which describes M given that there is a point with mark m at o. This means that

$$\mathbf{P}_L(M \in \mathcal{A}) = \int_L \mathbf{P}_{(o,m)}(M \in \mathcal{A}) f_{\mathcal{M}}(m) \mathrm{d}m.$$

Campbell–Mecke formula

Let $f(x, m, M)$ be a function depending on a point x in $\mathbb{R}^d$, a real number m and a stationary marked point process M. Then

$$\mathbf{E} \left(\sum_{[x; m(x)] \in M} f(x, m(x), M) \right) = \lambda \int_{\mathbb{R}^d} \int_{-\infty}^{\infty} \mathbf{E}_{(o,m)}(f(x, m, M_{-x})) f_{\mathcal{M}}(m) \mathrm{d}m \mathrm{d}x. \quad (5.1.22)$$

Here $\mathbf{E}_{(o,m)}$ denotes the expectation with respect to $\mathbf{P}_{(o,m)}$.

5.2   Summary characteristics

5.2.1   Introduction

Summary characteristics for stationary marked point processes serve a similar purpose as those for non-marked processes: they are numbers or functions that describe specific aspects of the point distribution. In the analysis of marked point patterns the marks of the points are also taken into account, resulting in a more interesting and diverse theory.

Summary characteristics for marked point processes are constructed based on the same ideas as those for non-marked point processes. Therefore this section assumes that the reader is familiar with Section 4.2, where summary characteristics for non-marked processes are treated in detail. In particular, this section does not discuss the various edge-correction methods again.

However, there are some problems which edge effects which do not occur in the analysis of non-marked processes. The mark information assigned to a point does not always originate directly from its location x as, for example, for tree marks $m_1(x)$ = species or $m_2(x)$ = dbh. Consider, for instance, marked points that describe particles, where x = centre and $m(x)$ = volume. To determine the volume, information from the neighbourhood of x has to be taken into account, which may be impossible to obtain for particles that are near the edge of the window W.

5.2.2   Intensity and mark-sum intensity

The *intensity* λ is the mean number of points per unit area or volume as discussed in Section 4.2.3, where the statistical estimation is also described in detail.

Usually it is not at all necessary to take the marks into consideration when assessing the intensity of a pattern. Indeed, if the points are well-defined, the marks do not play any role in the estimation of the intensity. Problems may arise with large objects or particles, e.g. microscopic objects such as in Figures 1.6 and 5.7, where the 'points' x_n have been constructed in the analysis as particle centres. In this case it is not always clear whether all particles in a window should be included in the analysis, e.g. if the particles cross the edge of the window W. Spatial statistics provides efficient methods for the estimation of λ in these cases, which are based on geometrical ideas: the Gundersen frame and the equation-system method, which uses, in the planar case, estimates of area, boundary length and Euler number of all particles in W; see Stoyan et al. (1995, p. 222).

For qualitative marks, the intensity λ_i of the individual sub-point process of the points with mark i is of particular interest. It is estimated by

$$\hat{\lambda}_i = \sum_{x \in W} \mathbf{1}(m(x) = i)/\nu(W), \qquad (5.2.1)$$

i.e. by counting the points with mark i in W and dividing this by the area or volume of W.

The *mark-sum intensity* λ_S is a first-order characteristic which takes the marks into account. It yields the mean of the sum of all marks of the points per unit area or volume and satisfies

$$\lambda_S = \lambda\mu,$$

where μ is the mean mark; see (5.1.19). The standard estimator of λ_S is

$$\hat{\lambda}_S = \sum_{x\in W} m(x)/\nu(W), \tag{5.2.2}$$

i.e. the marks of the points in the observation window W are summed and divided by window area or volume. This estimator simply follows the definition of λ_S and is unbiased. It is also consistent if the marked point process M is ergodic. The estimator assumes that it is possible to count all points in W and to determine their marks. For cases where this is impossible, methods such as those discussed in Section 4.2.3 may be used.

Voronoi cell weighting

This estimation method is based on two types of marks for the points x, the natural marks $m(x)$ and the constructed area marks $a(x)$, where $a(x)$ is the area of the Voronoi cell containing the point x. It is similar to the approach described on p. 191, but now for those points in the pattern that are closest to the measurement locations y_i the natural marks m_i are also used. λ_S may then be estimated as

$$\hat{\lambda}_{S,D} = \frac{1}{n}\sum_{i=1}^{n}\frac{m_i}{a_i}. \tag{5.2.3}$$

Its unbiasedness can be shown by analogy with that of $\hat{\lambda}_V$.

Line transect sampling

This estimation method is the same as on p. 192, but the marks m_i are considered in addition to the points in the strip $\mathcal{L}$. The estimator of λ_S is then

$$\hat{\lambda}_{S,H} = \frac{1}{2L}\sum_{i=1}^{n}\frac{m_i}{\varepsilon v_i}, \tag{5.2.4}$$

where L is the length of the strip and εv_i describes the visibility of point i as explained on p. 192.

Angle count sampling

In forestry, the following estimator has been applied to tree patterns described by marked points $[x; m(x)] = [\text{tree position; dbh}]$ to determine the 'basal area'

$$\lambda_S = \frac{\pi}{4}\lambda\mu_2,$$

where $\mu_2 = \sigma_m^2 + \mu^2$ is the second moment of the diameter mark, i.e. $\frac{\pi}{4}\mu_2$ is the mean cross-section area of the trees. The Bitterlich estimator (Bitterlich, 1947, 1984) for one observation point y_1 is

$$\hat{\lambda}_S = \sin^2\alpha \sum_{[x;m(x)]\in M} \mathbf{1}\left(\|x-y_1\| < \frac{m(x)}{2\sin\alpha}\right). \qquad (5.2.5)$$

Usually several points y_i are used and the results are averaged. In the estimator given by (5.2.5) all trees that can be seen from y_1 within an angle larger than a fixed size 2α are counted (see Figure 5.2). Analytically the counting condition is

$$\|x-y_i\| < \frac{m(x)}{2\sin\alpha},$$

which means that only trees that are sufficiently thick and not too far away from the point of observation are counted.

In theoretical terms, here the Bitterlich field plays a role, the shot-noise field $\{B(x)\}$ defined by

$$B(x) = \sum_{[x_i;m(x_i)]\in M} \mathbf{1}\left(\|x_i-x\| < \frac{m(x_i)}{2\sin\alpha}\right).$$

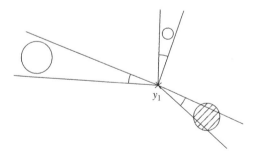

Figure 5.2   The Bitterlich count. The observer is positioned at y_1. All trees seen within an angle larger than 2α are counted, where α is a fixed angle. In the example shown only one tree (indicated by hatched shading) is counted.

Large marks

As noted above, the marks $m(x_n)$ of the points x_n are sometimes based on the configuration of the objects or their centres in some neighbourhood of x_n, for example when particles of considerable size are analysed or in the context of constructed marks such as the distance $d(x_n)$ from x_n to its nearest neighbour. In these cases it is possible that the marks of some of the points x cannot be determined if only information from inside the observation window W is available. Then *local plus sampling* and *local minus sampling* (or nearest-neighbour edge-correction) yield unbiased estimators:

$$\widehat{\lambda^+} = \sum_{[x; m(x)] \in M} \frac{\mathbf{1}_{W \oplus m(x)}(x)}{\nu(W \oplus m(x))} \tag{5.2.6}$$

and

$$\widehat{\lambda^-} = \sum_{[x; m(x)] \in M} \frac{\mathbf{1}_{W \ominus m(x)}(x)}{\nu(W \ominus m(x))}. \tag{5.2.7}$$

Here $m(x)$ denotes either the particle connected with x or a constructed numerical mark such as $d(x)$.

In the first case (which contradicts the general assumption in this book that the marks are numerical) $W \oplus m(x)$ and $W \ominus m(x)$ are interpreted in a set-theoretic sense (see Soille, 1999, and Stoyan et al., 1995, for the notation). In the second case, they are interpreted as $W_{\oplus m(x)}$ and $W_{\ominus m(x)}$ with

$$W_{\oplus m(x)} = W \oplus b(o, m(x)) \quad \text{and} \quad W_{\ominus m(x)} = W \ominus b(o, m(x)).$$

5.2.3 Mean mark, mark d.f. and mark probabilities

This section discusses the statistical estimation of the mark d.f. $F_{\mathcal{M}}(m)$, the mean mark μ and the mark probabilities p_i introduced in Section 5.1.5. These describe the probability distribution of the marks, i.e. their random fluctuations. These characteristics do not have a spatial nature, and hence classical statistical methods may be applied to estimate them. Problems occur only when the marks cannot be determined for all points if only data from inside the window can be used.

Recall that, as a distribution function, $F_{\mathcal{M}}(m)$ may be interpreted as the probability that the mark $\mathcal{M}$ of the typical point of M is smaller than (or equal to) m. Furthermore, p_i is the probability $\mathbf{P}(\mathcal{M} = i)$ that the typical point has mark i, in other words $\mathbf{P}_o(m(o) = i)$. The quantity μ is the mean of the mark of the typical point, i.e. the mean corresponding to $F_{\mathcal{M}}(m)$ or $\{p_i\}$.

The standard estimators are

$$\hat{F}_{\mathcal{M}}(m) = \sum_{x \in W} \mathbf{1}(m(x) \le m)/N(W), \tag{5.2.8}$$

$$\hat{\mu} = \sum_{x \in W} m(x)/N(W) \tag{5.2.9}$$

and

$$\hat{p}_i = \hat{\lambda}_i/\hat{\lambda}, \tag{5.2.10}$$

where $N(W)$ is the number of points of M in the observation window W (without the marks) and $\hat{\lambda}_i$ and $\hat{\lambda}$ are the intensity estimators given in (5.2.1) and (4.2.10). All three estimators are quite natural. $\hat{F}_{\mathcal{M}}(m)$ is simply the fraction of points x in W with $m(x) \le m$ and $\hat{p}_i$ the relative frequency of points with mark i. The mark variance σ_μ^2 as well as the standard deviation σ_μ may be estimated in a similar way.

Example 5.3.   Mark probabilities and mark probability density function for the amacrine cells and gold particles

The sample of amacrine cells in Section 1.2.1 consists of 294 points in total, 152 of which are on-cells ($i = 1$) and 142 are off-cells ($i = 2$). Consequently, the empirical mark probabilities are $\hat{p}_1 = 0.517$ and $\hat{p}_2 = 0.483$.

Figure 5.3 shows the estimated mark p.d.f. for the gold particles introduced on p. 6. The gold particle diameters have a nearly symmetric density function and the estimated mean mark and mark standard deviation are

$$\hat{\mu} = 6.42\,\mu\text{m} \quad \text{and} \quad \hat{\sigma}_\mu = 0.87\,\mu\text{m}.$$

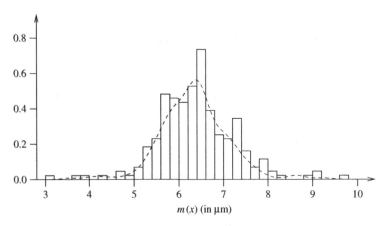

Figure 5.3   Mark p.d.f. for the gold particle diameters.

The three estimators above are not unbiased, since $N(W)$ is random. However, they are ratio-unbiased, since they are quotients of unbiased estimators. For example, $\hat{\mu}$ can be rewritten as

$$\hat{\mu} = \hat{\lambda}_S/\hat{\lambda}$$

with $\hat{\lambda}_S$ from (5.2.2) and $\hat{\lambda}$ from (4.2.10) or any of the other estimators in Section 4.2.3.

In the estimation of the mark d.f. it may be helpful to know that $F_{\mathcal{M}}(m)$ can be expressed in terms of a mark-sum intensity λ_S for a suitably constructed mark $c(x)$:

$$\lambda_S = \lambda F_{\mathcal{M}}(m),$$

where

$$c(x) = \begin{cases} 1 & \text{for } m(x) \le m, \\ 0 & \text{otherwise.} \end{cases}$$

Then the mark-sum intensity λ_S based on these c-marks is the same as $\lambda F_{\mathcal{M}}(m)$. In cases with constructed marks, the mark d.f.s may be estimated based on estimation approaches for the nearest-neighbour distance d.f. $D(r)$ as described in Section 4.2.6. Indeed, the neighbour distance d.f. can be regarded as a mark d.f. – specifically as a mark d.f. of constructed $d(x)$-marks.

Note an important consequence from the discussion on p. 209: λ_S and λ should be estimated based on the same principle. It is, for example, usually not helpful to combine an estimator of λ of very high precision with some non-edge-corrected estimator of λ_S.

5.2.4   Indices for stationary marked point processes

Indices are numerical summary characteristics that describe the distribution of point processes in a simple and elegant way. They were discussed for non-marked point patterns in Section 4.2.4. In the context of stationary marked point processes they have a similar role; again location- and point-related indices may be distinguished. However, the more complicated aim is to find indices that provide useful information on the relative positions of the points dependent on the marks and on the correlations of the marks.

Note that the construction of indices, including those considered in Chapter 4, may actually be described based on marked point process terminology. This provides a more systematic understanding of the nature of point-related indices, even for non-marked patterns.

Given a marked point process M (or a non-marked process N), new marks $c(x_n)$ are constructed for its points as follows. For all points x_n the $c(x_n)$ are determined according to some rule. This means that x_n is considered as the origin of a new coordinate system and $c(x_n)$ is then determined by some function f. In mathematical terms this means that the marked point process M is shifted such that point x_n is moved to the origin, i.e. $M_{-x_n} = \{[x_1 - x_n, m(x_1)], \ldots, [o; m(x_n)], \ldots\}$ is constructed and the numerical value $f(M_{-x_n})$ is calculated for the resulting marked point pattern by means of some function f. An example in the non-marked case is

$$c_1(x_n) = f(M_{-x_n}) = \text{distance from } o \text{ to the nearest point in } M_{-x_n},$$

which is simply $d(x_n)$, the distance from x_n to its nearest neighbour. An example in the marked case is the mark

$$c_2(x_n) = f(M_{-x_n}) = \text{sum of the marks of } M_{-x_n} \text{ in } b(o, r), \text{ not counting } o,$$

which is the number of points of M in the disc or sphere $b(x_n, r)$, not counting x_n.

In this way, the points x_n are allocated new marks $c(x_n)$. By construction, the marked point process $M_c = \{[x_n; c(x_n)]\}$ is stationary if M is stationary.

Valuable information is contained in the corresponding mark distribution, i.e. in the mark d.f., the mean mark or, if the $c(x_n)$ are integer-valued, the mark probabilities. A well-known example of such a mark d.f. is the nearest-neighbour distance d.f. $D(r)$, which is based on $d(x)$-marks.

Based on the mean (constructed) mark, indices may be derived in a straight-forward way. These are often standardised by considering the ratio with other point process characteristics, such as the intensity λ, to allow comparison between patterns with different intensities.

This subsection discusses several indices, which mainly originate in forestry and ecology, where they have been popular and useful, for example, in characterising, for a given plant, the strength of competition from its neighbours. However, these indices may be suitably applied in other areas as well, for instance in physics or materials science. The exposition commences with a location-related index and then presents point-related and distance-dependent indices related to the nearest neighbour, the k nearest neighbours or, finally, to all neighbours within some given distance r.

A location-related index

The index of dispersion defined by (4.2.22) characterises the variability of numbers of points in test sets. A similar index, the *index of mark-sum dispersion*, may be defined for marked point patterns:

$$IMD = \frac{\text{var} S(B)}{\lambda \mu \nu(B)}, \tag{5.2.11}$$

where B is some test set, for example a disc of radius r. In other words, IMD is the ratio of the variance of the sum of the marks of the points in B and the corresponding mean. Clearly, the numerical values depend on the set B or on the radius r.

Point-related indices for qualitative marks

Bivariate aggregation index. The Clark–Evans aggregation index CE defined by (4.2.24) may be generalised to the bivariate case as

$$CE_{12} = m_{D.1,2} \cdot 2\sqrt{\lambda_2}.$$

Here, $m_{D.1,2}$ is the mean distance from the typical point of type 1 (i.e. the typical point of the subprocess of points with mark 1) to its nearest neighbour of type 2, and λ_2 is the intensity of the subprocess of points of type 2. To facilitate the comparison of this mean distance to a similar mean for an independently marked Poisson process, the mean is multiplied by the factor $2\sqrt{\lambda_2}$.

To understand this, consider a Poisson process of intensity $\lambda = \lambda_1 + \lambda_2$ where the points are independently marked as type 1 and type 2 points yielding two subprocesses of points of type 1 and type 2 with intensities λ_1 and λ_2. For this process the mean distance from the typical type 1 point to its nearest type 2 point is $\frac{1}{2\sqrt{\lambda_2}}$, independent of λ_1.

Hence, values of CE_{12} greater than 1 indicate repulsion between the points of type 1 and the points of type 2, and values smaller than 1 indicate attraction.

Segregation and mingling. Consider now, in the bivariate case, the probabilities of all combinations of the mark of the typical point and the mark of its nearest neighbour, i.e. the joint probabilities p_{kl} that the typical point has mark k and its nearest neighbour has mark l, where $k, l = 1, 2$. These four probabilities may be summarised as in Table 5.2. p_1 and p_2 are the mark probabilities as discussed in Section 5.1.5, and $p_{.k}$ is the probability that the nearest neighbour has mark k, irrespective of the mark of the typical point. Note that p_i is not necessarily equal to $p_{.i}$.

Clearly, the nearest-neighbour probabilities p_{kl} describe important aspects of the distribution of marks within a pattern and therefore useful indices can be derived based on these.

Table 5.2 Nearest-neighbour table with probabilities p_{ij}.

		Mark of NN		
		1	2	
Mark of typical point	1	p_{11}	p_{12}	p_1
	2	p_{21}	p_{22}	p_2
		$p_{.1}$	$p_{.2}$	1

Coefficient of segregation. Pielou (1977) introduces an index that considers the ratio of the observed probability that the typical point and its nearest neighbour have different marks along with the same probability for independent marks, with fixed p_1 and p_2:

$$S = 1 - \frac{p_{12} + p_{21}}{p_1 p_{.2} + p_2 p_{.1}}. \qquad (5.2.12)$$

Consequently, $S = 0$ if the marks are independent. If the nearest neighbour always has the same mark as the typical point, then $p_{12} = p_{21} = 0$ and $S = 1$; if all neighbours have different marks, then $p_{11} = p_{22} = 0$, and S is negative, with a minimum of $S = -1$ for $p_1 = p_2 = \frac{1}{2}$.

Mingling index. The mingling index is defined here for general multivariate processes. For the special case of a bivariate process it can be expressed in terms of the p_{kl}. The index compares the mark of the typical point to those of its k nearest neighbours. Practical experience shows that in forestry applications it is often sufficient to consider the first three or four neighbours, $k = 3$ or 4; see Füldner (1995) and Aguirre et al. (2003). The mingling index is defined as

$$\overline{M}_k = \frac{1}{k} \mathrm{E}_o \left(\sum_{i=1}^{k} \mathbf{1}(m(o) \neq m(z_i(o))) \right), \qquad (5.2.13)$$

which is the mean fraction of points among the k nearest neighbours of the typical (or reference) point that have a mark different from that of the reference point. Thus $\overline{M}_k$ characterises the mixture of marks.

The mingling index is based on the constructed mark

$$\overline{M}_k(x_n) = \frac{1}{k} \sum_{l=1}^{k} \mathbf{1}(m(x_n) \neq m(z_l(x_n))). \qquad (5.2.14)$$

Here $z_l(x_n)$ is the lth neighbour of x_n and $m(z_l(x_n))$ its mark. That is, $\overline{M}_k(x_n)$ describes the fraction of the k nearest neighbours with a mark different from that of x_n.

When the marks of the typical point and its neighbours tend to be different, $\overline{M}_k$ has a large value; in the opposite case the index shows that points with different marks are segregated. In the bivariate case with independent marks,

$$\overline{M}_{ki} = 2 p_1 p_2. \qquad (5.2.15)$$

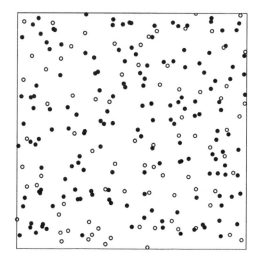

Figure 5.4   Oak (type 1,∘) and beech trees (type 2,•) in a square window *W* of side length 80 m. For a detailed description, see Pommerening (2002).

Example 5.4.   Segregation and mingling index for two bivariate patterns
This example considers the pattern of the amacrine cells introduced in Section 1.2.1 as well as a pattern of oaks and beeches with 80 oak trees (1) and 164 beech trees (2) in a square window *W* of side length 80 m from the Manderscheid 198 forest stand aged 118 years in the German federal state of Rheinland-Pfalz (see Figure 5.4). These trees form an interesting mixed pattern, which is the result of the clever work done by foresters whose aim is to harvest good-quality oak timber. For them, the most important tree species in the forest is oak, which needs light, while beech tolerates shade. Foresters use the beech trees to cast their shadow on the oak stems to prevent epicormic growth of oaks, which would otherwise reduce the timber quality. Beech trees that compete too strongly are removed to avoid dominance in the forest. In these managed forests the oak and beech trees may be considered having an equal degree of dominance (see Table 5.1) which is entirely due to the management strategies applied by the foresters.

Visual inspection reveals that the pattern is not completely randomly mixed. The beech trees appear in small clusters while the oaks have larger inter-tree distances (which again is a result of the foresters' work). The whole pattern (irrespective of the marks) is somewhat irregular and similar to a sample from a Poisson process. It was used in Mecke and Stoyan (2005) as a pattern which merits a test of CSR. It is difficult to decide whether the pattern exhibits CSR based on visual inspection alone and some tests of little power do not reject CSR. For this data set, the CSR hypothesis can only be rejected with very powerful tests.

The empirical nearest-neighbour probabilities $\hat{p}_{ij}$ for the amacrine cell pattern are as follows:

$\hat{p}_{ij}$		$\hat{p}_{i.}\hat{p}_{.j}$	
0.081	0.434	0.262	0.253
0.428	0.057	0.247	0.238

The $\hat{p}_{ij}$ indicate that on- and off-cells frequently have a cell of the other type as their nearest neighbour as $\hat{p}_{12}$ and $\hat{p}_{21}$ are relatively high. However, further analysis will show that the two subpatterns are actually independent and that the pattern is of class (a); see Table 5.1. Thus there is no attraction among type 1 and type 2 points, but it is the regularity of the two subpatterns which makes it unlikely that two points of the same type are nearest neighbours.

For the oaks and beeches the corresponding probabilities are as follows:

$\hat{p}_{ij}$		$\hat{p}_{i.}\hat{p}_{.j}$	
0.044	0.265	0.080	0.229
0.215	0.476	0.179	0.513

The $\hat{p}_{ij}$ show that there is some repulsion between trees of the same type and some mutual attraction between the two tree species, which is weaker than in the pattern of the amacrine cells. Further analysis will reveal that the two subpatterns are not independent and that the pattern is of class (b); see Table 5.1.

Table 5.3 shows the mingling and segregation indices for the two patterns above, as well as for the concrete pattern (see Example 5.5) and the mounds and palms pattern (see Example 5.7).

Table 5.3   The indices $\overline{M}_4$ and S (mingling and segregation) for four point patterns $\overline{M}_4$, in comparison with $\overline{M}_{4i}$ as in (5.2.15): (a) amacrine cells; (b) oaks and beeches; (c) concrete, see Example 5.5; (d) = mounds and palms, see Example 5.7.

Case	$\overline{M}_4$	$\overline{M}_{4i}$	S
(a)	0.86	0.50	−0.73
(b)	0.48	0.44	−0.18
(c)	0.36	0.36	0.07
(d)	0.37	0.43	0.09

The values in the third column of Table 5.3 are the mingling indices expected in the case of independent marks, with the observed frequencies of type 1 and type 2 points. For the amacrine cells clear differences between mark and nearest-neighbour mark are indicated, while for the oaks and beeches these are not so clear. In contrast, for the concrete pattern and the mounds and palms the marks of typical point and nearest neighbour seem to be independent.

Point-related indices for quantitative marks

Nearest-neighbour correlation indices. Stoyan and Stoyan (1992, 1994) introduce a general principle for the construction of nearest-neighbour indices of the form

$$\mathbf{E}_o(t(\text{mark of typical point, mark of nearest neighbour})),$$

where $t(m(o), m(z_1(o))$ is a suitable test function. In the following, three examples of this type are briefly discussed.

The *nearest-neighbour mark product index is given by*

$$\mathbf{n}_{mm} = \mathbf{E}_o(m(o) \cdot m(z_1(o)))/\mu^2, \tag{5.2.16}$$

where μ is the mean mark. A value of $\mathbf{n}_{mm}$ larger than 1 indicates that the mean mark product of the typical point and its nearest neighbour is above the average, i.e. it indicates some mutual stimulation. By definition, $\mathbf{n}_{mm} = 1$ if the marks are independent and $\mathbf{n}_{mm} < 1$ if there is mutual inhibition.

A related index is the *nearest-neighbour mark index,*

$$\mathbf{n}_{.m} = \mathbf{E}_o(m(z_1(o))/\mu. \tag{5.2.17}$$

This is the normalised mean mark of the nearest neighbour of the typical point. A value of $\mathbf{n}_{.m}$ larger than 1 means that points close to other points tend to have large marks. Again independent marks imply $\mathbf{n}_{.m} = 1$, and $\mathbf{n}_{.m} < 1$ indicates inhibition.

Finally, the *nearest-neighbour variogram index*

$$\mathbf{n}_\gamma = \frac{1}{2}\mathbf{E}(m(o) - m(z_1(o)))^2 \Big/ \sigma_\mu^2 \tag{5.2.18}$$

denotes half the squared deviation between the mark of the typical point and the mark of its nearest neighbour. It is related to the mark variogram $\gamma_m(r)$ discussed on p. 344. Theoretically, it can take any positive value. It characterises the variability of the marks of points that are close together. Constant marks imply that $\mathbf{n}_\gamma = 0$, while for independent marks $\mathbf{n}_\gamma = 1$ because of the normalisation by the mark variance σ_μ^2. Example 6.4 demonstrates the application of these indices in an example from forest ecology, which aims to show that the marks are independent, and in Example 5.8 they are used with the same aim for the gold particles.

Mark dominance index. Using an approach similar to the mingling index, Hui et al. (1998) introduce the index

$$Do_k = \frac{1}{k} \mathbf{E}_o \left(\sum_{l=1}^{k} \mathbf{1}(m(o) > m(z_l(o))) \right), \quad (5.2.19)$$

which includes the k nearest neighbours $z_l(o)$ of the typical point, for $l = 1, 2, \ldots, k$. It is the fraction of points of the k-neighbourhood of the typical point which have a smaller mark than the typical point.

The mark dominance index is based on constructed marks that are defined by

$$Do_k(x_n) = \frac{1}{k} \sum_{l=1}^{k} \mathbf{1}(m(x_n) > m(z_l(x_n))), \quad (5.2.20)$$

where $z_l(x_n)$ is the lth neighbour of point x_n. For example, if the points are tree positions and the marks tree size parameters such as dbh or height, $Do_k(x_n)$ is the proportion of trees in the k-neighbourhood that are dominated by the tree at x_n.

The approach applied in the mingling and dominance index can be used to construct further indices; see Füldner (1995), Hui et al. (1998) and Pommerening (2002).

The following two indices are based on r-neighbourhoods, i.e. on the points within a distance r from the reference tree.

Hegyi index. Hegyi (1974) introduced a competition index that reflects the strength of competition in a point pattern. It is based on the following constructed mark:

$$H(x_n) = \sum_{l=1}^{n_r(x_n)} \frac{m(z_l(x_n))}{m(x_n)(1 + \|x_n - z_l(x_n)\|)}, \quad (5.2.21)$$

where $\|x_n - z_l(x_n)\|$ is the distance between x_n and its lth nearest neighbour $z_l(x_n)$. The index for the whole pattern is again the corresponding mean mark H. Large values indicate a high degree of competition pressure on the points of the pattern.

Moravie index. The index introduced by Moravie et al. (1999) is again a competition index which aims to quantify the competition pressure resulting from close neighbours expressed in terms of their heights. It is based on the following mark:

$$Mo(x_n) = \sum_{l=1}^{n_r^>(x_n)} \frac{\pi - \arctan\left(\frac{m(x_n) - m(z_l(x_n))}{\|x_n - z_l(x_n)\|}\right)}{\pi}, \quad (5.2.22)$$

where $n_r^>(x_n)$ is the number of points in $b(x_n, r)$ with a mark larger than $m(x_n)$. This index was introduced in the context of trees, and the marks $m(x_n)$ were tree heights.

Further competition indices based on influence areas around trees and their mutual intersections were considered in Bella (1971); see Berger and Hildenbrandt (2000) for a more recent paper; see also the references on p. 159.

Statistical estimation of the indices

In the following, only naive estimators without edge-correction are presented, since Pommerening and Stoyan (2006) show that indices based on nearest neighbours should be estimated without any edge-correction. This means that it is only necessary to take into account the 'nearest neighbours *within* the window W'. The notation used in the estimators is simplified with reference to Table 5.4. As usual, $n = N(W)$ is the total number of points in W. Each of these points is used as a reference point. Here a is the number of all pairs of points where the reference point and its nearest neighbour in W both have mark 1. The frequencies b, c and d are defined analogously. Clearly $r = a + b$, and similarly for s, u and v.

Then the naive estimators of CE_{12}, $\overline{M}_k$ and S are

$$\widehat{CE} = 2\overline{d}_{12}\sqrt{\hat{\lambda}_2}, \tag{5.2.23}$$

$$\widehat{\overline{M}}_k = \frac{b+c}{n} \tag{5.2.24}$$

and

$$\hat{S} = 1 - \frac{n(b+c)}{rv+su}, \tag{5.2.25}$$

where

$$\overline{d}_{12} = \frac{1}{m}\sum_{i=1}^{m}d_{12,i}$$

Table 5.4 Nearest-neighbour frequency table showing the observed numbers.

		Mark of NN in W		
		1	2	
Mark of	1	a	b	r
reference point	2	c	d	s
		u	v	n

with $d_{12,i}$ the distance from the ith type 1 point to its nearest type 2 neighbour and summation over all type 1 points, and

$$\hat{\lambda}_2 = \frac{s}{\nu(W)}.$$

Finally,

$$\widehat{Do}_k = \frac{1}{n} \sum_{i=1}^{n} \frac{1}{k} \sum_{j=1}^{n} \mathbf{1}(m_i > m_{n(i,j)}) \tag{5.2.26}$$

and

$$\hat{\mathbf{n}}_{mm} = \frac{1}{n} \sum_{i=1}^{n} (m_i m_{n(i)}) / \hat{\mu}^2 , \tag{5.2.27}$$

where m_i is the mark of the ith point in W and $m_{n(i)}$ that of its nearest neighbour in W; $\hat{\mu}$ is the estimated mean mark as in (5.2.9).

5.2.5 Nearest-neighbour distributions

Functional summary characteristics based on nearest neighbours are as useful for marked point processes as they are for non-marked processes. Now, of course, they also take the marks into account. Processes with qualitative and quantitative marks have to be considered separately.

Qualitative marks

In the following, $d_i(x)$ denotes the distance from x to the nearest point of type i in M.

Spherical contact d.f. $H_{s,i}(r)$. For non-marked point patterns, the spherical contact distribution function $H_s(r)$ was defined in Section 4.2.5. The definition is very similar for processes with qualitative marks. The spherical contact d.f. is simply applied to each of the subpatterns, i.e. by considering $H_{s,i}(r)$ for N_i, the subprocess of points with mark i, for $i = 1, 2, \ldots$:

$$H_{s,i}(r) = \mathbf{P}(d_i(o) \leq r) \qquad \text{for } r \geq 0. \tag{5.2.28}$$

This is the d.f. of the distance from the origin o to the nearest point of type i. Note that the sum

$$\sum_{i=1}^{k} H_{s,i}(r)$$

may be usefully interpreted on its own – it is the mean number of different marks in a spherical test set of radius r.

Nearest-neighbour distance d.f. $D_{ij}(r)$.   Similarly, one may consider the d.f. of the distance from the typical point of type i to its nearest neighbour of type j. Thus

$$D_{ij}(r) = \mathbf{P}_{oi}(d_j(o) \leq r) \qquad \text{for } r \geq 0. \tag{5.2.29}$$

The distribution function $D_{ij}(r)$ may be used to reveal differences in neighbourhood relations in multivariate point patterns. The $D_{ij}(r)$ may indeed differ for different indices i and j. This is the case, for example, for the stand of oaks and beeches considered in Example 5.4.

In the case of random labelling, the $D_{ij}(r)$ are

$$D_{ij}(r) = 1 - \sum_{k=0}^{\infty} \mathbf{P}_o(n_r(o) = k)(1 - p_j)^k \tag{5.2.30}$$

or

$$D_{ij}(r) = 1 - \mathbf{E}_o((1 - p_j)^{n_r(o)}),$$

where, as above, $n_r(o)$ is the number of points in the sphere of radius r around the typical point and p_j the mark probability for points of type j.

For an independently marked Poisson process this formula simplifies to

$$D_{ij}(r) = 1 - \exp(-\lambda p_j \pi r^2) \qquad \text{for } r \geq 0.$$

Recall the J-function introduced in Section 4.2.7, which is a ratio of $D(r)$ and $H_s(r)$, and note that Van Lieshout (2006a) developed a theory of multivariate J-functions.

Mingling distribution $\{M_{i:k}\}$.   Consider the k closest neighbours $z_1(x_n), \ldots,$ $z_k(x_n)$ of the point x_n. Count the number $M_k(x_n)$ of those neighbours that have a mark different from that of x_n, i.e. $m(x_n) \neq m(z_l(x_n))$. The corresponding discrete mark distribution is the mingling distribution,

$$M_{i:k} = \mathbf{P}_o(M_k(o) = i) \qquad \text{for } i = 0, 1, \ldots, k; \tag{5.2.31}$$

$M_{i:k}$ is the probability that the typical point has i neighbours with a different mark among the k nearest neighbours. This distribution is more informative than the mingling index in Section 5.2.4, which is simply the mean of the distribution $\{M_{i:k}\}$, i.e.

$$\overline{M}_k = \frac{1}{k} \sum_{i=0}^{k} i M_{i:k}.$$

Quantitative marks

Distributions based on nearest-neighbour information may also be defined for patterns with quantitative marks. The following example shows an approach which uses the marks of the nearest neighbours rather than the distances to them.

Mark dominance distribution. Consider again the k nearest neighbours $z_1(x_n), \ldots, z_k(x_n)$ of a point x_n and count the number $ND_k(x_n)$ of these neighbours with a mark larger than x_n, i.e. $m(z_l(x_n)) > m(x_n)$. The corresponding mark distribution is the mark dominance distribution,

$$Do_{i:k} = \mathbf{P}_o(ND_k(o) = i) \qquad \text{for } i = 0, 1, \ldots, k. \qquad (5.2.32)$$

$Do_{i:k}$ is the probability that the typical point has i neighbours with a larger mark among the k neighbours. Large $Do_{i:k}$ for large i indicate a high competition load on the typical point. The mark dominance index Do_k given by (5.2.19) is the mean of the distribution $\{Do_{i:k}\}$ divided by k.

Estimation of the nearest-neighbour characteristics

The nearest-neighbour characteristics for qualitative marks can be estimated in a very similar way for univariate processes, i.e. processes without marks (see Chapter 4). This means that $H_{s,i}(r)$ is estimated in the same way as $H_s(r)$, by applying the estimator $H_s(r)$ to each of the sub-point processes N_i. Similarly, $D_{ij}(r)$ is estimated by analogy with $D(r)$. Its nearest-neighbour estimator is

$$\hat{D}_{ij}(r) = \hat{\mathcal{D}}_{ij,n}(r) \Big/ \hat{\lambda}_{i,nn} \qquad (5.2.33)$$

with

$$\hat{\mathcal{D}}_{ij,n}(r) = \sum_{x \in N_i} \mathbf{1}_{W_{\ominus d_j(x)}}(x) \mathbf{1}(0 < d_j(x) \le r) \Big/ \nu(W_{\ominus d_j(x)})$$

and

$$\hat{\lambda}_{i,nn} = \sum_{x \in N_i} \mathbf{1}_{W_{\ominus d_j(x)}}(x) \Big/ \nu(W_{\ominus d_j(x)}),$$

where $d_j(x)$ denotes the distance from x to the nearest point of type j.

The nearest-neighbour edge-correction method can also be applied to the other characteristics. It yields, for example,

$$\widehat{M}_{i:k} = \left(\sum_{[x; m(x)] \in M} \mathbf{1}_{W_{\ominus d^k(x)}}(x) \mathbf{1}(M_k(x) = i) \Big/ \nu(W_{\ominus d^k(x)}) \right) \Big/ \hat{\lambda}_{nn,k}$$

where $d^k(x)$ is the distance from x to the kth neighbour of x, $M_k(x)$ the number of neighbours of x with a mark different from that of x among the k nearest neighbours, and

$$\hat{\lambda}_{nn,k} = \sum_{[x;m(x)]\in M} \mathbf{1}_{W_{\ominus d_k(x)}}(x)/\nu(W_{\ominus d_k(x)}).$$

5.3 Second-order characteristics for marked point processes

5.3.1 Introduction

Second-order characteristics for marked point processes are as useful and as popular a tool as for non-marked point process. Here, the aim is to characterise not only the variability of the point distribution but also the variability of the marks and, further, to describe correlations among marks and points. A large variety of characteristics may be constructed for this purpose; this section presents some of the most useful ones. Section 7.5 presents significance tests for checking the significance of statistically detected mark correlations.

Again, the exposition first discusses characteristics for processes with qualitative marks, i.e. bi- and multivariate processes.

5.3.2 Definitions for qualitative marks

Recall that N_i is the sub-point process of points with mark i and intensity λ_i, $i = 1, 2, \ldots, m$, where m denotes the number of different marks. $\mathbf{E}_{oi}$ denotes the mean with respect to the Palm distribution for points of type i, i.e. the conditional mean given that the typical point, which is located at o, is of type i.

Multivariate or inter-type *K*-functions

The K-function as defined in Section 4.3.1 may be generalised for multivariate point processes in a straightforward way. Define K_{ij} by

$$\lambda_j K_{ij}(r) = \mathbf{E}_{oi}(N_j(b(o, r))) \qquad \text{for } r \geq 0. \tag{5.3.1}$$

Thus, $\lambda_j K_{ij}(r)$ is the mean number of points of type j in a disc (sphere) of radius r centred at the typical point of type i. Since $i \neq j$ the point at o is not counted. For $i = j$ the notation may be simplified to $K_i(r)$, which is the classical K-function for the subprocess N_i.

Note that

$$K_{ij}(r) = K_{ji}(r) \qquad \text{for } r \geq 0. \tag{5.3.2}$$

This important equation is discussed in detail below, in the context of partial pair correlation functions $g_{ij}(r)$.

Remark. It is often useful to also consider what one might call 'condensed' K-functions $K_{i\cdot}(r)$ and $K_{\cdot j}(r)$, defined by

$$\lambda K_{i\cdot}(r) = \mathbf{E}_{oi}(N(b(o, r)\backslash\{o\})) \qquad \text{for } r \geq 0 \qquad (5.3.3)$$

and

$$\lambda_j K_{\cdot j}(r) = \mathbf{E}_o(N_j(b(o, r)\backslash\{o\})) \qquad \text{for } r \geq 0 .$$

The interpretation of these functions is similar to that of the $K_{ij}(r)$: $\lambda K_{i\cdot}(r)$ is the mean number of points (irrespective of their marks) in a disc (sphere) of radius r centred at the typical point of type i, which itself is not counted. Similarly, $\lambda_j K_{\cdot j}(r)$ is the mean number of points of type j in a disc (sphere) of radius r centred at the typical point (irrespective of its mark). The condensed K-functions are given by

$$\lambda K_{i\cdot}(r) = \sum_{j=1}^{n} \lambda_j K_{ij}(r) \quad \text{and} \quad K_{\cdot j}(r) = \sum_{i=1}^{n} p_i K_{ij}(r)$$

with $p_i = \lambda_i/\lambda$.

The multivariate K-functions may be used to reveal information as to which marking model (see Section 5.1.3) may be suitable for a specific process.

If the marks are independent, i.e. if the marking may be regarded as random labelling, then all $K_{ij}(r)$ coincide, and

$$K_{ij}(r) = K(r) \qquad \text{for all } i \text{ and } j , \qquad (5.3.4)$$

where $K(r)$ is the K-function of the entire point process N, i.e. M irrespective of the marks. This is an effect of the definition of the K-functions by normalisation.

In the case of the random superposition model discussed on pp. 297 and 370,

$$K_{ii}(r) = K_i(r) \quad \text{and} \quad K_{ij}(r) = b_d r^d \qquad \text{for } i \neq j , \qquad (5.3.5)$$

where $K_i(r)$ is the K-function of the subprocess N_i.

Differences between the $K_{ij}(r)$ for different indices i and j and between the $K_{ij}(r)$ and $K(r)$ indicate correlations between the marks. Note, however, that the functions $g_{ij}(r)$ and $p_{ij}(r)$ introduced below are more suitable for the detection of these correlations in an exploratory analysis than the cumulative functions. This is analogous to the non-marked case as discussed in Section 4.3. And again the K_{ij}-functions should be used for the construction of statistical tests, which test the null hypothesis of independent marking vs. superposition. This is discussed in Section 7.5.

Again, as in Section 4.3, it is useful to transform the $K_{ij}(r)$ to *multivariate L-functions* via

$$L_{ij}(r) = \sqrt[d]{\frac{K_{ij}(r)}{b_d}} \qquad \text{for } r \geq 0. \tag{5.3.6}$$

Partial or cross-pair correlation functions $g_{ij}(r)$

The mark-independent product density $\varrho(r)$ introduced in Section 4.3.1 can be refined for multivariate point processes in a straightforward way by defining the product densities $\varrho_{ij}(r)$.

The term $\varrho_{ij}(r)\mathrm{d}x\mathrm{d}y$ describes the probability of finding a point of type i in an infinitesimally small sphere $b(x)$ of volume $\mathrm{d}x$ and a point of type j in an infinitesimally small sphere $b(y)$ of volume $\mathrm{d}y$, where the distance between the two sphere centres is r.

Clearly,

$$\varrho_{ij}(r) = \varrho_{ji}(r) \qquad \text{for all } i \text{ and } j \text{ and } r \geq 0. \tag{5.3.7}$$

As for K_{ij} above, $i = j$ yields the second-order product density of the subprocess N_i.

By analogy with the construction of the pair correlation function in (4.3.10), partial pair correlation functions $g_{ij}(r)$ may be derived from $\varrho_{ij}(r)$ by normalisation,

$$g_{ij}(r) = \varrho_{ij}(r)/\lambda_i\lambda_j \qquad \text{for } r \geq 0. \tag{5.3.8}$$

The symmetry of the $\varrho_{ij}(r)$ in (5.3.7) implies that

$$g_{ij}(r) = g_{ji}(r) \qquad \text{for all } i \text{ and } j \text{ and } r \geq 0. \tag{5.3.9}$$

And again, as in (4.3.8), the partial pair correlation function may also be defined as the derivative of the K_{ij}-function,

$$g_{ij}(r) = K'_{ij}(r) \big/ db_d r^{d-1} \qquad \text{for } r \geq 0. \tag{5.3.10}$$

It is now straightforward to show that (5.3.2) holds, by (5.3.10) together with (5.3.9) and since $K_{ij}(0) = 0$ for all i and j.

The global behaviour of the partial pair correlation functions $g_{ij}(r)$ is the same as the behaviour of the classical $g(r)$, i.e.

$$g_{ij}(r) \geq 0 \tag{5.3.11}$$

and

$$\lim_{r \to \infty} g_{ij}(r) = 1 \qquad\qquad (5.3.12)$$

for all i and j. They may be interpreted along the lines of $g(r)$; see the detailed discussion in Section 4.3.4 and the examples below. Most importantly, they are valuable tools for finding a suitable marking model (see Section 5.1.3) for a specific marked point pattern.

In the case of *random labelling*,

$$\varrho_{ij}(r) = p_i p_j \varrho(r)$$

and

$$g_{ij}(r) = g(r) \qquad \text{for } r \geq 0, \qquad\qquad (5.3.13)$$

where $g(r)$ is the pair correlation function of the whole non-marked point process N. The *random superposition model* yields

$$g_{ii}(r) = g_i(r) \quad \text{and} \quad g_{ij}(r) \equiv 1 \qquad \text{for } i \neq j,$$

where $g_i(r)$ is the pair correlation function of the subprocess N_i.

Example 5.5.   Partial pair correlation functions for bivariate point patterns

(a) The data for the amacrine cells shown in Figure 1.2 consist of locations of cells in the retina of a rabbit. Refer to Section 1.2.1 for more background to this data set and the nature of the different cells. Points of type 1 are on-cells and points of type 2 are off-cells. The relationship between the two subprocesses formed by points of type 1 and 2 is analysed here. Based on the biological function of the cells, one may assume that the two types of points do not dominate each other, thus one of the cases (a) or (b) in Table 5.1 is given.

The pattern is analysed with the partial pair correlation functions $g_{ij}(r)$ for i, $j = 1$, 2. Figure 5.5 shows estimates of these functions obtained with the estimator (5.3.49) with bandwidth $h = 13.2\,\mu\mathrm{m}$.

Apparently, $\hat{g}_{11}(r)$ and $\hat{g}_{22}(r)$, the estimates of the pair correlation functions for the sub-point processes of cells of type 1 and 2, are rather similar. The type 2 process (off-cells) is perhaps slightly more regular than the type 1 process, as indicated by the higher first maximum at around $r = 75$. The range of correlation in

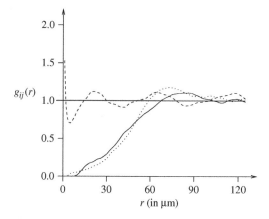

$g_{ij}(r)$

Figure 5.5   Empirical partial pair correlation functions $\hat{g}_{ij}(r)$ for the amacrine cells: $\hat{g}_{11}(r)$ (solid line), $\hat{g}_{12}(r)$ (dashed line), $\hat{g}_{22}(r)$ (dotted line).

both patterns seems to be around 100 μm. The soft-core form results from projection effects; the cells form hard-core patterns in their layers.

The function $\hat{g}_{12}(r)$, one of the functions describing inter-type correlations between the patterns of type 1 and type 2 points, fluctuates around the value 1. These fluctuations are irregular and show no clear pattern. One may thus assume that the 'true' $g_{12}(r)$ is constant and equal to 1, reflecting independence between the two patterns. A formal test of the corresponding independence hypothesis is presented in Section 7.5.

(b) Consider again the oaks and beeches introduced in Example 5.4. The aim is to identify the beech cluster size, the inhibition distance of oaks and the nature of the correlation between oaks (1) and beeches (2).

Figure 5.6 shows the estimated partial pair correlation functions $\hat{g}_{ij}(r)$. Clearly, these functions differ substantially: $g_{11}(r)$ indicates a hard-core process for the oaks (with minimum inter-point distance 2.0 m). It reflects some short-range order (reflected in the maxima at $r \approx 3$ m and 7 m, probably caused by the first and second neighbour of the typical oak).

In stark contrast, $\hat{g}_{22}(r)$ for the beeches resembles the pair correlation function of a cluster process with a cluster diameter of 1 m; the minimum inter-point distance for points of type 2 is 0.5 m. Note also the maximum of $\hat{g}_{22}(r)$ at $r \approx 6$ m, which may reflect the distance between beech clusters.

The function $\hat{g}_{12}(r)$, which describes the correlation between type 1 and type 2 points, is difficult to interpret in this example. It remains below 1 for almost all r, indicating repulsion between the two types of points. The range of correlation seems to be around 8 m for $(i, j) = (2, 2)$ and $(1, 2)$, whereas for $(1, 1)$ it appears to be larger, around 12 m.

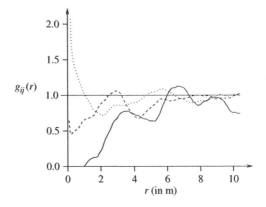

Figure 5.6   Estimated partial pair correlation functions $\hat{g}_{ij}(r)$ for the oak and beech forest obtained by (5.3.19) with bandwidth $h = 1\,\text{m}$: $\hat{g}_{11}(r)$ (solid line), $\hat{g}_{12}(r)$ (dashed line), $\hat{g}_{22}(r)$ (dotted line).

(c) Consider now the pattern of hard grains and air pores in a sample of concrete as shown in Figure 5.7. This is a planar section through a sample of self-flowing refractory castable, a specific type of concrete as described in Section 1.2.5. The nearly spherical corundium grains appear in white, while the matrix (the binding system) appears in grey; the air pores are shown in black. This material was produced for research purposes and investigated in Hubalková and Stoyan (2003) based on planar sections and in Ballani

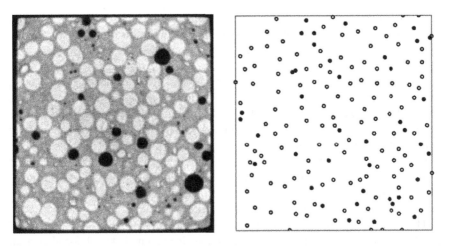

Figure 5.7   Planar section through a sample of concrete: (left) original image, 130 grain centres (2) in white and 41 air pores (1) in black, side length of the square around 10 mm; (right) point pattern of grain centres (o) and (•).

(2006) and Ballani et al. (2005) using three-dimensional information from computerised tomography. The number of air pores is unusually large.

Any statistical analysis of a planar section as shown in Figure 5.7 can only yield limited results due to the substantial amount of information lost through the reduction from three to two dimensions. Sophisticated stereological methods have been developed that may be used to determine the spatial density of particles per unit volume and the pair correlation function of sphere centres (see Stoyan et al., 1995) from planar sections. However, further stochastic modelling makes sense only if the aim is to yield information on the three-dimensional structure. Based on three-dimensional information, Ballani (2006) and Ballani et al. (2005) were able to show that a modified hard sphere Gibbs process model is the best model for the three-dimensional structure, better than an RSA model or a packing model.

There are, however, two questions that can probably be answered based on the planar data:

1. Are there any spatial correlations between the grains (points of type 2) and pores (points of type 1)? One would expect that type 2 points dominate as corundium is much harder than air.

2. Can the pattern of points of type 1 be considered completely random? The point distribution in the figure seems to confirm this assumption, but this should of course not be the case if the answer to question 1 was 'yes'.

On the one hand, the partial pair correlation function $g_{11}(r)$ in Figure 5.8 shows that the corundium particles exhibit a certain degree of clustering. The partial pair correlation function $g_{22}(r)$, on the other hand, indicates that the pores seem to form

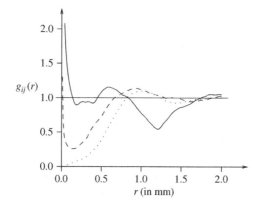

Figure 5.8   Estimated partial pair correlation functions $\hat{g}_{ij}(r)$ for the concrete sample: $\hat{g}_{11}(r)$ (solid line), $\hat{g}_{12}(r)$ (dashed line), $\hat{g}_{22}(r)$ (dotted line).

a soft-core process; stereological experience even suggests that the arrangement of grain centres in three-dimensional space is similar to that of centres of a random system of hard spheres as discussed in Section 6.5. This is indeed the case; see the discussion in Ballani et al. (2005).

There is probably some interaction among the pores, which causes neighbouring pores to coalesce to form larger and isolated pores. As a result, with reference to question 2, the pattern formed by the points of type 1 cannot be considered random.

The partial pair correlation function $g_{12}(r)$ has a shape similar to that of $g_{22}(r)$. It is difficult to interpret.

It is useful to combine g_{ij}-functions for a more refined correlation analysis. The quantity

$$\delta_{ij}(r) = \lambda_j g_{ij}(r) - \lambda_i g_{ii}(r) \qquad \text{for } r \geq 0 \qquad (5.3.14)$$

is a good indicator of correlations between i and j points. Positive values of $\delta_{ij}(r)$ indicate that, at distance r, there are more points of type j than points of type i around a point of type i.

Note that it is probably not advisable to consider the differences

$$g_{ij}(r) - g_{ii}(r)$$

or

$$K_{ij}(r) - K_{ii}(r).$$

Consider, for example, a bivariate point process with two independent subprocesses, the first of which is a cluster process with $g_{11}(r) > 1$. Due to the independence of the subpatterns, $g_{12}(r) \equiv 1$. Assume, in addition, that the intensity of N_1 is small but that the intensity of N_2 is large. As a result, there are mainly points of type 2 in the neighbourhood of a point of type 1, despite the clustering of the points of type 1. This is indicated by $\delta_{ij}(r)$, while the difference $g_{ij}(r) - g_{ii}(r)$ is negative and not influenced by the intensities.

In ecology, species diversity and species richness in ecosystems have long been discussed in the context of ecosystem functioning, where high degrees of species richness or diversity have often been associated with healthy and well-functioning ecosystems. Several indices have been derived in this context, of which the *Simpson index* and *Shannon index* have been the most prominent; see Krebs (1998). Neither of these indices take the spatial structuring in ecosystems into account.

Shimatani (2001) introduced a spatial Simpson index in the context of genetic marks, which may be written in terms of pair correlation functions as

$$\alpha(r) = 1 - \sum_{i=1}^{m} \lambda_i g_{ii}(r) \Big/ \lambda^2 g(r).$$

It can be interpreted as the probability that two arbitrarily chosen points occurring at a distance r have different marks.

Mark connection functions $p_{ij}(r)$

The mark connection functions $p_{ij}(r)$ are another refined tool for the correlation analysis of qualitative marks. The quantity $p_{ij}(r)$ can be interpreted as the conditional probability that two points at distance r have marks i and j, given that these points are in the point process N. Due to stationarity and isotropy the two points can be chosen as o and $\mathbf{r}$, where $\mathbf{r}$ is any point at distance r from o. Thus

$$p_{ij}(r) = \mathbf{P}_{or}(m(o) = i, m(\mathbf{r}) = j).$$

In practice, the $p_{ij}(r)$ are calculated by means of

$$p_{ij}(r) = \frac{\varrho_{ij}(r)}{\varrho(r)} \qquad \text{for } r > 0 \tag{5.3.15}$$

or

$$p_{ij}(r) = p_i p_j \frac{g_{ij}(r)}{g(r)} \qquad \text{for } r > 0. \tag{5.3.16}$$

This implies that $p_{ij}(r)$ is defined only for those r for which $g(r) > 0$. For $r = 0$,

$$p_{ii}(0) = 1 \quad \text{and} \quad p_{ij}(0) = 0 \qquad \text{for } i \neq j,$$

using the partial product density defined on p. 325. It is important to know the asymptotic behaviour of the mark connection function, i.e. the behaviour of $p_{ii}(r)$ and $p_{ij}(r)$ as r tends to infinity. In applications, one can use this to find the range of mark correlation:

$$\lim_{r \to \infty} p_{ii}(r) = p_i^2$$

and

$$\lim_{r \to \infty} p_{ij}(r) = 2p_i p_j \qquad \text{for } i \neq j.$$

If the marks are independent (random labelling), all $p_{ij}(r)$ are constant, it is

$$p_{ii}(r) \equiv p_i^2 \quad \text{and} \quad p_{ij}(r) \equiv 2p_i p_j \qquad \text{for } i \neq j, \tag{5.3.17}$$

while for the random superposition model

$$p_{ii}(r) = \frac{\lambda_i^2 g_i(r)}{\lambda_i^2 g_i(r) + \lambda_j^2 g_j(r) + 2\lambda_1 \lambda_2} \tag{5.3.18}$$

and

$$p_{ij}(r) = 2p_i p_j \qquad \text{for } i \neq j,$$

where $g_i(r)$ is the pair correlation function of the subprocess N_i.

Shimatani (2001) considers the function

$$\delta_i(r) = 1 - p_{ii}(r),$$

which yields the probability that two points with an inter-point distance of r have marks different from i, i.e. in the specific application discussed in the paper the probability that the trees represented by the points are from different species.

Note that $g_{ij}(r)$ and $p_{ij}(r)$ provide different information, similarly to the way the probability $\mathbf{P}(A \cap B)$ and the conditional probability $\mathbf{P}(A \cap B | C)$ provide different information for random events.

To illustrate this, assume that $g_{ij}(r)$ has a very small value for some $r = r'$ due to a small number of pairs of points of type i and j with an inter-point distance of r'. Assume further that, irrespective of the marks, pairs of points with a distance of r' are very rare in M in general. Then $g(r')$ is also small. However, $p_{ij}(r)$ can still be close to 1, if the few pairs of points with a distance of r' are mainly (i, j) pairs.

Example 5.6.   Mark connection functions for three bivariate point patterns
(a) The mark connection functions p_{ij} for the amacrine cells are shown in Figure 5.9. Since 51.7 % of the points are on-cells, the estimates of the long-range probabilities or theoretical limits of $\hat{p}_{ij}(r)$ for $r \rightarrow \infty$ are

$$\hat{p}_1 = 0.267, \qquad \hat{p}_2 = 0.233, \qquad \hat{p}_{12} = 0.500.$$

Clearly, the $\hat{p}_{ij}(r)$ in Figure 5.9 tend to converge towards these limits. Apparently, only some random fluctuations of the functions can be detected for distances from $r = 100 \, \mu$m onwards, and thus the correlation range of the pattern is around $100 \, \mu$m.

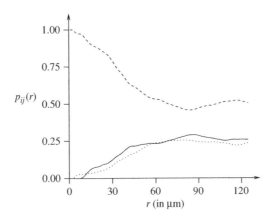

Figure 5.9   The estimated mark connection functions $\hat{p}_{ij}(r)$ for the amacrine data: $\hat{p}_{11}(r)$ (solid line), $\hat{p}_{12}(r)$ (dashed line), $\hat{p}_{22}(r)$ (dotted line).

This is perhaps not so clear in the graph of the partial pair correlation function in Figure 5.5.

The similarity of the curves for $\hat{p}_{11}(r)$ and $\hat{p}_{22}(r)$ again reflects the similarity of the distribution of the points of type 1 and 2; the increase in these functions reflects the same behaviour of the sub-point patterns as the increase in the partial pair correlation functions for $r \leq 60\,\mu$m.

The function $\hat{p}_{12}(r)$ suggests that there is a high probability that the points in pairs of points with small distances are of different types. This is less clear from the $\hat{g}_{ij}(r)$ in Figure 5.5, but simple reasoning shows that there is inhibition among points of type 1 as well as among points of type 2. In a pair of points that are close together those points are very likely to be of different type. It is natural that this effect disappears for $r > 100\,\mu$m, where the partial pair correlation functions begin to fluctuate around 1.

If the two sub-point processes N_1 and N_2 were really independent, the pattern may be interpreted as resulting from the superposition of two independent processes, as discussed in Section 6.2.3. If $g_1(r)$ and $g_2(r)$ are the corresponding pair correlation functions and λ_1 and λ_2 the corresponding intensities, this yields the pair correlation function $g(r)$ of the pattern resulting from the superposition as

$$g(r) = \left(\lambda_1^2 g_1(r) + \lambda_2^2 g_2(r) + 2\lambda_1\lambda_2\right) / (\lambda_1 + \lambda_2)^2 \qquad \text{for } r \geq 0. \qquad (5.3.19)$$

Figure 5.10 shows the pair correlation function estimate $\hat{g}(r)$ for all 294 cells in Figure 1.2 and for comparison the function resulting from (5.3.19) by plugging in $\hat{g}_{ii}(r)$ for $g_i(r)$ and $\hat{\lambda}_i$ for λ_i (the intensity estimates, e.g. $\hat{\lambda}_1 = 152/(1060 \times 662)\,\mu\text{m}^{-2}$).

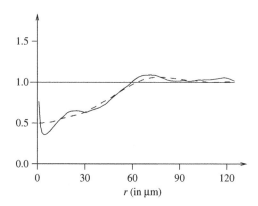

Figure 5.10   The estimated pair correlation function $\hat{g}(r)$ of all amacrine cells (ignoring the marks) (solid line) and the function obtained from (5.3.19) with $g_i(r) = \hat{g}_{ii}(r)$ from Figure 5.5 (dashed line). The similarity between the functions may be considered as 'evidence' in favour of the hypothesis that the pattern may be regarded as a superposition of two independent components.

The two curves are very similar which provides further evidence for the independent-superposition hypothesis.

This hypothesis can be more formally tested by significance tests based on simulation; see Example 7.4 for more details. It turns out that the random superposition hypothesis is accepted on the basis of the given data.

The amacrine data were analysed very carefully by means of the K_{ij}-functions in Diggle et al. (2006). The authors fitted a bivariate (i.e. marked) pairwise interaction point process (as in Section 3.6) to the data. The 'effective range' of correlation for the pattern was estimated as $90\,\mu$m. Diggle et al. (2006) refined the analysis by distinguishing 'functional' and 'statistical' independence: since the points represent cells, there is a natural minimum distance between points of type 1 and 2 (which is observed as $4.9\,\mu$m in the pattern) such that the patterns cannot be completely or statistically independent. However, there was evidence for functional independence.

(b) The empirical mark connection functions $p_{ij}(r)$ for the oaks and beeches are shown in Figure 5.11. They look less irregular than the $\hat{g}_{ij}(r)$. For large r they tend to converge to the values

$$\hat{p}_1 = 0.107, \qquad \hat{p}_2 = 0.452, \qquad \hat{p}_{12} = 0.441,$$

which result from the fact that 32.8 % of the trees are oaks.

The function $\hat{p}_{11}(r)$ reflects inhibition among the oak trees and $\hat{p}_{22}(r)$ mutual attraction among the beech trees. Both 'forces' seem to level out at a distance of 8 m. The maxima of $\hat{p}_{12}(r)$ and $\hat{p}_{22}(r)$ at $r=2$ m and $r=5$ m, respectively, are very prominent. The first maximum seems to reflect relations between clusters of beeches and their oak neighbours, while the second one may correspond to relationships between pairs of neighbouring clusters.

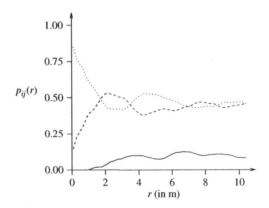

Figure 5.11   Empirical mark connection functions $\hat{p}_{ij}(r)$ for the oak and beech forest obtained by (5.3.50) with bandwidth $h = 1.0\,$m: $\hat{p}_{11}(r)$ (solid line), $\hat{p}_{12}(r)$ (dashed line), $\hat{p}_{22}(r)$ (dotted line).

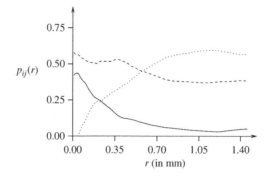

Figure 5.12   Empirical mark connection functions $\hat{p}_{ij}(r)$ for the concrete sample obtained with bandwidth $h = 0.4$ mm: $\hat{p}_{11}(r)$ (solid line), $\hat{p}_{12}(r)$ (dashed line), $\hat{p}_{22}(r)$ (dotted line).

The above analysis has provided a detailed picture of the spatial correlation in the oak and beech forest, both biologically and statistically: the oak trees form a hard-core pattern, whereas the beech trees form clusters in weak dependence on the oak trees.

(c) The mark connection functions $p_{ij}(r)$ for the pattern of hard grains and air pores in the concrete sample are shown in Figure 5.12. The functions reveal a range of correlation of around 0.8 mm. For small r, $p_{11}(r)$ indicates some attraction among points of type 1, which is further evidence of clustering of the pores. In contrast, $p_{22}(r)$ indicates repulsion among the grains, supporting the soft-core hypothesis. Finally, the values of $p_{12}(r)$ around $r = 0.4$ mm indicate attraction between pores and grains.

In summary, there is clearly some interaction between pores and grains, but this interaction cannot be simply classified as complete dominance of the grains over the pores.

Example 5.7.   Termite mounds and juvenile palms in the African tree savannah
Barot et al. (1999) and Barot and Gignoux (2003) studied ecological relationships in plant societies in the humid tree savannah of Lamto, Côte d'Ivoire. They focused on the analysis of the palm species *Borassus aethiopum*, which is a common tall palm tree of this savannah exhibiting a specific root foraging strategy: palm root density increases significantly in nutrient-rich patches, even far away from the actual stem of the palm. The savannah is a strongly heterogeneous environment, where the soil is globally nutrient-poor, but tree clumps and termite mounds constitute nutrient-rich patches. The palms were classified with respect to sex and age (seedlings, juveniles and adults). The life cycle of *B. aethiopum* starts with dense clusters of seedlings around mother trees. Seedlings that grow at some distance from these females are the most likely to survive and grow to become juveniles. The level of competition among the juveniles is very high and they are mainly found on

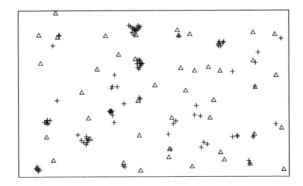

Figure 5.13   Pattern formed by 104 palms (+) and 48 termite mounds (△) in a 250 × 150 m window *W* from the tree savannah of Côte d'Ivoire. Courtesy of S. Barot. Reproduced by permission of the Ecological Society of America.

nutrient-rich patches. Some of them develop the above-mentioned root foraging strategy and develop into adults.

This example describes an analysis of one of the patterns, termed TS3, which consists of 104 juvenile palms (type 1, referred to simply as 'palms') and 48 termite mounds (type 2) in a 250 × 150 m rectangle of tree savannah, as shown in Figure 5.13. Visual inspection reveals the main properties of the pattern: it appears fairly random and looks statistically homogeneous. On the one hand, the pattern of mounds is rather regular, which may indicate that around each mound there has to be a mound-free region to provide enough room for the termite colonies. On the other hand, the palms exhibit clusters, and these clusters are often close to mounds.

This triggers the idea of a 'functional' relationship between the mounds and palms, in the sense of case (d) in Table 5.1, where the mounds 'control' the positions of palms by attraction. Point process statistics may be used to quantify this relationship.

Figure 5.14 shows the pair correlation function $g_2(r) = g_{22}(r)$ of the point pattern formed by the mounds. It looks like the pair correlation function of a soft-core point process, as discussed in Section 4.3.4. The first local maximum of $g_2(r)$ is at around 30 m, which is apparently the most frequent inter-mound distance. Some pairs of points with a distance of less than 10 m have given the function a *soft*-core form. The range of correlation r_{corr} is around 40 m, thus the window is perhaps slightly too small for a thorough analysis of the mound pattern.

The cluster behaviour of the palms is clearly reflected in the corresponding pair correlation function $g_1(r) = g_{11}(r)$ in Figure 5.14. If the mounds are regarded as cluster centres (with many 'empty' clusters) of a cluster process formed by the palms, the pattern cannot be a simple Neyman–Scott process, where, as explained in Section 6.3.2, the parent points form a Poisson process. In the given situation, the pattern formed by the mounds is clearly not CSR and hence the parent points or clusters centres cannot be assumed to form a Poisson process.

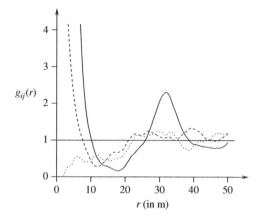

Figure 5.14   The estimated partial pair correlation functions for the mounds (dotted line) and the palms (solid line). For the mounds the pair correlation function looks like that of a soft-core process, whereas the function for the palms shows a behaviour similar to that of a cluster process. The function indicates a range of correlation of $r \approx 40$ m. Additionally, the figure shows the partial pair correlation function $\hat{g}_{12}(r)$ (dashed line).

The shape of the pair correlation function for the palms suggests a cluster diameter of around 15 m and a range of correlation of perhaps 40 m. The maximum at $r \approx$ 30 m may be easily interpreted: it is linked to the maximum of the pair correlation function $g(r)$ of mounds at $r \approx 25$ m.

The correlation between the palms ($i = 1$) and mounds ($i = 2$) may be characterised by both the partial pair correlation function $\hat{g}_{12}(r)$ and the mark connection functions $\hat{p}_{ij}(r)$ for $i, j = 1, 2$. Figure 5.14 shows $\hat{g}_{12}(r)$, which indicates strong attraction between points of type 1 (palms) and type 2 (mounds) for small r ($r \leq 10$). For larger values of r it almost coincides with $\hat{g}_2(r)$, the estimated pair correlation function for the mounds. This may also be a result of the clustering of points of type 1 (palms) around points of type 2 (mounds).

Figure 5.15 shows the three functions $p_{ij}(r)$. $p_{11}(r)$ is perhaps the most interesting mark connection function for this example. For small r, the values of $p_{11}(r)$ are very large. This is easy to understand since pairs of points that are close together are most likely to be two palms and the number of palms is high. The steep decline near $r \approx 10$ m might reflect the size of the clusters. There are not many pairs of points with a distances of around 17 m, and any pair at this distance is rarely a pair of palms. The distinct maximum at $r \approx 32$ m reflects pairs of palms from different clusters. The conditioning used in the definition of $p_{11}(r)$ provides clearer evidence of this second-cluster behaviour (which is not visible in $g_{11}(r)$), since the number of pairs of points with a distance around 32 m is small in comparison to that for 5 m.

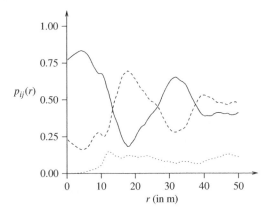

Figure 5.15   The estimated mark connection functions $\hat{p}_{ij}(r)$ for the palms and mounds: $\hat{p}_{11}(r)$ (solid line), $\hat{p}_{12}(r)$ (dashed line), $\hat{p}_{22}(r)$ (dotted line).

The behaviour of $p_{12}(r)$ is in some sense complementary to that of $p_{11}(r)$. There are of course pairs of points of type 1 and 2 at small values of r, but these are relatively rare since usually only a single mound forms the cluster centre with some palms surrounding it. Therefore, the pairs of points of type 1 and 2 are the most frequent among the few pairs of points with an inter-point distance of 17 m.

Finally, the behaviour of $p_{22}(r)$ is rather different from the behaviour of $p_{11}(r)$ above. It has small values for $r < 15$ m, then increases linearly and is nearly constant for larger r. The fluctuations are around 0.1, which equals $\hat{p}_2^2$, with $\hat{p}_2 = \frac{48}{104+48}$. Clearly, the small values for $r < 15$ m reflect the soft-core property of the mound pattern.

The statistical analysis shows clearly that the palms form a cluster process with the mounds as cluster centres or parent points. The diameter of the clusters is around 15 m.

In the following, the cluster size is estimated quantitatively. Note that here the aim is to estimate neither the number of points per cluster nor, more precisely, the probability p_k that the typical cluster consists of k points. (It is clear that p_0 is large due to the large number of mounds without any associated palms.) The aim is rather to approximately estimate the probability density $d_p(r)$ of the random distance of palms from 'their' mound. For this purpose the relationship between distance and point density given by formula (3.4.7) in Example 3.4 is used, where $p(r)$ corresponds to $g_{12}(r)$. Hence, for small r,

$$\hat{d}_p(r) \propto 2\pi r \hat{g}_{12}(r),$$

assuming that the pattern of mounds is regular and only one cluster of palms contributes. (Note that, of course, $\hat{g}_{12}(r) = \hat{g}_{21}(r)$.)

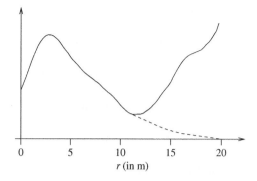

Figure 5.16   The function $2\pi r \hat{\lambda}_1 \hat{g}_{12}(r)$ for the palm and mound data. The additional dashed curve shows a rough non-normalised estimate of $d_p(r)$.

Figure 5.16 shows the result. Clearly, only the first part of the curve of $2\pi r \hat{\lambda}_1 \hat{g}_{12}(r)$ for small r should be seriously considered, since the curve for $r > 15\text{m}$ results from pairs of points of type 1 in different clusters. The probability density $d_p(r)$ is skewed and is similar to the distance probability density corresponding to the modified Thomas process considered in Section 6.3.2. In this case

$$d(r) = \frac{r}{\sigma^2} \exp\left(-\frac{r^2}{2\sigma^2}\right) \qquad \text{for } r \geq 0.$$

The deviation parameter is around $\sigma = 3\text{–}5\,\text{m}$.

The palm and mound data can also be used to demonstrate the potential of another statistical method. An important research issue in the context of cluster processes is the development of suitable algorithms that can be used to reconstruct (or estimate) the locations of cluster centres; see Van Lieshout (1995), Van Lieshout and Baddeley (1995, 2002) and many other papers. This is of interest, for example, in seismology where researchers aim to separate background seismicity from earthquake clusters; in this context the term 'declustering' has been used (see Zhuang et al., 2002).

A number of algorithms have been developed that can be applied to simulated data or to real clustered point patterns. However, for real data the true cluster centres, if there are any, are typically unknown and it is impossible to verify whether the method has identified the right centre. The palm and mound data provide a nice example which is well suited for testing the methods on real data if the mounds are assumed to be cluster centres.

M.N.M. van Lieshout kindly analysed the palm and mound data in this way. She used only the 108 palm locations (and did not know the locations of the mounds), considered them as points in a cluster process, reconstructed the cluster centres and compared them with the mound positions.

She assumed that the clusters are i.i.d with respect to their centres. For the cluster distribution she assumed that the numbers of points per cluster follow a Poisson distribution with mean μ and their locations are given by a radially symmetric Gaussian distribution with variance σ^2 (refer to the modified Thomas process in Section 6.3.2). In accordance with $d_p(r)$ as shown in Figure 5.16, σ was chosen to be 4 m.

Given the cluster centres $z_1, \ldots, z_c$, the joint density of the n points $x_1, \ldots, x_n$,

$$f(x_1, \ldots, x_n | \mathbf{z}) \qquad \text{with } \mathbf{z} = (z_1, \ldots, z_c),$$

can be given in closed from (Van Lieshout, 1995). The maximum likelihood estimate of $\mathbf{z}$ is the pattern $\hat{\mathbf{z}}$ maximising $f(\mathbf{x}|\mathbf{z})$,

$$\hat{\mathbf{z}} = \arg\max_{\mathbf{z}} f(\mathbf{x}|\mathbf{z}) ; \qquad (5.3.20)$$

here both c and the locations $z_1, \ldots, z_c$ have to be determined.

Numerical experience gathered by Van Lieshout shows that this naive approach does not work well and 'detects' too many cluster centres, which are close together. Therefore, a Bayesian approach was used which started from an a priori distribution $p(\mathbf{z})$ of $\mathbf{z}$ with sufficiently large inter-point distances. Van Lieshout used a nearest-neighbour Markov point process (see Section 3.6). Thus (5.3.20) is replaced by

$$\hat{\mathbf{z}} = \arg\max_{\mathbf{z}} f(\mathbf{x}|\mathbf{z}) p(\mathbf{z}) . \qquad (5.3.21)$$

Numerical methods as described in Van Lieshout and Baddeley (2002) were used to derive the estimate $\hat{c} = 39$ and the cluster centres shown in Figure 5.17. Since the approach is based entirely on the palm data, it is clear that the statistical method cannot detect empty clusters, i.e. mounds without palms, and hence $\hat{c} < 48$.

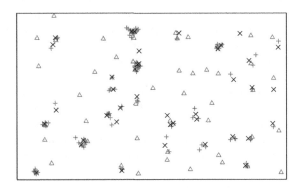

Figure 5.17 Reconstructed cluster centres shown as $\times$ for the point process of palms and the pattern of mounds. Both patterns are fairly similar, which indicates that the cluster reconstruction algorithm works very well.

5.3.3 Definitions for quantitative marks

This section discusses second-order characteristics for marked point patterns with quantitative marks. The spatial correlations considered here are of a very different nature than correlations in processes with qualitative marks. Whereas the latter processes are analysed for aggregation or repulsion among the subprocesses, the analysis of processes with quantitative marks typically focuses on questions such as the numerical differences among the marks dependent on the distances of the corresponding points. For instance, the marks of neighbouring points may tend to be smaller (larger) than the mean mark μ, i.e. there may be inhibition (stimulation) of the marks.

Another issue concerns the numerical similarity of the marks of neighbouring points. This means that these points may tend to have similar marks, but the opposite behaviour may be also observed, i.e. specific points may dominate the points in their vicinity and thus have a large mark, while the other points have small marks (see Example 6.4).

Second-order characteristics for point processes with quantitative marks can be introduced in an elegant, yet abstract and theoretical way. Before embarking on this, this subsection initially discusses a special case, which leads to the classical *mark correlation function* $k_{mm}(r)$.

Consider the mean

$$c_{mm}(r) = \mathbf{E}_{or}(m(o) \cdot m(\mathbf{r})) \qquad \text{for } r > 0. \tag{5.3.22}$$

In the notation used in this book, this is the conditional mean of the product of the marks of a pair of points in M with distance r, where o is the origin and $\mathbf{r}$ any point with distance r from o. The condition is that there are points in o and in $\mathbf{r}$. A naive statistical approach to determining such a mean is to consider all pairs of points in the marked point pattern of interest with an inter-point distance approximately equal to r and to calculate the mean of the products of the marks of these pairs of points.

A value of $c_{mm}(r)$ that is larger than the squared mean mark μ^2 is indicative of some form of mutual stimulation among the points resulting in increased marks at distance r.

This mean may be normalised by dividing it by μ^2 (assuming $\mu^2 \neq 0$) to allow comparison of the strength of mark correlation between different processes. This yields the *mark correlation function* $k_{mm}(r)$, which is defined by

$$k_{mm}(r) = \frac{c_{mm}(r)}{\mu^2} \qquad \text{for } r > 0.$$

Note that μ^2 can be interpreted as $c_{mm}(\infty)$, the value of $c_{mm}(r)$ for very large r, where the marks are independent and the mean of the product is equal to the product of means. The term 'mark correlation function' was introduced in Stoyan (1984).

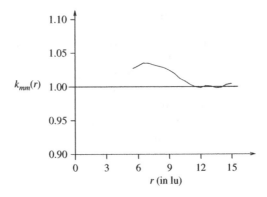

Figure 5.18   Estimated mark correlation function $\hat{k}_{mm}(r)$ for the gold particle data. The bandwidth $h = 3\,\mathrm{lu}$. The curve is shown only for $r \geq 5.65\,\mathrm{lu}$ since for the smaller distances the estimated pair correlation function $\hat{g}(r)$ vanishes.

It is also used in this book, even though it is not exactly a correlation function in the strict sense of the word, as discussed in Schlather (2001a). However, it is an elegant function describing important aspects of the spatial correlation of marks.

Example 5.8.   Gold particles: mark correlation function for the diameter marks
Figure 5.18 shows the empirical mark correlation function $\hat{k}_{mm}(r)$ for the diameter marks of the gold particles. The function is not defined for $r \leq 5.66$ since there $\hat{g}(r) = 0$. It reveals that the marks are indeed spatially correlated, but perhaps not very strongly. The empirical mark correlation function $\hat{k}_{mm}(r)$ is a little above 1 and decreases continuously. That there are indeed correlation between the marks is shown by the mark variogram $\gamma_m(r)$ in Figure 5.21. The range of mark correlation indicated by $k_{mm}(r)$ seems to be short, perhaps $r_{\mathrm{corr}} = 15\,\mathrm{lu}$.

For comparison consider also the empirical nearest-neighbour mark product index $\hat{\mathbf{n}}_{mm}$. Its value is 1.03, i.e. very close to 1. Hence one might conclude that the marks of points that are close together are independent. Similarly, the nearest-neighbour mark index $\mathbf{n}_{.m} = 1.02$. However, the nearest-neighbour variogram index $\hat{\mathbf{n}}_\gamma = 0.69$ is clearly smaller than 1. It indicates that the marks of points that are close together tend to be similar. In Example 5.10 the mark variogram provides further support for this initial impression.

The approach used in the derivation of the mark correlation function $k_{mm}(r)$ may be generalised in a natural way. This generalisation is discussed below to provide readers with a method that may be used to construct tailor-made correlation functions for specific questions relating to the analysis of marked point patterns.

The summary characteristics are constructed based on so-called *test functions* $t(m_1, m_2)$, which depend on two marks m_1 and m_2, where $m_1 = m(o)$ and $m_2 = m(\mathbf{r})$. For instance, for the mean $c_{mm}(r)$ the test function is

$$t(m_1, m_2) = m_1 m_2.$$

Different test functions may be used for this purpose; this book focuses on the following ones:

$$t_1(m_1, m_2) = m_1 m_2,$$

$$t_2(m_1, m_2) = m_1,$$

$$t_3(m_1, m_2) = m_2,$$

$$t_4(m_1, m_2) = \frac{1}{2}(m_1 - m_2)^2,$$

$$t_5(m_1, m_2) = (m_1 - \mu)(m_2 - \mu),$$

$$t_6(m_1, m_2) = \min\{|m_1 - m_2|, \pi - |m_1 - m_2|\}.$$

Note that $t_1(m_1, m_2)$ is the test function that is used to construct $c_{mm}(r)$.

In the following, the other test functions are explained, apart from t_6, which is discussed in Section 5.4.3 in the context of angular marks.

Non-normalised mark correlation functions $c_t(r)$

A number of second-order characteristics may be defined that can be subsumed under the term 'non-normalised mark correlation functions'. Analogously to the construction of the mean $c_{mm}(r)$, the summary function $c_t(r)$ given some test function $t(m_2, m_2)$ is defined as

$$c_t(r) = \mathbf{E}_{or}(t(m(o), m(\mathbf{r}))) \qquad \text{for } r > 0. \qquad (5.3.23)$$

This is the mean of $t(m(o), m(\mathbf{r}))$ given that there are points of M at locations o and $\mathbf{r}$. See below for a definition of $c_t(0)$ for $r = 0$.

In particular, for the test function $t = t_1$, $c_t(r)$ is the same as $c_{mm}(r)$ and is, as indicated, the mean of the product of the marks of a pair of points in M with inter-point distance r.

For $t = t_2$, $c_t(r)$ is denoted $c_{m \cdot}(r)$ and called the *r-mark function*. It is the mean of the mark of the first point of a pair of points in M with interpoint distance r. Note that, by definition, it is not the same as the mean mark μ introduced in Section 5.1. If the marks are not independent, it is likely that

$$c_{m \cdot}(r) \neq \mu,$$

since the existence of a partner point at distance r can influence the mean mark.

For $t = t_3$, $c_t(r)$ is of no interest since it leads to the same function as for t_2, with

$$c_{m\cdot}(r) = c_{\cdot m}(r) \qquad \text{for all } r > 0,$$

where $c_{\cdot m}(r)$ is simply the analogue of $c_{m\cdot}(r)$ based on the test function t_3.

For $t = t_4$, $c_t(r)$ is denoted $\gamma_m(r)$ and called the *mark variogram*. It characterises the squared differences between the marks of pairs of points with a distance of r. The mark variogram has small values if the marks of the points in a pair with inter-point distance r are similar and large values if the marks differ strongly. It is typically used in the non-normalised form. For large r it converges towards σ_μ^2, the variance of the marks:

$$\lim_{r \to \infty} \gamma_m(r) = \sigma_\mu^2.$$

Note that despite the similarity in name and character, the mark variogram $\gamma_m(r)$ and the geostatistical variogram differ strongly in their definition. Geostatistical variograms characterise regionalised variables, i.e. random fields as introduced in Section 1.8.3. They are variables that take values at any location; examples include nitrogen levels in the soil, water temperature and annual rainfall. The corresponding variograms satisfy some analytical conditions covered in the geostatistical literature (see also Appendix C), and statisticians have gained thorough experience in their interpretation. In contrast, the marks of a marked point process are given only at the points of the process, and otherwise are undefined. Thus there is no regionalised variable and only a mark variogram can be estimated. Statistical experience shows that the shapes of mark variograms would be considered rather unusual for geostatistical variograms (see Wälder and Stoyan, 1996; Kint et al., 2003). Attempts to interpret these with geostatistical logic in mind may fail. This is true, in particular, if a pattern exhibits inter-point interactions which favour point pairs with small and large marks in close proximity. Example 6.4 is an example of this case, whereas the mark variogram in Example 5.10 has a form similar to that of a typical geostatistical variogram. The confusion with variograms results from uncritical use of geostatistical software, misled by the similar structures of marked point pattern and geostatistical data, which both consist of points and Z-values.

For $t = t_5$, a characteristic is obtained that is akin to Moran's I-statistic and is defined as

$$I(r) = \mathbf{E}_{or}((m(o) - \mu)(m(\mathbf{r}) - \mu)) / \sigma_\mu^2 , \qquad (5.3.24)$$

which is a normalised $c(r)$; see Shimatani (2002). This is a marked point process analogue of the classical (Pearson) correlation coefficient.

The following assumes that the marks are positive, but the theory can be generalised to be applicable to negative as well as positive marks, where modifications are necessary if $\mu = 0$.

The value of the non-normalised mark correlation function for $r = 0$ is defined as

$$c_t(0) = \mathbf{E}_o(t(m(o), m(o)))\tag{5.3.25}$$

or

$$c_t(0) = \int\limits_0^\infty t(m, m) f_{\mathcal{M}}(m) \mathrm{d}m,$$

i.e. it is the mean of $t(m(o), m(o))$ where $m(o)$ is the mark of the typical point (Schlather, 2001b). Here $f_{\mathcal{M}}(m)$ is the mark probability density. Thus

$$c_{mm}(0) = \mu_2 = \int\limits_0^\infty m^2 f_{\mathcal{M}}(m) \mathrm{d}m, \qquad c_{m\cdot}(0) = c_{\cdot m}(0) = \mu, \qquad \gamma_m(0) = 0.$$

The definition of the $c_t(r)$ may appear rather complicated because of its conditional nature. It may thus be surprising that they can easily be estimated statistically; see (5.3.52). This is based on the following equation, which reflects clearly the conditional nature of $c_t(r)$ as a ratio of two infinitesimal probabilities:

$$c_t(r) = \frac{\varrho_t(r)}{\varrho(r)} \qquad \text{for } r > 0, \; \varrho(r) > 0.\tag{5.3.26}$$

Here $\varrho(r)$ is the second-order product density of the unmarked process N as defined in Section 4.3.1 and $\varrho_t(r)$ is another product density. Equation (5.3.26) implies that the general $c_t(r)$ and the special correlation functions $k_{mm}(r)$, $k_{m\cdot}(r)$ below and $\gamma_m(r)$ are defined only for those r for which $g(r) > 0$. The ratio in (5.3.26) may be interpreted as a conditional mean. The denominator contains the probability of the condition, based on the interpretation of $\varrho(r)\mathrm{d}x\mathrm{d}y$ (see p. 219). The term $\varrho_t(r)\mathrm{d}x\mathrm{d}y$ in the numerator is related to the mean of $t(m(x), m(y))$, where $m(x)$ and $m(y)$ are the marks of the points at x and y.

More technically, $\varrho_t(r)$ is the density of the factorial moment measure α_t defined by

$$\alpha_t(B_1 \times B_2) = \mathbf{E}\left(\sum_{\substack{[x_1; m(x_1)], \\ [x_2; m(x_2)] \in M}}^{\ne} \mathbf{1}_{B_1}(x_1) \mathbf{1}_{B_2}(x_2) t(m(x_1), m(x_2)) \right),$$

where B_1 and B_2 are subsets of $\mathbb{R}^d$.

Mark correlation function $k_t(r)$

In practice, it is helpful to normalise the functions $c_t(r)$ by dividing it by $c_t(\infty) = c_t$, the value the function takes for very large distances r, at which the marks are independent. This means that mark correlation functions consider correlations among marks relative to the case of independent marks. This facilitates their interpretation; the resulting mark correlation functions tend towards 1 for large r.

The mark correlation function $k_t(r)$ based on a test function t is defined as

$$k_t(r) = \frac{c_t(r)}{c_t} \qquad \text{for } r > 0 \tag{5.3.27}$$

and

$$k_t(0) = \frac{c_t(0)}{c_t} . \tag{5.3.28}$$

The normalising factor $c_t(\infty) = c_t$ is obtained from

$$c_t = \int_0^\infty \int_0^\infty t(m_1, m_2) f_{\mathcal{M}}(m_1) f_{\mathcal{M}}(m_2) dm_1 dm_2 . \tag{5.3.29}$$

For the test functions described on p. 343 the normalising factors are:

$$t_1 : \mu^2 ,$$

$$t_2 : \mu ,$$

$$t_3 : \mu ,$$

$$t_4, t_5 : \sigma_\mu^2 ,$$

$$t_6 : \frac{\pi}{4} \text{ for } d = 2 \text{ and } 1 \text{ for } d = 3 .$$

As above, μ is the mean mark, and σ_μ^2 the mark variance. As indicated, the mark variogram, like the geostatistical variogram, is typically not normalised. Important examples of mark correlation functions include the following functions:

- *mark correlation function,*

$$k_{mm}(r) = c_{mm}(r) / \mu^2 \quad \text{for } r > 0, \qquad k_{mm}(0) = \mu_2 / \mu^2 ,$$

  with $\mu_2 = \mu^2 + \sigma_m^2$;

- *r-mark correlation function,*

$$k_{m\cdot}(r) = c_{m\cdot}(r) / \mu, \qquad k_{m\cdot}(0) = 1 .$$

Mark correlation functions are valuable tools in the exploratory analysis of marked point processes. They may be used to detect correlations among marks and to identify suitable marking models (see Section 6.8).

If the marks are independent,

$$k_{mm}(r) = k_{m.}(r) \equiv 1 \qquad \text{for } r \geq 0 \tag{5.3.30}$$

and

$$\gamma_m(r) \equiv \sigma_\mu^2. \tag{5.3.31}$$

In applications, if the empirical correlation functions are not constant then there is reason to assume that the marks are not independent. Consider the following two main types of dependence among the marks (see also Figure 5.19):

- *Inhibition.* The objects represented by the points compete and thus have smaller than average marks if they are close together and

$$k_{mm}(r) < 1 \quad \text{and} \quad k_{m.}(r) < 1 \qquad \text{for small } r.$$

- *Mutual stimulation.* The points benefit from being close together and thus have on average larger marks than μ and then

$$k_{mm}(r) > 1 \quad \text{and} \quad k_{m.}(r) > 1 \qquad \text{for small } r.$$

For large r all of these functions converge to 1; empirical correlation functions fluctuate irregularly around 1 for $r > r_{\text{corr}}$. The latter quantity is the *range of correlation,* which is an important characteristic in the context of spatial correlation: r_{corr} is the distance at which the correlation function becomes constant. It is similar

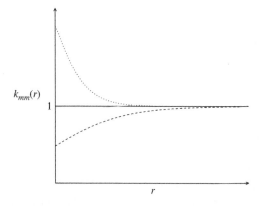

Figure 5.19   Idealised shapes of mark correlation functions: (dashed line) inhibition; (dotted line) mutual stimulation; (solid line) independence.

to the correlation range discussed in the context of pair correlation functions as defined on p. 220, but it is possible for the values of r_{corr} to differ for different test functions, e.g. that $g(r)$, $k_{mm}(r)$ and $\gamma_\mu(r)$ yield different correlation ranges.

The mark variogram and r-mark function are very useful in the context of the important random field model, see Section 6.8.3. For this model

$$k_{m.}(r) \equiv 1 \qquad \text{for } r \geq 0. \tag{5.3.32}$$

Multiplicatively weighted pair correlation function

The function

$$g_{mm}(r) = \frac{\varrho_{mm}(r)}{(\lambda\mu)^2}, \tag{5.3.33}$$

where $\varrho_{mm}(r)$ is a product density like $\varrho_t(r)$ in (5.3.26) with

$$t(m(x_1), m(x_2)) = m(x_1)m(x_2),$$

may be called the multiplicatively weighted pair correlation function. It can be interpreted as a second-order version of the mark-sum intensity λ_S. It is a second-order characteristic of the mark-sum measure, which describes the spatial distribution of the mark *mass* rather than the point distribution. Since points with small marks get small weights, $g_{mm}(r)$ for a cluster process may resemble the pair correlation function of a Poisson process, if the points in dense clusters have small marks and isolated points have large marks. In this case, the function reflects the uniform distribution of mass.

Example 5.9. A three-dimensional sample of concrete
This example analyses the three-dimensional sample of hard spheres shown in Figure 1.6. The marks are the radii of the spheres. Since the radii determine the smallest possible nearest-neighbour distance for each sphere, it is clear that marks and nearest-neighbour distances are closely related. Figure 5.20 shows estimates of the pair correlation function $g(r)$, the mark correlation function $k_{mm}(r)$ and the r-mark function $k_{m.}(r)$.

The shape of the pair correlation function resembles the functions in Figure 4.27 and 6.11, which correspond to random dense systems of hard spheres with random radii. Its maximum is at $r = 0.8$, which corresponds to the mean diameter of the spheres; cf. the corresponding estimated mark p.d.f. shown in Figure 5.3.

The functions $k_{mm}(r)$ and $k_{m.}(r)$ are not defined for values of $r \leq 0.5$, since for these $g(r) = 0$. For the larger r, $k_{mm}(r)$ indicates inhibition among the marks for $r \leq 0.8$: pairs of the hard spherical particles can only be close together if their radii are smaller than the mean radius. This changes for $0.8 \leq r \leq 1.1$: these distances are common in the pattern, and pairs of points with these distances can also include large spheres. Finally, for $r \leq 1.5$, $k_{m.}(r)$ deviates clearly from the value of 1, hence neither a marking model assuming independent marks nor the random field model hold, which is also clear because of the nature of the structure investigated. All three curves suggest a range of correlation $r_{corr} \leq 2$; this may be interpreted as some form of short-range order.

For this pattern models for systems of hard spheres are natural and were studied in Ballani (2006).

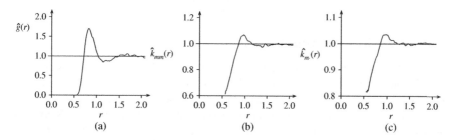

(a)   (b)   (c)

Figure 5.20   Estimates of second-order characteristics for the concrete sample: (a) pair correlation function $\hat{g}(r)$; (b) mark correlation function $\hat{k}_{mm}(r)$; (c) r-mark function $\hat{k}_{m\cdot}(r)$. Courtesy of F. Ballani. With kind permission of Springer Science and Business Media.

Example 5.10.   Gold particles: mark correlation analysis, continued
This example continues the mark correlation analysis of the gold particle pattern in Example 5.8. Figure 5.21 shows $\hat{k}_{m\cdot}(r)$ and the mark variogram $\hat{\gamma}_m(r)$. Both functions indicate that the marks are indeed correlated, but perhaps not very strongly.

The behaviour of $\hat{k}_{m\cdot}(r)$ is similar to that of $\hat{k}_{mm}(r)$ with values just below 1 for small r.

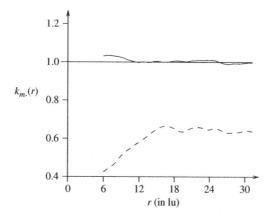

Figure 5.21   Estimated r-mark correlation function $\hat{k}_{m\cdot}(r)$ (solid curve) and mark variogram $\hat{\gamma}_m(r)$ (dashed curve) for the gold particles. The bandwidth h is 3 lu for $\hat{k}_{m\cdot}(r)$ and 8 lu for the mark variogram. The curves are shown only for $r \geq 5.65$ lu since for the smaller distances $\hat{g}(r)$ vanishes. Note that the ordinate starts at 0.4.

The empirical mark variogram $\hat{\gamma}_m(r)$ resembles a geostatistical variogram if determined by a larger bandwidth than that used for the estimation of $g(r)$ and the other correlation functions. It increases monotonously and converges to the empirical mark variance $\hat{\sigma}_\mu^2$, which is $0.757\,\mu m^2$. The range of correlation of the variogram is around $r_{\text{corr}} = 50\,\text{lu}$. For this example, the random field model with variogram $\gamma(r) = 0.4 + 0.24(1 - \exp(-0.13r))$ is suitable, as a simulation test has shown.

Cumulative characteristics for marks

The mark correlation functions discussed so far provide an excellent insight into correlations among marks in the exploratory analysis of spatial point patterns. However, they are not the only second-order characteristics that should be considered in the context of marked point processes with quantitative marks. Statistical tests of correlation hypotheses should not be based on mark correlation functions but should rather be constructed based on generalised L-functions. These result from mark-weighted K-functions by the usual square root transform. The following paragraphs discuss these functions.

Like the K_{ij}-function in Section 5.3.2, the mark-weighted K-function $K_{mm}(r)$ is a natural generalisation of Ripley's K-function. Hence, K_{mm} is constructed in a very similar way and has a similar interpretation to K and K_{ij}. Recall that Ripley's classical K-function can be written as

$$\lambda K(r) = \mathbf{E}_o \left(\sum_{x \in N} \mathbf{1}_{b(o,r)}(x) \right)$$

(see (4.3.1)). The mark-weighted K-function has a very similar form but the marks are now also taken into account, i.e.

$$\lambda K_{mm}(r) = \mathbf{E}_o \left(\sum_{[x;m(x)]} m(o)m(x)\mathbf{1}_{b(o,r)}(x) \right) \bigg/ \mu^2. \qquad (5.3.34)$$

This means that $\lambda\mu^2 K_{mm}(r)$ is the mean of the sum of the products formed by the mark of the typical point and the marks of all points in the disc (sphere) of radius r centred at the typical point. Due to stationarity the typical point can be chosen as the origin o. Because of its cumulative nature, Shimatani and Kubota (2004) call the function the 'cumulative mark product function'.

Like the K-function, $K_{mm}(r)$ is also normalised by dividing it by the intensity λ, but also by the squared mean mark μ^2. If the marks are independent, this normalisation yields

$$K_{mm}(r) = K(r) \qquad \text{for } r \geq 0. \qquad (5.3.35)$$

The derivative $K'_{mm}(r)$ reveals the close relationship to the mark correlation function $k_{mm}(r)$:

$$K'_{mm}(r) = db_d r^{d-1} k_{mm}(r), \tag{5.3.36}$$

by analogy with (4.3.8) for $K(r)$ and $g(r)$.

The concept of the mark-weighted K-function can be easily extended to the K_t-function for any test function t. Note first that the weighting in the construction of $K_{mm}(r)$ may be regarded as a weighting based on the test function t_1 as defined on p. 343. Other test functions may of course be used in a similar way and thus more general mark-weighted K-functions may be defined as

$$\lambda K_t(r) = \mathbf{E}_o \left(\sum_{[x;m(x)]\in M} t(m(o), m(x)) \mathbf{1}_{b(o,r)}(x) \right) \Big/ c_t. \tag{5.3.37}$$

In particular, the test function

$$t_3(m_1, m_2) = m_2$$

yields

$$\lambda K_{.m}(r) = \mathbf{E}_o \left(\sum_{[x;m(x)]\in M} m(x) \mathbf{1}_{b(o,r)}(x) \right) \Big/ \mu. \tag{5.3.38}$$

This is a Palm mean corresponding to the mark-sum measure,

$$\lambda \mu K_{.m}(r) = \mathbf{E}_o \left(S(b(o, r)) \right) / \mu \qquad \text{for } r \geq 0.$$

Similarly, the test function $t_2(m_1, m_2) = m_1$ yields

$$\lambda K_{m.}(r) = \mathbf{E}_o \left(\sum_{[x;m(x)]\in M} m(o) \mathbf{1}_{b(o,r)}(x) \right) \Big/ \mu,$$

and the right-hand side can be rewritten as

$$\mathbf{E}_o(m(o) N(b(o, r))) / \mu.$$

In general, this mean is not equal to $\lambda K(r)$, since the mark of the typical point and the number of points in its neighbourhood may be correlated.

However, if the marks are independent or if the random field model is a suitable marking model, then

$$K_{m.}(r) = K_{.m}(r) = K_{t_4}(r) = K(r). \tag{5.3.39}$$

A test of the hypothesis of the random field model may be based on this equation, see Section 7.5.

A generalisation of (5.3.36) yields

$$K'_t(r) = db_d r^{d-1} k_t(r) g(r) \qquad \text{for } r > 0, \tag{5.3.40}$$

since $K_t(r)$ can be expressed in terms of 'two-point Palm means' as

$$\lambda \mu K_t(r) = \int_0^\infty \mathbf{E}_{ou}(t(m(o), m(\mathbf{u}))) db_d u^{d-1} g(u) du, \tag{5.3.41}$$

where $g(u)$ is the pair correlation function of N with variable u instead of the usual r and $u = \|\mathbf{u}\|$. By (5.3.23), the right-hand side becomes

$$\int_0^\infty c_t(u) db_d u^{d-1} g(u) du,$$

which yields the formula for $K'_t(r)$.

As mentioned above and discussed in Chapter 4, the L-function as a square-root transformed version of the classical K-function should be used. Similarly, $K_{mm}(r), \ldots, K_t(r)$ may be transformed in the same way yielding the mark-weighted L-function

$$L_{mm}(r) = \sqrt[d]{\frac{K_{mm}(r)}{b_d}} \qquad \text{for } r \geq 0 \tag{5.3.42}$$

and

$$L_t(r) = \sqrt[d]{\frac{K_t(r)}{b_d}} \qquad \text{for } r \geq 0. \tag{5.3.43}$$

5.3.4   Estimation of second-order characteristics

This subsection presents estimators of the most important second-order characteristics for marked point processes. The order is similar to that of Section 4.3.3.

Mark-weighted K-functions

The K-functions $K_{ij}(r)$, $K_{mm}(r)$ and $K_t(r)$ for marked point processes may be estimated in a similar way to Ripley's K-function as discussed in Section 4.3.3. Recall that this implies first calculating characteristics similar to $\hat{\kappa}_{st}(B)$ and $\hat{\kappa}_i(r)$ and then normalising these by estimators of intensities and suitable constants.

In the following, estimators are introduced based on stationary edge-correction, by analogy with $\hat{\kappa}_{st}(b(o, r))$; the formulas for isotropic edge-correction can be derived analogously to $\hat{\kappa}_i(B)$ on p. 229.

An unbiased estimator of $\lambda_i \lambda_j K_{ij}(r)$ is given by

$$\hat{\kappa}_{st,ij}(r) = \sum_{x_1,x_2 \in W}^{\neq} \frac{\mathbf{1}(m(x_1) = i)\mathbf{1}(m(x_2) = j)\mathbf{1}(\|x_1 - x_2\| \leq r)}{\nu(W_{x_1} \cap W_{x_2})} \tag{5.3.44}$$

for $r \geq 0$. Ratio-unbiased estimators of $K_{ij}(r)$ result from division by intensity estimators

$$\hat{\lambda}_i = \frac{N_i(W)}{\nu(W)} \quad \text{and} \quad \hat{\lambda}_j = \frac{N_j(W)}{\nu(W)}$$

or, better, by $\hat{\lambda}_{i,V}(r)$ and $\hat{\lambda}_{j,V}(r)$, which are constructed for N_i and N_j as $\hat{\lambda}_V(r)$ in Section 4.2.3.

The following is an unbiased estimator of $\lambda^2 c_{mm} K_{mm}(r)$:

$$\hat{\kappa}_{st,mm}(r) = \sum_{x_1,x_2 \in W}^{\neq} \frac{m(x_1)m(x_2)\mathbf{1}(\|x_1 - x_2\| \leq r)}{\nu(W_{x_1} \cap W_{x_2})} \quad \text{for } r \geq 0; \tag{5.3.45}$$

see Stoyan and Stoyan (1994, p. 303). Again ratio-unbiased estimators of $K_{mm}(r)$ result from division, now by estimators of λ^2 and c_{mm}.

λ^2 is chosen as in Section 4.3.3, either by (4.3.34) or as $(\hat{\lambda}_V(r))^2$. Since $c_{mm} = \mu^2$, c_{mm} may be estimated by

$$\hat{c}_{mm} = \hat{\mu}^2, \tag{5.3.46}$$

using $\hat{\mu}$ in (5.2.9).

Finally, an unbiased estimator of $\lambda^2 c_t K_t(r)$ is given by

$$\hat{\kappa}_{st,t}(r) = \sum_{x_1,x_2 \in W}^{\neq} \frac{t(m(x_1), m(x_2))\mathbf{1}(\|x_1 - x_2\| \leq r)}{\nu(W_{x_1} \cap W_{x_2})} \quad \text{for } r \geq 0, \tag{5.3.47}$$

and ratio-unbiased estimators of $K_t(r)$ can be derived by dividing by estimators of λ^2 and c_t. The characteristic c_t may be estimated by

$$\hat{c}_t = \sum_{i=1}^{n} \sum_{j=1}^{n} t(m_i, m_j)/n^2, \tag{5.3.48}$$

where $m_1, \ldots, m_n$ are the marks of the points of M in W and n is the number of points in W.

Partial pair correlation functions and mark connection functions

Estimators of $g_{ij}(r)$ and $p_{ij}(r)$ are based on (5.3.8) and (5.3.15). Ratio-unbiased estimators of $g_{ij}(r)$ and $p_{ij}(r)$ are

$$\hat{g}_{ij}(r) = \frac{\hat{\varrho}_{ij}(r)}{\hat{\lambda}_i \hat{\lambda}_i} \qquad \text{for } r > 0 \tag{5.3.49}$$

and

$$\hat{p}_{ij}(r) = \frac{\hat{\varrho}_{ij}(r)}{\hat{\varrho}(r)} \qquad \text{for } r > 0, \tag{5.3.50}$$

where $\hat{\varrho}_{ij}(r)$ and $\hat{\varrho}^{(2)}(r)$ are estimators of the product densities $\varrho_{ij}(r)$ and $\varrho^{(2)}(r)$ and $\hat{\lambda}_i$ and $\hat{\lambda}_j$ are intensity estimators. For $\varrho(r)$ the estimator in (4.3.29) is recommended and for $\varrho_{ij}(r)$,

$$\hat{\varrho}_{ij}(r) = \sum_{x_1, x_2 \in W}^{\neq} \frac{\mathbf{1}(m(x_1) = i)\mathbf{1}(m(x_2) = j)k(\|x_1 - x_2\| - r)}{2\pi r \nu(W_{x_1} \cap W_{x_2})} \tag{5.3.51}$$

for $r > 0$; see Stoyan and Stoyan (1994, p. 291). The denominator $2\pi r$ in the above equation is appropriate for the planar case, i.e. $d = 2$. For $d = 3$ it has to be replaced by $4\pi r^2$.

Mark correlation functions

Estimators of mark correlation functions are based on (5.3.26). Therefore,

$$\hat{k}_t(r) = \frac{\hat{\varrho}_t(r)}{\hat{\varrho}(r)} \Big/ \hat{c}_t \qquad \text{for } r > 0 \tag{5.3.52}$$

is a ratio-unbiased estimator of $k_t(r)$. Here

$$\hat{\varrho}_t(r) = \sum_{x_1, x_2 \in W}^{\neq} \frac{t(m(x_1), m(x_2))k(\|x_1 - x_2\| - r)}{2\pi r \nu(W_{x_1} \cap W_{x_2})}. \tag{5.3.53}$$

For the special case of $t_1(m_1, m_2) = m_1 m_2$,

$$\hat{\varrho}_{mm}(r) = \sum_{x_1, x_2 \in W}^{\neq} \frac{m(x_1)m(x_2)k(\|x_1 - x_2\| - r)}{2\pi r \nu(W_{x_1} \cap W_{x_2})}. \tag{5.3.54}$$

The quantity c_t (see (5.3.29)) used for normalisation is estimated by (5.3.48).
   The values for $r = 0$, $k_{mm}(0)$ and $k_t(0)$, can be estimated by

$$\hat{k}_{mm}(0) = \frac{1}{n} \sum_{i=1}^{n} m_i^2 \Big/ \hat{\mu}^2 \tag{5.3.55}$$

and

$$\hat{k}_t(0) = \frac{1}{n} \sum_{i=1}^{n} t(m_i, m_i) \bigg/ \hat{c}_t, \qquad (5.3.56)$$

where $m_1, \ldots, m_n$ are the marks of the points of M in W and n is the number of these points.

Note that the estimators of mark correlation functions are ratios of estimators which have a rather similar structure. Therefore, they yield precise estimates since fluctuations cancel out. Practical experience shows that for the estimation of mark correlation or connection functions the estimators of $\varrho_{ij}(r)$, $\varrho_t(r)$ and $\varrho(r)$ do not have to be of a particularly good quality. They can be estimated without edge-correction, if both functions are estimated based on the same estimation principle. As a result, estimates of $g_{ij}(r)$ might be of a lower quality than those of $p_{ij}(r)$. This information may be useful in the context of patterns observed in irregular windows, where edge-correction is difficult.

Note that for the estimation of $p_{ij}(r)$ and $k_t(r)$, intensity estimation is not necessary.

Mark correlation functions for quantitative marks were first defined and discussed in Stoyan (1984) and Isham (1987). Further development and applications may be found in Penttinen and Stoyan (1989), Cressie (1993), Schlather (2001a), Schlather et al. (2004), Stoyan et al. (1995) and Stoyan and Stoyan (1994).

5.4　Orientation analysis for marked point processes

5.4.1　Introduction

Many geometrical structures are anisotropic, either globally or locally. In point process applications, objects which are represented by points may have their own orientations. In the planar case, using angles as marks is a natural choice in this context. This section introduces statistical methods that may be applied to data sets with angular marks or for which angular marks can be constructed. (This is different from Section 4.5, which only discusses constructed marks, since there marks that were given a priori were not considered.) Figure 5.22 shows a pattern of this type, where the points are centres of particles the orientation of which is described by main axis orientation. This structure is globally anisotropic.

Angular marks are called *directions* if the angles are between 0° and 360° and correspond to rays or directed lines, or *orientations* if the angles are between 0° and 180° and correspond to axes or undirected lines.

In the case of anisotropic processes, the main aim is the estimation of the directional distribution, i.e. the mark distribution. This distribution is usually different from the uniform distribution on $[0, 360°]$ or $[0, 180°]$.

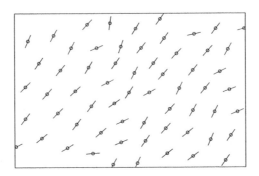

Figure 5.22   A point pattern obtained from a planar section through a Cd-Zn eutectic with rod-shaped Zn in a $171 \times 111\,\mu$m rectangle. The centres of gravity of the section profiles form the point pattern. For the orientation analysis the directions of maximum projection lengths are shown.

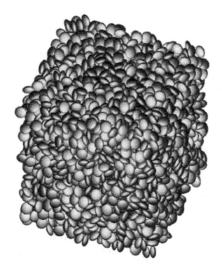

Figure 5.23   A packing of 4000 oblate spheroids of aspect ratio 1:2 and packing density (= volume fraction of spheroids) 0.55.

Note that the situation becomes more complicated if the pattern is isotropic, which implies that it has a uniform direction distribution. Even in this case directional statistics can be usefully applied. However, the analysis now aims to detect *inner orientations* or *local non-isotropies* such as a tendency to local parallelity (or anti-parallelity). For example, Figure 5.23 shows a (nearly) isotropic system of spheroids with a tendency to local parallelity. The spheroids are arranged in parallel due to their shape, whereas their centres form a nearly isotropic point process.

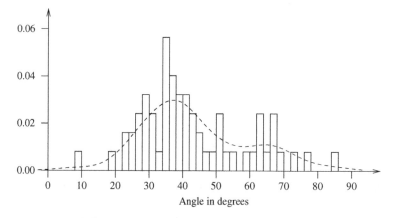

Figure 5.24   Empirical orientation probability density function for the particle axes of Figure 5.22. The apparent non-isotropy of the pattern is well described by this density; the 'main' direction is at 38°, corresponding to the mode in the density function.

5.4.2   Orientation analysis for anisotropic processes with angular marks

This subsection discusses anisotropic marked point patterns with angular marks, i.e. data sets as in Figure 5.22. The statistical analysis is based on the general approach to the estimation of mark distributions as described in Section 5.2.3.

The analysis of marked point patterns with angular marks applies methods that have already been developed in an area of statistics called 'directional statistics'; see Upton and Fingleton (1989) and Fisher et al. (1987) for the two- and three-dimensional case, respectively. It is beyond the scope of this book to provide full coverage of these methods, but it suffices to understand that the angular marks are regarded as a sample of angular data and analysed by directional statistics. Figure 5.24 shows a kernel estimate of the orientation probability density function for Figure 5.22, reflecting the frequencies of the different orientations in the pattern.

The second-order analysis of Section 4.5.3 can also be applied to bivariate data, as demonstrated by Haase (2001). In an ecological study of shrub positions in dryland he uses four quadrant functions $K_{12}(r, \alpha)$ corresponding to 90° sectors and shows that there is a preference for the more sheltered quadrants north-west to north-east.

5.4.3   Orientation analysis for isotropic processes with angular marks

This subsection introduces methods to detect potential inner orientations based on the *orientation correlation function* $k_d(r)$. This function is a mark correlation

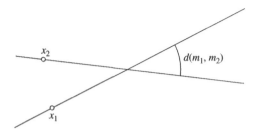

Figure 5.25   Angles $d(m_1, m_2)$ between points marked by orientations.

function as discussed in Section 5.3.3 that is based on a specific test function denoted here as $d(m_1, m_2)$.

For two orientation marks m_1 and m_2, $d(m_1, m_2)$ is the smaller angle between the orientations m_1 and m_2. In the planar case the function is simply

$$d(m_1, m_2) = \min\{|m_1 - m_2|, 180° - |m_1 - m_2|\},\qquad(5.4.1)$$

where m_1 and m_2 are angles between 0° and 180° (see Figure 5.25).

In the spatial case two undirected lines through the origin o having directions $m(x_1)$ and $m(x_2)$, respectively, are considered for the marked points $[x_1; m(x_1)]$ and $[x_2; m(x_2)]$. These lines form a plane, and $d(m(x_1), m(x_2))$ is the angle between the two lines in this plane.

To obtain $k_d(r)$, the function

$$c_d(r) = \frac{\varrho_d(r)}{\varrho(r)}\qquad(5.4.2)$$

is defined analogously to $c_t(r)$ in (5.3.26). $\varrho_d(r)$ is the same as $\varrho_t(r)$ there with $t_6 = d$ with $d(m_1, m_2)$ defined in (5.4.1). Then $c_d(r)$ is normalised by $c_d(\infty)$, which corresponds to the case of independent uniform marks on $[0, 180°]$,

$$k_d(r) = \frac{c_d(r)}{c_d(\infty)}.\qquad(5.4.3)$$

For the planar case

$$c_d(\infty) = 45° = \frac{\pi}{4} \text{ rad},$$

and for the spatial case

$$c_d(\infty) = 57.3° = 1 \text{ rad}.$$

In the spirit of Section 5.3.3, $c_d(r)$ may be interpreted as the conditional mean of the angle between the orientations of two points of distance r. (Note that

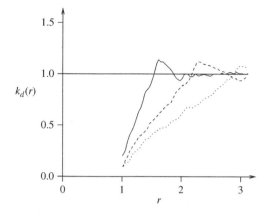

Figure 5.26   Orientation correlation functions of three packings of oblate spheroids: (solid line) aspect ratio $1:2$, Figure 5.23; (dashed line) aspect ratio $1:3$, packing density 0.45; (dotted line) aspect ratio $1:4$, packing density 0.37 (see Bezrukov and Stoyan, 2006). Reproduced by permission of John Wiley & Sons, Ltd.

this mean is based on the usual arithmetic mean and not on the mean definition of directional statistics.) Values of $k_d(r)$ smaller than 1 indicate a tendency for local parallel alignment at distance r; in the case of independent orientations $k_d(r) \equiv 1$.

The function $k_d(r)$ is estimated in a similar way as $k_t(r)$ (see Section 5.3.4), since the $d(m_1, m_2)$ in (5.4.1) is just the test function $t_6(m_1, m_2)$. Based on this argument, a K_d-function can also be defined and estimated, by analogy with the constructions discussed above.

Figure 5.26 shows the empirical orientation correlation function $\hat{k}_d(r)$ for the system of spheroids in Figure 5.23, where the marks are the orientations of the axes of rotational symmetry. It indicates a tendency towards local parallelity. Such local parallelity is quite natural for the patterns, where non-spherical objects were forced to arrange very densely in the space, which implies the need to be locally parallel. If the density increases further, even global anisotropy is necessary; see, for example, Bezrukov and Stoyan (2006). The statistical methods for the three-dimensional case are described in Low (2002).

5.4.4   Orientation analysis with constructed marks

The orientation correlation function $k_d(r)$ can also be easily applied to constructed marks and hence is a useful method in the context of non-marked point processes. Note that this approach was not discussed in Section 4.5 as some knowledge of the theory of marked point processes is necessary to fully understand it.

A natural mark construction principle uses the orientation $o(x_n)$ of the line connecting the point x_n with its nearest neighbour $z_1(x_n)$. In the planar case, where $x_n = (\xi_n, \eta_n)$ and the nearest neighbour of x_n, $z_1(x_n) = (\xi'_n, \eta'_n)$, the mark is computed by

$$o(x_n) = \arctan \left(\frac{\eta_n - \eta'_n}{\xi_n - \xi'_n} \right) \tag{5.4.4}$$

with the convention that

$$o(x_n) := o(x_n) - 180° \qquad \text{if } o(x_n) > 180°.$$

Remarks

1. The ray starting at x_n and pointing towards $z_1(x_n)$ defines a direction. But its use seems unnatural, since often x_n and $z_1(x_n)$ are mutual nearest neighbours and would then have diametrically different directional marks. Working with orientations of connection lines results in the same mark for two points that are mutually nearest neighbours.

2. In empirical data sets it is possible that $\xi_n = \xi'_n$, which implies that $o(x_n)$ cannot be defined. The resulting numerical difficulties can be avoided by adding very small random independent numbers to the coordinates.

Example 5.11.   Sea anemones on a rock
Figure 5.27 shows the locations of 231 beadlet anemones in a 280×180 cm rectangle on a rock. The pattern looks like an isotropic pattern, but there seem to be chains of points, i.e. the animals have a tendency to be arranged in lines. These chains have been already analysed by Upton and Fingleton (1985), who constructed marks, the nearest-neighbour orientations $o(x_n)$, and then used the corresponding

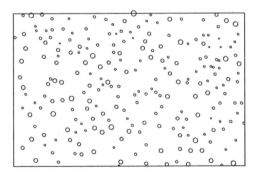

Figure 5.27   Locations of 231 beadlet anemones in a 280×180 cm rectangle on a stone. The diameters of the circles are related to animals' sizes. (After S. A. L. Kooijman.) Courtesy of G. Upton. Reproduced by permission of John Wiley & Sons, Ltd.

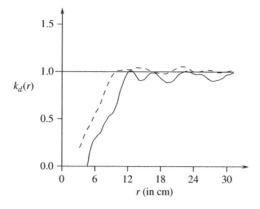

Figure 5.28   The orientation correlation function $\hat{k}_d(r)$ for the anemones (solid line). The function indicates close orientational correlations for short distances (up to $r = 4$, it is $k_d(r) = 0$) and a range of correlation of around $r_{corr} = 10$ cm. The dashed line is $\hat{k}_d(r)$ for 1012 simulated completely random points in a larger window. There is also some orientation correlation, but it is clearly weaker.

mark distribution as discussed in Section 4.5.2. This method failed to reveal the structure, but the approach with mark correlation is slightly more successful.

In the first step of the analysis, marks $o(x_n)$ were constructed for all 231 points following (5.4.4). The analysis accepts that the true nearest neighbours cannot be determined for points close to the boundary, but only the nearest neighbour inside the window. In the second step the method introduced in Section 5.4.3 is applied, i.e. the estimation of the orientation correlation function $k_d(r)$. The result is shown in Figure 5.28. This function indicates some orientation correlation: for short distances ($r \leq 4$ cm), $\hat{k}_d(r) = 0$. This means that at these distances pairs of points are considered which are mutual nearest neighbours. For larger distances the correlation decreases, which is indicated by increasing $\hat{k}_d(r)$, and the range of orientation correlation is around $r_{corr} = 10$ cm, as the empirical mark correlation function $\hat{k}_d(r)$ has only statistical fluctuations around 1 for $r > 10$ cm.

Note that the mark construction method introduces some artificial correlation, which has to be taken into account. It is demonstrated in Figure 5.28, which shows $\hat{k}_d(r)$ also for a simulated pattern of 1012 completely random points in a 470×470 cm window. (This is a larger point pattern with the same point density as the pattern of anemones.) There is also some spatial correlation, but it is apparently weaker than in the anemone pattern. The observed orientation correlation results partly from pairs of points which are mutual nearest neighbours.

6

Modelling and simulation of stationary point processes

Models are indispensable in all areas of statistics, and point process statistics is no exception. Models aid in the interpretation and understanding of empirical patterns and their summary characteristics, as they show possible theoretical forms and give information on the possible statistical fluctuations. And, of course, simulations of spatial point processes are typically based on models.

This chapter first discusses some operations on point patterns, which represent construction principles and are useful for constructing and modifying given patterns and often mimic the natural generation of point patterns. Then several classical point process model classes – cluster, Cox, hard-core and Gibbs processes – are introduced. Some marking models are also presented.

In addition, three very important and rather diverse topics are discussed. The first topic is a method for simulating point patterns without any explicit model, the second are space–time point processes, and finally the text covers problems from spatial statistics that are more general than point process theory by studying correlations between point processes and random fields or fibre systems.

Statistical Analysis and Modelling of Spatial Point Patterns   J. Illian, A. Penttinen, H. Stoyan and D. Stoyan
© 2008 John Wiley & Sons, Ltd

6.1   Introduction

A large part of classical statistics deals with fitting statistical models and theoretical probability distributions to data as well as with their interpretation and the evaluation of their suitability. Well-known examples of distributions that may be suitably fitted to specific data structures include the Gaussian, the Poisson and the exponential distribution. Choosing such a distribution in itself constitutes 'modelling'. It is particularly satisfying if such a distribution is not only formally chosen but can also be motivated by knowledge of mechanisms underlying the observed data.

Both distributions and statistical models typically depend only on a small number of parameters. These parameters often have an intuitive meaning and interpretation. For this reason the characteristics of a specific phenomenon can easily be described based on the type of distribution or model and the parameters.

In point process statistics, models have a similar role. Some basic models have already been discussed in earlier chapters, such as the Poisson process (Chapter 2) and finite point process models (Chapter 3). This chapter presents further models for stationary point patterns. As in classical statistics, these models help to characterise the given data sets:

- Given a model, a small number of parameters provide a unique description of the process of interest. Ideally, it should be clear how these parameters control the structure of the corresponding point patterns. In other words, different patterns that are samples from the same model can be distinguished entirely on the basis of different parameters. For example, in an evolution process, these may be point density and object size.

- Statistical analysis is greatly simplified if it is based on a parametric model. Then it suffices to estimate the parameters, i.e. a parametric statistical approach is taken.

- The model construction may reflect the processes that have caused the pattern. Hence, the model construction itself provides an improved understanding of these underlying processes.

- Section 1.9 discussed the many issues that can be addressed with simulation approaches and the reasons why they are indispensable in the context of spatial point processes. Models are very important as simulations are typically based on models.

6.2   Operations with point processes

Many point process models can be derived from less complex models by applying one or several operations to simpler models such as the Poisson process. This section describes three fundamental operations that may be used to generate new point processes from given processes. The resulting models, and many others, are described in later sections of this chapter. The three fundamental operations described here are thinning, clustering and superposition.

6.2.1   Thinning

A thinning operation uses a specifed rule determining which points in a basic process N_b are deleted, yielding the *thinned point process* N. Regarded as a random set, N is a subset of N_b:

$$N \subset N_b.$$

There are many different thinning rules.

   p-thinning. The simplest form of thinning is *p-thinning:* each point in N_b is deleted with probability $1 - p$. The deletion is independent of the location of the point and the deletion or non-deletion of other points in N_b. The constant p is called the *retention probability*.

   p(x)-thinning. $p(x)$-thinning generalises the simple approach of p-thinning in so far as that the retention probability p now depends on the location x of the point. Thinning is based on a deterministic function $p(x)$ on $\mathbb{R}^d$, with $0 \le p(x) \le 1$. A point x in N_b is deleted with probability $1 - p(x)$, and again its deletion is independent of deletion or non-deletion of any other points. The $p(x)$-thinning of a stationary point process is a second-order intensity-reweighted point process, i.e. an important example of a non-stationary point process as discussed in Section 4.10.

   P(x)-thinning. In a further generalisation, the thinning function $p(x)$ itself is random. More formally, thinning is based on a random field $\{P(x)\}_{x \in \mathbb{R}^d}$ which is independent of N_b. A realisation of the thinned process N is constructed by taking a realisation of N_b and applying $p(x)$-thinning to it, where $p(x)$ is a sample $\{p(x)\}$ of the random field $\{P(x)\}$.

The thinning approaches discussed so far are *independent thinnings*. In other words, the deletion or non-deletion of any particular point is not correlated with the operation on the other points, i.e. the thinning functions (which are independent of N_b) completely determine the operation. However, more general thinning approaches may be considered where the thinning operation depends on the configuration of N_b, thus yielding the class of *dependent thinnings*. For example, thinning operations in this class drive the evolution of plant communities. Often, the spatial distribution of plants becomes more regular over time due to competition-induced mortality. Humans apply thinning to plants, for example by weeding (as gardeners) or thinning-out trees (as foresters). Clearly, in applications thinnings are often dependent thinnings, but independent thinnings are nevertheless useful approximations.

If the characteristics of the basic process N_b (denoted by the subscript b) are known, the characteristics of the point processes N resulting from independent thinning may be derived.

First-order characteristics

If $\lambda_b(x)$ is the intensity function of the basis process N_b, the intensity function of the thinned process is

$$\lambda(x) = p(x)\lambda_b(x). \tag{6.2.1}$$

If N_b is stationary, the p-thinned process N is also stationary. Note that this is not true for a $p(x)$-thinned process. The intensity of a p-thinned processes is given by (6.2.1), i.e.

$$\lambda = p\lambda_b. \tag{6.2.2}$$

A $P(x)$-thinned process from a stationary N_b is stationary if $P(x)$ is a stationary random field. Equation (6.2.2) generalises to

$$p = \mathbf{E}(P(o)), \tag{6.2.3}$$

i.e. p is the mean of the random thinning field.

Second-order characteristics

The product density of a p-thinned process N is given by

$$\varrho^{(2)}(r) = p^2 \varrho_b^{(2)}(r) \tag{6.2.4}$$

and the K-functions as well as the pair correlation functions of N and N_b coincide:

$$K_p(r) = K(r), \tag{6.2.5}$$

$$g_p(r) = g(r). \tag{6.2.6}$$

The product density of a $p(x)$-thinned stationary and isotropic point process with pair correlation function $g(r)$ is

$$\varrho^{(2)}(x, y) = \lambda(x)\lambda(y)g(r) \qquad \text{for } r = \|x - y\|, \tag{6.2.7}$$

with $\lambda(x) = \lambda p(x)$.

$P(x)$-thinning based on a stationary and isotropic thinning field $\{P(x)\}$ with mean p and covariance function

$$k(r) = \mathbf{E}\left(P(o)P(r)\right) - p^2 \qquad \text{where } r = \|\boldsymbol{r}\|$$

yields

$$\varrho^{(2)}(r) = \left(k(r) + p^2\right) \varrho_b^{(2)}(r). \tag{6.2.8}$$

Consequently, Ripley's *K-function* is

$$K(r) = \int_0^r \left(k(t) + p^2\right) dK_b(t) \bigg/ p^2 = \int_0^r (k(t) + p^2) db_d t^{d-1} g_b(t) dt.$$

The intuitive interpretation of the product density $\varrho^{(2)}(r)$ on p. 32 shows that (6.2.4) is plausible: if two points both contribute to the product density then both of them have to survive the thinning.

Thinning a Poisson process

Consider a Poisson process N_b with intensity function $\lambda_b(x)$. The $p(x)$-thinned process N is an inhomogeneous Poisson process with intensity function given by (6.2.1). This fact is implemented in the context of simulation of inhomogeneous Poisson processes as described in Section 3.4.1. If N_b is a homogeneous Poisson process with intensity λ_b then the p-thinned process is a homogeneous Poisson process with intensity $p\lambda_b$. Note that the points that are *deleted* by this thinning operation form another Poisson process, and the two processes are stochastically independent. A $P(x)$-thinned Poisson process is a doubly stochastic Poisson processes (or Cox process) as discussed in Section 6.4.

Example 6.1.   Thinning a forest of Sitka spruce
Figure 6.1 shows again the pattern of 294 Sitka spruce trees that was discussed in Example 4.14 in its November 2003 state. In June 2004 this stand was thinned by a forester, resulting in a more regular pattern of 74 trees. The objectives of the intervention were to release approximately 80 windfirm trees and to uniformly open the crown canopy to provide more light for Sitka spruce regeneration. Clearly, this was not a random p-thinning with $p = \frac{74}{294} = 0.252$, but the neighbourhood situation of each tree influenced the forester's decision.

Simple point process methodology was applied to analyse the thinning scheme, with the idea in mind that trees with small marks were more likely to be felled. The following marks (natural or constructed) were considered:

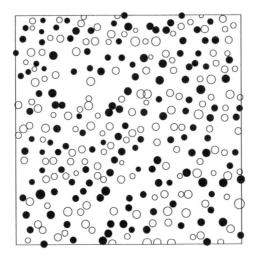

Figure 6.1   Positions of 294 Sitka spruce trees: o = thinned out, • = remaining tree. Data courtesy of University of Wales, Bangor (Arne Pommerening), and Forest Research (Forestry Commission).

- dbh (diameter);
- height;
- nearest-neighbour distance;
- mean of the distances to the four nearest neighbours;
- mean-direction index $IR_4(x)$;
- dbh dominance, $Do_4(x)$ for dbh;
- height dominance, $Do_4(x)$ for height (see (5.2.20)).

The clearest relationship between thinning policy and marks was found for dbh dominance and dbh.

Table 6.1 shows that a large percentage (69) of trees with dbh dominance index 1 were considered as windfirm trees, while a large percentage (57) of trees with index 0 were thinned out. According to Table 6.2 many trees with small dbh were thinned out.

6.2.2 Clustering

In a clustering operation every point x in a given point process N_p is replaced by a cluster N^x of points. The clusters N^x are finite point processes and their set-theoretic union is the *cluster point process*

$$N = \bigcup_{x \in N_p} N^x.$$

Table 6.1   Percentages of tree types (0 = normal, 1 = windfirm, 2 = thinned out) for the dbh dominance indices 0, 0.25, ..., 1.

0			0.25			0.5			0.75			1		
0	1	2	0	1	2	0	1	2	0	1	2	0	1	2
42	1	57	58	1	41	67	13	20	45	42	13	29	69	2

Table 6.2   Stem-and-leaf diagram of dbh (thinned trees in **bold**, 2|4 represents 24 cm)

2	**4445**
2	**667** 77777
2	**8888** 8 **99999** 999
3	**00000000000** 000000 **111111111** 11111111
3	**2222** 222222222222222 **3333333333** 333333
3	**44444444** 4444444444444 **55555** 55555555555555555555
3	**66** 666666666666666 **7** 7777777777
3	**88** 888888888888888 **99** 99999999999999999999999999
4	000000000000 1111111111
4	22222222 **3** 3333333333333333
4	444444444 5555555
4	**6** 666 **7** 777
4	8888 999
5	0001
5	
5	
5	6

In some models the original point x is included in the resulting process N^x, but usually this is not the case. Figure 6.3 on p. 379 shows a sample from a simulated cluster point process.

Cluster processes have been used as models for many natural phenomena. A typical interpretation regards N_p as a collection of 'parent points', e.g. representing the locations of plants. The points in the clusters may then be regarded as the 'daughters' of parent points, e.g. representing the locations of dispersed seeds or young plants. Often, however, the locations of the parent points are unknown or even fictitious. As a result of this and also because the clusters typically overlap, statistical methods for cluster processes are rather complicated, perhaps even more difficult than for Gibbs processes. An important special case is the Neyman–Scott process, which is discussed in detail in Section 6.3.2.

In the following, formulas for a slightly more general model are given. Suppose that the parent point process $N_p = \{x_1, x_2, \ldots\}$ is stationary with intensity λ_p and second-order product density $\varrho_p^{(2)}(\mathbf{r})$ and that the clusters N^x are of the form

$$N^{x_n} = N_n + x_n$$

for each x_n in N_p. The N_n form a sequence of i.i.d. finite point sets centred at the origin with the same distribution, independent of the parent point process N_p. The $+x_n$ term means that N^{x_n} is centred at x_n.

The 'typical cluster' N_c is a further set with the same distribution. The origin o is not included in N_c, which has the consequence that the parent points are not included in N. The random total number of points of N_c is denoted by c, and the corresponding mean is $\bar{c}$. The intensity function of the finite point process N_c is denoted by $\lambda_c(x)$ and its second-order product density by $\varrho_c^{(2)}(x, y)$. This approach is termed *homogeneous independent clustering*. If N_p is a Poisson process the resulting process N is called a *Poisson cluster process*.

Irrespective of the form of N_p in this model, the intensity λ of N is given by

$$\lambda = \lambda_p \bar{c}. \tag{6.2.9}$$

The second-order product density $\varrho^{(2)}(\mathbf{r})$ of N is (Felsenstein, 1975; Shimatani, 2002)

$$\varrho^{(2)}(\mathbf{r}) = \lambda_p \int_{\mathbb{R}^d} \varrho_c^{(2)}(\mathbf{r} + x, x) dx$$

$$+ \int_{\mathbb{R}^d} \int_{\mathbb{R}^d} \varrho_p^{(2)}(\mathbf{r} - x + y) \lambda_c(x) \lambda_c(y) dx dy. \tag{6.2.10}$$

Several consecutive clustering operations may be applied to a basis process, for example mimicking a pattern of plants where the offspring become the parents of the next generation; see Liemant et al. (1988) and Shimatani (2002).

6.2.3 Superposition

In a superposition operation two or more point processes are superimposed onto each other such that the resulting process is formed by the set-theoretic union of these processes.

Let N_1 and N_2 be two point processes with intensities λ_1 and λ_2, pair correlation functions g_1 and g_2, etc. Consider the union

$$N = N_1 \cup N_2,$$

assuming that none of the points of N_1 and N_2 coincide. The pattern in Figure 1.2 may be regarded as an example of a process resulting from the superposition of two independent point processes of • and ○ points, as discussed in Examples 5.5 and 5.6.

The usual point process characteristics can be given analytically. Regardless of whether N_1 and N_2 are independent or stationary, the intensity function of the superposition is

$$\lambda(x) = \lambda_1(x) + \lambda_2(x) \qquad \text{for } x \in W.$$

If N_1 and N_2 are stationary and stationarily connected (but not necessarily independent), then $N = N_1 \cup N_2$ is stationary and

$$\lambda = \lambda_1 + \lambda_2.$$

If N_1 and N_2 are mutually *independent*, stationary and isotropic the second-order characteristics of $N = N_1 \cup N_2$ are

$$g(r) = \left(\lambda_1^2 g_1(r) + \lambda_2^2 g_2(r) + 2\lambda_1\lambda_2\right) / (\lambda_1 + \lambda_2)^2,$$

$$\lambda K(r) = \frac{\lambda_1}{\lambda}\left(\lambda_1 K_1(r) + \lambda_2 b_d r^d\right) + \frac{\lambda_2}{\lambda}\left(\lambda_2 K_2(r) + \lambda_1 b_d r^d\right) \quad \text{for } r \geq 0.$$

The superposition of n i.i.d. stationary and isotropic point processes has the pair correlation function

$$g_n(r) = \frac{ng(r) + n^2 - n}{n^2}.$$

For $n \to \infty$ this converges towards $g(r) \equiv 1$, the pair correlation function of the Poisson process.

The nearest-neighbour distance d.f. of the superposition of two independent stationary and isotropic point processes takes the form

$$D(r) = 1 - \left(\frac{\lambda_1}{\lambda}\left(1 - D_1(r)\right)\left(1 - H_{s,2}(r)\right) + \frac{\lambda_2}{\lambda}\left(1 - D_2(r)\right)\left(1 - H_{s,1}(r)\right)\right) \text{ for } r \geq 0,$$

and the spherical contact d.f. can be expressed as

$$H_s(r) = 1 - (1 - H_{s,1}(r))(1 - H_{s,2}(r)) \qquad \text{for } r \geq 0.$$

6.3 Cluster processes

6.3.1 General cluster processes

In the natural world, clustered patterns are very common, perhaps more common than regular or random patterns. For example, young trees in natural forests and galaxies in the universe form clustered patterns. The point process literature describes many examples of clustering; see Lawson and Denison (2002) or any book on point process statistics. In clustered patterns the point density varies strongly through space. In some areas the point density is very high, i.e. the points form clusters and these are surrounded by areas of low point density, perhaps even by empty space.

The definition of a cluster is somewhat subjective: clusters are groups of points with an inter-point distance that is below the average distance in the pattern. This local aggregation is not simply a result of random point density fluctuations. There is a 'fundamental ambiguity between heterogeneity and clustering, the first corresponding to spatial variation of the intensity function $\lambda(x)$, the second to stochastic dependence amongst the points of the process', and these are 'difficult to disentangle' (Diggle et al., 2007).

In general, spatial clustering may have been caused by a number of different processes:

1. The objects represented by the points were originally scattered in the entire region of interest, but remain only in some subregions that are distributed irregularly in space. A classical example of this are plants with wind-dispersed seeds which germinate only in areas where the environmental conditions are suitable. These patterns are most suitably modelled by Cox processes (see Section 6.4).

2. The cluster pattern is a result of physical processes which cause objects to move in space. Physical laws determine the formation of clusters, as in the case of galaxies. Models for this type of patterns are beyond the scope of this book.

3. The pattern is a result of a mechanism that involves 'parent points' and 'daughter points', where the daughter points are scattered around the parent points, as discussed in Section 6.2. This results in a pattern which is most suitably modelled by the classical cluster model. A specific type of cluster model is the Neyman–Scott model where the parent points form a Poisson process. This model is discussed in detail in Section 6.3.2. It is of course possible to construct similar models based on non-Poisson parent processes, but closed-form expressions for summary characteristics might be difficult to find. Nevertheless, these models are of practical interest, and they may be simulated in a way that is similar to Neyman–Scott processes. By the way, the clusters of the final cluster process are usually the single 'primary clusters' (all daughters of the same parent) as expected; however, sometimes they result from the superposition of primary clusters, in particular when these are thin and large.

4. There is positive interaction among the points. Over time, the point density increases in areas where the density was initially slightly higher. In areas

where the density was initially lower, it decreases further. Inter-species cooperation among plants may be an example of this.

In each of these cases the behaviour of the summary characteristics is similar and it is difficult to distinguish the four cases based on statistical approaches alone.

Second-order characteristics

In a clustered pattern more points occur at short distances than would be expected if the pattern were CSR and hence the summary characteristics take larger values than a Poisson process at these distances. More specifically, the K-function is greater than $b_d r^d$, the L-function is greater than r, and the pair correlation function is greater than 1. Note that even $g(0) = \infty$ is possible, i.e. that $g(r)$ has a pole at $r = 0$. The order of such a pole cannot exceed $d - 1$, i.e. it is only possible that $g(r) \propto r^{-\beta}$ with $\beta \leq d - 1$.

Distance characteristics

The nearest-neighbour distance d.f. $D(r)$ for a clustered pattern takes larger values than for Poisson patterns with equal intensity. By construction, this d.f. mainly reflects the form of the clusters: the typical point is likely to be part of a cluster and its nearest neighbour is frequently in the same cluster. Thus $D(r)$ provides little more than some information on the cluster geometry.

In contrast, the spherical contact d.f. $H_s(r)$ takes smaller values than both $D(r)$ and the spherical contact d.f. for a Poisson process with equal intensity. It describes the empty space between the clusters.

Detection of clusters

In the analysis of clustered patterns one issue concerns the detection of clusters. This may be easy for isolated clusters, but in general the problem is very hard and computationally intensive. Okabe et al. (2000, Section 8.6), present an early systematic approach to this issue. Simple approaches to cluster detection are described in Fotheringham and Zhan (1996). These are based on counts of points within test circles with radii that are systematically varied and centred at either lattice points or randomly distributed points. Clusters are indicated by extremely large numbers of points within a circle. A more refined approach is discussed in Stoica et al. (2007).

Many cluster-detection methods may be described in terms of graph theory. A simple classical method is the so-called friends-of-friends algorithm. It uses the sphere graph $G(N, r)$ with a distance r chosen according to some idea of inter-point distances in clusters. The connected components of the graph are the clusters constructed by the algorithm. More recent algorithms use local point densities attached to the points, e.g. based on the Voronoi tessellation using (3.3.7). An example is the HOP algorithm (Eisenstein and Hut, 1998). In this algorithm the point with the highest local density is determined among the nearest neighbours of each point. The point and its 'nearest' neighbour are then linked by an edge. Again, the connected components are the clusters; cluster centres are those points which are their own nearest neighbour. Neyrinck et al. (2005) improved this algorithm by eliminating Poisson noise, based on the observation that in a homogeneous Poisson process, density maxima occur at 1/13.6 of the points.

In some applications, the shape of clusters is analysed, e.g. isolated points, lineaments or compact clusters are of interest. In the analysis of the highly clustered system of galaxies one looks for halos, blobs, lineaments and walls.

Modern MCMC-based (Bayesian) approaches are also used to detect clusters or lineaments. Stoica et al. (2005) present a method for the detection of filaments in cluster patterns, while Guillot et al. (2006) and Skare et al. (2007) consider clusters of points along the edges of tessellations.

6.3.2   Neyman–Scott processes

Neyman–Scott processes are the result of a specific type of homogeneous independent clustering (see p. 370) applied to a stationary Poisson process. The parent points form a stationary Poisson process with intensity λ_p and the daughter points in the typical cluster N_c are random in number and scattered independently and with identical distribution around the origin. The parent points are only auxiliary constructs; they are not observable and do not form part of the final point pattern, which consists exclusively of the daughter points.

Neyman–Scott processes have been applied in many contexts. They were introduced by Neyman and Scott (1952) to model patterns formed by the locations of galaxies in space; see the discussion in Section 1.3. The review by Neyman and Scott (1972) contains further examples: the distribution of insect larvae in fields and the geometry of patterns reflecting the impact of bombing. In the pattern of insect larvae the parent points are locations of egg masses and the daughter points are the locations of the larvae. In the bombing pattern, the parent points represent the target locations at bomb release and the daughter points are the locations of impact of individual bombs. Cluster processes of Neyman–Scott type may also be appropriate models for patterns of trees such as pines in natural forests; see Penttinen et al. (1992) and Tanaka et al. (2008).

This subsection discusses several general properties of the Neyman–Scott processes. Due to the distributional assumptions the resulting cluster process N is stationary. If the scattering distribution is isotropic, then so is N.

An isotropic Neyman–Scott process is based on the following *process parameters:* the number of cluster points and the distribution of the distance between a parent and a daughter point. The first parameter is an integer random variable c with mean $\bar{c}$ and probabilities

$$p_i = \mathbf{P}(c = i) \qquad \text{for } i = 1, 2, \ldots .$$

The p.d.f. of the distances from the cluster centre is denoted by $\delta(r)$.

In some formulas, another distance distribution has to be considered, the distribution of the random distance between two independent points in the same cluster. The corresponding d.f. is denoted by $F_d(r)$; its p.d.f. is $f_d(r)$. Stoyan and Stoyan (1994, p. 310) show how $F_d(r)$ may be calculated given $\delta(r)$.

General formulas for Neyman–Scott processes

Since Neyman–Scott processes are constructed in a rather simple way based on Poisson processes and independent clusters, it is possible to derive simple formulas for first- and second-order characteristics. However, such formulas cannot be derived for other characteristics such as nearest-neighbour distances.

The general formula (6.2.9) for the intensity discussed in the context of clustering operations yields

$$\lambda = \lambda_p \bar{c}.$$

The K-function, pair correlation function and distance d.f. respectively are given by

$$K(r) = b_d r^d + \frac{1}{\lambda \bar{c}} \sum_{n=2}^{\infty} p_n n(n-1) F_d(r) \qquad \text{for } r \geq 0, \tag{6.3.1}$$

$$g(r) = 1 + \frac{1}{\lambda \bar{c}} \sum_{n=2}^{\infty} p_n n(n-1) \frac{f_d(r)}{d b_d r^{d-1}} \qquad \text{for } r \geq 0, \tag{6.3.2}$$

$$D(r) = 1 - (1 - H_s(r)) \overline{D}_{cl}(r) \qquad \text{for } r \geq 0. \tag{6.3.3}$$

Here $\overline{D}_{cl}(r)$ is the probability that there is no other point of N_c in a disc of radius r centred at an arbitrary (randomly chosen) point of the typical cluster N_c. Since it is difficult to determine $\overline{D}_{cl}(r)$ the last equation is rarely used (for some calculations, see Stoyan and Stoyan, 1994, p. 313; Tanaka et al., 2008; Daley and Vere-Jones, 2008, Example 15.1 (a) and Exercise 15.1.3). Furthermore,

$$H_s(r) = 1 - \exp(-\lambda_p \mathbf{E}(\nu(N_c \oplus b(o, r)))) \qquad \text{for } r \geq 0. \tag{6.3.4}$$

Thus, in order to calculate $H_s(r)$ the mean area (volume) of the union set of discs (spheres) of radius r centred at the points of the typical cluster has to be found. This is an elementary but very difficult problem (again see Stoyan and Stoyan, 1994, p. 313).

By definition and by (6.3.3) it is clear that the J-function is given by

$$J(r) = \overline{D}_{cl}(r). \tag{6.3.5}$$

It is decreasing in r and smaller than 1.

The distribution of the number $N(W)$ of points in a set W of volume V may be approximated by

$$\mathbf{P}(N(W) = n) = \frac{\lambda V(1-b)}{n!}(\lambda V(1-b) + nb)^{n-1}\exp(-\lambda V(1-b) - nb)$$

for $r \geq 0$, with b given by

$$(1-b)^2 = 1 + \lambda V \xi_2$$

and

$$\xi_2 = \frac{1}{V}db_d\int\limits_0^\infty \overline{\gamma}_W(r)r^{d-1}(g(r) - 1)dr;$$

see Sheth and Saslaw (1994).

Two examples of Neyman–Scott processes

In the following models, the representative cluster N_c is an isotropic centred Poisson process with mean total number $\overline{c}$, which implies that c has a Poisson distribution.

Matérn cluster process   (Matérn, 1960). Here, the points in N_c are independently uniformly scattered in the disc (ball) $b(o, R)$, where R is a further model parameter. For this model the density function for the distance from the cluster centre is

$$\delta(r) = \frac{dr^{d-1}}{R^d} \qquad \text{for } 0 \leq r \leq R$$

and

$$f_d(r) = \begin{cases} \frac{4r}{\pi R^2}\left(\arccos\frac{r}{2R} - \frac{r}{2R}\sqrt{1 - \frac{r^2}{4R^2}}\right) & \text{for } d = 2, \\ \frac{3}{2}\frac{r^2}{R^6}\left(R - \frac{r}{2}\right)^2\left(2R + \frac{r}{2}\right) & \text{for } d = 3, \end{cases}$$

for $0 < r < 2R$; otherwise $f_d(r) = 0$.
    Consequently,

$$g(r) = 1 + \frac{f_d(r)}{\lambda_p db_d r^{d-1}} \qquad \text{for } r \geq 0 \qquad (6.3.6)$$

and

$$g(0) = 1 + \frac{1}{\lambda_p b_d R^d}.$$

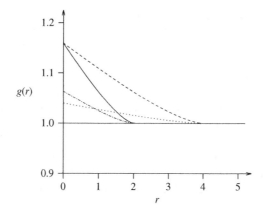

Figure 6.2   Pair correlation functions of Matérn cluster processes of equal intensity $\lambda = 10$ for different cluster radii R and mean numbers of points $\bar{c}$: (solid line) $R = 1$, $\bar{c} = 5$; (dot-dashed line) $R = 1$, $\bar{c} = 2$; (dashed line) $R = 2$, $\bar{c} = 20$; (dotted line) $R = 2$, $\bar{c} = 5$.

Figure 6.2 shows the pair correlation functions for four Matérn cluster processes for varying values of R and $\bar{c}$ to demonstrate the impact of the parameters R and $\bar{c}$ on the form of $g(r)$. For $R = 1$, $\bar{c} = 5$ and $R = 2$, $\bar{c} = 20$ the point density in the clusters is the same and thus the values of $g(0)$ coincide; $2R$ is that value r_{corr} of r at which $g(r)$ becomes constant.

Modified Thomas process, planar case   (Thomas, 1949). In Thomas processes the distribution of the daughter points around the parent points is the symmetric normal distribution with parameter σ. Here

$$\delta(r) = \frac{r}{\sigma^2} \exp\left(-\frac{r^2}{2\sigma^2}\right)$$

and

$$f_d(r) = \frac{r \exp\left(-\frac{r^2}{4\sigma^2}\right)}{2\sigma^2} \qquad \text{for } r \geq 0.$$

Consequently,

$$g(r) = 1 + \frac{1}{4\pi\lambda_p\sigma^2} \exp\left(-\frac{r^2}{4\sigma^2}\right) \qquad \text{for } r \geq 0$$

and

$$g(0) = 1 + \frac{1}{4\pi\lambda_p\sigma^2}.$$

Both the Matérn cluster process and the Thomas process can be generalised by replacing the Poisson distribution of the random number c of cluster points by any other discrete distribution; the formulas for $f_d(r)$ can be plugged into the general formula (6.3.2). Tanaka et al. (2008) generalise the model by introducing two types of clusters with σ_1 and σ_2.

Note that many Neyman–Scott processes are also Cox processes, which are considered in Section 6.4. This is the case if the number of points per cluster has a Poisson distribution. Examples are the processes considered above, the Matérn cluster process and the modified Thomas process.

Non-stationary or inhomogeneous variations of Neyman–Scott processes have also been used in the literature; see Provatas et al. (2000), Fleischer et al. (2006), Møller and Waagepetersen (2007) and Waagepetersen (2007).

Simulation of Neyman–Scott processes

The simulation of Neyman–Scott processes in a window W is straightforward and follows the model construction. First, the parent Poisson process is simulated in the enlarged window $W \oplus b(o, R)$, where R is a radius that is large enough for the influence of parent points outside W on the pattern in W to be (almost) completely eliminated. In the simulation of a Matérn cluster process, the radius R is the same as a the model parameter, whereas for other models 99th percentiles of the distance distribution may be used. For example, for modified Thomas processes, R may be chosen as $R = \sqrt{-2\sigma^2 \cdot \ln(\alpha)}$, yielding the $1 - \alpha$ quantile of the parent–daughter distance. (A more exact way of handling the problem is presented in Brix and Kendall, 2002, and is based on perfect simulation.)

In the second step, a cluster distributed as N_c is generated for each parent point x in $W \oplus b(o, R)$ and shifted towards x. This means that the following quantities have to be generated:

(1) the number of cluster points, following the p_i;

(2) the directions of the daughter points relative to the parent points, following the uniform distribution;

(3) the distances between parent points and daughter points, following the p.d.f. $\delta(r)$.

In the simple cases of a Matérn cluster and a modified Thomas process it is more efficient to simulate directly from either the uniform distribution in the sphere of radius R or the bivariate Gauss distribution, respectively. Figure 6.3 shows a simulated Matérn cluster process.

Conditional simulation is not difficult either. Assume that the aim is to simulate a sample from a Neyman–Scott process with exactly n points in $\boxed{1}$. This requires the following steps:

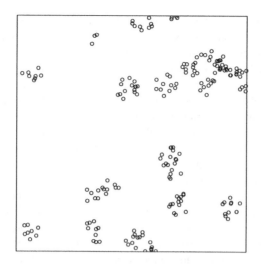

Figure 6.3   Simulated planar Matérn cluster process in $\boxed{1}$. The intensity is $\lambda = 200$, the cluster radius is $R = 0.05$ and the mean number of points per cluster is $\bar{c} = 10$.

1. Generate parent points, just as in the unconditional case. Denote their number by n_p.

2. Choose a random integer i between 1 and n_p (following the discrete uniform distribution).

3. Generate a daughter point for parent point i. If the daughter point is in $\boxed{1}$, it belongs to the sample, otherwise it is rejected.

Continue steps 2 and 3 until n daughter points have been generated.

The parametric statistics of cluster processes follows the general approach of Section 7.2; see Examples 7.1 and 7.2, where a Matérn cluster process is fitted to the pattern in Figure 1.4.

6.4   Stationary Cox processes

6.4.1   Introduction

Cox processes are a class of spatial point process models describing aggregation or clustering resulting from environmental variability. Finite Cox processes have already been briefly discussed in Section 3.4.2; this section now considers stationary Cox processes. Due to their elegant construction, Cox processes are sufficiently general to be applied to a large variety of different scientific questions but are still amenable to calculations. Cox processes were first systematically studied in Cox (1955).

In the following, the general construction of a Cox process and a number of examples of specific Cox processes are discussed. Towards the end of this section a specific Cox process is fitted to the pattern in Figure 1.4.

As a motivation for the construction of a Cox processes, recall the inhomogeneous Poisson process, based on a non-constant intensity function $\lambda(x)$. It models point patterns with variable point density and even clusters in the sense that the pattern is more clustered in areas of high relative intensity. Hence an inhomogeneous Poisson process might be regarded as a nice and simple model of spatial pattern formation in the presence of environmental heterogeneity. However, the inhomogeneous Poisson process is never stationary since the intensity function is not constant, and hence many of the methods developed for stationary processes are not applicable.

In a stationary Cox process the intensity function is replaced by a stationary random field with non-negative values. The realisations of this random field are functions which are treated as intensity functions of inhomogeneous Poisson processes. All distributional properties of the point process generated are inherited from the stationary random field, which is called here the *intensity field,* yielding a stationary point process model.

Formally, the Cox point process model is defined in two steps:

- Consider a stationary non-negative valued random field $\{\Lambda(x)\}$.

- Given a realisation of the random field, i.e. given that $\Lambda(x) = \lambda(x)$ for all $x \in \mathbb{R}^d$, the points of the corresponding realisation of the Cox process form an inhomogeneous Poisson process with intensity function $\lambda(x)$.

The resulting process N is called a *Cox process* and sometimes also a *doubly stochastic Poisson process* due to the construction described above. By construction, a Cox process is a hierarchical point process model with two levels. The stationary random field forms the bottom level and in applications typically represents unobserved (environmental) local heterogeneity. The second level is formed by the points, which are independently scattered given the intensity function. Thus a Cox process is a clever construction in which independence (as in the Poisson process) is replaced by conditional independence.

Note that Cox processes can be also constructed based on more general random processes than random fields (see p. 383 and Example 6.8).

There are many ways of constructing the random field underlying Cox processes. As a consequence, the class of Cox processes is very flexible yet yields computationally tractable models; the following provides several examples of Cox process models in which this flexibility is well reflected.

Examples of Cox processes

Mixed Poisson process.   The mixed Poisson process is a very simple example of a Cox process, presented here only for the sake of exposition. It may be regarded as a stationary Poisson process with randomised intensity parameter. More specifically, the intensity λ is now a non-negative random variable rather than a fixed value. An individual sample from this process looks like a sample from a stationary Poisson process, but the intensities differ from sample to sample. It is not possible to investigate the Cox nature of the process based on a single sample; it can be analysed only as a sample from a Poisson process.

Matérn cluster process.   Let N_p be a homogeneous Poisson process with intensity λ. Each point x in N_p is used as the centre of a disc (ball) $b(x, R)$ with radius R, and a finite Poisson process with intensity λ_c is generated within the disc (ball). The superposition of all these circular (spherical) clusters is the Matérn cluster process as discussed in Section 6.3.2. For this process the random field is

$$\Lambda(x) = \lambda_c \cdot \sum_{y \in N_p} \mathbf{1}_{b(y,R)}(x).$$

It is a shot-noise field as introduced in Section 1.8.3 and considered further in Section 6.9.

Log-Gaussian Cox process.   A random field $\{Z(x)\}$ is Gaussian if, for any finite collection of locations $x_1, \ldots, x_k$, any linear combination $b_1 Z(x_1) + \cdots + b_k Z(x_k)$ with real $b_1, \ldots, b_k$ has a one-dimensional normal distribution; see Lantuéjoul (2002) and Wackernagel (2003). If $\{Z(x)\}$ is stationary and isotropic, its distribution is determined completely by the mean μ and the (spatial) covariance function $k(r)$. However, this type of field cannot be used as the intensity field of a Cox process since it can take negative values. Thus, a suitable transformation has to be applied to the field $\{Z(x)\}$ to yield a Cox process. A very elegant transformation, resulting in a mathematically tractable model, is

$$\Lambda(x) = \exp(Z(x)) \qquad \text{for } x \in \mathbb{R}^d.$$

The corresponding process is termed a *log-Gaussian Cox process*. It was first described in Rathbun (1996) and Møller et al. (1998) and has been widely used in the modelling of environmental heterogeneity. It was independently defined in astronomy; see Coles and Jones (1991). The log-Gaussian Cox process can be seen as a link between point process statistics and geostatistics. A simulated realisation of an LGC process is presented in Figure 6.4(a).

Poisson-gamma random field Cox process.   This model is based on an intensity field which is generated by a homogeneous Poisson process N_I of impulse centres in $\mathbb{R}^d$, a sequence of i.i.d. gamma-distributed impulses $\{w_i\}$ for marks of the points in N_I, and a d-dimensional p.d.f. (kernel function) $k_s(x)$, leading to a shot-noise random field as described in Section 1.8.3, i.e.

$$\Lambda(x) = \sum_{x_i \in N_I} w_i k_s(x - x_i).$$

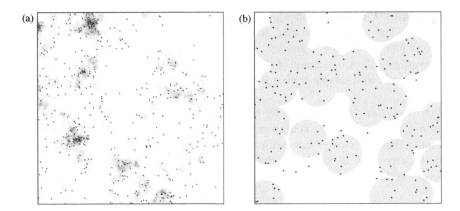

Figure 6.4   Two simulated patterns generated from Cox processes. (a) A sample from a log-Gaussian Cox process. The (unobservable) intensity field is indicated by the grey shading in the background. Note that in applications usually only the points can be observed. (b) A sample from a random-set generated Cox process where the intensities in the two phases determined by the random set differ. Again, typically only the points can be observed.

This model was introduced in Ickstadt and Wolpert (1997) and Wolpert and Ickstadt (1998). Note the difference from the Matérn cluster process seen as a shot-noise random-field-generated Cox process, where the impulses are constant ($= \lambda_c$) and the kernel function is the uniform distribution in $b(o, R)$.

The Matérn cluster and Poisson-gamma random field Cox processes are examples of so-called *shot-noise Cox processes*; see Brix (1999), Møller (2003), Møller and Waagepetersen (2004) and Møller and Torrisi (2005).

Random-set-generated Cox process.   A stationary random closed set X divides $\mathbb{R}^d$ into two parts or phases: the set X and its complement X^c. In both phases, a Poisson process is generated with intensities λ_1 and λ_2, respectively. This process was first introduced for a data set of a pattern of seedling locations in a commercial tree plantation where the soil had been treated in two different ways. The random set reflects these two different soil treatments (Penttinen and Niemi, 2007). The random intensity of this Cox process is simply

$$\Lambda(x) = \lambda_1 \mathbf{1}_X(x) + \lambda_2(1 - \mathbf{1}_X(x)),$$

where $\mathbf{1}_X(x)$ denotes the indicator function of the set X. A simulated example is shown in Figure 6.4(b).

Fibre-process-generated Cox process.   Assume that $S = \bigcup_l S_l$ is a so-called stationary fibre process, i.e. a mathematical model for a random system of fibres or curves (see Stoyan et al. 1995), where S_l is a randomly located, oriented and shaped fibre in $\mathbb{R}^d$, for example a line segment. Construct a Poisson process on each fibre. The number of points on the fibre S_l follows the Poisson distribution with mean $\lambda_f \ell(S_l)$, related to the fibre length $\ell(S_l)$, and the points are uniformly and independently distributed along the fibre. This point process is also a Cox process, which may be used to model alignments of points. Fibre-process generated Cox processes may also be applied in the statistics of random fibre systems (see Section 6.11.3). This example shows that Cox processes can be more general than in the construction introduced above that uses an intensity field; see Stoyan et al. (1995, p. 155).

Thinning of Poisson processes.   In this case the random process is generated from a Poisson process with intensity λ through random independent thinning, where the location-dependent retention probabilities are based on a random field $\{P(x)\}$, with values in the interval $[0, 1]$. Note also that independent thinning of a Cox process yields a new Cox process.

6.4.2   Properties of stationary Cox processes

General formulas

Due to the construction and its close relationship to the Poisson process, relatively simple formulas can be derived for Cox processes. Assume that the intensity field $\{\Lambda(x)\}$ is a second-order random field. Then the first-order and second-order intensity functions of the Cox process are, respectively,

$$\lambda(x) = \mathbf{E}(\Lambda(x))$$

and

$$\varrho^{(2)}(x_1, x_2) = \mathbf{E}(\Lambda(x_1)\Lambda(x_2)).$$

The latter can be generalised to the kth-order intensity in a straightforward way.

If the generating random field is stationary the Cox process is also stationary, and if the field is isotropic the process is also isotropic. The intensity of the Cox process is

$$\lambda = \mathbf{E}(\Lambda(o)) = \mu, \tag{6.4.1}$$

where μ is the mean of the intensity field.

In the stationary and isotropic case the second-order product density may be written as

$$\varrho^{(2)}(r) = \mu^2 + k(r) \tag{6.4.2}$$

with $r = \|x - y\|$, and the pair correlation function is

$$g(r) = 1 + \frac{k(r)}{\mu^2}, \tag{6.4.3}$$

where $k(r)$ is the covariance function of the random field,

$$k(r) = \mathbf{cov}(\Lambda(x), \Lambda(y)) \qquad \text{for } r = \|x - y\|.$$

If the random field has positive autocorrelations, then $g(r) \geq 1$, i.e. the resulting point pattern is clustered.

The void probability of a Cox process is

$$v_K = \mathbf{P}(N(K) = 0) = \mathbf{E}\left(\exp\left(-\int_K \Lambda(x)\mathrm{d}x\right)\right)$$

for any (compact) set K of $\mathbb{R}^d$. Here the expectation is with respect to the intensity field $\{\Lambda(x)\}$.

Formulas for some specific models

Of the many Cox process models the log-Gaussian Cox process and Poisson-gamma random field Cox process are perhaps the most widely applicable processes. The random-set-generated Cox process is also a natural model in many applications. The following lists some important point process characteristics for these three models.

Log-Gaussian Cox process.   Suppose that $\{Z(x)\}$ is a stationary and isotropic Gaussian random field with mean μ_Z, variance σ_Z^2 and covariance function $k_Z(r)$. Recall that the random intensity is $\Lambda(x) = \exp(Z(x))$. Then the first- and second-order point process characteristics are

$$\lambda = \exp\left(\mu_Z + \frac{1}{2}\sigma_Z^2\right)$$

and

$$g(r) = \exp(k_Z(r)) \qquad \text{for } r \geq 0.$$

Poisson-gamma random-field Cox process. The generating shot-noise random field has the following parameters: intensity λ_p for the Poisson process locations of the impulses, gamma distribution with shape α and inverse scale parameter β for the random impulses, and smoothing kernel $k_s(x)$. Then the intensity and the pair correlation function of the generated Cox process are

$$\lambda = \lambda_p \alpha / \beta$$

and

$$g(r) = 1 + \frac{1+\alpha}{\alpha} \int \int k_s(u) k_s(u+x) \mathrm{d}u$$

with $r = \|x\|$; both integrals are over $\mathbb{R}^d$.

Random-set-generated Cox process. The first- and second-order characteristics of a stationary and isotropic random set X are area fraction (volume fraction in $\mathbb{R}^3$)

$$p = \mathbf{P}(o \in X)$$

and (non-centred) covariance

$$C(r) = \mathbf{P}(o \in X \text{ and } \mathbf{x} \in X) \qquad \text{for } \mathbf{x} \in \mathbb{R}^d \text{ with } r = \|\mathbf{x}\|.$$

These two characteristics, together with the Poisson process intensities λ_1 and λ_2 of the two phases, determine the intensity and pair correlation function, which are

$$\lambda = p\lambda_1 + (1-p)\lambda_2$$

and

$$g(r) = \frac{1}{\lambda^2} \left(C(r)(\lambda_1 - \lambda_2)^2 - 2p(\lambda_1 - \lambda_2)\lambda_2 + \lambda_2^2 \right) \qquad \text{for } r \geq 0.$$

Simulating Cox processes

Since samples from a Cox process are samples from inhomogeneous Poisson processes given a realisation of the intensity field, the main issue concerns the simulation of the intensity field model. Once this is achieved the algorithm is the same as for the inhomogeneous Poisson process. For example, a simulation algorithm for the log-Gaussian Cox process can be found in Møller and Waagepetersen (2002, 2004), and conditional simulation is discussed in Lantuéjoul (2002). A detailed discussion of these methods is beyond the scope of this book.

6.4.3   Statistics for Cox processes

For the estimation of the parameters of Cox processes the standard methods described in Section 7.2 can be used, i.e. the minimum contrast method and likelihood methods, based on either MCMC (Møller and Waagepetersen, 2004) or the approximate method of Tanaka et al. (2008). Also goodness-of-fit tests follow the usual pattern in Section 7.4.

Example 6.2.   Phlebocarya *pattern: fitting a log-Gaussian Cox process*
The statistical analyses in Chapter 4 for the *Phlebocarya* pattern suggest analysing it with methods for stationary point processes and using models with spatially variable point density. Thus, a Cox process is fitted to the data here. (In Example 7.1 a cluster process is also used to model the data.)

The minimum contrast method explained in Section 7.2 based on the pair correlation function yields the following estimates:

$$\hat{\mu}_Z = 0.428, \qquad \hat{\sigma}_Z^2 = 1.89,$$

and

$$\hat{k}_Z(r) = 1.89 \exp(-\alpha r) \qquad \text{with } \alpha = 6.89.$$

Figure 6.5 shows the empirical and model pair correlation functions, which are close together. Simulation tests based on $D(r)$ and $H_s(r)$ also indicate that the log-Gaussian Cox model is acceptable (see Figure 6.6).

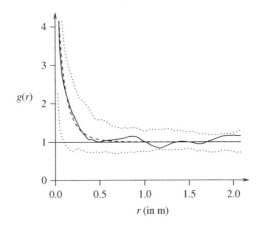

Figure 6.5   The empirical (solid line) and model (dashed line) pair correlation function for the *Phlebocarya* pattern. Additionally, envelopes resulting from 99 simulations of the log-Gaussian Cox model are shown (dotted lines).

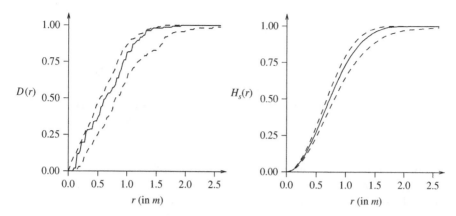

Figure 6.6   Results of simulation tests for the log-Gaussian Cox process for the *Phlebocarya* pattern based on (left) $D(r)$ and (right) $H_s(r)$. In both cases a good fit is indicated, since the empirical curve is well between the simulated envelopes.

6.5   Hard-core point processes

6.5.1   Introduction

A hard-core point process is a point process in which there are no points at a distance smaller than a specific minimum distance denoted by r_0. Hard-core processes describe patterns representing the locations of centres of non-overlapping objects, typically circles or spheres with radius $R \leq r_0/2$. Consequently, the pair correlation function $g(r)$ and the nearest-neighbour distance d.f. $D(r)$ satisfy

$$g(r) = 0 \quad \text{and} \quad D(r) = 0 \qquad \text{for } 0 \leq r \leq r_0.$$

In reality, of course, there are hard-core distances in all point patterns, since the objects represented by the points take up some space themselves. However, the space around the objects is often large enough that the objects' size can be ignored. The measurement process itself may also lead to patterns with a hard core, for example if the point coordinates are measured on a grid; see Section 1.2.3.

Hard-core processes are typical examples of processes with a tendency towards regularity, which results from repulsion among the points. A number of suitable models have been considered, which can be simulated. However, formulas for summary characteristics have been derived only for the simplest and least interesting ones. Hence, in most applications simulation approaches have been used to find these characteristics.

There are two main types of hard-core processes:

- Processes resulting from *thinning operations*. These operations may remove points that are close to other points or eliminate points in clusters to producing a pattern of isolated points. Consider, for example, ecological processes in plant communities, where dense patterns of seedlings develop into hard-core patterns as a result of competition. Different thinning rules generate different patterns and models.

- Processes resulting from *interaction of hard objects*. Here the objects represented by the pattern are hard and non-penetrable such that if they are randomly distributed in space they cannot be closer together than permitted by their sizes. These objects can exist either simultaneously from the very beginning or can appear over time.

This section describes some common hard-core models which have been used in many applications. The first of these processes are the Matérn hard-core processes, for which explicit formulas have been derived. These models are based on very strict thinning rules resulting in rather sparse processes, which are of particular interest in biological applications. The simple sequential inhibition or RSA models result in less sparse patterns which are generated in an iterative way. Gibbs hard-core processes as discussed in Section 6.6.1 and the packing models discussed below may result in rather dense patterns.

6.5.2   Matérn hard-core processes

Matérn (1960, 1986) suggested two hard-core models. Here the model (of his type 2) yielding a higher final intensity of points is described along with two other models which generalise Matérn's original idea.

Matérn hard-core process

The model is essentially based on dependent thinning applied to a stationary Poisson process N_b with intensity λ_b. The points in N_b are marked independently by random numbers uniformly distributed in $(0, 1)$. The dependent thinning retains a point x in N_b with mark $m(x)$ if the sphere $b(x, r_0)$ contains no points in N_b with marks smaller than $m(x)$.

In applications, the points in N_b may be the locations of seeds, while the marks are points in time at which they germinate. Then N consists of the locations x of those plants the seeds of which were the earliest to germinate within an area $b(x, r_0)$, the area the plants need for nutrient uptake.

The intensity λ of N is given by

$$\lambda = p\lambda_b.$$

Here p is the *Palm retention probability* of the 'typical point' in N_b, which is given by

$$p = \int_0^1 r(t)\mathrm{d}t = (1 - \exp(-\lambda_b V))/(\lambda_b V) \qquad \text{for } V = b_d r_0^d,$$

where $r(t) = \exp(-\lambda_b Vt)$ is the probability of retention of a point with mark t with $0 \leq t \leq 1$. This formula results from the observation that the sub-point process of N_b that consists of the points with a mark smaller than t is simply a t-thinning of a Poisson process, which is itself a Poisson process of intensity $\lambda_b t$. Hence $r(t)$ is the probability that a sphere of radius r_0 contains no points in the t-thinned process. Consequently,

$$\lambda = (1 - \exp(-\lambda_b V))/V. \tag{6.5.1}$$

If $r_0 = 1$ the maximum intensity λ is approximately 0.318 if $d = 2$, and 0.239 if $d = 3$. These maxima result from letting λ_b tend to infinity.

The second-order product density $\varrho^{(2)}(r)$ is

$$\varrho^{(2)}(r) = \begin{cases} 0, & \text{for } r \leq r_0, \\ \frac{2\Gamma_{r_0}(r)(1-e^{-\lambda_b V})-2V(1-e^{-\lambda_b \Gamma_{r_0}(r)})}{V\Gamma_{r_0}(r)(\Gamma_{r_0}(r)-V)} & \text{for } r > r_0, \end{cases} \tag{6.5.2}$$

where

$$\Gamma_{r_0}(r) = v(b(o, r_0) \cup b(\mathbf{r}, r_0)) = 2V - \gamma_{r_0}(r) \qquad \text{for } r > r_0, \tag{6.5.3}$$

in which $\mathbf{r}$ is any point with $\|\mathbf{r}\| = r$; $\gamma_{r_0}(r)$ is the area (volume) of the intersection of two discs (spheres) of radius r_0 and midpoint distance r.

In the important low-dimensional cases for $r \leq 2r_0$,

$$\gamma_{r_0}(r) = \begin{cases} 2r_0^2 \arccos \frac{r}{2r_0} - \frac{r}{2}\sqrt{4r_0^2 - r^2} & \text{for } d = 2, \\ \frac{4\pi}{3}r_0^3 \left(1 - \frac{3r}{4r_0} + \frac{r^3}{16r_0^3}\right) & \text{for } d = 3. \end{cases}$$

Clearly, $\gamma_{r_0}(r)$ vanishes for $r > 2r_0$.

These formulas result from using $\varrho^{(2)}(r) = \lambda_b^2 \kappa(r)$ where $\kappa(r)$ is the two-point Palm probability that two points in N_b a distance r apart are both retained. It is straightforward to check that $\varrho^{(2)}(r) = \lambda^2$ for $r \geq 2r_0$. Refer to Stoyan and Stoyan (1985) for a mathematical derivation of the formulas.

The following two models have variable hard-core radii and may be regarded as generalisations of the Matérn hard-core process.

Random competition model

Månsson and Rudemo (2002) study the following model. Each of the points x in a basic Poisson process N_b with intensity λ_b are assigned two independent marks, $r(x)$ and $c(x)$, by independent marking, where $r(x)$ is a radius mark and $c(x)$ a competition mark. The $r(x)$ follow the p.d.f. $f(r)$ and d.f. $F(r)$, whereas the $c(x)$ are uniformly distributed on $(0,1)$; points with small c-marks are regarded as strong.

For this model the *thinning rule* is as follows. The point x in N_b is deleted if there is another point y in N_b with

$$\|x - y\| \le r(x) + r(y) \quad \text{and} \quad c(y) \le c(x).$$

In other words, x is eliminated if its sphere $b(x, r(x))$ intersects with the sphere of a stronger point y.

The intensity of the thinned process N is

$$\lambda = \lambda_b \int_0^\infty p(r) f(r) \mathrm{d}r \tag{6.5.4}$$

with

$$p(r) = \left(1 - \exp\left(-\lambda_b b_d \int_0^\infty (r+y)^d f(y) \mathrm{d}y \right) \right) \bigg/ \left(\lambda_b b_d \int_0^\infty (r+y)^d f(y) \mathrm{d}y \right).$$

Furthermore, the radius (or mark) distribution of the resulting process is

$$F_{\mathcal{M}}(r) = 1 - \int_r^\infty p(s) f(s) \mathrm{d}s \bigg/ \int_0^\infty p(s) f(s) \mathrm{d}s \qquad \text{for } r \ge 0.$$

This distribution function usually differs from $F(r)$, since those points in the marked Poisson process that have larger radius marks are less likely to survive. Therefore, $F(r)$ is called the 'proposal' d.f. and $F_{\mathcal{M}}(r)$ 'resulting' d.f. For the pair correlation function only complicated formulas can be given which contain multiple volume integrals.

Dominance competition model

Stoyan (1988) studies the following model: the points in a Poisson process N_b with intensity λ_b are independently assigned radius marks $r(x)$, which follow the p.d.f. $f(r)$ or d.f. $F(r)$.

The point x in N_b is deleted if there is another point y in N_b with

$$\|x - y\| < 2r(y) \quad \text{and} \quad r(y) > r(x).$$

In other words, x is eliminated if it is within the distance $2r(y)$ of a point with larger mark $r(y)$, i.e. points with larger marks are less likely to be deleted.

The intensity of the thinned process is

$$\lambda = \lambda_b \int_0^1 r(t) \mathrm{d}t \tag{6.5.5}$$

with

$$r(t) = \exp(-\lambda_b V(t))$$

and

$$V(t) = \nu\left(\left\{(x, y, z) : \sqrt{x^2 + y^2} \leq 2F^{-1}(z), t \leq z \leq 1\right\}\right),$$

where $F^{-1}(z)$ is the inverse of $F(r)$. The mark d.f. is

$$F_{\mathcal{M}}(r) = 1 - \int\limits_{F(r)}^{1} r(t)\mathrm{d}t \left/ \int\limits_{0}^{1} r(t)\mathrm{d}t \right. .$$

Stoyan (1988) also presents a formula for the pair correlation function $g(r)$ as well as several examples.

Simulation of the Matérn models

In general, the simulation of a classical Matérn hard core process is straightforward. First, a Poisson process with intensity λ_b is simulated in $W \oplus b(o, r_0)$ and the points are marked independently with random uniform marks in $(0, 1)$. Second, the specific thinning rule is applied to each pair of points. Note that points deemed to be deleted still cause the deletion of other points. This means in practice that those points that are not contained in the final pattern are initially marked as 'to be deleted' but are removed only after all points have been considered; the points in W that have not been deleted form the final sample. Figure 6.7 shows a simulated sample of a Matérn hard core process. For models with random radii an analogous procedure may be applied, replacing r_0 by some quantile of $F(r)$.

For the pattern of gold particles, Glasbey and Roberts (1997) use a hard-core model which modifies a Poisson process by shifting particular points: if points are too close together, they are pushed apart.

6.5.3 The dead leaves model

Matheron (1968) introduced an interesting model of non-overlapping discs or spheres, discussed here for the planar case. Place discs (with constant or random radii) randomly and uniformly on the plane for a long time, within the time interval $(-\infty, 0]$, such that

(1) a new disc may cover discs that have already been placed in the plane, and

(2) infinitely many discs are placed in any subset of the plane of positive area. This generates a pattern as shown in Figure 6.8. The pattern may be interpreted as a layer of circular dead leaves seen from above, from the air (or, equivalently, from below, e.g. from the point of view of a mole).

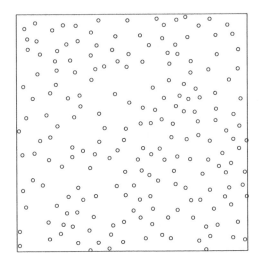

Figure 6.7   A simulated sample from a planar Matérn hard-core process in $\boxed{1}$. The intensity λ is 200 and $r_0 = 0.039$.

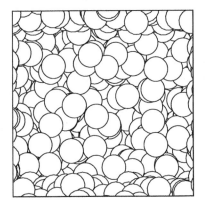

Figure 6.8   A simulation of Matheron's dead leaves model with identically sized discs. The intact discs form the sample of the dead leaves model. Data courtesy of C. Lantuéjoul.

Now consider the non-intersecting or intact discs in the uppermost layer; their centres form a hard-core process that has the same distribution as Matérn's hard core process of Section 6.5.2 in the limit as $\lambda_b \to \infty$.

For the case of constant radii $R = r_0/2$, the formulas are

$$\lambda = 1/(b_d r_0^d), \qquad (6.5.6)$$

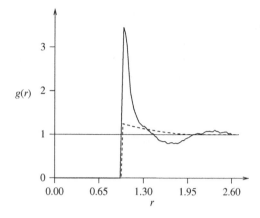

Figure 6.9 Pair correlation functions of two planar hard-core processes with r_0: dead leaves model (dashed line); RSA model (solid line).

$$g(r) = \begin{cases} 0 & \text{for } r \le r_0, \\ \frac{2b_d r_0^d}{\Gamma_{r_0}(r)} & \text{for } r > r_0, \end{cases} \qquad (6.5.7)$$

with $\Gamma_{r_0}(r)$ as in (6.5.3); see also Figure 6.9.

Note that the simulation of the dead leaves process (also called 'falling leaves process') is sometimes used as an illustration of the application of the idea of exact or perfect simulation; see Kendall and Thönnes (1999). In order to generate a sample in the window W the process of falling leaves is simulated in $W_{\oplus R}$ until the whole window is covered by discs. At this point the simulation can be stopped and the sample may be considered as resulting from a simulation with infinitely many steps. This is because infinitely many other invisible discs may be assumed underneath the layer that is currently visible.

6.5.4 The RSA process

RSA is an abbreviation for 'random sequential adsorption', a term used in physics and chemistry. The RSA process is known as the 'simple sequential inhibition' model in the statistical literature. Since the model is used much more frequently in physics and chemistry, the name RSA is used here.

The RSA process is usually considered within a finite set W and is hence a finite point process model (see Figure 3.9). The pattern is constructed by placing iteratively and randomly spheres in W with radii following the d.f. $F(r)$. If a new sphere intersects with an existing sphere, the new sphere is rejected and another sphere, with a different centre and radius, is generated etc. (Note that the process may be modified by generating a new sphere after rejection with a different centre but the same radius.) The process is stopped if either the required number of spheres is placed or if it is impossible to place any new sphere (termed *jamming*). The pattern formed by the sphere centres is a sample of the point process to be generated, and the radii may be considered as marks.

This book only discusses this finite RSA process, but note that Stoyan and Schlather (2000) discuss a stationary version of the model. They show that for constant radii the intensity of the RSA process is higher than for the dead leaves model. In the case of random radii, a proposal and a resulting d.f. have to be distinguished; see details in Stoyan and Schlather (2000). All numerical information for the RSA process, (see Evans, 1993), has been obtained by simulation.

The intensity of the RSA process is usually expressed in terms of the area or volume fraction, A_A or V_V, of space occupied by the spheres. For a process with constant radii the maximum values are

$$A_A = 0.547 \quad \text{and} \quad V_V = 0.382.$$

Figure 6.9 shows the pair correlation function for a sample from a particular RSA process, which was obtained by simulation. The RSA process is simulated along the lines of the model description. For an efficient simulation close to jamming, the search for potential locations for new spheres should be organised with an efficient search algorithm; see Döge (2001).

Statistical methods for the RSA process are described in Van Lieshout (2006c). Provatas et al. (2000) consider closely related models where the spheres are replaced by fibres.

Remarks.   For the simulation of random systems of hard spheres, the RSA process is a popular simple *ad hoc* choice, but much better models with higher area or volume fractions exist and will be discussed in the following sections. Note also that in the RSA process the spheres never touch.

6.5.5   Random dense packings of hard spheres

This subsection describes simulation methods for generating random dense packings of spheres; only the three-dimensional case is considered. It is assumed that these methods correspond to unknown but correct point process models.

Sedimentation algorithm

This algorithm simulates the process of sedimentation of hard and heavy spheres, as introduced by Jodrey and Tory (1979). It generates a random system of hard spheres in a parallelepipedal container. Periodic boundary conditions in the horizontal direction ensure that the simulated sphere system may be considered a sample with a statistically homogeneous structure, if the container is large enough. A typical implementation of the concept of sedimentation is as follows.

First, some initial configuration of spheres is generated, a layer of spheres at the bottom of the container. Every subsequent iteration puts a new sphere into the system. It moves downwards (following 'gravitation') until it hits another already existing sphere in the system. Then, the sphere rolls across the surface of the existing packing until it reaches a stable position, usually given by contact with three spheres. (If a stable position cannot be found after a long time, the algorithm starts another attempt with a new sphere.) This process of filling the container with new spheres continues until all spheres that were intended to be placed have been packed or until the container has been filled up.

Typically, the packings obtained with this method are not as dense as natural random dense packings of hard spheres. With identical spheres the maximum volume fraction is $V_V = 0.58$. Some gradient, i.e. some inhomogeneity in the vertical direction, can be observed in the packing.

Force-biased algorithm

This algorithm generates very dense random packings of hard spheres with constant or random radii. It has, similar to other algorithms with the same aim, little to do with the real processes leading to packings of hard spheres.

The algorithm is based on an old idea due to Jodrey and Tory (1985) and was later refined and generalised to random radii; see Moscinski et al. (1989), Bargiel and Moscinski (1991) and Bezrukov et al. (2002). Its main idea may also be applied to non-spherical objects such as ellipsoids (Bezrukov and Stoyan, 2006) or polyhedra. The name of the algorithm originates from physical terms such as 'force' and 'potential'.

The initial configuration of the algorithm is a system of n spheres $b(x_{i,\text{start}}, r_{i,\text{start}})$. The number n is fixed for the entire simulation. The centres $x_{i,\text{start}}$ are uniformly distributed in the parallelepipedal container, while the radii $r_{i,\text{start}}$ are independent random numbers generated from the radius d.f. In the initial configuration, the spheres may overlap and this configuration is the only stochastic aspect of the algorithm; the rest is completely deterministic.

While the algorithm is running, the centres and radii are changed and denoted by x_i and r_i. At every step of the algorithm, all r_i are uniformly reduced by multiplication with some factor smaller than 1; this ensures that the proportions between them remain fixed; for identical r_i the radii of all spheres are identical but decrease with the number of iterations. A process parameter ϱ describes the current radii, given by

$$\varrho = r_i / r_{i,\text{start}};$$

the parameter is 1 at the start and decreases gradually.

The x_i are changed in order to reduce overlapping. To avoid the size of the spheres being reduced more quickly than necessary, the overlapping is reduced faster than the radii. In the end, only one pair of spheres is in direct contact, while many others are very close together, so close that the small gaps between them can be considered numerical errors. However, the algorithm also produces some 'rattlers', i.e. completely isolated spheres; for systems with random radii these are typically small spheres.

At each iteration step of the algorithm all x_i and r_i are recalculated and then simultaneously replaced by the new values. The new centre of the ith sphere is given by

$$x_i \leftarrow x_i + \frac{1}{r_i} \sum_{j \neq i} F_{ij}, \qquad (6.5.8)$$

i.e. results from a small shift. The F_{ij} are so-called 'repulsion forces' acting between two spheres $b(x_i, r_i)$ and $b(x_j, r_j)$ if these overlap. They are written as

$$F_{ij} = \varphi \mathbf{1}_{ij} p_{ij} e_{ij}, \qquad (6.5.9)$$

where φ is a scaling factor, $\mathbf{1}_{ij}$ is 1 if $b(x_i, r_i) \cap b(x_j, r_j) \neq \emptyset$ and 0 otherwise, p_{ij} is a 'potential value' and e_{ij} the unit vector pointing from x_i to x_j,

$$e_{ij} = \frac{x_j - x_i}{\|x_j - x_i\|}.$$

The p_{ij} depend on the degree of overlapping of $b(x_i, r_i)$ and $b(x_j, r_j)$. In Bezrukov et al. (2002) the form

$$p_{ij} = r_i r_j \left(\frac{\|x_i - x_j\|^2}{(r_i + r_j)^2} - 1 \right) \qquad (6.5.10)$$

is recommended. (If the spheres overlap, p_{ij} is negative and the spheres are moved further apart.) This explains why it is much more difficult to model non-spherical particles as it is not a trivial matter to determine analogues of $\mathbf{1}_{ij}$ and p_{ij}.

The radii shrink according to

$$\varrho \leftarrow \varrho - 2^{-\delta} \tau, \qquad (6.5.11)$$

where τ is another scaling factor and

$$\delta = -\log_{10}(V_{V,\text{nom}} - V_{V,\text{act}}). \qquad (6.5.12)$$

Here

$$V_{V,\text{nom}} = \frac{4}{3} \pi \sum_{i=1}^{N} r_i^3 \bigg/ (\text{volume of container})$$

and

$$V_{V,\text{act}} = \eta^3 V_{V,\text{nom}},$$

where η is the largest number (< 1) such that all spheres $b(x_i, \eta r_i)$ do not intersect. The system formed by these spheres is already a random system of non-overlapping spheres with the right radius proportions, but perhaps of minor quality, since $V_{V,\text{act}}$ is smaller than the target value. The aim of the choice of δ in (6.5.12) is to adapt the radius reduction to the current quality of the packing.

The algorithm stops if during the simulation $V_{V,\text{act}} \geq V_{V,\text{nom}}$ (or $\eta \geq 1$) or after a fixed number of steps have been carried out; in the latter case the spheres $b(x_i, \eta r_i)$ are the output.

Bezrukov et al. (2002) discuss the choice of φ and τ in detail. If one wishes to generate very dense packings, an iterative procedure may be useful: the packing resulting from one iteration step (a run of the force-biased algorithm) is used as the starting configuration for the next step, with the same x_i but increased r_i, $cr_i \rightarrow r_i$, with $c > 1$ (such that again spheres overlap).

Figure 6.10 shows a random packing of 10 000 spheres with lognormal radii. Figure 6.11 shows the corresponding pair correlation function, which looks like the pair correlation function of a soft-core process, along with the pair correlation function for a case of constant spheres; see also Figure 4.27.

Collective rearrangement algorithms

The force-biased algorithm is part of a larger family of algorithms used to generate systems of non-overlapping spheres, called collective rearrangement or concurrent

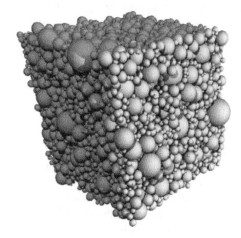

Figure 6.10   A random dense packing of 10 000 spheres with lognormal radii in a cubic container. The ratio $\mu : \sigma$ of the parameters of the basic Gauss distribution is $\mu : \sigma = 4 : 1$, and the volume fraction of the packing is $V_V = 0.7$.

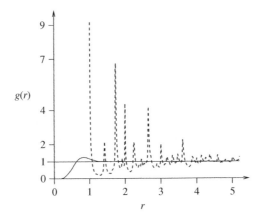

Figure 6.11   Pair correlation functions for the system of sphere centres in Figure 6.10 (solid line) and for a system corresponding to a packing with constant radii and $V_V = 0.7$ (dashed line).

algorithms. Here the number of spheres is fixed for the entire simulation, and many spheres are moved, i.e. the sedimentation algorithm is not a collective rearrangement algorithm. A successful algorithm of this type is the Stillinger–Lubachevsky algorithm, where the spheres grow during the simulation and move more frequently and more strongly than in the force-biased algorithm; see Stillinger et al. (1964) and Lubachevsky and Stillinger (1990).

Another very important approach is *molecular dynamics* where the spheres do not change size, but move randomly in space, following Newton's laws of motion, having contact with other spheres (assuming elasticity) and the container's boundaries. A successful algorithm of this type is the SPACE algorithm; see Stroeven and Stroeven (1999, 2001).

6.6   Stationary Gibbs processes

6.6.1   Basic ideas and equations

In Section 3.6 finite Gibbs processes were discussed, which form an important class of point processes that model the interaction among points in an elegant way. This section now considers their stationary analogue, stationary Gibbs processes. These have infinitely many points, distributed over the whole of $\mathbb{R}^d$. They may be used to adequately model large point patterns with interesting forms of interaction, if an approximation by a finite Gibbs process is considered unsuitable. Again, similar to Section 3.6, the following exposition only considers a subclass of Gibbs processes, those with pairwise interaction. The reader is advised to read Section 3.6 prior to embarking on this section.

The formal mathematical definition of stationary Gibbs processes is more complicated than that of Gibbs processes with a finite number of points. It was developed by the mathematicians Georgii, Glötzl, Kallenberg, Kerstan, Matthes, Mecke, Nguyen, Warmuth and Zessin; see the references in Stoyan et al. (1995, Section 5.5.3) and Georgii (1976). Note that in these theories the process definition is not based on limit procedures with $W \uparrow \mathbb{R}^d$ where the window increases to the whole space (physicists term this the 'thermodynamic limit').

Two basic facts are important for this section. First, stationary and isotropic Gibbs processes are based on two characteristics, the *chemical activity* α and the *pair potential* $\phi(r)$. The latter plays the same role as in the finite case, determining the character of the point distribution (i.e. exhibiting hard cores, soft cores, weak clustering, ...), whereas α influences the intensity λ of the process. For a fixed pair potential, λ increases with decreasing α. (Note that α can also have negative values.) In the context of stationary Gibbs processes, the number of points in the window W is assumed to be the realisation of a random variable.

The conditional intensity of the process is

$$\lambda(x|N) = \exp\left(-\alpha - \sum_{y \in N \setminus \{x\}} \phi(\|x - y\|)\right); \qquad (6.6.1)$$

the term $\alpha + \sum_{y \in N} \phi(\|x - y\|)$ is often referred to as the 'local energy'.

Second, the distribution of a stationary and isotropic Gibbs process satisfies two important mean-value relations. The first of these is the Georgii–Nguyen–Zessin formula:

$$\mathbf{E}\left(\sum_{x \in N} h(x, N \setminus \{x\})\right)$$

$$= \mathbf{E}\left(\int h(x, N) \exp\left(-\alpha - \sum_{y \in N} \phi(\|x - y\|)\right) dx\right), \qquad (6.6.2)$$

where $h(x, N)$ is any function that assigns a non-negative number to a point x and the point process N. The second relation concerns a function $f(N)$ that depends only on N:

$$\lambda \mathbf{E}_o(f(N \setminus \{o\})) = \mathbf{E}\left(f(N) \exp\left(-\alpha - \sum_{x \in N} \phi(\|x\|)\right)\right). \qquad (6.6.3)$$

The left-hand side contains a Palm mean, as introduced in Section 4.1, and the right-hand side a normal mean with respect to the process distribution. An example of such a function is $f(N) = N(b(o, r))$, the number of points in N contained in the ball $b(o, r)$.

It is quite difficult to calculate the summary characteristics discussed in Chapter 4 for Gibbs processes in terms of α and $\phi(r)$. Much work has been done in this context by physicists, who are particularly interested in the pair correlation function $g(r)$ for patterns in $\mathbb{R}^3$. One well-known result is the *Percus–Yevick approximation:*

$$\phi(r) \approx \frac{g(r)}{g(r) - c(r)} \qquad \text{for } r \geq 0 \qquad (6.6.4)$$

(see Hansen and McDonald, 1986; Diggle et al., 1987). Here, $c(r)$ is a further function, the so-called 'direct correlation function', which is defined implicitly as the solution of the Ornstein–Zernike equation,

$$c(r) = h(r) - \lambda(c^*h)(r), \qquad (6.6.5)$$

with $h(r) = g(r) - 1$ and

$$(c^*h)(r) = \int\limits_0^{2\pi} \int\limits_0^{\infty} c(s)h\left(\left(r^2 + s^2 - 2rs\cos\xi\right)^{1/2}\right) ds\,d\xi.$$

Combining (6.6.4) and (6.6.5) yields an integral equation for $g(r)$, which may be used to approximate the pair correlation function.

To get a better idea of the nature of the problem, refer to (3.6.4) and (6.6.7) for the pair potential of the Gibbs hard-core process and the pair correlation functions shown in Figures 4.27 and 6.11. (Note that these functions have been computed for random dense packings of hard spheres, which are not samples of Gibbs processes, since in three-dimensional Gibbs hard-core processes the hard spheres do not touch. Nevertheless, the global behaviour of the pair correlation functions of both process classes is quite similar.) The pair potential $\phi(r)$ is a simple step function, but $g(r)$ has many waves and no formula is known.

Stoyan–Grabarnik residuals

The issue of model diagnostics in the context of point processes is discussed in Section 4.6.5. As in classical statistics, the suitability of a model for a particular data set is assessed based on residuals, which describe deviations between model and data.

Equation (6.6.3) may be used to define 'residuals' based on constructed marks, which may be used both for model and simulation diagnostics. To do so, Stoyan and Grabarnik (1991b) mark the stationary Gibbs process with 'exponential energy marks' defined by

$$m(x) = \frac{1}{\lambda(x|N)}. \qquad (6.6.6)$$

The mean mark corresponding to $m(x)$ is λ, the intensity. This is a direct consequence of (6.6.3), since for

$$f(N) = \frac{1}{\lambda(o|N)} = \exp\left(\alpha + \sum_{x \in N} \phi(\|x\|)\right)$$

the right-hand side of (6.6.3) is 1, and the left-hand side is $\lambda \times$ mean mark.

These constructed marks have two main applications. First, the marks may be used to control iterative simulation algorithms such as those discussed in Section 3.6. As indicated, these algorithms generate samples from a point process model only after a certain number of iterations, the burn-in period. The number of iterations required for burn-in is typically found by experimentation. However, the marks defined above may be used to check whether the Monte Carlo chain has reached the end of the burn-in phase. One can simply take a sample, compute the exponential energy marks $m(x)$ as in (6.6.6) and determine their sum for all x in W, which should be approximately $\nu(W)$ if the burn-in phase is over.

The second use of the constructed marks is in model-fitting. With $\hat{\alpha}$ and $\hat{\phi}(r)$ estimated from the data, the marks

$$\hat{m}(x) = \exp\left(\hat{\alpha} + \sum_{y \in W} \hat{\phi}(\|x - y\|)\right)$$

are used as residuals which should approximately sum to $\nu(W)$ if the model is correct.

Note that these residuals have the disadvantage that $\lambda(x|N)$ must be positive. This causes a problem with Gibbs hard-core processes for which $\lambda(x|N)$ can be zero so that the residuals above cannot be defined.

Gibbs hard-core process

Stationary Gibbs processes with the very simple pair potential

$$\phi(r) = \begin{cases} \infty & \text{for } r \le r_0, \\ 0 & \text{for } r > r_0, \end{cases} \tag{6.6.7}$$

form another class of hard-core point processes, called Gibbs hard-core processes. These processes and their variations have been extensively investigated in statistical physics; see Löwen (2000). If the intensity is very high (or the chemical activity α 'very negative') these processes are similar to random dense packings of identical spheres, but, as mentioned above, the spheres do not touch. Corresponding pair correlation functions are shown in Figures 4.24 and 6.11.

Note that all finite Gibbs process models discussed in Section 3.6 have stationary counterparts; the pair potentials on pp. 141–142 can also be used in the stationary case. In Example 6.3 two interesting pair potentials are used.

6.6.2   Simulation of stationary Gibbs processes

In practice, stationary Gibbs processes, like all stationary processes, can only be simulated within a finite window. This can be done by applying methods derived for finite Gibbs processes with the pair potential $\phi(r)$ of interest. There are two main approaches:

- Simulate a finite Gibbs process in an extended window W_{sim}, which is much larger than W such that the influence of the edge of W is negligible. Of course, the number of points in W is variable, determined by α and $\phi(r)$. To achieve this, the algorithm described in Section 3.6.3 with a variable number of points is applied in W_{sim}.

- Simulate a finite Gibbs process with periodic boundary conditions in W, if W is a rectangle or parallelepiped. This is usually done with a birth-and-death or Metropolis-Hastings algorithm with a variable number of points.

Note that the simulation of Gibbs hard-core processes of high intensity is a particularly difficult problem; Mase et al. (2001) and Döge et al. (2004) present a suitable approach to this issue.

6.6.3   Statistics for stationary Gibbs processes

As noted, in applications the patterns that could be regarded as samples from stationary point processes are often treated as samples from finite Gibbs processes and thus the methods introduced in Section 3.6 are applied. Diggle et al. (1994) show that this approach yields acceptable results for the pair potential $\phi(r)$, while estimation of α is beyond this approach.

Stoyan and Grabarnik (1991a) apply the cusp-point method as discussed in Section 3.6.4, which is appropriate for stationary processes if $K_{\text{fin}}(r)$ is replaced by $K(r)$. While this can only be applied to hard-core Strauss processes, the following method may be used for all Gibbs processes.

The *Takacs–Fiksel method* (Takacs, 1986; Fiksel, 1984, 1986) is a general parametric estimation approach that is based on (6.6.3) in the spirit of the method of moments. It yields estimates of the parameter α as well as of parameter θ in the pair potential $\phi(r, \theta)$. For example, in the case of a hard-core Strauss process, $\theta = (r_0, \beta, r_{\max})$.

The basic idea of the approach is to find values for α and θ such that estimates $\hat{L}_k(\alpha, \theta)$ and $\hat{R}_k(\alpha, \theta)$ of the left- and right-hand side of (6.6.3) for suitable test functions $f = T_k$ are as similar as possible. In other words, one seeks to minimise the sum of squared differences

$$S(\alpha, \theta) = \sum_{k=1}^{m} \left(\hat{L}_k(\alpha, \theta) - \hat{R}_k(\alpha, \theta) \right)^2.$$

The number m of these test functions should be larger than the number of parameters of the pair potential, i.e. the dimension of θ. Note that for stationary Gibbs processes α is an additional parameter, which can be estimated by the Takacs–Fiksel method in the same way as the pair potential parameters. For finite Gibbs processes, however, the number of points n is typically known and does not have to be estimated.

Since Takacs' first paper in 1983, various test functions T_k have been considered. Two versions appear to be particularly useful:

$$T'_k(N) = N(b(o, r_k)) \exp\left(\alpha + \sum_{x \in N} \phi(\theta; \|x\|)\right)$$

and

$$T''_k(N) = N(b(o, r_k)) = \text{number of points in } b(o, r_k),$$

where the r_k are suitable radii.

For planar patterns, $L_k(\alpha, \theta)$ and $R_k(\alpha, \theta)$ are estimated as follows. If T'_k is used, $\hat{R}_k(\alpha, \theta)$ has the simple form

$$R_k(\alpha, \theta) = \lambda \pi r_k^2.$$

Thus there is no (direct) dependence on α and θ. The estimator is simply

$$\hat{R}_k(\alpha, \theta) = \hat{\lambda} \pi r_k^2. \tag{6.6.8}$$

An estimator of the left-hand side of (6.6.3) is then

$$L_k^*(\alpha, \theta) = \frac{1}{\nu(W)} \sum_{i=1}^{n} N(W \cap b(x_i, r_k) \setminus \{x_i\})$$

$$\times \exp\left(\alpha + \sum_{\substack{j=1 \\ (j \neq i)}}^{n} \phi(\theta; \|x_i - x_j\|) w(x_i, x_j)\right). \tag{6.6.9}$$

Here $n = N(W)$ is the number of points observed and

$$\alpha_i = \frac{\pi r_k^2}{\nu(W \cap b(x_i, r_k))},$$

whereas $w(x_i, x_j)$ is defined as on p. 188.

For the second test function T''_k, $L_k(\alpha, \theta)$ is

$$\lambda^2 K(r_k),$$

since

$$\lambda E_o(N(b(o, r_k) \setminus \{o\})) = \lambda^2 K(r_k),$$

where $K(r)$ is Ripley's K-function. It can be estimated by the methods described in Section 4.3.3. To estimate $R_k(\alpha, \theta)$ use a point lattice $\{y_j\}$ for $j = 1, 2, \ldots, l$ in W and calculate:

$$\hat{R}_k(\alpha, 0) = \frac{1}{l} \sum_{j=1}^{l} N(b(y_j, r_k)) c_j \exp\left(-\alpha - \sum_{i=1}^{n} \phi(\|x_i - y_j\|; \theta) w(x_i, x_j)\right),$$

with

$$c_j = \frac{\pi r_k^2}{\nu(W \cap b(y_j, r_k))}.$$

It is also possible to use $c_j = 1$ if all lattice points are far away from the boundary of W. (For example, if $\phi(r) = \infty$ for $r \le r_0$ the distance of the y_j from the boundary of W should be at least r_0.)

As far as the choice of test function T_k is concerned, experience shows that T_k'' yields better results for 'repulsive' pair potentials, i.e. if $\phi(r)$ is non-negative. For a pair potential that models attraction, i.e. $\phi(r)$ is negative for some r, $T_k'(N)$ is preferable.

The radii r_k should be chosen as

$$r_k = k\frac{R}{m} \qquad \text{for } k = 1, \ldots, m,$$

where R is a relatively large number with $R \approx 1.3 r_{\max}$ and $r_{\max}$ is the range of interaction, i.e. the smallest r-value such that

$$\phi(r) = 0 \quad \text{for all } r > r_{\max}.$$

Tomppo (1986) has developed a variant of this method based on nearest-neighbour distances. It may be used when data on the coordinates of the objects' location is not available but the distances among them have be recorded. Ripley (1988), Särkkä (1993) and Goulard et al. (1996) show that a specific choice of T_k yields estimation equations where the Takacs–Fiksel and so-called pseudo-likelihood estimation equations coincide. This extends the use of the pseudo-likelihood method; see Särkkä and Tomppo (1998).

Example 6.3.   Estimation of a pair potential for trees in a spruce forest
This example considers the pattern of 134 Norwegian spruce trees shown in Figure 6.12. The data were collected in a forest research area in the Tharandter

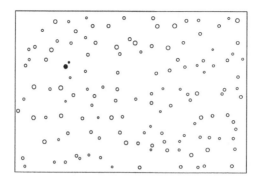

Figure 6.12   Pattern formed by 134 Norwegian spruce trees in a $56 \times 38\,\mathrm{m}$ rectangle in the Tharandter Wald. The black dots represent the pair of points with a small inter-point distance and largest difference in their marks.

Wald (near Dresden, Germany) and the trees were 60 years old at the time of data collection done by G. Klier in the 1960s.

They were originally planted and were later thinned by the forester with the aim of obtaining a stand of trees of similar size; in particular, those trees that were dominated by other trees were removed. Since this stand is part of a larger forest, a stationary point process model seems to be preferable; the trees close to the window's boundary are influenced by trees outside the window.

In the following, these data are analysed on the assumption that they are a subset of a sample from a stationary Gibbs process. The pattern formed by the trees in the stand as well as in the entire forest has been influenced by growth, thinning by foresters and natural competition, such that interactions among the trees are difficult to interpret. Fitting a Gibbs model to the data is a form of data analysis. It leads to an estimated pair potential, which can be regarded as a further summary characteristic. Furthermore, it enables simulation of such patterns. Unlike the approach taken in Møller and Waagepetersen (2007) the marks are ignored here since they appear to be independent (see Example 6.4) and do not show large fluctuations. (Without the forester's work the marks would probably be strongly correlated since the trees that have previously been suppressed would still be present in the plot.)

Following Fiksel (1984, 1986), two families of pair potentials are considered:

$$\phi_1(r) = \phi_{h_1,\beta,R_1}(r),$$

the hard-core Strauss potential as on p. 142, and

$$\phi_2(r) = \begin{cases} \infty & \text{for } r \le h_2, \\ a^{-\zeta r} & \text{for } h_2 < r \le R_2, \\ 0 & \text{for } r > R_2. \end{cases}$$

Here $r_{\max}$ is denoted as R_i. The parameters are estimated as

$$\hat{h}_1 = 1\,\text{m}, \qquad \hat{R}_1 = 2.7\,\text{m}, \qquad \hat{\beta} = 0.85\,\text{m}$$

and

$$\hat{h}_2 = 1\,\text{m}, \qquad \hat{R}_2 = 3.5\,\text{m}, \qquad \hat{a} = 1.47\,\text{m}, \qquad \hat{\zeta} = 1.0\,\text{m}^{-1},$$

and the chemical activities are estimated as

$$\hat{\alpha}_1 = 0.99, \qquad \hat{\alpha}_2 = 1.89.$$

The estimates were obtained by the Takacs–Fiksel method with the test functions $T_k''(N) = N(b(o, r_k))$ for

$$r_k = 0, 0.25, 0.50, \ldots 5.00$$

and suggest inhibition among trees up to a distance of approximately 3 m.

Testing the goodness-of-fit of both models shows that the second model is better. The inhibition between the trees is slightly stronger at distances between 1 and 2 m than that for the points in the model, as can be seen in Figure 6.13. (A goodness-of-fit test using $D(r)$ yielded the same result.) Another pair potential leads to a better result.

The pair potential $\phi_2(r)$ may be interpreted in the following way in forestry terms: the pattern may be assumed to result from two thinning operations. The aim of the first of these was to obtain a pattern with 1600 trees per hectare, which corresponds to an inter-tree distance of 2.5 m, but shorter distances close

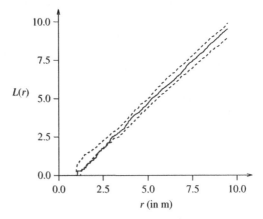

Figure 6.13   Empirical L-function for the spruces with envelopes derived from 99 simulations of the stationary Gibbs process model with pair potential $\phi_2(r)$.

to gaps. The aim of the second was to obtain 600 trees per hectare, which is approximately the intensity of the final spruce forest.

Note that $\hat{r}_{max} = 2.7$ m for the hard-core Strauss potential and 3.5 for $\phi_2(r)$ mark the ranges of interaction of the pair potentials, while the range of correlation r_{corr} corresponding to the pair correlation function is close to 8 m.

6.7 Reconstruction of point patterns

6.7.1 Reconstructing point patterns without a specified model

Simulation is an important tool in point process statistics as it may be used to test statistical hypotheses, to visualise patterns, to evaluate statistical procedures (e.g. assess estimation variances) and to continue an empirical point pattern beyond the edges of the original observation window W (see p. 185).

All simulation methods discussed so far generate samples from explicit parametric point process models. These models are elegant since they are based on a small number of numerical parameters only. However, the underlying mathematical theory and methodology are rather complicated, which may make them less popular among applied researchers. Fortunately, a simple and universal simulation method has been developed which can be used in many situations, when point patterns similar to an empirical pattern have to be generated in some window. In other words, based on an empirical point pattern in an observation window W, one would like to simulate an irregular pattern with a similar point distribution in a new window W_{rec} which may be different from W.

The main idea underlying the approach described in this section is to generate a simulated pattern such that appropriate summary characteristics of the simulated and the empirical pattern are as similar as possible. Note that in some theoretical studies the summary characteristics may be known analytically. Note also that this reconstruction approach yet again demonstrates the power of summary characteristics to extract important features in spatial point pattern data sets.

Different summary characteristics may be used in the algorithm, e.g. the intensity λ, the L-function $L(r)$, the pair correlation function, the spherical contact distribution function $H_s(r)$ or some of the indices discussed in Sections 4.2.4 and 5.2.4. The algorithm tries to find point patterns that minimise the difference between the summary characteristics of the simulated and the empirical pattern.

The patterns resulting from the algorithm may be regarded as having been derived from a general class of point processes that are similar to Gibbs processes. However, these have rather unusual or intractable energies, which cannot be motivated from the geometry of the pattern or from background knowledge on the pattern formation, but are simply treated as an empirical phenomenon. Fortunately, and despite the fact that the true form of these unusual energies might be very complicated, the reconstruction algorithm is actually rather easy to understand.

Note, however, that the approach has its clear limitations. As discussed in Section 4.3.6, different point processes might have the same intensity and the same *L*-function. They might also have the same intensity and the same $H_s(r)$, or even different λ, $L(r)$ and $H_s(r)$. In other words, rather different patterns might be generated based on the algorithm, which might sometimes be similar to those generated from one specific point process model and sometimes to another point process model. In many cases, the simulated pattern might be somewhere 'in between' the two proesses. All these patterns, though, always yield almost the same estimates of λ, $L(r)$ and $H_s(r)$. Note also that it is sometimes impossible to construct a point process based on specific presribed parameters, if these yield contradicting characteristics which cannot be satisfied simultaneously. (A simple example is a process with a large hard-core distance r_0 and an intensity λ that is too high.)

It is very important to carefully choose the summary characteristics such that the variability among the generated patterns is as small as possible. This will be discussed in the theoretical example below.

This approach was originally developed in physics by Rice in 1945, and refined by Joshi, Quiblier and Adler, who used it to simulate two-phase structures such as random media; see Torquato (2002, p. 295). To the authors' knowledge the idea of applying the method to point processes was first discussed in Tscheschel (2001).

The reconstruction algorithm

The algorithm starts with an arbitrary pattern in the window W_{rec}. This pattern has n points, where n is

$$\lambda \cdot \nu(W)$$

rounded to the nearest integer. This is done to fix the correct point density as a minimum requirement.

The number of points n does not change during the simulation. (That is to say, the algorithm in some way simulates a Gibbs process with a fixed number of points.) The initial pattern may be, for example, a sample from a binomial process with n points in W_{rec}.

In each of the simulation steps, one point in the current pattern is randomly chosen and a new candidate point is generated, to potentially replace the chosen point. The values of the summary characteristics, e.g. $L(r)$ and $H_s(r)$, are determined for both the pattern without the chosen point and with the candidate point and compared to those for the empirical pattern. If the values for the modified pattern are closer to these than for the old pattern, the candidate point replaces the chosen point. Otherwise a new candidate is chosen. The following text and the subsequent examples describe the method in more detail.

The terminology used to describe the algorithm is derived from physics, and hence the deviation of the characteristics for a simulated pattern from the characteristics for the empirical pattern is measured by an *energy* $E(X)$ (with arguments $X = N$, N_k, etc.). The aim of the algorithm is to simulate a pattern with as small an energy as possible.

Let the summary characteristics be some functional summary characteristics $f_i(r)$ for $i = 1, 2, \ldots, I$, and some numerical summary characteristics n_j for $j = 1, 2, \ldots, J$. For example, f_1 may be the L-function and f_2 the spherical contact distribution function, and n_1 may be some numerical index. (Note that the intensity is not used in this way as it is directly included and fixed in the simulation.) As mentioned above, the $f_i(r)$ and n_j result from statistical analysis of an empirical point pattern in the original window W. For the sake of the exposition, the text below considers the case where a theoretical model is known and hence explicit formulas for the summary characteristics may be used.

Note further that even though $f_i(r)$ and n_j are statistical estimates, the 'hats' $\widehat{}$, which usually indicate estimation, are omitted, to simplify the notation.

During the reconstruction, a series $\{N_k\}$ of point patterns in W_{rec} is generated. For each N_k statistical estimates of the $f_i(r)$ and n_i, denoted by $f_i^{(k)}(r)$ and $n_i^{(k)}$, are determined. Then the *energy of the pattern* N_k is

$$E(N_k) = \sum_{i=1}^{I} E_{f,i}^{(k)} + \sum_{j=1}^{J} E_{n,j}^{(k)} \tag{6.7.1}$$

with

$$E_{f,i}^{(k)} = k_i \int_0^{R_i} \left(f_i(r) - f_i^{(k)}(r) \right)^2 \mathrm{d}r \qquad \text{for } i = 1, 2, \ldots, I \tag{6.7.2}$$

and

$$E_{n,j}^{(k)} = c_j \cdot (n_j - n_j^{(k)})^2 \qquad \text{for } j = 1, 2, \ldots, J, \tag{6.7.3}$$

where the R_i are suitably chosen limits of the integrals and the k_i and c_j are positive weights.

The energy controls the simulation. Assume that N_k is given and let N_k' be the pattern in which one of the points has provisionally been randomly chosen to be deleted and a candidate point has been uniformly placed in the window W_{rec}. Note that 'randomly chosen' means that each of the n points in N_k has the same chance $1/n$ of being selected. If

$$E(N_k') < E(N_k), \tag{6.7.4}$$

the candidate point is now part of the pattern and the chosen point is definitely deleted and N_k' becomes N_{k+1}. Since the energy has been reduced, the algorithm has produced an improved pattern.

The algorithm can be modified such that candidate points may also be accepted in the main simulation step with some probability if the energy increases as a consequence of this. Then this simulation step resembles the Metropolis-Hastings step in Gibbs process simulation as described in Section 3.6.3. If the acceptance rate of candidates which result in a higher energy is decreased in each step until it is almost 0, the simulation follows the principle of the well-known *simulated annealing algorithm,* which is a global optimisation method. Tscheschel (2001) and Tscheschel and Stoyan (2006) compare the simulated annealing method with the (faster) algorithm that only accepts points that reduce the energy as described above. They find some evidence that the latter is sufficient for good point pattern reconstruction, although it usually finds only a local minimum. Note that all these approaches are based on modern heuristic optimisation algorithms as discussed in Winkler and Gilli (2004) and that the particular choice of algorithm is not important.

However, it is very important to calculate the $f_i^{(k)}(r)$ and $n_j^{(k)}$ *efficiently.* Rather than completely recalculating the estimators at each step only some 'minor pieces' of the estimators should be recalculated. The example below outlines an approach for $L(r)$ and $H_s(r)$.

The iteration $N_k \rightarrow N_{k+1}$ is carried out until

$$E(N_k) - E(N_{k+1}) < \varepsilon$$

for some chosen small ε. Note, however, that this might actually never happen in some cases. If so, it is possible that there is no point process which has the specific 'parameters' – recall that it is possible to choose summary characteristics which are contradictory.

It is not unlikely that the energy as a function of point patterns has many local minima. Different starting configurations will lead to different local minima of the energy functional and so the output of the reconstruction algorithm will be additionally randomised. If different models have the same summary characteristics, different models may relate to different local minima.

6.7.2 An example: reconstruction of Neyman–Scott processes[*]

By way of illustration, this subsection describes the reconstruction of a theoretical pattern. Readers who are mainly interested in applications may want to skip most of this section but should read the conclusions on p. 415.

Consider a model for which the summary characteristics are theoretically known and such that it is possible to assess the quality of the results of reconstruction. To this end, a planar Neyman–Scott process is used. The reconstruction procedure is applied twice, each time with different sets of summary characteristics; a third approach is described in Tscheschel and Stoyan (2006).

In the *first case* only the L-function is considered (in addition to the intensity λ), i.e. $I = 1$ and $J = 0$ with

$$f_1(r) = L(r).$$

The L-function is derived from the K-function, which is known for Neyman–Scott processes and has the general form

$$K(r) = \pi r^2 + \frac{F_d(r)}{\lambda \bar{c}} \sum_{i=2}^{\infty} p_i i(i-1) \qquad \text{for } r \geq 0, \tag{6.7.5}$$

where $F_d(r)$ is the distribution function of the distance between two random points in the typical cluster (see p. 376). Furthermore, the intensity λ is the intensity of the cluster process, which is given by

$$\lambda = \lambda_p \cdot \bar{c}, \tag{6.7.6}$$

where λ_p is the intensity of the parent Poisson process and $\bar{c}$ the mean number of points in the typical cluster. The distribution of the number of points in the cluster is given by $\{p_i\}$, where p_i is the probability that the cluster consists of i points.

For the example considered here, $F_d(r)$ is given by

$$F_d(r) = \frac{1}{2\pi R^3} \left(4R^3 \arcsin\left(\frac{r}{2R}\right) + 4Rr^2 \arccos\left(\frac{r}{2R}\right) \right.$$
$$\left. - r(2R^2 + r^2)\sqrt{1 - \frac{r^2}{4R^2}} \right) \tag{6.7.7}$$

for $0 \leq r < 2R$ and $F_d(r) = 1$ for $r \geq 2R$. Formula (6.7.7) corresponds to the case of a Matérn cluster process (see p. 376); however, not only this case is considered. The text below also discusses cases where the number of points per cluster does not follow a Poisson distribution.

The intensity of all cluster processes considered here is assumed to be

$$\lambda = 100$$

and the equality

$$\sum_{i=2}^{\infty} i(i-1)p_i / \bar{c} = 3$$

always holds. Three possible variations of cluster processes which all result in the same intensity and L-function are

$$\lambda_p = 25 \qquad \text{with } p_4 = 1 \text{ and } p_i = 0 \text{ for } i \neq 4,$$

$$\lambda_p = 70 \qquad \text{with } p_1 = \frac{20}{21}, \ p_{10} = \frac{1}{21} \text{ and } p_i = 0 \text{ for all other } i$$

and

$$\lambda_p = \frac{100}{3} \qquad \text{with } p_i = \frac{3^i}{i!} e^{-3} \text{ for } n = 0, 1, \ldots.$$

The first variant corresponds to the fixed number of cluster points ($= 4$) in every cluster, whereas in the second variant most clusters consist of one point and a small number of clusters have 10 points. In the third variant the number of daughter points is Poisson-distributed, i.e. this is a Matérn cluster process with mean number $\bar{c} = 3$ of cluster points and cluster radius R. Figure 6.14 shows two simulated patterns to illustrate the first and the second variants.

What type of pattern will result from the reconstruction in the first approach, which uses only the L-function and is therefore unable to discriminate between the two Neyman–Scott processes? Simply a cluster pattern with an empirical L-function similar to that constructed by (6.7.5), (6.7.7) and the combinations of λ_p and $\{p_i\}$ above. However, it is not clear which geometry the patterns constructed from the algorithm will have, i.e. whether the resulting patterns are similar to one of the two variants shown in Figure 6.14 or have yet another behaviour.

Simulations in the window $W_{\text{rec}} = [0, 2] \times [0, 2]$, i.e. a square of side length 2 with periodic boundary conditions and a start configuration of 400 uniformly distributed points, produced the following results.

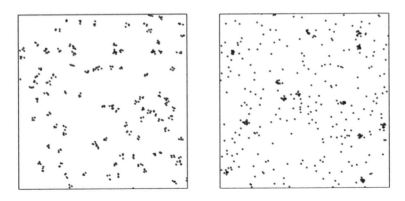

Figure 6.14   Samples from two different Neyman–Scott processes with the same K- and L-function as explained in the text. In the pattern on the left all clusters consist of exactly four points, whereas on the right the clusters usually consist of one point only and with a small probability there are 10-point clusters. The simulations were performed using the usual cluster process simulation algorithm. Reproduced by permission of Elsevier Publishers.

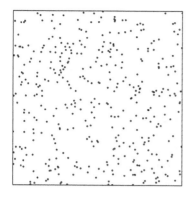 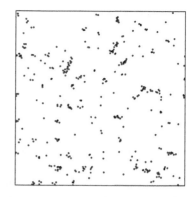

Figure 6.15   Start configuration (left) used in the reconstruction, consisting of 400 points randomly scattered in the 2×2 square. Result of the reconstruction (right) based on the intensity and the L-function only. Because the two models have the same intensity and L-function, the same reconstruction is obtained for each. Compare with Figure 6.14. Reproduced by permission of Elsevier Publishers.

After $k = 40\,000$ steps the pattern shown in Figure 6.15 (right) was obtained. This looks very much like a cluster process. Of course, the number of points per cluster is random, and one might say that the pattern is somewhere 'in between' the patterns in Figure 6.14. Hence, the algorithm worked well, as well as might have been be expected. The final energy $E(L)$ is very small, less than $5 \cdot 10^{-7}$.

Remarks on the implementation. The integral in (6.7.2) was replaced by a sum,

$$E(L) = \sum_{i=1}^{50} \left(L^{(k)}(r_i) - L_{\text{th}}(r_i) \right)^2 \qquad (6.7.8)$$

with $r_i = 0.004, 0.008, \ldots, 0.196, 0.2$ and

$$L_{\text{th}}(r) = \sqrt{\frac{K_{\text{th}}(r)}{\pi}},$$

where $K_{\text{th}}(r)$ is given by (6.7.5). For the estimation of the K-function an estimator adapted to the torus conditions and fixed number of points is used:

$$\hat{K}(r) = \frac{\nu(W_{\text{rec}})}{n^2} \sum_{i=1}^{n} \sum_{j=1, j \neq i}^{n} \mathbf{1}(\|x_i - x_j\| \le r). \qquad (6.7.9)$$

The distances are redefined based on the torus (see p. 184). No edge-correction is necessary.

Note that when moving from point pattern N_k to N_{k+1} only one point is deleted and replaced by a new one. Thus only a small part in the double sum (6.7.9) has to be recalculated: the partial sums related to those two points. If the deleted point is x_l and the new point x_m then the following term has to be added to the double sum:

$$-2 \sum_{i=1, i \neq m}^{n} \mathbf{1}(\|x_m - x_i\| \leq r) + 2 \sum_{i=1, i \neq m}^{n} \mathbf{1}(\|x_l - x_i\| \leq r).$$

Consider now the *second case*, which uses a larger number of summary characteristics. Because the L-function does not discriminate the point processes corresponding to the two patterns shown in Figure 6.14, an additional functional characteristic has to be considered which differs for the two Neyman–Scott processes. In order to realise this aim, the spherical contact distribution function $H_s(r)$ is included as a second functional summary characteristic $f_2(r)$, for two reasons:

(a) The nature of $L(r)$ and $H_s(r)$ is rather different, $L(r)$ is point-related, while $H_s(r)$ is location-related.

(b) It is well known that $H_s(r)$ is a good summary characteristic for cluster processes.

If only two functional summary characteristics are used, the nearest-neighbour distance d.f. $D(r)$ would be a bad choice for $f_2(r)$, as it is also point-related and is not very informative for cluster processes, see also Section 4.2.6.

Figure 6.16 shows the result of the reconstruction with two functional summary characteristics and weights $k_1 = k_2 = 1$ obtained after 40 000 iterations. The spherical contact distributions differ and correspond to the two cases

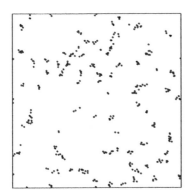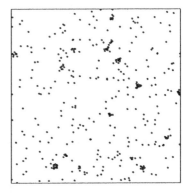

Figure 6.16   Results from the reconstruction based on L-function and spherical contact d.f. $H_s(r)$. Two different patterns corresponding to different spherical contact d.f.s are obtained for the two models. The reconstructions resemble the simulated point patterns in Figure 6.14. Reproduced by permission of Elsevier Publishers.

in Figure 6.14, and indeed different patterns are obtained. The similarity of the simulated patterns to both the left- and the right-hand side of Figure 6.14 is apparent.

The energies $E(L)$ and $E(H_s)$ are both smaller than $6 \cdot 10^{-6}$ for the final patterns; in the simulated pattern the L- and H_s-functions of the original and simulated patterns almost coincide. Thus, the reconstruction algorithm worked very successfully in both cases.

Clearly, a reconstruction based on $L(r)$ and $H_s(r)$ is not the only possibility. In a *third approach*, Tscheschel and Stoyan (2006) successfully tried the d.f.s of the distance to the kth neighbour $D_k(r)$ for $k = 1, 2, \ldots, 17$. This led to an excellent reconstruction. Other energies can also be used. For example, Pommerening (2006) used some of the indices discussed in the Sections 4.2.4 and 5.2.4 to reconstruct forest patterns.

Conclusions

The reconstruction should be based on summary characteristics which either describe different aspects of the distribution of the point pattern or yield a very precise description if large numbers I and/or J of summary characteristics are used.

6.7.3 Practical application of the reconstruction algorithm

Non-marked point processes

The example discussed in the previous section is of a theoretical nature. In practical applications, the aim is to reconstruct empirical patterns. In these cases, the reconstruction is based on *estimates from an empirical pattern,* rather than on known theoretical functions and parameters. For instance, the empirical L-function, the empirical spherical contact distribution function, or estimates of $D_l(r)$ for $l = 1, 2, \ldots$ may be used. Thus it is not necessary to find explicit formulas for $L(r)$, $H_s(r)$ and other summary characteristics.

Figure 6.17 shows a reconstruction of the amacrine cell point pattern of Figure 1.2. Here the marks ('on' and 'off') have been ignored. Whereas the original pattern of 294 points was given in a rectangular window of side lengths 1060 and 662 μm, the window size is now chosen to be $1060 \times 1060\,\mu$m and the resulting number of points is $n = 470$.

The reconstruction is based on the L-function and the spherical contact distribution function, where the L-function was reconstructed for $r = 5.3, 10.6, \ldots,$ $100.7, 106.0\,\mu$m and the spherical contact distribution function for $r = 4.6, 9.2, \ldots,$ $88.1, 92.7\,\mu$m. The energy of the final configuration after 47 000 steps shown in Figure 6.17 is rather small, $E(L) = 1.8 \times 10^{-7}$ and $E(H_s) = 8.4 \times 10^{-7}$.

Of course, simulated point patterns such as those shown in Figure 6.17 can also be obtained from the Gibbs process model constructed by Diggle et al. (2006), i.e. based on a classical point process that depends on a small number of parameters with a clear probabilistic interpretation. Thus theoreticians may prefer this model, which is very elegant and straightforward to interpret. However, it requires a lot of statistical expertise, and those who simply want to generate a realistic series of 'amacrine cell patterns' in windows different from $1060 \times 662 \, \mu$m might prefer the reconstruction method.

Note that the pattern shown in Figure 6.17 is not a 'complete' reconstruction of the amacrines cell pattern. The original pattern in Section 1.2.1 is marked with • (off) and ○ (on). Therefore, the reconstruction should also assign marks to the points. An approach to this is sketched in the following.

Marked point processes

Marked point patterns may be reconstructed in a similar way. One approach simply considers the marked points as points in a higher-dimensional space and uses the same method as above, with adapted summary characteristics which also include the marks. Another approach reconstructs the pattern in two steps. The first step yields the point positions as described above and the second the marks, where the locations generated in the first step are fixed. The marks are reconstructed in a similar way as the locations, based on mark-related summary characteristics such as $L_{mm}(r)$, $L_{ij}(r)$ and $p_{ij}(r)$. These characteristics are used to construct *mark energy functions,* which depend only on the marks, and then the marks are changed in a stepwise fashion to find minima of the deviation energy.

Pommerening (2006) considers marked point patterns of forests, where the marked points correspond to trees. The trees carry marks $m = (l, s)$, where the l are qualitative (discrete) marks characterising tree species and the s are quantitative (real-valued) marks, which characterise tree size, e.g. stem diameter (dbh).

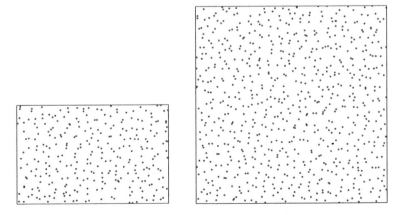

Figure 6.17   The original amacrine cell pattern (left) and the result of a reconstruction (right) based on $L(r)$ and $H_s(r)$. The window W_{rec} is now a square. Reproduced by permission of Elsevier Publishers.

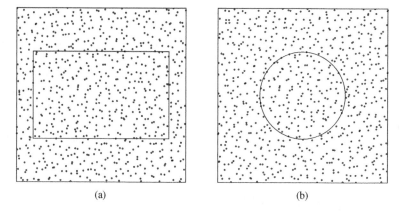

(a)   (b)

Figure 6.18   Conditional reconstruction of amacrines patterns based on $L(r)$ and $H_s(r)$. The points in (a) the inner rectangle and (b) the inner disc are fixed, whereas the others were generated with the reconstruction algorithm. The window W_{rec} is now a square. Reproduced by permission of Elsevier Publishers.

Pommerening carries out the simulations in three steps: construction of the positions; construction of the l-marks; and, finally, construction of the s-marks. Note that this approach should be considered as a recipe that works well rather than as a general principle that one has to adhere to in all applications.

Evaluation of summary characteristics

Pommerening (2006) uses the reconstruction idea to *evaluate* summary characteristics, specifically some of the indices mentioned in Section 5.2.4, with the aim of finding out which characteristics describe the distribution of trees in forests particularly well. Characteristics which provided a good reconstruction of given forest patterns were considered as 'good' and 'informative'.

Conditional reconstruction

Figure 6.18 shows the results of two *conditional* reconstructions for the amacrines pattern. The points in the small rectangle (a) and in the disc (b) are fixed, while the other points are reconstructed as above. The positions of the fixed points have some influence on the new points.

6.8   Formulas for marked point process models
6.8.1   Introduction

Chapter 5 was entirely dedicated to marked point patterns but, apart from a short treatment of marking models at the beginning of the chapter (Section 5.1.3), it mainly focused on exploratory data analysis based on summary characteristics. This section now discusses three types of marked point process models. The first model is a model with independent marks, which may be regarded as a null model. The next model, the random field model, considers dependent marks, but point density and marks are still uncorrelated. In the last model, a marked Cox process, these are correlated; the marks depend linearly on local point density. Clearly, other more complex models may be constructed – in particular, models with marks constructed in relation to the neighbourhood configurations of the points, as sketched towards the end of Section 5.1.3. These, however, are not considered here any further.

6.8.2   Independent marks

It is important to know the behaviour of the various summary characteristics for marked point processes if the marks are independent. The independent marking model serves as a null model, and empirical patterns may be classified as patterns with positive and negative correlation of the marks relative to this null model. However, the model is also interesting in its own right as the correlations among marks may be rather weak in some applications.

In Chapter 5 formulas for many summary characteristics were given, along with the values for independent marks. These values are repeated here for some of the summary characteristics.

- Coefficient of segregation (p. 314):

$$S = 0.$$

- Mingling index (p. 314, bivariate case):

$$\overline{M}_k = 2p_1 p_2.$$

- Nearest-neighbour-correlation indices (p. 317):

$$\mathbf{n}_{mm} = \mathbf{n}_{.m} = \mathbf{n}_\gamma = 1.$$

- Mark correlation functions (Section 5.3.3):

$$k_{mm}(r) \equiv 1,$$
$$k_{m.}(r) \equiv 1,$$
$$\gamma_m(r) \equiv \sigma_\mu^2.$$

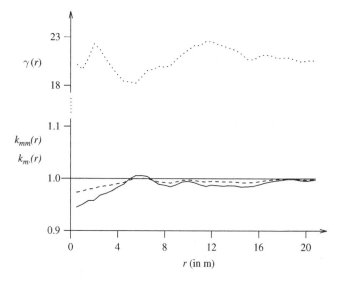

Figure 6.19   The empirical mark correlation functions $k_{mm}(r)$ (solid line), $k_{m.}(r)$ (dashed line) and $\gamma_m(r)$ (in normalised form, dotted line) for the spruce pattern, obtained with bandwidth $h = 1$ m.

Example 6.4.   Correlation of spruce diameters
This example continues the analysis of the pattern of Norwegian spruce trees shown in Figure 6.12 and now also considers the dbh marks. Figure 6.19 shows the empirical mark correlation functions; the nearest-neighbour correlation indices are

$$\hat{\mathbf{n}}_{mm} = 0.99, \qquad \hat{\mathbf{n}}_{.m} = 1.00, \qquad \hat{\mathbf{n}}_{\gamma} = 1.13.$$

The behaviour of the functions $k_{mm}(r)$ and $k_{m.}(r)$ and the indices $\mathbf{n}_{mm}$ and $\mathbf{n}_{.m}$ is similar to that for independent marks, which is not surprising in view of the history of the forest stand. The behaviour of the nearest-neighbour variogram index $\mathbf{n}_{\gamma}$ and $\gamma_m(r)$ is different: the mark variogram $\hat{\gamma}_m(r)$ has large values for small r and thus its shape deviates from that of a geostatistical variogram, and $\hat{\mathbf{n}}_{\gamma}$ is clearly larger than 1. This behaviour indicates large mark differences for trees at close distances. However, inspection of the marks shows that there are a few pairs of trees (the most important one is shown as filled circles in Figure 6.12) which have caused this behaviour. Whereas the mean mark is $\hat{\mu} = 25.0$, the marks of the extreme tree pair are 18 and 35. If the smaller mark of that pair were increased to 25, $\hat{\gamma}_m(r)$ would clearly decrease for small r and $\hat{\mathbf{n}}_{\gamma}$ would become 1.08, while $k_{mm}(r)$, $k_{m.}(r)$, $\mathbf{n}_{mm}$ and $\mathbf{n}_{.m}$ would not change much.
    The estimated dominance index for the diameters is $\widehat{Do}_4 = 0.493$, whereas the dominance index for independent marks would be 0.466. The difference is rather small, which might indicate that there are very few dominating trees in the stand.

The conclusion of the statistical analysis is that the stand may be regarded as a sample from an independently marked point process; a formal test is presented in Example 7.8. The fact that the variation of the mark of a single point changes $\hat{\mathbf{n}}_\gamma$ and $\hat{\gamma}_m(r)$ so drastically shows that the sample is rather small, perhaps too small for a serious correlation analysis.

6.8.3 Random field model

The random field model was introduced in Section 5.1.3. In this model, the points x_n of a non-marked point process N are allocated marks from a random field $\{Z(x)\}$, which is independent of the points, by

$$m(x_n) = Z(x_n),$$

i.e. the mark of point x_n is simply the value of the random field at the location of x_n. This type of marking is termed *geostatistical marking*. The mean, variance and variogram of $\{Z(x)\}$ are denoted by m_Z, σ_Z^2 and $\gamma_Z(r)$.

In this marking model the marks are correlated, i.e. marks of points close together are typically similar. However, there is no correlation between marks and point density. The mean mark and mark variance are

$$\mu = m_Z \quad \text{and} \quad \sigma_\mu^2 = \sigma_Z^2.$$

The mark d.f. $F_\mu(m)$ is the same as the one-dimensional d.f. of the random field. Thus point process estimates $\hat{\mu}$, σ_μ^2 and $\hat{F}_\mu(m)$ can be used to determine the random field characteristics.

Due to the independence between N and $\{Z(x)\}$ the relevant correlation characteristics exhibit rather simple behaviour. The mark correlation functions discussed in Section 5.3.3 are

$$k_{m\cdot}(r) \equiv 1, \tag{6.8.1}$$

$$k_{mm}(r) = 1 + \sigma_{\mu/\mu^2}^2 - \gamma_\mu(r)/\mu^2, \tag{6.8.2}$$

$$\gamma_m(r) = \gamma_Z(r). \tag{6.8.3}$$

Thus, estimates of $k_{mm}(r)$ and $\hat{\gamma}_m(r)$ both separately yield estimates of $\gamma_Z(r)$.

Due to its simple form $\hat{k}_{m\cdot}(r)$ is an attractive candidate use in for testing the hypothesis that a marked point process follows the random field model as suggested by Schlather (2001a) and Schlather et al. (2004); it is applied in Section 7.5. The mark variogram $\gamma_m(r)$ is the same as the random field variogram $\gamma_Z(r)$, which means in particular that $\gamma_m(r)$ is a geostatistical variogram as in Appendix C. This is an important property since if the points and marks do interact, i.e. in situations where the random field model is not suitable, mark variograms can differ strongly from geostatistical variograms; see Wälder and Stoyan (1996).

An interesting example was discussed in Dimov et al. (2005). In a forestry application geostatistical variograms only resulted from the correlation analysis after suppressed trees had been removed. Trees close to dominant trees, i.e. in short-range interaction, result in large values of $\hat{\gamma}_m(r)$ for small r, which in turn produce non-geostatistical variograms like that in Figure 6.19.

The nearest-neighbour correlation indices are

$$\mathbf{n}_{mm} = \int_0^\infty k_{mm}(r)d(r)\mathrm{d}r, \tag{6.8.4}$$

$$\mathbf{n}_{.m} = 1 \tag{6.8.5}$$

and

$$\mathbf{n}_\gamma = \int_0^\infty \gamma(r)d(r)\mathrm{d}r/\sigma_\mu^2, \tag{6.8.6}$$

where $d(r)$ is the p.d.f. of nearest-neighbour distance as defined in Section 4.2.6.

6.8.4   Intensity-weighted marks

As an example of a model with correlations between point density and marks the intensity-weighted log-Gaussian Cox (ILGC) model introduced in Ho and Stoyan (2008) is considered. Here, the point process is a stationary log-Gaussian Cox process as introduced in Section 6. 4. This means that the local density $\Lambda(x)$ of the point process is

$$\Lambda(x) = \exp(S(x)),$$

where $\{S(x)\}$ is a Gaussian field with mean m_S, variance σ_S^2 and variogram $\gamma_S(r)$. Then $\{\Lambda(x)\}$ is also a random field with mean

$$m_\Lambda = \exp\left(\mu_S + \frac{\sigma_S^2}{2}\right).$$

Clearly, the intensity λ of the log-Gaussian Cox process is

$$\lambda = m_\Lambda.$$

The points x_n of the log-Gaussian Cox process are assigned marks $m(x_n)$ according to

$$m(x_n) = a + b\Lambda(x) + \varepsilon(x_n) \tag{6.8.7}$$

to yield the ILGC process. Here a and b are real-valued model parameters and $\varepsilon(x_n)$ is normally distributed with mean 0 and variance σ_ε^2; for different x_n the $\varepsilon(x_n)$ are independent. For positive b, the marks are large in areas of high point density and small in areas of low point density. For negative b the marks are small in regions of high point density and vice versa.

The relationship between the characteristics of $\{\Lambda(x)\}$ and those of the ILGC process is not as simple as between $\{Z(x)\}$ and the random field model; therefore it is not true that $\mu = a + b\lambda$. This is because the marks are big in regions of high point density if $b > 0$, which increases the mean mark. The mean mark of the ILGC process is

$$\mu = a + b\lambda \exp(\sigma_S^2), \tag{6.8.8}$$

which is larger than λ for $b > 0$ and smaller for $b < 0$.

The mark correlation functions are as follows:

$$k_{mm}(r) = \begin{cases} \dfrac{a^2 + 2ab\lambda\exp(\sigma_S^2 + \sigma_S^2\varrho_S(r)) + b^2\lambda^2\exp(2\sigma_S^2 + 3\sigma_S^2\varrho_S(r))}{(a + b\lambda\exp(\sigma_S^2))} & \text{for } r > 0, \\[3mm] \dfrac{a^2 + 2ab\lambda\exp(\sigma_S^2) + b^2\lambda^2\exp(3\sigma_S^2) + \sigma_\varepsilon^2}{(a + b\lambda\exp(\sigma_S^2))^2} & \text{for } r = 0, \end{cases}$$

$$k_{m\cdot}(r) = \begin{cases} \dfrac{a + b\lambda\exp(\sigma_S^2 + \sigma_S^2\varrho_S(r))}{a + b\lambda\exp(\sigma_S^2)} & \text{for } r > 0, \\[3mm] 1 & \text{for } r = 0, \end{cases}$$

$$\gamma_m(r) = \begin{cases} b^2\lambda^2\exp(2\sigma_S^2 + 2\sigma_S^2\varrho_S(r)) \\ \times(\exp(\sigma_S^2) - \exp(\sigma_S^2\varrho_S(r))) + \sigma_\varepsilon^2 & \text{for } r > 0, \\ 0 & \text{for } r = 0. \end{cases}$$

In patterns generated from this model, the ranges of correlation corresponding to the points (described by $g(r)$) and the marks (described by $k_{mm}(r)$, $k_{m\cdot}(r)$ and $\gamma_m(r)$) are equal; all correlation functions depend on $\gamma_S(r)$, the variogram of the Gaussian field $\{S(x)\}$.

Myllymäki and Penttinen (2007) study intensity-weighted marks which allow also the mark variance to depend on the intensity.

6.9   Moment formulas for stationary shot-noise fields

Shot-noise fields were introduced in Section 1.8.3. This section provides some formulas for this important class of models. Stationary shot-noise fields are constructed from a marked point process M and an impulse function $s(x, m)$. In the stationary case the mean, variance and variogram can be derived analytically for these random fields, but the relevant equations are unfortunately only simple for the mean. Despite their complexity the reader may get an impression of the structure of these equations by considering the volume integrals for the second-order characteristics in this section, taken from Cox and Isham (1980) and Schmidt (1985).

In order to avoid excessively complex formulas, assume that the marks are quantitative (real-valued) and independent, that the point process N is motion-invariant and that the impulse function is rotation-invariant with respect to o, which is in some cases used to replace $s(x, m)$ by $s(r, m)$.

With these assumptions the field $\{S(x)\}$ is stationary and isotropic,

$$S(x) = \sum_{[x_n; m(x_n)] \in M} s(x - x_n, m(x_n)) \quad \text{for } x \in \mathbb{R}^d.$$

This means that the value of the random field at location x is a sum of impulses $s(x - x_n, m(x_n))$ at the points x_n and depends on their marks $m(x_n)$.

The *mean value* at $x = o$ is given by

$$\mathbf{E}(S(o)) = \lambda \int_{\mathbb{R}^d} e(x) dx \qquad (6.9.1)$$

with

$$e(x) = \int_0^\infty s(x, m) f_{\mathcal{M}}(m) dm,$$

where $f_{\mathcal{M}}(m)$ is the mark p.d.f. Due to stationarity, the mean of $S(x)$ is the same for all $x \in \mathbb{R}^d$, therefore only the value for $x = o$ has to be considered. This equation is a simple consequence of the Campbell theorem for marked point processes.

Example 6.5.   Adler competition field
Adler (1996) discusses the following shot-noise field in the context of ecological competition modelling in plant communities using the impulse function

$$s(x, m) = m^\alpha \exp\left(-\delta \frac{\|x\|}{m^\beta}\right),$$

where α, β and δ are model parameters, the points are plant locations and the marks size parameters. In this planar case, the mean $E(S(o))$ of the competition field can be derived based on polar coordinates as

$$E(S(o)) = 2\pi\lambda \int\limits_0^\infty \int\limits_0^\infty m^\alpha r \exp\left(-\delta \frac{r}{m^\beta}\right) dr f_{\mathcal{M}}(m) dm$$

$$= \frac{2\pi\lambda}{\delta^2} E\left(\mathcal{M}^{\alpha+2\beta}\right),$$

where $E\left(\mathcal{M}^{\alpha+2\beta}\right)$ denotes the $(\alpha+2\beta)$th moment of the marks.

The variance, again at $x = o$, is

$$\mathbf{var}(S(o)) = \lambda \int\limits_{\mathbb{R}^d} e^{(2)}(x)dx + \lambda^2 \int\limits_{\mathbb{R}^d} \int\limits_{\mathbb{R}^d} e(x)e(x+h)dx \mathcal{K}(dh)$$

$$- (E(S(o)))^2 \qquad\qquad (6.9.2)$$

with

$$e^{(2)}(x) = \int\limits_0^\infty \left(s(x,m)\right)^2 f_{\mathcal{M}}(m)dm.$$

Here $\mathcal{K}$ is the reduced second-moment measure of the point process N (i.e. M without the marks), as introduced on p. 224. For a homogeneous Poisson process, (6.9.2) simplifies to

$$\mathbf{var}(S(o)) = \lambda \int\limits_{\mathbb{R}^d} e^{(2)}(x)dx, \qquad\qquad (6.9.3)$$

since then $\mathcal{K} = \nu$. For the Adler field in Example 6.5,

$$\mathbf{var}(S(o)) = \frac{2\pi\lambda}{4\delta^2} E\left(\mathcal{M}^{2(\alpha+\beta)}\right).$$

The variogram is given by

$$\gamma(r) = \mathbf{var}(S(o)) + (E(S(o)))^2 - \lambda \int\limits_{\mathbb{R}^d} e_r(x)dx$$

$$- \lambda^2 \int\limits_{\mathbb{R}^d} \int\limits_{\mathbb{R}^d} e(x)e(x+h-r)dx \mathcal{K}(dh) \qquad\qquad (6.9.4)$$

with

$$e_r(x) = \int_0^\infty s(x, m)s(x - \mathbf{r}, m)f_{\mathcal{M}}(m)\mathrm{d}m,$$

where $\mathbf{r}$ is any point at a distance r from the origin o.

Mean value of the shot–noise field at the typical point

In ecological modelling, the mean competition load at the typical point is of substantial interest, since it reflects, for individuals in the community, the strength of competition from other individuals, while $\mathbf{E}S(o)$ is a spatial mean reflecting the strength of the competition faced by a potential individual at an arbitrary position. In the planar case, again based on polar coordinates,

$$\mathbf{E}_o(S(o)) = 2\pi\lambda \int_o^\infty rg(r) \int_o^\infty s(r, m)f_{\mathcal{M}}(m)\mathrm{d}m\mathrm{d}r. \tag{6.9.5}$$

For the Adler example (Example 6.5)

$$\mathbf{E}_o(S(o)) = 2\pi\lambda \int_0^\infty \int_0^\infty m^\alpha rg(r) \exp\left(-\delta\frac{r}{m^\beta}\right) f_{\mathcal{M}}(m)\mathrm{d}r.$$

For a homogeneous Poisson process the pair correlation function $g(r)$ is equal to one and consequently

$$\mathbf{E}(S(o)) = \mathbf{E}_o(S(o)),$$

which is again a special case of the Slivnyak–Mecke theorem (see p. 78).

6.10 Space–time point processes

6.10.1 Introduction

So far, this book has only discussed patterns observed at just one point in time or where temporal changes are not taken into account. The space–time point processes which are now considered are models for time-dependent, dynamic point patterns. The 'points' in these pattern represent events which take place at random times and at random locations (such as earthquakes), or objects which move through space or objects which appear at random instants at random locations and remain there for a random length of time (such as trees in a forest or forest fires).

From the point of view of time-dependent processes, classical spatial point processes are merely snapshots, and a space–time description of a phenomenon typically provides much more information and a deeper understanding of underlying biological and physical processes.

In fact, ignoring time dependence may result in serious misinterpretations, when the pattern formed by all points that appear in some time interval is analysed. For example, within a specific short time interval, points (e.g. individuals from a specific plant species that mainly occurs in spring) may be clustered in a certain area. Within another short time interval (e.g. in summer), however, there might be a gap in the same area. As a consequence, aggregating the point pattern over a longer time interval that includes both these shorter time intervals may 'cancel out' the clusters and the gaps. This provides the wrong impression on the phenomenon.

Section 1.4 has already shown that collecting point process data can be rather laborious. On top of this, space–time data have to be collected over time, by definition, aggravating the situation even further. This is particularly difficult when phenomena are being analysed that change only very slowly over time, such as developments in forests or ecosystems. This, along with the computational problems arising in the analysis of the huge data sets for space–time point patterns, may explain the fact that the methodology for space–time point processes is still underdeveloped. No systematic theory has been developed so far, which may be also explained by the large number of different types of time dependence. Therefore, this section can only try to give some ideas and examples.

The following five cases are perhaps particularly important:

1. The points (t_i, x_i) represent events of zero (or negligible) duration at random instants t_i at random locations x_i and form spatial point patterns if the process is observed within a time interval $T_1 \le t_i \le T_2$. A typical example of this case are earthquakes.

2. At the start, at some point in time, the space–time process is a dense pattern of points representing objects with random lifetimes. Over time the pattern gets thinner and thinner, because the objects die and are not replaced. A planted forest without self-reproduction may serve as typical example of this. Similarly, the opposite situation may occur which starts with no points. Subsequently, new points appear and remain in the pattern. Examples include activated fault points in fracture processes or the points in a simulation of an RSA process.

3. The two situations described in 2 are two extremes in a class of space–time processes in which the points appear randomly, remain within the pattern for a random length of time and subsequently disappear. Forest fires and plant communities with self-reproduction and competition may be considered as examples.

4. The points are moving in space, with or without interaction. They may represent physical particles or objects such as rainfall cells, storm centres or animals.

5. The point pattern of interest develops in discrete time-steps. An example may be a sequence of cluster patterns, where the offspring of the last generation become parents of the new one, etc.

In the following, processes as described in 1 are referred to as *event processes* (Figure 6.20) and the resulting spatial point processes collected over time as (T_1, T_2)-*summary processes*. Furthermore, the processes discussed in 2 or 3 are called *birth-and-death processes* (Figure 6.20), the examples in 2 being *pure birth* or *pure death processes*. (Note that in this book the term 'birth-and-death process' usually refers to time-dependent processes that are used for simulating Gibbs processes; see p. 144. These processes are interesting space–time processes as well and can be used as models in statistical analyses.) All points of these processes that exist at time t form a *t-snapshot*, which is a spatial point process as considered in the other chapters of this book.

Space–time processes can have various invariance properties such as time stationarity, space stationarity or homogeneity, and complete stationarity. Definitions will be given for the special case of space–time Poisson processes in Section 6.10.2. In Section 6.10.3 second-order characteristics are considered for the case of complete stationarity. Finally, two applications are sketched, which present ideas from earthquake statistics and ecological modelling. More information can be found in Daley and Vere-Jones (2008), Vere-Jones (2009) and Diggle (2007). Spatio-temporal extensions of log-Gaussian Cox processes are considered in Brix and Møller (2001) and Brix and Diggle (2001), and Gibbs processes with a time component in Renshaw et al. (2007). Brix and Møller (2001) consider a birth process (to describe the spread of weeds in a plot) backwards in time, by assuming that the pattern at time t_1 results by independent thinning of the pattern at time t_2, where $t_1 < t_2$.

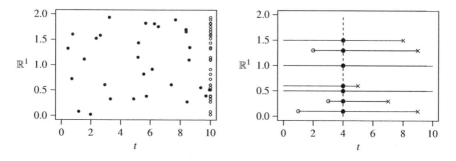

Figure 6.20 Illustration of an event process (left) and a birth-and-death process (right). The figures show the entire histories of the processes. The abcissa is the time axis, and the ordinate shows the locations, which are one-dimensional in these examples. In the event process the open circles show the (0,10) summary. In the birth-and-death process the 4-snapshot consists of seven points (filled circles) on the vertical line, and it is a point process.

6.10.2   Space–time Poisson processes

Event processes

Space–time Poisson processes are defined in a similar way to inhomogeneous Poisson processes in Section 3.4.1. While the latter processes live in $\mathbb{R}^d$, the points (t, x) of space–time processes live in $\mathbb{R} \times \mathbb{R}^d$, where t is the instant and x the location. They have the following properties:

(1) *Poisson distribution of point counts.* The number of points of the process N in any bounded set B in $\mathbb{R} \times \mathbb{R}^d$ has a Poisson distribution with mean $\int\int_B \lambda(t, x) dt dx$, where $\lambda(t, x)$ is a non-negative function called the intensity function. A special case is $B = T \times A$ with T a time interval $[T_1, T_2]$ and A a bounded subset of $\mathbb{R}^d$. The mean then represents the number of points appearing between T_1 and T_2 in the set A.

(2) *Independent scattering.* The random numbers of points of N in k disjoint subsets of $\mathbb{R} \times \mathbb{R}^d$ are independent random variables, for arbitrary k.

Any (T_1, T_2)-summary of a space–time Poisson process is an inhomogeneous Poisson process; the corresponding intensity function is given by

$$\lambda_{T_1 T_2}(x) = \int_{T_1}^{T_2} \lambda(t, x) dt. \tag{6.10.1}$$

Analogously, the instants t_i at which the points (t_i, x_i) appear with x_i in a fixed bounded set A form a (temporal) one-dimensional Poisson process with intensity function

$$\lambda_A(t) = \int_A \lambda(t, x) dx. \tag{6.10.2}$$

Space–time Poisson processes may be *time-stationary*, which means that $\lambda(t, x)$ depends only on x, $\lambda(t, x) = \lambda(x)$. This is equivalent to the property that the point process consisting of time points t_i of the points (x_i, t_i) with x_i in any fixed bounded subset of $\mathbb{R}^d$ is time-stationary. (The latter property is 'stationarity' in the sense defined in Section 1.6, but now for the one-dimensional point process of t_i.) The (T_1, T_2)-summaries of time-stationary Poisson processes are inhomogeneous Poisson processes with intensity function

$$\lambda_{T_1 T_2}(x) = (T_2 - T_1)\lambda(x). \tag{6.10.3}$$

Space-stationary (or *homogeneous*) processes, i.e. processes for which $\lambda(t, x) = \lambda(t)$, may be defined similarly as above, this is equivalent to the property that the point process of locations x_i of the points (x_i, t_i) with t_i in any fixed time interval is stationary in the sense as defined for spatial point processes, but now in the $(d + 1)$-dimensional case. The (T_1, T_2)-summaries of space-stationary Poisson processes are homogeneous Poisson processes with intensity

$$\lambda_{T_1 T_2} = \int_{T_1}^{T_2} \lambda(t) \, \mathrm{d}t. \tag{6.10.4}$$

The combination of both invariance properties results in *complete stationarity*, $\lambda(t, x) \equiv \lambda$. Then

$$\lambda_{T_1 T_2} = (T_2 - T_1)\lambda. \tag{6.10.5}$$

The statistical analysis follows Section 3.3.3, for $(d + 1)$-dimensional processes.

For time-stationary processes the points x_i of the (T_1, T_2)-summary in some window W are recorded and the corresponding estimator $\widehat{\lambda}(x)$ is calculated, e.g. by means of (3.3.6). Then the intensity function estimator of the space–time process is

$$\widehat{\lambda}(t, x) = \frac{\widehat{\lambda}(x)}{T_2 - T_1}. \tag{6.10.6}$$

Analogously, for space-stationary processes the instants t_i for the points (t_i, x_i) with x_i in some space window W are recorded in the time interval of interest. The corresponding estimator $\widehat{\lambda}(t)$ is calculated e.g. by means of (3.3.6) with t and t_i instead of x and x_i. Then the intensity function estimator of the space-time process is

$$\widehat{\lambda}(t, x) = \frac{\widehat{\lambda}(t)}{\nu(W)}. \tag{6.10.7}$$

Birth-and-death processes

Pure death process. Consider a homogeneous Poisson process N_0 at time $t = 0$ with intensity λ_0. The points in N_0 have i.i.d. lifetimes with d.f. $L(t)$. Then the point process N_t, the t-snapshot, is a homogeneous Poisson process with intensity

$$\lambda_t = \lambda_0[1 - L(t)]. \tag{6.10.8}$$

The process N_t is a result of p-thinning, with $p = 1 - L(t)$. If the starting process is an arbitrary stationary and isotropic point process with intensity λ_0 and pair correlation function $g(r)$, then (6.10.8) holds and $g(r)$ is the pair correlation function of N_t for all t. In any of these cases, the intensity of the death process decreases towards zero with increasing t.

Pure birth process.   At time $t = 0$ the process has no points, but new points appear with intensity function $\lambda(t)$. This means that the random number $N_t(B)$ of points in any bounded subset B of $\mathbb{R}^d$ at time t has a Poisson distribution with mean

$$\mu(t) = \int_0^t \lambda(s)\mathrm{d}s \cdot \nu(B) \qquad (6.10.9)$$

and the numbers of points in disjoint sets are independent. Thus the point process N_t formed by the points that are present at time t is a Poisson process with intensity $\lambda_t = \int_0^t \lambda(s)\mathrm{d}s$.

Birth-and-death process.   This model is based on a completely stationary event process with intensity λ. Each of its points (t, x) represents a point at position x, which exists for a specific length of time, $[t, t + l]$. l is a random lifetime and all random lifetimes are assumed to be i.i.d. with d.f. $L(t)$ and mean m_L.

This model may be regarded as a spatial version of the $M/G/\infty$ queue, a system with infinitely many servers and thus no waiting times; see Tijms (2003). The properties and formulas for the $M/G/\infty$ queue imply that

- the snapshot process at any time is a stationary Poisson process with intensity λm_L;

- the covariance of $N_t(B)$ and $N_{t+u}(B)$, i.e. of the numbers of points in a fixed set B at times t and $t + u$, is given by

$$\mathbf{cov}(N_t(B), N_{t+u}(B)) = \lambda m_L \left(1 - \frac{1}{m_L} \int_0^u (1 - L(s))\,\mathrm{d}s \right), \qquad (6.10.10)$$

using a result due to Beneš (1957). If the lifetimes are bounded by l_0 (i.e. $L(l_0) = 1$), then $N_t(B)$ and $N_{t+u}(B)$ are uncorrelated for $u \geq l_0$.

The statistical analysis of these three processes is not difficult if the entire history of the processes has been observed for a sufficiently long time in a large window.

6.10.3 Second-order statistics for completely stationary event processes

The classical statistical methods developed for stationary processes may often be straightforwardly generalised to space–time processes. In order to illustrate this, this section discusses an example of a statistical method for space–time processes – an estimator for the pair correlation function of a completely stationary event process is derived. By way of illustration, the method is applied here to seismological data.

A space–time point process $N = \{[t_1, x_1], [t_2, x_2], \ldots\}$ is called *completely stationary* ('homogeneous and stationary') if, for all $x \in \mathbb{R}^d$ and all real t,

$$N \stackrel{d}{=} N_{t,x}, \tag{6.10.11}$$

where $N_{t,x}$ is the translated process $N_{t,x} = \{[t_1 + t, x_1 + x], [t_2 + t, x_2 + x], \ldots\}$. The definition is an analogue of the stationarity definition in (1.6.1).

Note the difference from the case of marked point processes, where the marks remain unchanged in translations but the times do not. Furthermore, event processes may also be treated as point processes in $\mathbb{R}^{d+1}$. Nevertheless, the explicit notation of x and t is typically preferred due to the different roles of time and space including different scaling of spatial and temporal coordinates and the fact that there is no concept in time that is equivalent to the concept of isotropy.

As stationary point processes, completely stationary point processes have an *intensity* λ as first-order characteristic that does not vary in time or space. It is defined as

$$\lambda = \mathbf{E}(N(\boxed{1} \times \boxed{1})), \tag{6.10.12}$$

i.e. λ is the mean number of points per unit area (volume) and time unit. The first $\boxed{1}$ denotes the unit square (cube) and the second the unit interval.

The intensity λ is best estimated along the lines of (4.2.10),

$$\hat{\lambda} = \frac{N(W \times T)}{\nu(W)\nu(T)}, \tag{6.10.13}$$

where W is the spatial observation window and T the temporal observation window. $N(W \times T)$ is the total number of points $[t, x]$ with $x \in W$ and $t \in T$. Note that it does not make sense to define a spatial or temporal intensity, since completely stationary point processes have infinitely many points in space and time; for example, the total number of points $[x, t]$ with $x \in \boxed{1}$ is infinite.

The second-order structure of completely stationary point processes can be described in a similar way to that of stationary point processes. It is possible to define K- and L-functions of type $K(r, t)$ (see Diggle et al., 1995) and $L(r, t)$, but in the following only the corresponding product density and pair correlation function are considered for the special case where N is spatially isotropic. Then the second-order product density is a function $\varrho(r, u)$ where r denotes spatial distance as before and u is the time lag. The quantity

$$\varrho(r, u)\mathrm{d}x\mathrm{d}y\mathrm{d}t\mathrm{d}s$$

may be interpreted heuristically as the probability that there is a point of N in each of the two infinitesimally small sets with volumes $dxdt$ and $dyds$ with distance r and time lag u. The corresponding pair correlation function is defined by

$$g(r, u) = \varrho(r, u)/\lambda^2 \qquad \text{for } r \geq 0 \text{ and } u \geq 0. \qquad (6.10.14)$$

An estimator for $\varrho(r, u)$ and $g(r, u)$ is derived analogously to the stationary case (cf. (4.3.29)):

$$\hat{\varrho}(r, u) = \frac{1}{2db_d r^{d-1}} \sum_{[x,t],[y,s] \in W \times T} \frac{k_{sp}(r - \|x - y\|)k_{ti}(u - \|t - s\|)}{\nu(W_x \cap W_y)\nu(T_t \cap T_s)}, \qquad (6.10.15)$$

where $k_{sp}(z)$ and $k_{ti}(z)$ are kernel functions with bandwidths h_{sp} and h_{ti}. (Experience with pair correlation function estimation recommends box kernels, see Section 4.3.3.) The estimator is unbiased in the same sense as $\hat{\varrho}(r)$ in (4.3.32).

The pair correlation function is estimated by

$$\hat{g}(r, u) = \hat{\varrho}(r, u) \Big/ \hat{\lambda}^2 . \qquad (6.10.16)$$

This estimator can probably be improved by the use of adapted intensity estimators.

Example 6.6.   Earthquakes with a magnitude of more than 4.5 in central Japan
Figure 6.21 shows the positions of earthquakes with a magnitude of more than 4.5 in central Japan in the years 1926–2005. There are 2646 of them in total. The pattern looks like a sample from a stationary point process and is here initially analysed with methods for those processes. This is done despite an apparent large-scale inhomogeneity in the pattern since the aim is to analyse short-range correlation (spatial as well as temporal). However, the approach produces invalid results for large distances. (The point density in the north–east corner is clearly smaller than average and for a larger window the spatial inhomogeneity would become even more pronounced. This is indicated further by the fact that the spatial pair correlation function $\hat{g}(r)$ for the pattern in Figure 6.21, shown in Figure 6.22, exceeds the value of 1.1 even for very large r.)

To simplify the statistical analysis, the spherical quadrangle is approximated by a planar rectangle W of side lengths 543 km and 556 km. The time interval T of observation has a length of 29 100 d. This leads to the estimate $\hat{\lambda} = 3 \times 10^{-7} \, \mathrm{km}^{-2} \mathrm{d}^{-1}$ for the intensity λ. Figure 6.22 shows the estimate $\hat{g}(r, u)$ of the space–time pair correlation function. The bandwidths have been chosen as $h_{sp} = 100$ km and $h_{ti} = 100$ d. The pair correlation function shows the strong clustering of earthquakes in space and time. It is greater than 2 (a value of 1 would have been expected for random patterns) for distances $r \leq 40$ km (and time lags up to 200 d) and time lags $u \leq 50$ d (and distances up to 100 km).

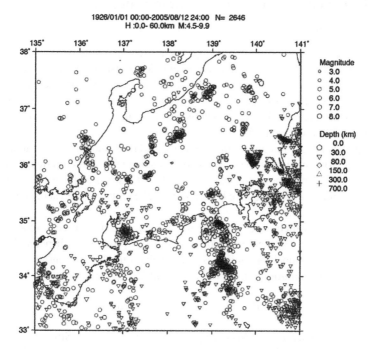

Figure 6.21   Positions of earthquakes in central Japan in the years 1926–2005 with a magnitude of more than than 4.5. Data courtesy of Y. Ogata.

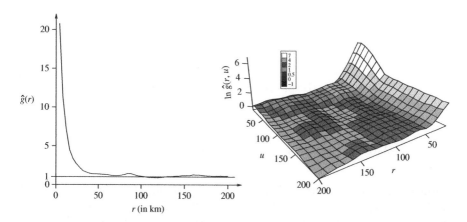

Figure 6.22   Empirical pair correlation function $\hat{g}(r)$ for the positions of Japanese earthquakes, ignoring the time dependence (left), and logarithm of empirical time-dependent pair correlation function $\hat{g}(r, u)$ (right).

6.10.4 Two examples of space–time processes

Statistics for earthquake data

Point process methods have frequently been used in the statistical analysis of earthquakes. Typically, enough data are available to justify a space–time analysis. In order to illustrate some of the methods and to encourage the reader to apply these to other event processes in different scientific contexts, the following sketches one successful approach. For more details, see Zhuang et al. (2006) and Daley and Vere-Jones (2008, Chapter 15).

Earthquakes appear in clusters in both space and time and are usually understood as resulting from superposition of background seismicity and earthquake clusters. These two cases correspond to objectives in the statistical analysis, i.e. long-term prediction (zoning and earthquake hazard potential estimation) and short-term earthquake prediction, respectively.

The approach, which is now standard, applies a point process model, more specifically a marked space–time process (an event process) with points $[t, x, m]$, where t is time, x is position, and m is a real-valued mark representing magnitude. It is based on the conditional intensity $\lambda(t, x, m|\mathcal{H}_t)$, given by

$$\lambda(t, x, m|\mathcal{H}_t)\mathrm{d}t\mathrm{d}x\mathrm{d}m = \mathbf{E}(N(\mathrm{d}t\mathrm{d}x\mathrm{d}m)|\mathcal{H}_t), \qquad (6.10.17)$$

where $\mathcal{H}_t$ is the observational history up to time t (excluding t). Similarly to $\lambda(x|N)$ on p. 28, the right-hand side can be interpreted as the probability of observing a point around time t and location x with mark around m, given the history $\mathcal{H}_t$.

A model of shot-noise nature is the epidemic-type aftershock sequence (ETAS) model (Ogata, 1998). Here

$$\lambda(t, x, m|\mathcal{H}_t) = \lambda_0(t, x)J(m) \qquad (6.10.18)$$

with

$$J(m) = \beta \exp(-\beta(m - m_0)), \qquad \text{for } m \geq m_0,$$

where β is a seismic parameter, m_0 a magnitude threshold and

$$\lambda_0(t, x) = \mu(x) + \sum_{i:t_i < t} k(m_i)h(t - t_i)f(x - x_i, m_i), \qquad (6.10.19)$$

in which $\mu(x)$ is the background intensity, which is assumed to be time-independent but location-dependent, $k(m)$ is the mean number of aftershocks resulting from an event of magnitude m, $h(t)$ the p.d.f. of the occurrence times of aftershocks following an event at time point 0, and $f(x, m)$ is the p.d.f. of the locations of events following an event at location o. Explicit equations have been derived from seismological theory for all of these functions. These equations depend on parameters which have to be estimated statistically.

The data for the statistical analysis are marked points

$$(t_i, x_i, m_i) \quad \text{for } i = 1, \ldots, n, \text{ with } T_1 \le t_i \le T_2 \text{ and } x_i \in W.$$

If $\mu(x)$ is given in the form $\mu(x) = \mu_0 u(x)$ with known $u(x)$, the parameters in a vector θ (which includes μ_0) can be estimated by means of the maximum likelihood method. The log-likelihood is of the form

$$\log L(\theta) = \sum_{j:T_1 \le t_j \le T_2, x_j \in W} \log \lambda_0(t_j, x_j) - \int_W \int_{T_1}^{T_2} \lambda_0(t, x) \mathrm{d}t \mathrm{d}x \qquad (6.10.20)$$

and has to be maximised with respect to θ.

Zhuang et al. (2006) describe an approach of 'stochastic declustering', which aims to separate background and cluster points. This may be used to estimate $\mu(x)$ iteratively. Zhuang et al. (2006) apply this method to earthquake data from the Japan and Taiwan regions. Furthermore, they describe a method of testing the goodness of fit of their model which uses residuals.

Self-thinning through local competition

Consider an even-aged plant community without self-reproduction such that only growth and mortality are important for community dynamics. Both processes are controlled by local competition: plants experiencing strong competition grow more slowly and have higher mortality. These processes and their modelling are discussed extensively in the literature; the following sketches only one of many possible approaches which follows the ideas of Adler (1996) and Berger and Hildenbrandt (2000).

The behaviour of the plant community may be described as a pure death process, with some Markov structure, i.e. it is enough to consider a t-snaphot to predict the further development. Any t-snaphot M_t is assumed to be a stationary marked point process $\{[x_i, m(x_i)]\}$. Here the mark $m(x_i)$ is a size parameter of the plant at x_i at time t. The competition pressure on a plant at location x is

$$c(x) = \sum_{x_i \in M_t, x_i \ne x} f(x - x_i, m(x_i)) \qquad (6.10.21)$$

with some 'impulse' function $f(z, m)$ such as

$$f(z, m) = m^\alpha \exp\left(-\frac{\delta \|z\|}{m\beta}\right),$$

where α, β and δ are model parameters. Equation (6.10.21) means that the competition pressure results from additive superposition of individual competition pressures from the other plants, as in a shot-noise field; for plants with many close neighbours the pressure will be larger than for isolated plants.

The growth rate $r(x)$ of a plant at x is

$$r(x) = h(m(x))(\gamma_o - \gamma c(x))_+, \qquad (6.10.22)$$

where γ and γ_0 are further model constants and h some function of the mark. x_+ is equal to x if $x > 0$ and 0 otherwise. Formula (6.10.22) has been derived from biological theory of plant growth; $r(x)$ decreases with $c(x)$, i.e. growth decreases with increasing local competition.

Finally, a plant at x dies if its growth rate is too small, e.g. if

$$r(x) < \varepsilon \mathbf{E}_o(r(o)) \qquad (6.10.23)$$

for $0 < \varepsilon < 1$ (in Berger and Hildenbrandt, 2000, $\varepsilon = \frac{1}{2}$). $\mathbf{E}_o(r(o))$ is the mean growth rate in the point process.

This model can be easily simulated and hence the process of self-thinning can be conveniently studied. Figure 6.23 shows six steps of such as process which aims

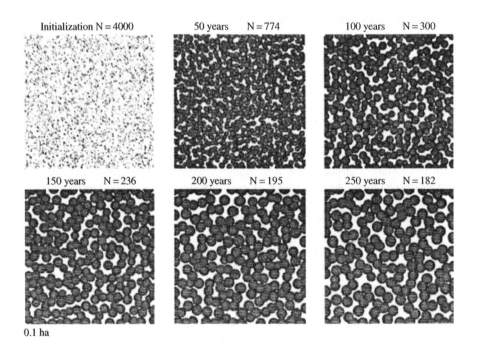

0.1 ha

Figure 6.23 Simulation of the development of a mangrove tree cohort with an initial density of 1000 individuals per hectare. The panels show the distribution of trees for six consecutive time-steps. Data courtesy of U. Berger.

to model the growth of a mangrove forest. The underlying model is more complex than the model described above.

Note that the means $\mathbf{E}_o(c(o))$ and $\mathbf{E}_o(r(o))$ are of the same nature as $\mathbf{E}_o(S(o))$ in (6.9.5) for shot-noise fields and estimated in a similar way to the point-related indices in Section 5.2.4.

Similar, more general models, which also include immigration, are discussed in Särkkä and Renshaw (2006), Renshaw et al. (2007) and Comas and Mateu (2007).

The statistical methodology for this type of model is still in its infancy. These days, the model parameters are usually determined by trial and error with the aim to carry out simulations which yield plausible results. Least-squares methods are the natural numerical approach.

Spatio-temporal processes in the context of disease spread are studied in Diggle (2007). One of the main focuses in these studies is the detection of spatial clustering.

6.11 Correlations between point processes and other random structures

6.11.1 Introduction

Often a point process N is controlled by or correlated with another random structure S. For example, the local point density of N may depend on a covariate. The tree density in a forest may depend on a spatially varying soil property. Or the marks of the points in N may depend on the distances to curves scattered in space, e.g. the diameters of trees in a forest may depend on the trees' distances from a river. The aim of this section is to present non-parametric methods that may be used to detect and quantify these correlations. Acquiring knowledge on these correlations may be of interest in its own right, and, in addition, it may be used as a starting point for more complex spatial modelling. The approach includes three cases where the partner structure S may be a random field, random set or fibre process. This section does not discuss the case where S is a point process itself. Methods covered in Section 5.2 for discrete marks may be applied, by considering the union of N and S and marking the points of N by 1 and those of S by 2. Then the correlation functions for bivariate point processes, such as $g_{12}(r)$, $L_{12}(r)$, $D_{12}(r)$ etc., may be used.

A rigorous general theory for the corresponding correlation analysis based on the theory of random measures is detailed in Stoyan and Ohser (1982, 1984). In its original form, geometrical measurement of areas, curve lengths and angles are necessary. In 1982 this was still done manually but today computerised image analysis and geographical information systems techniques simplify the work.

This section presents two approaches which can be applied with standard software. The first of these exclusively uses point process methodology and exploits the information contained in the data very well. The structure S is replaced by a point process, more specifically a Cox process the points of which are scattered on the segments of S. Methods for bivariate point processes can then be applied, where the points of N are marked by 1 and those of S by 2.

The other approach can be applied in the case where S is a random field or a regionalised variable. Then N is also regionalised, by one of the methods discussed in Section 1.8.3, to obtain a second random field such as $\{Z_N(x)\}$ given by (1.8.2) or (1.8.3), and the correlations between the two fields are analysed by geostatistical methods. Note that, as a result of the regionalisation, some loss of information has to be accepted. Both approaches have the advantage of being easy to understand and of being based on standard software.

In the following, methods for stationary and isotropic processes are used, assuming that N and S are both stationary and isotropic and stationarily connected. As before, the intensity of the point process N is denoted by λ.

6.11.2 Correlations between point processes and random fields

Consider the situation where S is a random field $Z = \{Z(x)\}$. Its mean $\mathbf{E}(Z(o))$ is denoted by m_Z. Note that the case of a random set can be treated as a special case, where $Z(x)$ is the indicator function of the set.

The correlation between a point process N and a random field Z can be described by various summary characteristics. Here only a point-process statistical approach is described, and the partial pair correlation function $g_{12}(r)$ is applied as the summary characteristic.

The basic idea is to replace Z by a point process the intensity of which is determined by Z, i.e. the point density is high in regions of large values of Z, etc. If Z is positive (and this is often the case in applications), it can be used as the intensity field function of a Cox process, perhaps multiplied by a scaling factor f. If Z is derived from values Z_i at the points y_i in a grid G, the point densities within the grid cells C_i are chosen to be constant, i.e. fZ_i in C_i. Based on this, it is straightforward to simulate the corresponding Cox process.

Now, the given points of the empirical point process N are assigned the mark 1, and the constructed Cox points are assigned the mark 2. This yields a bivariate point process, which can be analysed by the methods in Section 5.3.2. In particular, the use of the corresponding partial pair correlation function $g_{12}(r)$ can be recommended.

What information does $g_{12}(r)$ provide about the relationship between N and Z? Values larger than 1 indicate positive correlation, i.e. there are large values of Z at a distance r from the typical point in N. In other words, in areas where Z has large values the point density of N is high, and for values of Z less than 1 it is low.

An alternative to the point-process-based characteristic $g_{12}(r)$ is the geostatistical cross-variogram $\gamma_{12}(r)$. The idea is to regionalise N, i.e. to replace it by a random field Z_N and then to analyse the correlations of both fields. This approach is recommended when correlations among long-range fluctuations of Z and of the point density in the window W are of primary interest, and the short-range interactions of points are negligible or meaningless.

The new point-process-related random field $\{Z_N(x)\}$ (or Z_N for short) can simply be constructed as in (1.8.2), i.e.

$$Z_N(y_i) = N(b(y_i, R)) \qquad \text{for } y_i \in G, \qquad (6.11.1)$$

where the same grid G as for Z is used. The value of the random field Z_N at location y_i is simply the number of points of N in the sphere $b(y_i, R)$ centred at y_i; the radius R is a numerical parameter which plays a role similar to that of the bandwidth h in kernel estimators.

An alternative is to set

$$Z_N(y_i) = N(C(y_i)) \qquad \text{for } y_i \in G,$$

where $C(x)$ is the cell associated with grid point y_i, i.e. $C(y_i)$ is one of the cells C_i above. Thus $Z_N(y_i)$ is now the number of points in this cell.

The data set

$$Z(y_i), Z_N(y_i) \qquad \text{for } y_i \in G,$$

can be analysed by geostatistical software, i.e. by estimating the cross-variogram given by

$$\gamma_{12}(r) = \frac{1}{2}\mathbf{E}\left((Z(o) - Z(\mathbf{r}))(Z_N(o) - Z_N(\mathbf{r}))\right) \quad \text{for } r \geq 0. \qquad (6.11.2)$$

Here, as before, $\mathbf{r}$ is any point at distance r from o.

Some of the properties of $\gamma_{12}(r)$ for the case of stationary fields Z and Z_N are as follows:

$$\gamma_{12}(0) = 0,$$

$$\gamma_{12}(\infty) = \mathbf{cov}(Z(o), Z_N(o)),$$

$$\gamma_{12}(r) \equiv 0 \qquad \text{if } Z \text{ and } Z_N \text{ are independent.}$$

Correlations between point processes and random fields may be also assessed with *parametric methods*. In the case of a Cox process with random intensity function $\lambda(x)$, the random field may control $\lambda(x)$, e.g.

$$\log \lambda(x) = \beta Z(x) + \varepsilon(x),$$

where β is a model parameter and the $\varepsilon(x)$ are i.i.d. with mean 0 and variance σ^2; see Møller and Waagepetersen (2007) and Waagepetersen (2007).

Example 6.7.   Correlation among flowers, stoniness and coverage by another plant
Figure 6.24 shows 341 positions of the arctic-alpine flower glacier buttercup

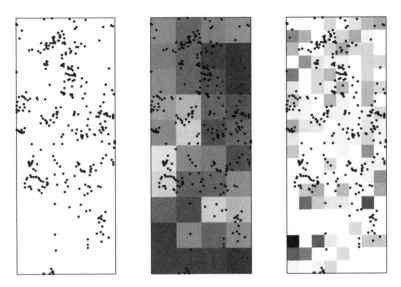

Figure 6.24   Positions of glacier buttercup in a $10 \times 4\,\text{m}$ rectangle and, in the same rectangle, stoniness (%) in $1 \times 1\,\text{m}$ squares and coverage (cm^2) of mountain sorrel in $0.5 \times 0.5\,\text{m}$ squares; dark grey squares correspond to high values. The data were collected in 1982 at Mount Saana in Lapland (Finland). Data courtesy of A. Järvinen (Kilpisjärvi Biological Station, University of Helsinki).

(Ranunculus glacialis) in a $10 \times 4\,\text{m}$ rectangle. The figure also shows two covariates in discretised form: the stoniness (percentage of area covered by stones) in $1 \times 1\,\text{m}$ squares and the coverage (cm^2) of area by mountain sorrel *(Oxysia diguna)* in $0.5 \times 0.5\,\text{m}$ squares.

Visual inspection of Figure 6.24 reveals irregular fluctuations with no clear spatial trend in any of the three patterns; and it seems sensible to assume similar fluctuations outside the sampling rectangle. The small number of flowers near the point $(1,1)$ may be interpreted simply as a random fluctuation. Based on these assumptions and findings, methods for stationary and isotropic structures as discussed above may be applied. In other words, the point pattern is regarded as a sample from a stationary and isotropic point process and the covariates are samples of two stationary and isotropic random fields.

However, it is rather difficult to draw any conclusions on the correlation between point density and covariates based on visual inspection alone. Biology might suggest some negative correlation in the sense that in areas of high stoniness (or coverage by mountain sorrel) there are few buttercups, as around the point $(1,1)$, but this may just be a misleading subjective impression.

The correlation analysis for the relationship between flower distribution and the two covariates uses both approaches described above. Before reporting the result, simple methods from classical statistics are applied to avoid unreasonable

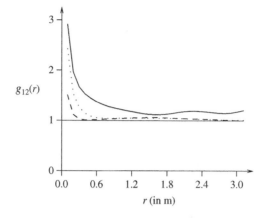

Figure 6.25 Partial pair correlation function estimates $\hat{g}_{12}(r)$ for flower density and stoniness (solid line) and coverage (dashed line). The bandwidth is $h = 1\,\text{m}$. Both curves indicate independence between flower density and covariates. The empirical pair correlation function for the flowers (dotted line) shows the cluster or Cox process nature of the pattern.

expectations on the magnitude of a potential effect. Consider the values of the covariates in the grid squares and the corresponding numbers of flowers. The coefficients of correlation are -0.146 for stoniness and -0.116 for coverage, indicating a very weak negative correlation. Perhaps the resolution of the covariates is simply too low. Nevertheless, the statistical analysis is continued here.

Figure 6.25 shows the Cox $\hat{g}_{12}(r)$ for both covariates, stoniness and coverage. The bandwidth h of the box kernel in (5.3.51) is chosen as $1\,\text{m}$, and the scaling factor f on p. 438 is 0.40 for stoniness and 0.15 for coverage. This resulted in 1226 and 1297 points in the simulated samples, the results of which are discussed here.

Both curves are decreasing in r, with values larger than 1 for very small r, and then are close to 1. The conclusion of this part of the analysis is that there is no significant correlation between the buttercups and either stoniness or sorrel coverage, since the large values for small r are interpreted as statistical artefacts.

In the analysis of the correlation between point density and the covariates coverage and stoniness, one may be prepared to accept the idea that the locations of individual flowers are not very relevant and thus it makes sense to regionalise the point field of buttercup positions.

This yields the cross-variograms for Z_N and Z (coverage or stoniness). The grid G is different for coverage and stoniness, adapted to Figure 6.24. Figure 6.26 shows the resulting cross-variograms, where Z_N was constructed using the numbers of points in the grid cells. In agreement with the negative correlation coefficients, the $\hat{\gamma}_{12}(r)$ are also negative. For stoniness it is close to zero for small r and for larger r approximately equal to the estimated covariance (which can be derived

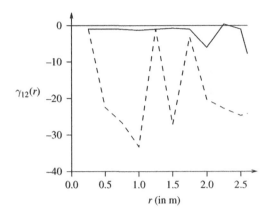

Figure 6.26   Cross-variograms for flower density and stoniness (solid line) as well as coverage (dashed line). These are based on the numbers of points in grid cells and indicate a weak negative correlation between point density and covariates.

from the correlation coefficient and the sample variances of Z and Z_N), which is -12.0. This may indicate independence between stoniness and flower density.

For sorrel coverage $\hat{\gamma}_{12}(r)$ is also close to the estimated covariance for small r, which is -24.1. This indicate a weak negative correlation, which has, however, no spatial component.

Formal simulation tests discussed in Section 7.5 show that none of these correlations is significant; see Table 7.2. Hence it may be assumed that there is no correlation between flower density and the two covariates, which may appear somewhat surprising.

6.11.3   Correlations between point processes and fibre processes

A fibre process F is a mathematical model for a random system of curves or line segments in space; see Stoyan et al. (1995). Examples include water courses in a forest or geological lineaments in a landscape. The locations of these 'fibres' and the points in a point process N may be correlated. For example, the objects represented by the points are attracted by the fibres in the sense that there are more points in the vicinity of the fibres than further away from them. In Stoyan and Ohser (1982) these correlations are analysed based on an extensive use of geometrical methods, including measurements of lengths and of angles between the fibres and test circles. Here, a simpler approach is presented which is based on an approximation of the fibres by points and F by a point process N_F and applies statistical methods for bivariate point processes.

The approximation is discussed here for the case where the fibres are line segments, certainly a common method for the digitalisation of the fibres. Points are distributed either regularly or randomly on these line segments. Due to the varying length of the line segments, a regular distribution may be problematic, and therefore a completely random distribution is considered here. In other words, points in a Cox process are constructed where the so-called leading measure is the length measure of the fibre process, as discussed on p. 383. This means the following: a scatter intensity λ_s is chosen and n_l points are scattered on a line segment of length l, where n_l is a random number with a Poisson distribution with mean $\lambda_s l$. These n_l points are independently and uniformly scattered on the line segment. This is done for all line segments and yields a second (constructed) point pattern, in addition to the sample of N, the empirical point pattern. The intensity λ_s should be chosen high enough that the essential details of the fibre system are reflected by N_F; visual comparison of F and N_F is recommended.

Both point patterns are now analysed as one qualitatively marked point process; the original points are assigned the mark 1 and the fibre points the mark 2. This bivariate pattern can then be analysed by the methods discussed in Sections 5.3.2 and 5.3.4 (discrete marks). In particular, $g_{12}(r)$ and $D_{12}(r)$ are useful in this context.

The function $g_{12}(r)$ is interpreted along the lines of Section 5.3.1. It is $g_{12}(r) = 1$ if N and F are independent. Values of $g_{12}(r)$ larger than 1 indicate attraction between the fibres and points, values smaller than 1 repulsion. $D_{12}(r)$ is an approximation of the d.f. of the nearest distance from the typical point to a fibre point.

Example 6.8. Copper deposits and lineaments
Figure 6.27 shows the pattern of 57 copper deposits (black dots) and 90 geological lineaments (sequences of small points) in Queensland (Australia). The potential correlation between the point and line segment pattern has been studied in various papers; see Berman (1986) and Baddeley and Turner (2006). Based on several models and nearest-neighbour distance summary characteristics such as the distance from deposit to nearest lineament, these authors conclude that there is no correlation. This example presents the results of a second-order analysis based on the ideas above.

In the analysis, the line segments are approximated by sequences of points; the intensity λ_s is $30\,\text{km}^{-1}$. A simulation is used to estimate $g_{12}(r)$ and 1724 points represent all line segments. Thus a pattern of a total of 1781 points is

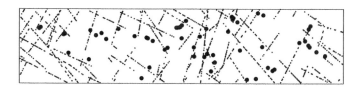

Figure 6.27 Copper ore deposits (large dots) and lineaments (sequences of small dots) in a rectangular region of central Queensland of side lengths 35.335 km and 158.043 km. Data courtesy of A. Baddeley and J. Huntington.

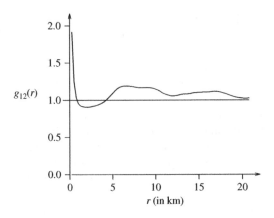

Figure 6.28   The empirical partial pair correlation function for the copper data. The curve indicates a weak positive yet non-significant correlation (attraction).

analysed, 57 with mark 1 (deposits) and 1724 with mark 2. Figure 6.28 shows the corresponding estimated partial pair correlation function $\hat{g}_{12}(r)$. It indicates a weak positive correlation between deposits and lineaments, as $\hat{g}_{12}(r)$ is around 1.2 for $4 \leq r \leq 20$. Some deposits very close to lineaments cause the very large values of $\hat{g}(r)$ for small r.

Since $\hat{g}_{12}(r)$ is not far away from 1, the value for independent patterns, the correlation might be just an artefact and was therefore tested more rigourously. A formal test using the methods in Section 7.5 shows that the correlation is not significant; see Table 7.2. However, some doubts do remain as several of the deposits are on the lineaments. This contradicts any stochastic model of independence between a point process and a fibre process, where the probability of an event like this is zero.

7

Fitting and testing point process models

Once a point pattern has been successfully analysed by the exploratory methods and the analysis has suggested a class of suitable models, in the last step these results are synthesised, i.e. a model is fitted to the data. This includes 'model criticism', consisting of various forms of model evaluation – in particular, formal goodness-of-fit testing. This final chapter presents methodology that may be used for this purpose. First, fundamental methods for parameter estimation are covered, the maximum likelihood method in an approximate form and the method of moments. Second, the use of simulation tests for testing distributional hypotheses is discussed, which includes tests of the hypotheses of independent and geostatistical marking.

7.1   Choice of model

One of the most important steps in the statistical analysis of a point pattern or a series of patterns is constructing and fitting a model. In some ways this is the synthesis of the knowledge gained from a thorough exploratory analysis. Due to the strong link between the methods for model fitting and the material covered

Statistical Analysis and Modelling of Spatial Point Patterns   J. Illian, A. Penttinen, H. Stoyan and D. Stoyan
© 2008 John Wiley & Sons, Ltd

in previous chapters of this book, this chapter assumes that the reader is familiar with such methods, in particular with exploratory data analysis and point process models.

This section briefly discusses those properties that make a good model and then provides some advice on a suitable strategy for identifying a suitable model for a data set.

In most cases, the aim of fitting a model to empirical data is to gain an understanding of the pattern(s). Of particular interest are the nature and spatial extent of the interaction or correlation between the points, the influence of the marks and the relation of covariates to the (marked) point pattern. A good statistical model is one that is easy to understand and interpret. In other words, a good model depends on only a small number of parameters but is still complex enough to suitably reflect those aspects in the data that are of scientific interest. It is important that the model is simple enough that both estimation and simulation are still feasible – models that are too complex easily result in algorithms that are prohibitively expensive to run.

However, it is not obvious how to identify a suitable model for a specific data set. Usually the data will originally have been collected with the aim of proving or disproving some hypotheses about the spatial distributions of the objects that are represented by the points, i.e. in the analysis of a spatial pattern the model is directly linked to some scientific question. In other cases, however, the aim is simply to describe the spatial behaviour in the pattern without reference to underlying mechanisms.

In most applications, some a priori knowledge is available, which often supports the choice of model. For example, seed dispersal may suggest that a cluster process model is suitable for a data set describing a plant pattern. However, further analysis may indicate that the modelling approach has to be refined. For example, airborne seeds disperse over large distances. Seeds within the observation window might stem from mother plants that are far away from the observation window but also from mothers close to or within the window. Hence, a superposition with a Poisson process to model global dispersal in addition to local dispersal may be more suitable.

Sometimes the structure of the model and the structure of the underlying process that formed the pattern are not directly linked. These models are fitted to merely describe or explain the statistical fluctuations and to use this information for the simulation of similar point patterns or to determine the precision of statistical estimates – in particular, of the intensity. However, these models do not explain any true natural mechanisms underlying the data. An example of this is the Matérn cluster process, which may be used to model clustering (around randomly located centres) or, alternatively, environmental heterogeneity. The variation of local point density can be suitably modelled with this process, but the structure of the model deviates strongly from the true structure in any realistic pattern.

As indicated in Section 2.7, the first step in the analysis of a spatial point pattern should be a test of the CSR hypothesis. Accepting the CSR hypothesis may be a disappointing result since it rules out any interesting correlations in the pattern. On the other hand, modelling and further scientific calculations are clearly simplified and can be based on the elegant Poisson process model.

Only once the CSR hypothesis has been rejected, a suitable class of models should be sought. Often, visual inspection and experience from the CSR test aid decision-making on more fundamental questions such as whether the pattern is finite or not finite, stationary or non-stationary, and clustered or regular.

In the exploratory analysis below, the methods discussed in Chapters 3–5 are applied using some of the summary characteristics described there. Quite often, this may be sufficient since these characteristics reveal details on the range and strength of correlation in the pattern. If one wishes to gain a deeper understanding of the pattern, one should look for a suitable model based on the results of the exploratory analysis.

If the data can be considered stationary, the pair correlation function $g(r)$ is probably the best exploratory tool. The detailed discussion of this function and its interpretation in Section 4.3.4, along with the description of the models in Chapter 6, can help the reader to identify appropriate classes of models. Recall, for example, that large values of $g(r)$ for small r indicate clustering and small values of $g(r)$ for small r indicate regularity. This simple information on clustering or regularity at different scales, combined with a priori knowledge, may already indicate which classes of models are more suitable than others. As a general rule, in the first instance every modelling approach should start with basic classical models such as Cox, Neyman–Scott or Gibbs processes, since these can easily be fitted and simulated. Cox and Neyman–Scott processes are suitable for modelling fluctuating point density, while Gibbs processes may be used to describe interesting repulsive interactions between the points. If these models turn out not to be suitable, they can be modified by relaxing some of the underlying Poisson process assumptions, such as by choosing a non-Poisson parent process for a cluster process or by including hard cores, i.e. minimum inter-point distances. Any of these modifications should still be simple enough that the model can be simulated.

Once a suitable model has been identified, the next step involves the estimation of model parameters. These model parameters, such as λ (for the Poisson process), λ, μ and R (for the Matérn cluster process) or β and $r_{\max}$ (for the Strauss process), are of course of a different nature for different models. However, to simplify the notation and language, the general symbol θ is used throughout this chapter to represent any of these parameters. Note that parameter estimation is based on the same principles as the estimation of μ and σ^2 in classical statistics, even if it appears to be more complicated in the context of point processes.

Researchers often consider the analysis as finished once a model has been fitted to the data, supported by good agreement of some graphics of theoretical and empirical summary characteristics. However, a goodness-of-fit test should be performed to formally assess the suitability of a model. It is not difficult to do so if it is possible to simulate from the fitted model. The goodness-of-fit test may on the one hand confirm that the model is suitable, but on the other hand it may help to identify and eliminate any weaknesses of the model.

The following sections describe approaches to parameter estimation and discuss model tests, assuming that a model has been identified for a particular data set. All model tests considered here are simulation tests, which are very common in point processes statistics. Note that, unfortunately, classical tests such as the t-test or χ^2 goodness-of-fit test cannot be applied here. Other approaches to assessing the model fit use, for example, residuals; see Section 4.6.5 and Baddeley et al. (2005). Mathematically more sophisticated tests (e.g. likelihood-ratio tests) can be found in Geyer (1999) and Møller and Waagepetersen (2004).

7.2 Parameter estimation

A number of different approaches to parameter estimation have been used in the context of spatial point processes. These are based on the same ideas as in classical statistics. Which estimation method is used for a specific data set depends on the model and the nature of the parameters and is to a certain extent also a matter of taste.

As far as the performance of the estimators is concerned, these are usually required to be *unbiased*, and to have a small *mean squared error* (mse). Another requirement is that the estimators are *consistent*, i.e. that their increases with increasing window size. This is the case for many parameter estimators of stationary point processes if these are ergodic.

7.2.1 Maximum likelihood method

Maximum likelihood methods are widely used in classical statistics, and many statisticians believe that they should also be preferred in point process statistics. Indeed, famous theorems by Fisher, Rao and Cramér show that in classical statistics maximum likelihood estimators represent the 'hard currency' among the estimators as they are efficient, sufficient and consistent. Those readers familiar with estimation methods in classical statistics may know that maximum likelihood method techniques can only be applied if the likelihood function – describing the probability of observing the data given the model – is known. This probability is maximised (with fixed data and variable parameters), yielding parameter estimators that best fit the data. However, often and particularly for stationary point processes, it is extremely difficult, even impossible, to find the likelihood function. As a result, the maximum likelihood method can only be applied to specific classes of models. These are Poisson processes (pp. 80 and 121), Cox processes (with approximative likelihoods; see Møller and Waagepetersen, 2004, and below) and finite Gibbs processes (p. 161). In addition to the models' flexibility and interpretability in applications, this might account partly for the popularity of finite Gibbs (and Markov) point processes in the statistical point process literature.

It is possible to apply the maximum likelihood method to spatial point patterns where the likelihood function is not known explicitly by (heuristically) approximating the likelihood function. This can be done in many ways.

An interesting example of this approach is a method developed by Tanaka et al. (2008) for stationary point processes. It is based on the model pair correlation function $g(r)$ and works well if $g(r)$ is known explicitly and contains the parameters of interest. The idea is to analyse the finite point process N_δ in addition to the original point process N. N_δ consists of all the difference points

$$\delta = x - y \qquad \text{for } x \neq y,$$

where x and y are the points of the process N in the window W, which is assumed to be convex. If the original pattern has n points then the difference pattern N_δ has $n(n-1)$ points. N_δ is central-symmetric in $W \oplus \check{W} = \{z : z = x - y, x \in W, y \in W\}$ since it contains $x - y$ as well as $y - x$. Figure 7.1 shows an example of a small point pattern and the corresponding pattern of difference points.

For the intensity function of the difference pattern a formula can be given in which λ and $g(r)$ appear. Consider the mean number $\Lambda_\delta(r)$ of points of N_δ in the disc $b(o, r)$. The derivative $\Lambda'_\delta(r) = \lambda_\delta(r)$ can be calculated as

$$\lambda_\delta(r) = \lambda^2 \overline{\gamma}_W(r) g(r) \qquad \text{for } r \geq 0,$$

where $\overline{\gamma}_W(r)$ is the set covariance of the window W.

Tanaka et al. (2008) then assume that N_δ can be approximated well by an inhomogeneous Poisson process with intensity function $\lambda_\delta(r, \theta)$. This function depends on θ through $g(r)$, and the authors use the maximum likelihood approach for inhomogeneous Poisson processes to estimate θ. Using (3.4.4) and polar coordinates, the corresponding log-likelihood function becomes

$$\ln L(\theta) = \sum_{x, y \in N \cap W} \ln \left(\lambda_\delta(\|x - y\|, \theta) \right) - \int_0^R \lambda_\delta(r, \theta) db_d r^{d-1} dr$$

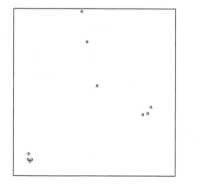
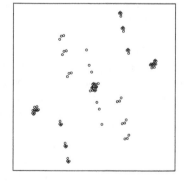

Figure 7.1   A pattern of 10 points and the corresponding pattern of difference points.

for $R = \min\{r: \overline{\gamma}_W(r) = 0\}$. Tanaka et al. (2008) simplify the likelihood function further and use numerical methods to compute the integral and find the maximum. They show that the method works well for Neyman–Scott cluster processes, for which the exact maximum likelihood method usually cannot be applied. The same approach can also be used for other Cox processes.

Other approximate likelihood methods include the pseudo-likelihood method (see Besag, 1975, 1978; Møller and Waagepetersen, 2004, p. 171) and its improvement by Huang and Ogata (1999); see also Section 3.6.

In classical statistics as well as in point process statistics often several potential models with different parameters and numbers of parameters may initially be considered appropriate for a specific data set. A common approach to the comparison of different models where parameter estimation has been done using maximum likelihood methods is the Akaike information criterion (Akaike, 1974). The *AIC* is defined as

$$AIC = -2 \cdot \ln(L(\hat{\theta})) + 2k,$$

where $L(\hat{\theta})$ is the likelihood function evaluated at the maximum likelihood estimator $\hat{\theta}$, and k is the number of independently fitted parameters. Tanaka et al. (2008) use the AIC for model comparison.

7.2.2   Method of moments

The method of moments has not been very popular in classical statistics in general as it lacks some of the desirable properties of the maximum likelihood method. Nevertheless, it often provides very good estimators that are unbiased or ratio-unbiased and consistent. It has many applications in the context of spatial point processes, especially when the likelihood function is not available.

Note that the term 'method of moments' is used here somewhat loosely since the approaches described here are all based on the same general idea but this idea is applied to moments or moment-measure-related characteristics as well as to other summary characteristics that are not moment-related. The general idea is to find parameters that minimise the difference between a 'suitable' summary characteristic S that is known analytically (or from simulations) and the summary characteristic $\hat{S}$ as estimated from the data. It is important that S depends on the unknown parameter θ; to emphasise this dependence, the characteristic is denoted by S_θ. The methods discussed in Chapter 3, 4 and 5 are then applied to the data to yield an empirical $\hat{S}$. The value θ for which S_θ and $\hat{S}$ are 'as similar as possible' is used as an estimator. The term 'as similar as possible' means here 'similar in the sense of a specific approximation method', such as the least-squares approach.

As indicated by the vague expression 'suitable', different summary characteristics S may be used. Which of these is deemed suitable depends on the context. A first criterion for the choice of summary characteristics is often whether a formula for

S_θ is known. However, if this is not the case simulation approaches may be used instead. Another criterion should be that S_θ is sensitive to variation in θ.

In its simplest form, the method of moments is applied to numerical summary characteristics. If, for example, the intensity λ is a model parameter, $\hat{\lambda}$ is its estimator as in (4.2.10). Similarly, if the hard-core distance r_0 is a model parameter, as in Gibbs hard-core or Matérn hard-core processes, the estimator is simply the minimum inter-point distance in the sample pattern, see (4.2.46).

In many other examples the method of moments is based on a functional summary characteristic, i.e. the S above is a function $S(r)$. Sometimes it is enough simply to plot $\hat{S}(r)$ and identify specific points, e.g. cusp points or points where $S_\theta(r)$ becomes constant. This approach may be used to find estimators of particular distances such as r_{corr}. In the context of Neyman–Scott processes the radius of the clusters may be found in this way. The parameters r_1 and β of a Strauss process may be estimated using the cusp-point method.

Typically, however, the method of moments for functional summary characteristics usually applies a least-squares approach that is often referred to as the *minimum contrast method*. This is based on the simple idea of minimising

$$\Delta(\theta) = \int_{s_1}^{s_2} |\hat{S}(r) - S_\theta(r)|^\beta \mathrm{d}r \qquad (7.2.1)$$

with respect to θ. The value of θ that yields the minimum is the estimator $\hat{\theta}$.

Here, the parameters s_1, s_2 and β as well as the summary characteristic $S(r)$ can in principle be chosen arbitrarily; often $\beta = 2$ is used. In the literature most applications use either

$$S(r) = L(r) \quad \text{or} \quad S(r) = H_s(r).$$

For these summary characteristics, the integral limits can be chosen as $s_1 = 0$ and $s_2 = s$ where s is some suitable maximum distance as on p. 95. In Stoyan and Stoyan (1996), Møller and Waagepetersen (2004, p. 183) and Taylor et al. (2001) the pair correlation function $g(r)$ is used as the summary characteristic $S(r)$. The reader is advised to follow this example in order to avoid dependence among the residuals in the sum of squares. However, because of the well-known inaccuracy of $\hat{g}(r)$ for small r, a positive value for s_1 should be chosen, somewhere in the region of $\hat{m}_D$, the estimated mean nearest-neighbour distance, or of $\hat{r}_0$, the estimated hard-core distance. For finite point processes $L_{\mathrm{fin}}(r)$ (see p. 131) may be a good choice for $S(r)$.

In practice, the integral in (7.2.1) is replaced by a sum,

$$\int_{s_1}^{s_2} |\hat{S}(r) - S_\theta(r)|^\beta \mathrm{d}r \approx \sum_{i=0}^{k} |\hat{S}(\varrho_i) - S_\theta(\varrho_i)|^\beta \qquad (7.2.2)$$

with $\varrho_0 = s_1$, $\varrho_i = r_1 + i\delta$, $\varrho_k = s_2$ and $\delta = \frac{s_2 - s_1}{k}$ for some integer k.

Jolivet (1986) and Heinrich (1992, 1993) investigate the statistical properties of $\hat{\theta}$ for special cases. In reasonable cases the minimum contrast estimator is 'consistent' in the sense of probability theory, i.e. it converges to the true value as the size of the window W increases until it is the whole space.

7.2.3 Trial-and-error estimation

If all else fails, parameters can be estimated by trial and error, as demonstrated by Ripley (1977), even if there is a possibility that the results it yields are of dubious quality. The idea is to try several estimates $\hat{\theta}_1, \hat{\theta}_2, \ldots$, and use a goodness-of-fit test from Section 7.4 with each of these. The first estimate $\hat{\theta}_i$ for which the hypothesis that the model containing $\hat{\theta}_i$ fits the data is accepted, is used as the estimate. An extension of this method may even be used to construct confidence intervals for θ, using the well-known duality between significance tests and confidence intervals: the set of all θ_i for which the test leads to acceptance determines such an interval.

Clearly, the dimension of θ, i.e. the number of parameters, should be small.

Example 7.1.   Phlebocarya *pattern: fitting a cluster process model*
After a successful statistical analysis of the pattern of 207 *Phlebocarya filifolia* plants in a 22×22 m square (see Figure 1.4) in Chapters 4 and 6, a cluster process model is now fitted to the data. Indeed, the pattern looks like a sample from a cluster process and the pair correlation function in Figure 4.19 also suggests this. The fact that in Example 6.2 a Cox process could be fitted to the pattern may trigger the idea of also trying a simpler cluster model. This model is rather primitive and likely to be too simplistic. It is discussed here mainly as an illustration. If such a model could really be fitted to the data this may lead to a clearer statement on the range of correlation in the pattern. Figure 4.19 does not give a clear answer on the question of the range of correlation: it may be that the range is smaller than 1 m and that the values of $g(r)$ larger than 1 for $r > 1$ m have to be considered irrelevant or that the range is large, up to 10 m, but for distances between 1 and 10 m the correlations are very weak.

In this example, a Matérn cluster process is used as the cluster process model, which depends on the parameters cluster radius R, mean number of points $\bar{c}$ per cluster, and intensity λ, as explained on p. 376. Two estimation methods are applied: the minimum contrast method with $S(r) = g(r)$, $s_1 = 0$ and $s_2 = 3.0$, yielding the estimates $\hat{R} = 0.12$ m, $\hat{\bar{c}} = 0.078$; and approximate maximum likelihood method, yielding $\hat{R} = 0.19$ m, $\hat{\bar{c}} = 0.134$. For both models the intensity estimate is $\hat{\lambda} = 0.428$ m^{-2}.

Although the methods yield similar results, in the following the minimum contrast estimates are preferred as their behaviour in goodness-of fit tests is a little better. Figure 7.2 shows the empirical pair correlation function and envelopes from 99 simulations of the fitted model, which show good agreement. This was to be expected since a second-order characteristic was used for parameter estimation. So

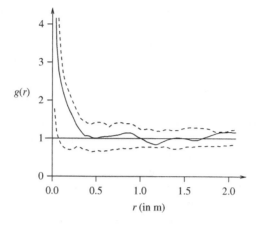

Figure 7.2 Empirical pair correlation function for the *Phlebocarya* pattern and the envelopes from 99 simulations of the fitted Matérn cluster process corresponding to the minimum contrast estimates.

using a similar characteristic to assess the fit of the model is likely to indicate a (potentially spurious) fit and should not be considered a useful goodness-of-fit test, but see Example 7.2.

Note that for this example the estimates of R deviate greatly from values which visual inspection of the empirical pair correlation function in Figure 4.19 may suggest.

In many applications, different estimation methods are combined: different components of the parameter vector θ are estimated by different methods, often in an iterative way. Consider, for example, a pair potential parameter $\theta = (\theta_1, \theta_2)$ where θ_1 is the range of interaction and θ_2 determines the strength of interaction. Then θ_1 may be estimated by the cusp-point method and θ_2 by the partial maximum likelihood method given θ_1. Another example is the profile likelihood method explained on p. 167 in Section 3.6.

7.3 Variance estimation by bootstrap

In all areas of statistics, parameter estimates are difficult to interpret without additional information on their accuracy. Accuracy is usually characterised by mse, estimation variance or confidence interval, but in point process statistics it is rather difficult to calculate these. Simulation and resampling methods are often used in practice to overcome this problem. A very popular method is the bootstrap; see Davison and Hinkley (1997) and Manly (2004) for details.

The bootstrap approach is based on empirical data, but in some cases simulated data are also used for the same purpose. The method typically does not make any distributional assumptions. In classical statistics, a large number of new artificial samples are resampled from the data $x_1, x_2, \ldots$ with replacement. These samples are analysed separately and the fluctuations of the estimates are considered and used in order to characterise the variability of the estimator of interest.

The analogue of the artificial data in point process statistics are independent point pattern *samples*, rather than samples that consist of points from one original pattern as this is a misuse of the bootstrap idea; see the discussion in Martínez and Saar (2002, p. 91) and Snethlage (1999). Methods discussed in the point process literature, where new patterns are constructed by 'block resampling', i.e. random subsamples of the pattern in W are taken and combined to form new patterns within W, may be useful for data on the real axis, in time series analysis. However, for spatial patterns these are unsuitable as they ignore existing correlations along the edges of the subwindows and generate configurations that do not exist or may even be impossible in the original pattern.

Therefore, in point process statistics bootstrapping can only be successfully used in the context of replicated patterns, where a series of point patterns are given in windows $W_1, \ldots, W_k$, which are of equal size and shape and k is not very small; see Diggle et al. (2000) and Schladitz et al. (2003). These windows may be subwindows of a large window or separate windows which observe the same spatial phenomenon. It is important that the patterns in the W_i are (practically) independent.

This section provides a sketch of bootstrapping for the estimation of the variances of estimators of the intensity λ and the pair correlation function $g(r)$ of replicated stationary point processes.

The intensity estimator (4.2.10) yields the values $\hat{\lambda}_i$ for $i = 1, 2, \ldots, k$ for the windows W_i. Equation (4.7.1) then yields the global value $\overline{\lambda}$,

$$\overline{\lambda} = \frac{1}{k} \sum_{i=1}^{k} \hat{\lambda}_i.$$

The aim of bootstrapping is to evaluate the precision of $\overline{\lambda}$ based on the values $\hat{\lambda}_i$. A new sample of k $\hat{\lambda}_i$-values is generated by randomly resampling from the $\hat{\lambda}_i$ with replacement. If $k = 5$, it may for example consist of the values $\hat{\lambda}_2, \hat{\lambda}_2, \hat{\lambda}_4, \hat{\lambda}_5$, and $\hat{\lambda}_5$ or $\hat{\lambda}_1, \hat{\lambda}_2, \hat{\lambda}_3, \hat{\lambda}_3$ and $\hat{\lambda}_3$. These values are used to calculate a new global $\overline{\lambda}$. The whole procedure is repeated m times to obtain m global means $\overline{\lambda}_1, \overline{\lambda}_2, \ldots, \overline{\lambda}_m$. The corresponding sample variance s_λ^2 is used as an approximation of the estimation variance σ_λ^2 of the intensity.

A confidence interval for λ results from rearranging the values $\overline{\lambda}_1, \overline{\lambda}_2, \ldots, \overline{\lambda}_m$ in increasing order and using the values with indices closest to $\frac{\alpha}{2} m$ and $(1 - \frac{\alpha}{2})m$ as the bounds of a confidence interval for λ of level $1 - \alpha$. (If $\alpha = 0.05$ and $m = 200$ the numbers are 5 and 195.)

Using a similar approach, the hypothesis $H_0 : \lambda_1 = \lambda_2$ i.e. that the intensities of two point processes are the same, can be tested by simulation. If the null hypothesis H_0 is rejected the data show evidence against the hypothesis of equal intensities.

Consider k_1 windows which yield the intensity estimates $\hat{\lambda}_{11}, \ldots, \hat{\lambda}_{1k_1}$ and k_2 with $\hat{\lambda}_{21}, \ldots, \hat{\lambda}_{2k_2}$ corresponding to λ_1 and λ_2 yielding $\overline{\lambda}_1$ and $\overline{\lambda}_2$ respectively as well as $\Delta\lambda = \overline{\lambda}_1 - \overline{\lambda}_2$. This $\Delta\lambda$ is compared to m values $\Delta\lambda_1, \ldots, \Delta\lambda_m$ which are obtained by resampling. Both samples of $\hat{\lambda}_{1i}$ and $\hat{\lambda}_{2j}$ are merged to form a unique sample of $k_1 + k_2$ values. Then m new pairs of samples of k_1 and k_2 values are generated from this sample, by randomly resampling with replacement, and labelled as 1 and 2 even though they have been drawn from the same set of λ-values. Then the differences of λ_1- and λ_2-estimates are calculated, yielding $\Delta\lambda_1, \ldots, \Delta\lambda_m$. H_0 is rejected if $\Delta\lambda$ has an extreme position in the series of ordered values $\Delta\lambda_i$. If the error probability is α, then values at positions smaller than $\frac{\alpha}{2}m$ and larger than $(1 - \frac{\alpha}{2})m$ are considered extreme.

The same procedure as for λ may be applied for all r separately to the pair correlation function $g(r)$, i.e. instead of $\hat{\lambda}_i$ the $\hat{g}_i(r)$ are used now for all values of r of interest.

Another common method is based on Monte Carlo simulation of models with estimated parameters. It is sometimes called the *parametric bootstrap*, although it is not a conventional bootstrap method. Assume that θ and $\hat{\theta}$ are the model parameter and its estimator, respectively. The model with parameter $\hat{\theta}$ is simulated m times independently, and the estimators $\tilde{\theta}_k$ are determined for $k = 1, 2, \ldots, m$ for the resulting data. The variance of these values is used as an approximation of the variance of the estimator $\hat{\theta}$; see Efron and Thibshirani (1993) and Givens and Hoeting (2005). This method can be applied regardless of the method that was used to derive $\hat{\theta}$. The parametric bootstrap is clearly computationally intensive, as it requires both simulation and parameter estimation for a large number of samples.

7.4   Goodness-of-fit tests

This section discusses goodness-of-fit tests for point processes, i.e. tests that are analogues of the Kolmogorov–Smirnov test familiar from classical statistics. For all these tests, the null hypothesis is simply H_0: 'the model fits the data'. In point process statistics these tests are usually based on simulations, with the exception of some tests that are used in the context of the Poisson process, as discussed in Section 2.7. These tests are special cases of Monte Carlo tests as described in Chapter 4 of Davison and Hinkley (1997); see also Ripley (1988) and Robert and Casella (2005).

Two main approaches are discussed here. The first is a method that has been very popular since its introduction in Ripley (1977), even though it does not produce a formal significance test with known and predefined error probability α. For this reason, Loosmore and Ford (2006) refer to the method as 'incorrect'. However, this book warmly recommends the approach in applications.

7.4.1   Envelope test

This method is based on some functional summary characteristic $S(r)$ such as $g(r)$, $L(r)$, $L_{\text{fin}}(r)$, $H_s(r)$ or $D(r)$, as discussed in Chapters 3, 4 and 5. The idea of the test is to compare the empirical summary characteristic estimated from a point pattern in the observation window W to estimates of the summary characteristic for simulations from the model using the estimated parameters generated in the same window. The model is simulated k times and the estimate of $S(r)$, $\hat{S}_i(r)$ for $i = 1$, $2, \ldots, k$, is determined for each sample. Then the extreme values

$$S_{\min}(r) = \min_{(i)} \hat{S}_i(r) \quad \text{and} \quad S_{\max}(r) = \max_{(i)} \hat{S}_i(r)$$

are determined. Finally, three curves showing $S_{\min}(r)$, $\hat{S}(r)$ and $S_{\max}(r)$ are plotted, as in Figure 7.3 as well as in many other figures in this chapter. Since $S_{\min}(r)$ and $S_{\max}(r)$ are envelopes of the $\hat{S}_i(r)$, the name 'envelope method' is often used, which leads to pointwise confidence bands.

The values $k = 19$ and $k = 99$ are often used for k, where $k = 19$ may appear to be rather small. On the other hand, the choice $k = 999$ might be considered as rather large. In the literature, values of k satisfying $\alpha k \geq 5$ are recommended, where α is the error probability.

If the inequality

$$S_{\min}(r) \leq \hat{S}(r) \leq S_{\max}(r) \tag{7.4.1}$$

holds for all r, the model is accepted, otherwise it is rejected. If the model is rejected, the values r for which (7.4.1) is violated provide some information on the nature and reason for the deviations of the data from the model.

This test is often regarded and interpreted as a significance test. Indeed, if a fixed $r = r^*$ has been chosen prior to the simulation, the test which rejects the model if the inequality (7.4.1) is not satisfied for $r = r^*$ is a correct simulation test. Its error probability in one-sided testing is $1/(k+1)$. Thus $k = 19$ corresponds to $\alpha = 0.05$ and $k = 99$ to $\alpha = 0.01$. In a two-sided test these values of α should be multiplied by 2. However, since 'all' r are considered simultaneously, the probability of rejecting H_0 is increased and the true error probability is larger than 0.05 and 0.01, respectively. On the other hand, it is to be expected that a test based on single r^* is rather conservative, i.e. the null hypothesis is rather unlikely to be rejected. This is because the model is simulated with parameters that have been estimated from the same data as those that were used for the test. Note that this problem has also been discussed in classical statistics in the context of the Kolmogorov–Smirnov test; see Conover (1999, pp. 443 and 448) and Armitage et al. (2001, p. 373).

To address this issue it is recommended to carefully choose the summary characteristic $S_{\text{test}}(r)$ used in the test. It should be of a different nature than $S_{\text{est}}(r)$, which is used to estimate the model parameters. A suitable approach, for example, is to estimate θ through $S_{\text{est}}(r) = g(r)$ (by the minimum contrast or approximate maximum likelihood methods) and to use $D(r)$ or $H_s(r)$ as $S_{\text{test}}(r)$ for the test rather than another second-order characteristic; see the discussion in Diggle (2003, p. 89). The rejection of a null hypothesis that is based on a test that applies the same summary characteristic as for parameter estimation casts particular doubt on the null hypothesis.

This book suggests accepting that the classical choices of $\alpha = 0.05$ and 0.01 are only conventions and are not given by first principles. In other words, in the context of point processes a point process convention may be to work with $k = 19$ and $k = 99$ and to interpret these as '$\alpha = 0.05$' and '$\alpha = 0.01$', respectively.

Note that the test above can be modified and larger or smaller values can be used, rather than the minimum and maximum values. For $\alpha = 0.05$, $k = 999$ may be chosen and $S_{\min}$ is replaced by the 25th of the ordered $\hat{S}_i(r)$ values and $S_{\max}$ by the 975th.

7.4.2 Deviation test

Simulation tests with an exact error probability α can be constructed based on the general recipe described on p. 54.

Each simulated pattern l is assigned a global deviation measure Δ_l, by analogy with D_l on p. 54, i.e.

$$\Delta_l = \max_{0 \leq r \leq s} |S_{\hat{\theta}}(r) - \hat{S}_l(r)|$$

or

$$\Delta_l = \int_0^s |S_{\hat{\theta}}(r) - \hat{S}_l(r)|^\beta dr \qquad \text{for } l = 1, 2, \ldots, k,$$

where $S_{\hat{\theta}}(r)$ is the theoretical $S(r)$ with the estimated parameter $\hat{\theta}$ and s a maximum r-value as on p. 96. In practice, the integrals are of course replaced by finite sums.

The values Δ_l and the Δ-value for the original data,

$$\Delta = \max_{0 \leq r \leq s} |S_{\hat{\theta}}(r) - \hat{S}(r)|$$

or

$$\Delta = \int_0^s |S_{\hat{\theta}}(r) - \hat{S}(r)|^\beta dr,$$

are arranged in increasing order. If Δ has an extremely high position among these values, H_0 is rejected. For $\alpha = 0.05$ and $k = 19$, 99 and 999 the critical positions are those larger than 19, 95 and 950, respectively; for $\alpha = 0.01$ and $k = 99$ and 999 they are 99 and 990.

Summary characteristics commonly chosen as $S(r)$ are $L(r)$, $L_{\text{fin}}(r)$, $D(r)$ and $H_s(r)$. In accordance with common practice in classical statistics the density functions $g(r)$, $d(r)$ and $h_s(r)$ are not used here, because the estimation of $L(r)$, $D(r)$ and

$H_s(r)$ is more standardised than that of $g(r)$, $d(r)$ and $h_s(r)$. Again, as mentioned in the discussion of the envelope approach, a deviation test should be based on a different summary characteristic $S_{\text{test}}(r)$ than the one that was used as $S_{\text{est}}(r)$.

The observed P-value of the deviation test can be approximatively calculated as

$$\hat{p} = \frac{1 + \sum_{l=1}^{k} \mathbf{1}(\Delta_l > \Delta)}{k+1}. \tag{7.4.2}$$

Loosmore and Ford (2006) discuss the variation of $\hat{p}$ as a function of k.

Practical experience has indicated that the deviation test is probably less powerful than the envelope test. To rectify this, the deviation test may be improved by using two summary characteristics, $S_1(r)$ and $S_2(r)$, of different nature (e.g. $L(r)$ and $D(r)$) and constructing a combined deviation measure, given here for the max case,

$$\Delta = \max_{0 \leq r \leq s} |S_{1,\hat{\theta}}(r) - \hat{S}_1(r)| + \max_{0 \leq r \leq s} |S_{2,\hat{\theta}}(r) - \hat{S}_2(r)|.$$

Deviation tests based on simulation approaches as before may be also applied if an explicit formula for $S_{\hat{\theta}}(r)$ is not known. Then m additional independent patterns with the parameter $\hat{\theta}$ are generated, the corresponding $S_l(r)$ are calculated and averaged over l. Diggle (2003, p. 89), recommends using the k samples above to also derive an estimate $S_{\hat{\theta}}(r)$.

Example 7.2.   Phlebocarya *pattern: testing a Matérn cluster process hypothesis*
This example continues the analysis of the *Phlebocarya* pattern in Example 7.1 and tests the hypothesis that the Matérn cluster process with the minimum contrast estimates given in Example 7.1 fits the data.

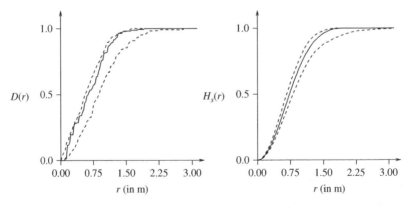

Figure 7.3   Empirical nearest-neighbour distance d.f. $\hat{D}(r)$ (solid line, left) and spherical contact d.f. $\hat{H}_s(r)$ (solid line, right) for the *Phlebocarya* pattern and envelopes from 99 simulations of the Matérn cluster process with minimum contrast estimates (dashed lines).

Since the parameter estimation is based on the second-order characteristic $K(r)$, different summary characteristics are used for the test, $D(r)$ and $H_s(r)$. Figure 7.3 shows the empirical d.f. $\hat{D}(r)$ and the envelopes resulting from 99 simulations of the Matérn cluster process with the two parameter sets derived in Example 7.1. The empirical nearest-neighbour distance d.f. $\hat{D}(r)$ is completely within the envelopes for both sets and the test confirms that the models with the very small values of $\hat{R}$ fit the data. For the spherical contact d.f. $H_s(r)$ the situation is the same.

The result of the test is that the Matérn cluster model is also acceptable. However, the log-Gaussian Cox process performs better and seems to be a more realistic model. The Matérn cluster models suggest a very short range of correlation of 20 cm. This may show that the pattern is globally close to a Poisson process; however, an L-test of the CSR hypothesis leads to rejection.

Example 7.3.   Testing a Gibbs process hypothesis for the Spanish towns
This example continues the analysis of the Spanish town pattern in Example 3.14 and tests the hypothesis that the hard-core Strauss process with the parameters obtained via the maximum likelihood method given in Example 3.14 fits the data. Since parameter estimation was not based on second-order characteristics the test applies the finite L-function. The parameters of the model are $\hat{r}_0 = 0.83$ miles, $\hat{r}_{max} = 3.5$ miles, $\hat{\alpha} = 2.08$ and $\hat{\beta} = 0.76$. Here the simulation is performed using the random variable $N(W)$ and that particular value of $\hat{\alpha}$ which corresponds to the mean value $\mathbf{E}(N(W)) = 69$, the number of towns in W. (The value of $\hat{\alpha}$ was found by trial-and-error and simulation.) Figure 7.4 shows the empirical L-function and the envelopes resulting from each 99 simulations of the Strauss process. Since the empirical L-function is completely within the envelopes, the Strauss process hypothesis is accepted. By the way, for the distance summary characteristics the result is the same.

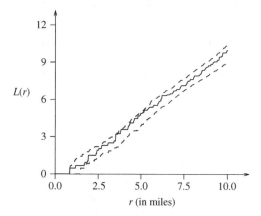

Figure 7.4   The empirical finite L-function for the Spanish towns pattern (solid line) and the envelopes from 99 simulations of the hard-core Strauss process (dashed lines).

7.5   Testing mark hypotheses

7.5.1   Introduction

The discussion in Section 7.4 has not explicitly covered examples with marked point patterns, but in the context of marked point processes similar tests to those described above may also be used, based on suitable summary characteristics for marked point processes. Two fundamental hypotheses are particularly important:

(1) the marks are independent;

(2) the marks result from geostatistical marking, i.e. the random field model of Section 5.1.3 and 6.8 can be applied.

In the first case, the aim may be to find out whether the marks are positively or negatively correlated and in the second case, whether marks and points are correlated, e.g. whether marks depend on local point density. In some applications, the analysis may reveal that the marks can be considered independent. This might appear to be an unexciting result and be disappointing in a specific context where scientific theory would have suggested some correlations. Nevertheless, showing independence among the marks is also an important and useful result. If the marks turn out to be spatially correlated, a further question concerns the interaction among points and marks or whether the simple random field model without interaction is suitable.

The test procedures discussed in the following are all non-parametric approaches. They use functions of the type $L_t(r)$ with a suitable test function t as introduced in Section 5.1.3, e.g. $L_{mm}(r)$, $L_{m\cdot}(r)$ or $L_{ij}(r)$.

7.5.2   Testing independent marking, test of association

All tests of independence hypotheses for marked point processes apply some form of resampling. This means that k new samples of marked point patterns with 'independent' marks are generated and the corresponding summary characteristics are compared to those of the original pattern. This is explained here in detail for L_t-functions as defined in Section 5.4.1.

Qualitative marks

Tests for qualitative marks are discussed here only for the most important bivariate case. This implies that, in a specific pattern, two marks i and j are selected from among the m different qualitative marks for an analysis of the correlation of the i- and j-marks, and points of type i a assigned the new mark 1 and points of type j the mark 2. Or, more generally, two groups of marks $(i_1, \ldots, i_{m_1})$ and $(j_1, \ldots, j_{m_2})$ are selected and all points of the first group get mark 1 and those of the second group mark 2. For example, in a forest conifers and deciduous trees may be considered, while the individual species are ignored.

The test considered here is based on the L_{12}-function as introduced in Section 5.3. Note that there are two different interpretations of the null hypothesis of 'independent marking'. These need to be carefully distinguished and are discussed below: random labelling, and random superposition or population independence.

Random labelling means that the points in an originally non-marked point process are independently marked 1 or 2. Goreaud and Pélissier (2003) therefore refer to this as *a posteriori* marking. Typical examples of this may be forests in which the trees are infected by some disease or damaged by wind or frost (mark 1) or not (mark 2).

Random superposition means that a priori there are two independent point patterns in the same window W, one consisting of points of type 1, the other of type 2. These two patterns are combined, yielding the bivariate pattern. Typical examples include plant communities where the locations of plants from the different species within W result from different dispersal mechanisms.

The summary characteristics for the two cases show a different but characteristic behaviour, as shown in Table 7.1. The quantities mentioned in the table are mainly defined in Section 5.3.2. $D_{12}(r)$ is the d.f. of the random distance from the typical type 1 point to its nearest type 2 neighbour and $H_{s,2}(r)$ is the spherical contact d.f. of the subprocess of type 2 points.

If the hypothesis of 'independent marking' is to be tested, the right null hypothesis has to be chosen first. In some cases, one of the two null hypotheses is completely inadequate (see Example 7.4) and thus can never be accepted. In many cases the choice is clearly determined by the study question and the origin and nature of the data. Below, a number of examples are used to illustrate this point. In more complicated cases where it is not clear which one of the two hypotheses seems more appropriate, statistical estimates of the pair correlation functions $g_{11}(r)$, $g_{22}(r)$ and $g_{12}(r)$ may be helpful. Certainly, $L_{11}(r)$, $L_{22}(r)$ and $L_{12}(r)$ may also be used,

Table 7.1 Summary characteristics with index '12' and simulation method for random labelling and random superposition. This table and the structure of the text referring to it were inspired by Goreaud and Pélissier (2003).

Random labelling	Random superposition
$L(r) = L_{11}(r) = L_{22}(r) = L_{12}(r)$	$L_{12}(r) = r$
$g(r) = g_{11}(r) = g_{22}(r) = g_{12}(r)$	$g_{12}(r) \equiv 1$
$p_{12}(r) \equiv 2p_1 p_2$	$p_{12}(r) \equiv 2p_1 p_2$
$p_{ii}(r) \equiv p_i^2$	$p_{ii}(r)$ given by (5.3.18)
$D_{12}(r)$: (5.2.30)	$D_{12}(r) = H_{s,2}(r)$
reallocation	shifting

but due to the cumulative nature of the *L*-functions their use is not recommended for exploration; see the discussion in Section 4.3. Further, it is important also to assess $g_{11}(r)$ and $g_{22}(r)$, and not just $g_{12}(r)$.

Under the null hypothesis of random labelling, all three functions are equal and their estimates will be similar. For random superposition, however, $g_{12}(r) \equiv 1$ but $g_{11}(r)$ and/or $g_{22}(r)$ may differ substantially and this should be also reflected in statistical estimates of the three functions. If it is not clear which of the hypotheses is suitable for a given data set, both null hypotheses may be tested and the above functions may be used for clarification. Note that the two cases cannot be distinguished for Poisson processes.

Once the appropriate choice of null hypothesis has been determined, a test of this null hypothesis may be carried out. This is again a simulation test, where *k* marked point patterns are generated with new marks for each of these. If the null hypothesis is a random labelling hypothesis, *random reallocation* is used to generate these new marks, i.e. the marks are permuted while the points are fixed. This implies that for all simulated patterns the numbers of points of type 1 and type 2 are the same as for the data.

If the null hypothesis is a random superposition hypothesis, *random shifts* are used. Assume that *W* is a rectangle or parallelepiped. The pattern of points of type 1 is fixed and the entire pattern of points of type 2 is shifted. If these points, which are x_i with $m(x_i) = 2$, are re-denoted by $y_1, y_s, \ldots, y_m$, a 'shift' implies

$$y_i \rightarrow y_i + u,$$

where *u* is a random uniform location in *W*, the same for all *i*, but different for different *l*. The operation $y_i \rightarrow y_i + u$ is based on an idea similar to periodic edge-correction (see p. 184).

Envelopes $L_{12,\min}(r)$ and $L_{12,\max}(r)$ are constructed based on the *k* simulated patterns. If the empirical L_{12}-function derived from the data $\hat{L}_{12}(r)$ is not completely within the envelopes or the corresponding confidence band the respective independence null hypothesis is rejected.

If one or both null hypotheses have been rejected one should reconsider the three partial pair correlation functions $g_{11}(r)$, $g_{22}(r)$ and $g_{12}(r)$ in order to find the right explanation. Considering only $\hat{L}_{12}(r)$ and the envelopes may be risky, as demonstrated in Goreaud and Pélissier (2003). It can lead to wrong conclusions as to the reason for rejection, i.e. on the nature of correlations among the two types of points.

Example 7.4.   Testing the independence of amacrine on- and off-cells
Figure 1.2 shows the bivariate pattern of on- and off-cells (type 1 and type 2 points) and Figure 5.5 the partial pair correlation functions $g_{ij}(r)$. The functions $g_{11}(r)$ and $g_{22}(r)$ are quite similar and seem to correspond to point processes with some

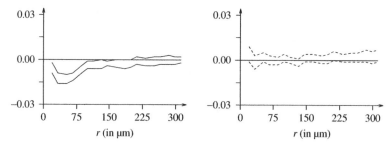

Figure 7.5   The envelopes $L_{12,\max}(r) - \hat{L}_{12}(r)$ and $L_{12,\min}(r) - \hat{L}_{12}(r)$ resulting from 99 simulations obtained by random reallocation (solid lines) and random shifts (dashed lines), for the amacrine cells. Since the null line is outside the confidence band corresponding to random reallocation for many values of r, the hypothesis of random marking is rejected. The fact that the null line is completely within the confidence band corresponding to random shifts suggests that the two subpatterns of type 1 and type 2 points are independent.

tendency towards regularity. In contrast to this, $g_{12}(r)$ fluctuates randomly around 1. The statistical analysis in Example 5.5 suggests that the random superposition model is appropriate.

Here, the random labelling hypothesis is also considered, but only to point out the difference between the two hypotheses. In this particular context it is clear from the start that this hypothesis will not be accepted. The simulation tests yield the expected results. Figure 7.5 shows the differences of the empirical L_{12}-function and the L_{12}-envelopes from each 99 simulated patterns obtained by random allocations and shifts. Clearly, the random superposition hypothesis is accepted. In contrast, the random labelling hypothesis is rejected, since the null line is outside the confidence band for many values of r. Apparently, random allocations completely destroy the correlation structure in the pattern, in particular that in the patterns of type 1 and type 2 points. Consequently, the independence of the systems of type 1 and type 2 points appears to be proved and the random superposition model seems to be the appropriate model for the amacrine cells.

Example 7.5.   *Frost shake in an oak forest*

This example reconsiders the distribution of trees damaged by frost shake in a young oak forest (*Quercus petraea*) at Allogny in France, as shown in Figure 7.6. 'Frost shake is a split in a tree trunk, produced by the interaction of frost and sun, that leads to a lowering of timber quality' (Goreaud and Pélissier, 2003). Visual inspection of Figure 7.6 shows that this example presents a difficult statistical problem as there is only some weak clustering of damaged trees. In the following, the question whether the pattern exhibits independent or dependent marking is addressed.

Figure 7.7 shows the three partial pair correlation functions $g_{11}(r)$, $g_{21}(r)$ and $g_{22}(r)$. There are no clear differences between the three functions. $g_{12}(r)$ is only

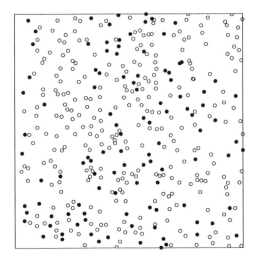

Figure 7.6   Positions of 392 oak trees in a 100×100 m square at Allogny, France. The pattern consists of 285 sound (o, type 1) and 107 damaged (•, type 2) trees. Courtesy of F. Goreaud and R. Pélissier. Reproduced by permission of Opulus Press.

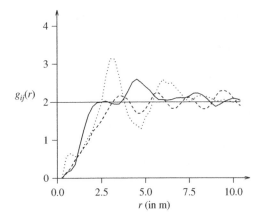

Figure 7.7   The empirical partial pair correlation functions for the oak pattern: $g_{11}(r)$ (solid curve), $g_{12}(r)$ (dashed curve) and $g_{22}(r)$ (dotted curve). The curves show that the pattern is close to a randomly labelled pattern with some tendency to clustering of trees of damaged type 2.

smaller than $g_{11}(r)$ and $g_{22}(r)$ for r between 2.0 m and 3.0 m, which indicates some weak repulsion among the trees of type 1 and 2. Between 2.5 and 3.5 m $g_{22}(r)$ has larger values, which indicates some form of clustering of damaged trees.

In this example it is clear which null hypothesis is appropriate. The biological problem clearly dictates that the random labelling hypothesis is the right one. But even in the absence of biological background information it is clear that the random labelling hypothesis should be chosen, since the similarity of the $g_{ij}(r)$ indicates that there is some probability that the null hypothesis may be rejected and it is unlikely that the clusters have been generated by superposition. Hence, the random superposition hypothesis is tested here only to demonstrate the difference between the two hypotheses. Indeed, Figure 7.8 shows that it is rejected; for r-values smaller than 5.5 m the empirical L_{12}-function is too small. For the random labelling hypothesis the situation is not that clear: only for r-values around 5 m is $\hat{L}_{12}(r)$ smaller than $L_{12,\min}(r)$. Note that, using data from a larger window (125×180 m, containing the window considered here), Goreaud and Pélissier (2003) concluded that frost shake *is* a clustered phenomenon; damaged trees appear in clumps. This is a nice example that shows the strong impact of the choice of the observation window.

Example 7.6. *Testing the randomness of the distribution of oaks and beeches*
Figure 5.11 on p. 334 shows the mark connection functions $p_{ij}(r)$ which suggest some weak correlation between oaks (1) and beeches (2). Figure 5.6 shows the partial pair correlation functions, which do not indicate clearly which hypothesis for independence testing is appropriate here. Whereas $g_{11}(r)$ and $g_{22}(r)$ differ clearly ($g_{11}(r)$ looks like the pair correlation function of a hard-core process, $g_{22}(r)$ like that of a cluster process), $g_{12}(r)$ is smaller than 1 and does not indicate independence. Thus, both versions are considered in the following in an attempt to clarify the situation.

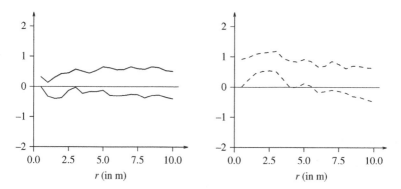

Figure 7.8 The envelopes $L_{12,\max}(r) - \hat{L}_{12}(r)$ and $L_{12,\min}(r) - \hat{L}_{12}(r)$ resulting from 99 simulations of random allocations (solid lines) and random shifts (dashed lines) for the Allogny oak forest. The envelopes indicate varying degrees of deviation from the independence hypothesis: for the random labelling hypothesis it is not clear, whereas the random superposition hypothesis (which is nonsense for these data) is clearly rejected.

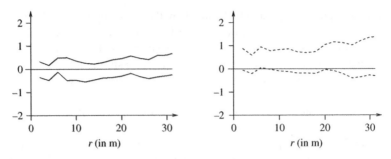

Figure 7.9   The envelopes $L_{12,\text{max}}(r) - \hat{L}_{12}(r)$ and $L_{12,\text{min}}(r) - \hat{L}_{12}(r)$ resulting from 99 simulations of random allocations (solid lines) and random shifts (dashed lines) for the pattern of oaks and beeches. The results of the two tests differ: whereas the null line is entirely within the confidence band for the random allocations, this is not true for the random shifts.

Figure 7.9 shows the differences in the empirical L_{12}-function and the L_{12}-envelopes resulting from 99 simulations, for random allocations and random shifts. For the random superposition test the null line is not completely within the confidence band. There are deviations for values around $r = 6.0\,\text{m}$ and hence the conclusion is, as expected, that the marks correlate weakly. For the random allocation method the result is different: the null line is completely within the confidence band. This might suggest the conclusion that there is no correlation between oaks and beeches; however, this has been proved wrong by the random shift test and the correlation functions. The discrepancy in the result based on the random allocation test may be due to the fact that relabelling does not destroy the global structure in the sample, which consists of clusters with many type 2 trees and few type 1 trees.

Note that for the example of palms and mounds on p. 335 the random labelling test also rejects the independence hypothesis, whereas the random shift test accepts it.

Quantitative marks

For quantitative marks, the test functions

$$t_2(m_1, m_2) = m_1 m_2 \quad \text{and} \quad t_4(m_1, m_2) = \frac{1}{2}(m_1 - m_2)^2$$

can be recommended for defining $L_{mm}(r)$ and $L_\gamma(r)$, respectively. If the marks are independent,

$$L(r) = L_{mm}(r) = L_\gamma(r).$$

For the k L_t-functions $L_{t,l}(r)$ which result from simulated samples for $l = 1, 2, \ldots,$ k, the envelopes based on $L_{t,\min}(r)$ and $L_{t,\max}(r)$ may be determined. An $\hat{L}_t(r)$ that is outside these envelopes indicates non-independence at specific distances. In graphical presentation it is useful to plot differences $L_{t,\max}(r) - \hat{L}_t(r)$ and $L_{t,\min}(r) - \hat{L}_t(r)$ and to assess whether the null line remains within the confidence band.

Marked point patterns with quantitative marks are resampled by *random reallocation* (or random marking or labelling), i.e. the points are fixed but are allocated new marks. There are two different approaches to this: permutation of the marks, and resampling the marks with replacement from the empirical mark distribution. Both approaches ignore existing mark correlations, but if the marks are independent these approaches do not change the mark correlation functions significantly. Note that permutation of the marks guarantees that all samples have the same empirical mark d.f.; this is not the case for independent marking using the empirical mark d.f.

The following examples discuss independence tests for two different data sets.

Example 7.7. Gold particles: testing the independence of diameter marks
Figure 5.18 on p. 342 shows the mark correlation function $k_{mm}(r)$ for the pattern of gold particles. For small r its values are below 1, which seems to indicate that there is some inhibition among the particles. The aim of this example is to test whether the relationship is significant using the random reallocation method and the L_{mm}- and L_γ-functions.

Figure 7.10 shows the differences in the empirical L_{mm}-function and the L_{mm}-envelopes resulting from 99 simulations. The null line is not completely within the confidence band, indicating that the marks are indeed correlated. For L_γ the situation is similar but much clearer, also shown in Figure 7.10. Considering the results of the tests here and of the exploratory analysis in Section 5.3.3, one may

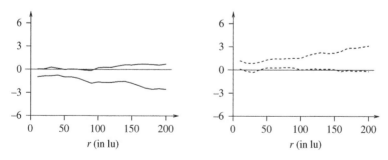

Figure 7.10   The envelopes $L_{mm,\max}(r) - \hat{L}_{mm}(r)$ and $L_{mm,\min}(r) - \hat{L}_{mm}(r)$ (solid lines) and $L_{\gamma,\max}(r) - \hat{L}_\gamma(r)$ and $L_{\gamma,\min}(r) - \hat{L}_\gamma(r)$ (dashed lines) resulting from 99 simulations with random allocation for the pattern of gold particles. Since the null line is not completely in the confidence bands, the marks are considered correlated.

conclude that there is a weak correlation among the diameter values with a tendency for diameters of particles at short distances to be similar.

Example 7.8.   Testing the independence of the marks in the spruce stand in the Tharandter Wald

Figure 6.19 shows the mark correlation function $k_{mm}(r)$ and the mark variogram for the 134 spruce trees as discussed in Example 6.4. The two correlation functions lead to different conclusions: whereas $k_{mm}(r)$ indicates independence, $\gamma_m(r)$ indicates that close pairs of big and small trees appear slightly more frequently than would have been expected for independent marks.

The random allocation method is used and L_{mm} and L_γ are considered. Perhaps slightly surprisingly, the null hypothesis of independent marking is not rejected by the envelope test with L_γ. The results are different for L_{mm}. Figure 7.11 shows the differences in the empirical L_{mm}-function and the L_{mm}-envelopes resulting from 99 simulations. The null line is not completely within the confidence band and the null hypothesis of independence is rejected. In conclusion, a weak correlation among the diameter marks can indeed be assumed.

The contradictory results for the spruce stand might be a result of the small size of the point pattern, which consists of only 134 trees. Probably the most realistic assumption is that the marks are simply independent.

7.5.3   Testing geostatistical marking

Tests of geostatistical marking are based on the $L_{m\cdot}$-function and random allocation, and strongly resemble tests of independent marking. Under geostatistical marking,

$$L(r) = L_{m\cdot}(r). \tag{7.5.5}$$

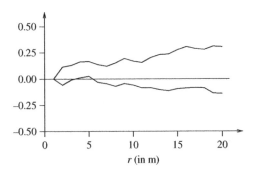

Figure 7.11   The envelopes $L_{mm,\max}(r) - \hat{L}_{mm}(r)$ and $L_{mm,\min}(r) - \hat{L}_{mm}(r)$ from 99 simulations resulting from random reallocation for the stand of spruces. Since the null line is not completely inside the confidence band, the marks are considered correlated.

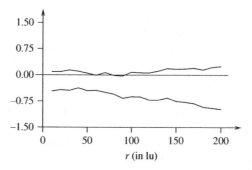

Figure 7.12   The envelopes $L_{m\cdot,\max}(r) - \hat{L}_{m\cdot}(r)$ and $L_{m\cdot,\min}(r) - \hat{L}_{m\cdot}(r)$ resulting from 99 simulations with random reallocation for the gold pattern. Since the null line is not completely within the confidence band, the hypothesis of geostatistical marking is rejected. But note that the upper envelope only slightly crosses the null line.

Example 7.9.   Gold particles: testing for geostatistical marking
This example continues Example 7.7 and uses the $L_{m\cdot}$-function to test for geostatistical marking. Figure 7.12 shows the differences the empirical $L_{m\cdot}$-function and the $L_{m\cdot}$-envelopes from 99 simulations with random reallocation. The null line is not completely within the envelopes, but nevertheless the hypothesis may be accepted, as a simulation test with an estimated variogram shows (see p. 349).

Example 7.10.   Testing for geostatistical marking in the pattern of spruce trees from the Tharandter Wald
This example continues Example 7.8 and tests for geostatistical marking, again using the $L_{m\cdot}$-function. Figure 7.13 shows the differences in the empirical $L_{m\cdot}$-function and the $L_{m\cdot}$-envelopes resulting from 99 simulations with random reallocation. The

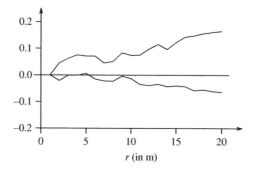

Figure 7.13   The envelopes $L_{m\cdot,\max}(r) - \hat{L}_{m\cdot}(r)$ and $L_{mm\cdot,\min}(r) - \hat{L}_{m\cdot}(r)$ resulting from 99 simulations with random reallocation for the spruces. Since the null line is not completely within the confidence band, the hypothesis of geostatistical marking is rejected. But note that the lower envelope only slightly crosses the null line.

null line is not completely within the confidence band, which indicates that the hypothesis of geostatistical marking might have to be rejected. Since independent marking is a special case of geostatistical marking, this result may suggest rejecting the independence hypothesis as well. However, for this example this result may be again due to the small size of the sample, and the authors tend to assume that the marking here is independent.

Note, by the way, that the pattern in this stand is the result of a forester's work. However, since the aim was to grow trees of a similar size the result is certainly a pattern of trees with similar diameters and only random fluctuations around the mean, where any effect of competition among the trees can hardly be identified.

Table 7.2 lists the results of mark independence tests for a number of examples.

Table 7.2   Results from different tests of independence hypotheses of marked point patterns by simulation tests.

Point pattern	L_t used	Envelope test	
Gold particles	L_{mm}	−	
	$L_{m\cdot}$	o	
	L_γ	−	
Spruces in Tharandter Wald	L_{mm}	o	
	$L_{m\cdot}$	o	
	L_γ	+	
Amacrine cells	L_{12}	+	tr
	L_{12}	−	pe
Oaks and beeches	L_{12}	o	tr
	L_{12}	+	pe
Oak frost shake	L_{12}	−	tr
	L_{12}	o	pe
Concrete sample (planar section)	L_{12}	o	tr
	L_{12}	+	pe
Palms and mounds	L_{12}	+	tr
	L_{12}	−	pe
Buttercup			
coverage	L_{12}	+	tr
stoniness	L_{12}	+	tr
Copper deposits	L_{12}	+	tr

− Rejection of independence hypothesis, empirical $L_t(r)$ for many r-values outside the envelope strip.
o The same, but only for few (≤ 3) r-values.
+ No rejection.
tr Random shift.
pe Random allocation.

Note that all tests are based on 99 replications (simulations) and 20 equidistant r-values with maximum distance approximately equal to half the smaller side length of the window W.

7.6 Bayesian methods for point pattern analysis

The Bayesian approach to statistical inference has recently become very popular, especially in the analysis of complex data sets. This is mainly a result of the development of Markov chain Monte Carlo methods, which has made it possible to apply Bayesian methods, since the early 1990s, to a much wider range of situations than had previously been possible. Not surprisingly, the Bayesian approach, together with MCMC methodology, has also made its way into point process statistics. In Section 3.2.1 the Bayesian approach was briefly mentioned in the context of statistics for the binomial distribution, and in Example 5.7 in the context of identifying mother points in a cluster process, but it may also be applied in many other situations, including intensity estimation, reconstruction of tessellations from point process data, and estimation of pair potential functions for Gibbs processes.

The Bayesian approach follows a different philosophy than the classical frequentist approach to statistical inference; the parameters and other unknown quantities, such as unobserved variates (covariates) and missing data, are considered random variables that are assumed to follow some probability distribution. In other words, the analysis involves formulating a model for both the data and the parameters. All uncertainties are expressed in terms of probability distributions and probabilities.

This implies the following modelling steps:

- As in a non-Bayesian context, the observations $\mathbf{x}$ are modelled by a probability distribution $f(\mathbf{x}|\theta)$. However, in the Bayesian setting the distribution is interpreted as a conditional distribution of the data $\mathbf{x}$ given the vector θ of all unknown and unobserved quantities in the model, including the parameters but possibly also unobserved variates and missing data.

- As mentioned above, the unknowns θ are further modelled with a *prior distribution* $\pi(\theta)$. This is a specific distribution that has to be chosen a priori and may be multivariate and may even have a complex dependence structure. This distribution reflects the investigator's beliefs on the unknown quantities prior to the analysis.

- The investigator's uncertainty on θ given the data $\mathbf{x}$ is expressed by the *posterior distribution*

$$\pi(\theta|\mathbf{x}) = \frac{\pi(\theta)f(\mathbf{x}|\theta)}{\int \pi(\theta)f(\mathbf{x}|\theta)\,d\theta}. \qquad (7.6.6)$$

Equation (7.6.6) is an application of *Bayes' theorem*. It relates unknown quantities given the data to the prior distribution and the distribution of the data.

- The posterior distribution is often difficult to handle and an analytical calculation of its characteristics is impossible in most cases due to the normalising integral in the denominator. However, MCMC methods can be applied to simulate from the posterior distribution. Refer to the literature (see below) for details on how this is done in practice.

- Once the posterior distribution has been determined, the analysis typically focuses very much on the characteristics of the marginal posterior distributions $\pi(\theta_i|\mathbf{x})$ for $\theta = (\theta_1, \ldots, \theta_p)$. This includes, in particular, the summary characteristics familiar from classical statistics, such as mean, standard deviation and quantiles.

There is an ample literature on the general principles of the use of MCMC in the context of Bayesian statistics; see Clark and Gelfand (2006), Gelman et al. (2004) and Gilks et al. (1996). Banerjee et al. (2004) concentrates on the Bayesian approach in spatial statistics, and Møller and Waagepetersen (2004) and some of the chapters in Lawson and Denison (2002) cover point process statistics.

To illustrate Bayesian methods, Example 3.4 is reconsidered here as this is based on a probability distribution and on a modelling approach that is familiar from classical statistics.

Example 7.11.   Analysis of fruit dispersal around an ash tree
Consider the estimation of the intensity function $\lambda(r)$ for the data in Example 3.4 on p. 122. In this example, the locations of fruit dispersed by an ash tree are modelled using an inhomogeneous Poisson process. The data set $\{n_i, r_i\}$ consists of trap counts n_i and of the distances r_i of 66 traps from the mother tree, where the traps are located around the tree as shown in Table 3.1.

Again, as in Example 3.4, the observations are modelled by a Poisson distribution

$$f(\{n_i\}|\{\mu_i\}) = \prod_{i=1}^{n} \frac{\mu_i^{n_i}}{n_i!} e^{-\mu_i}. \tag{7.6.7}$$

The parameter $\{\mu_i\}$ of the Poisson distribution results from the intensity function $\lambda(r)$ on p. 122 and models the mean number of fruit in trap i. It depends on the total number m of fruits and the random dispersal, which is assumed to follow a lognormal distribution with density function

$$d(r) = \frac{1}{\sigma r \sqrt{2\pi}} \exp\left(-\frac{(\ln r - \mu)^2}{2\sigma^2}\right) \qquad \text{for} \quad r \geq 0,$$

as in Example 3.4. Given m and the dispersal parameters μ and σ, the mean number of fruits in trap i of area a is

$$\mu_i = \frac{amd(r_i)}{2\pi r_i}.$$

The analysis now applies the Bayesian philosophy and treats the unknowns m, μ and σ as random variables, with $\theta = (m, \mu, \sigma)$, requiring the specification of their joint prior distribution. It is natural to assume that the total number of fruit (m) and the dispersal parameters (μ, σ) are independent of each other a priori. Hence,

$$\pi(m, \mu, \sigma) = \pi(m)\pi(\mu, \sigma).$$

In order to avoid complex notation $\pi(\cdot)$ is used for multivariate distributions as well as for the corresponding marginal distributions. The specific distributions can be identified on the basis of the arguments.

Prior information on the number of fruit is not available here – it is not clear, for example, whether certain values are more likely than others. In many applications, however, researchers do have expert knowledge on this and would probably be able to specify a specific distribution. In the absence of prior information, a natural choice of a prior distribution for m is the uniform distribution on $\{0, 1, \dots\}$, which is well approximated by its continuous counterpart, the uniform distribution on $[0, \infty)$. For this prior $\int_\infty^0 \pi(m)dm$ is infinite and referred to as *improper*. This implies that $\pi(m, \mu, \sigma)$ is also improper. However, such a prior may be used, provided that the posterior (7.6.6) is a proper distribution, i.e. that the integral of the posterior over the whole space is 1.

The parameters μ and σ determine the distances r of the dispersion of fruits from the mother tree. Accordingly, the mean distance is

$$\mathrm{E}(r) = \exp\left(\mu + \frac{1}{2}\sigma^2\right) \tag{7.6.8}$$

and the mode

$$r_{\mathrm{mode}} = \exp(\mu - \sigma^2). \tag{7.6.9}$$

The mode of the dispersion distribution is an intuitive characteristic for a priori reasoning. Assume now that some prior knowledge on the seed dispersal mechanism of the tree and hence prior information on the μ and σ can be used to derive the prior distribution of these $\pi(\mu, \sigma)$. In other words, the analysis now assumes that the maximum is between $0\,\mathrm{m}$ and $6\,\mathrm{m}$ with a high probability, i.e. $\ln(r_{\mathrm{mode}})$ has a Gaussian distribution where the parameters are chosen as $\mu = 3$ and $\sigma = 1.6$ and σ can be modelled as

$1/\sigma^2 \sim$ the gamma distribution with mean 1 and variance 10.

Note that in Bayesian statistics the precision parameter $1/\sigma^2$ is commonly modelled as a gamma distribution. A weaker prior such as the uniform distribution may be chosen if less information is available. Finally, the prior $\pi(\mu, \sigma)$ can be specified based on these two distributional assumptions and (7.6.9).

The posterior distribution $\pi(m, \mu, \sigma, \{\mu_i\}|\{n_i\})$ is rather complex but can be explored by simulation. Figure 7.14 gives a graphical description of the dependence structure. Note that complex hierarchical models are often displayed in this way or even constructed with reference to a similar graph.

The parameters of the marginal posterior distributions are shown in Table 7.3, and have been calculated using a large number of MCMC simulation runs, which were necessary to obtain stable results. This corresponds to the experience in Example 3.4 that the likelihood function was rather flat.

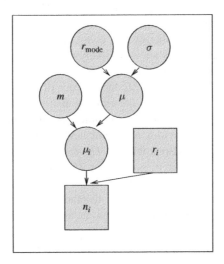

Figure 7.14   Directed acyclic graph showing the hierarchical Bayesian model describing the dependence structure among the variables. Observed variables are represented by squares and unobserved variables by circles.

Table 7.3   Description of the marginal posterior distributions for m, μ and σ extracted from MCMC simulations: mean, standard deviation and 95 % credible intervals (2.5 % and 97.5 % quantiles).

Parameter	Mean	Standard deviation	$q_{0.025}$	$q_{0.975}$
m	219 800	84 760	138 700	440 700
μ	4.11	0.35	3.65	5.04
σ	0.95	0.13	0.77	1.28

The quantiles $q_{0.025}$ and $q_{0.975}$ are the bounds of the 95% credible intervals for the parameters. For example, the posterior probability that the mean number of fruits from the tree is between 138 700 and 440 300 is 0.95. This interval is wide, which means that the experiment provides little information on the number of fruit. However, this is more valuable than the simple $\hat{m} = 179\,800$ obtained in Example 3.4.

Note that in this approach, which is often referred to as 'fully Bayesian', the estimation also provides information on the precision (in terms of standard deviation and credible interval) of the parameters. Unlike in the frequentist approach, which often depends on large sample size assumptions, these estimates are 'exact' in the sense that they do not rely on any such assumptions. Recall that the maximum likelihood estimates in Example 3.4 were: $\hat{m} = 179\,800$, $\hat{\mu} = 3.93$ and $\hat{\sigma} = 0.94$. These are in good agreement with the results of the fully Bayesian analysis.

Bayesian estimation of the intensity for a Poisson process

The following discusses the Bayesian approach, now applied to a simple point process model, the inhomogeneous Poisson process. Bayesian approaches for more complex models are briefly discussed below.

Assume that N is an inhomogeneous Poisson process with intensity function $\lambda(x)$ and a sample $x_1, \ldots, x_n$ of points in the observation window W is given. In the Bayesian approach, $\lambda(x)$ is considered a random variable, i.e. a random field. Consequently, the conditional distribution of the data given $\{\lambda(x)\}$ is the likelihood of an inhomogeneous Poisson process, given by (3.4.4), which in the notation used here is

$$f(x_1, \ldots, x_n | \{\lambda(x)\}, \theta) = \prod_{i=1}^{n} \lambda(x_i; \theta) \exp\left(-\int_W \lambda(x; \theta)\mathrm{d}x\right).$$

Several kinds of prior distributions for the intensity function, i.e. distributions of the random field $\{\lambda(x)\}$, have been suggested such as the log-Gaussian Cox process (Beneš et al., 2005), the gamma–Poisson process (Wolpert and Ickstadt, 1998), and the partition process (Heikkinen and Arjas, 1998; Ferreira et al., 2002).

Note that two of these processes have been discussed in this book in the context of Cox processes (refer to p. 380), but the partition process has not. The intensity of the partition process is

$$\lambda(x) = \sum_{k=1}^{K} \lambda_k \mathbf{1}_{E_k}(x),$$

where $\{E_k\}$ is a random partition of W, typically the Voronoi tessellation generated by some point process. Here, the $(\lambda_1, \ldots, \lambda_K)$ are the unknown parameters θ. The function $\lambda(x)$ is not continuous, but the average of several simulated realisations from the posterior distribution of $\lambda(x)$ results in a smooth surface.

Any of the three processes listed above may be chosen as the prior for the random field and contains unknown hyperparameters θ, i.e. for each of these a similar hierarchical model structure may be used and the posterior $\pi(\theta, \{\lambda(x)\}|N)$ is proportional to

$$\pi(\theta)\,\pi(\{\lambda(x)\}|\theta)\,f(x_1, \ldots, x_n|\{\lambda(x)\}).$$

Note that in this application the denominator in (7.6.6) may be 'ignored' in the MCMC simulation, i.e. it suffices to use the numerator to characterise the posterior distribution.

The estimation of the random intensity $\{\lambda(x)\}$ of point pattern data assumes a priori information on the smoothness of the result. There is some similarity to kernel smoothing of intensity functions as discussed on p. 115 and sometimes the term 'Bayesian smoothing' is used.

Other applications

Blackwell (2001), Blackwell and Møller (2003) and Skare et al. (2007) model badger territories. Badgers (*Meles meles*) of a clan create latrines near their territorial borders. These markings form a point pattern with increased intensity near the borders of territories. A natural tool in modelling territories is the Voronoi tessellation (see Section 1.8.4) generated as a secondary structure of a prior point pattern with one 'central' point for each territory. Bayesian modelling may be used to provide estimates of badger territories and their centre points, and posterior simulations produce measures of the uncertainty in these reconstructions. Similar approaches have been applied to image segmentation (Byers and Raftery, 2002), and delineation of homogeneous regions from categorical and quantitative soil measurements for precision farming purposes (Guillot et al., 2006).

Illian et al. (2008) apply both a non-Bayesian and a Bayesian approach to a multi-type pattern of 24 species of Australian plants (Armstrong, 1991; see also Section 4.9 above) and they notice that the Bayesian approach proves to be rather flexible in particular in the context of complex patterns and a large number of parameters.

For finite Gibbs processes (see Section 3.6) the likelihood is known and, in theory, a Bayesian approach could be applied to the estimation of the pair potential function. However, a big difficulty here is the fact that the normalising factor of the Gibbs likelihood (see (3.6.11)) remains in the MCMC algorithm for posterior simulation and also requires simulation, except in the Poisson case. Usually MCMC within MCMC is regarded as intractable. However, some progress has recently been made on this issue. It is possible to find the maximum of the posterior density; see Heikkinen and Penttinen (1999). Hence a Bayesian approach can be taken here, with the advantage that maximum a posteriori estimation is more useful than the maximum likelihood solution if regularisation of the estimates is important. This is the case, for example, if the pair potential is parametric but very flexible, such as a multi-scale model with many steps. Møller et al. (2006) suggest an ingenious method based on an auxiliary variable for a fully Bayesian estimation of the pair potential.

Marked point process models have been used as *regularisation priors* in image analysis and computer vision; see Stoica et al. (2004).

For Bayesian methods to be applied to point pattern analysis the likelihood of the point process has to be known. However, many important point processes are defined by construction, and the likelihood is not readily available. For these models the Bayesian methods are, at least currently, not a competitive alternative to the frequentist approach. The inhomogeneous Poisson process and the finite Gibbs process are two of the few exceptions. Indeed, many Bayesian applications for point pattern analysis rely on the inhomogeneous Poisson process as the model for observations. This model has been frequently applied, for example, in epidemiology, most often to counts of points in small compartments or cells; see Elliott et al. (2000) and Diggle (2003). Currently, Bayes methods in the context of point process statistics are very topical and many new methods are being developed.

Appendix A

Fundamentals of statistics

This appendix presents some general fundamental ideas of mathematical statistics that are not specific to spatial point processes and may be found in many general books on statistical theory. These are included here in order to make the book self-contained. For more information and further details, refer to any good book on statistics or a statistical encyclopedia.

A.1 Mean and variance

This book uses the notation of mathematical statistics, in particular its notation for the mean (or expected value) of a random variable X, i.e. EX, which denotes the same as $\langle X \rangle$ or $\overline{X}$ as used by physicists or engineers. Using the $\mathbf{E}$-notation, the variance is

$$\mathbf{var}X = \mathbf{E}(X - \mathbf{E}X)^2,$$

and the covariance of two random variables X and Y

$$\mathbf{cov}(X, Y) = \mathbf{E}(X - \mathbf{E}X)(Y - \mathbf{E}Y).$$

Statistical Analysis and Modelling of Spatial Point Patterns   J. Illian, A. Penttinen, H. Stoyan and D. Stoyan
© 2008 John Wiley & Sons, Ltd

A.2   Properties of estimators

Unbiasedness is a quality criterion that is frequently used in point process statistics as well as in classical statistics. For instance, $\bar{x}$ and s^2 as estimators of mean and variance are unbiased. Estimators in point process statistics are often required to be at least unbiased, i.e. to yield the true value on average. Technically, an *unbiased* estimator $\hat{\theta}$ of a parameter or characteristic θ has the property that

$$\mathbf{E}\hat{\theta} = \theta. \qquad (A.1)$$

A less stringent property is *ratio-unbiasedness*. An estimator $\hat{\theta}$ of θ is called ratio-unbiased if θ can be written as

$$\theta = \frac{\theta_1}{\theta_2}$$

and $\hat{\theta}$ has the form of a quotient

$$\hat{\theta} = \frac{\hat{\theta}_1}{\hat{\theta}_2},$$

where $\hat{\theta}_1$ and $\hat{\theta}_2$ are unbiased estimators of θ_1 and θ_2, respectively. This is done with the aim that $\hat{\theta}$ is asymptotically unbiased, i.e. that the bias of $\hat{\theta}$ vanishes for large samples by the law of large numbers. These estimators often have small variances if fluctuations of $\hat{\theta}_1$ and $\hat{\theta}_2$ cancel out.

A.3   Bias, errors and estimation variance

The bias of an estimator $\hat{\theta}$ of the parameter θ is the quantity

$$\text{bias}(\hat{\theta}) = \hat{\theta} - \theta,$$

which might also be called systematic error. Clearly, the bias of an unbiased estimator is zero.

The *mean squared error* (mse) is the mean of the squared difference of the estimator minus the true value,

$$\text{mse}(\hat{\theta}) = \mathbf{E}(\hat{\theta} - \theta)^2.$$

The *estimation variance* is the variance of $\hat{\theta}$, i.e.

$$\mathbf{var}(\hat{\theta}) = \mathbf{E}(\hat{\theta} - \mathbf{E}\hat{\theta}).$$

Note that

$$\text{mse}(\hat{\theta}) = \mathbf{var}(\hat{\theta}) + \text{bias}(\hat{\theta})^2.$$

A.4　*P*-value

For a statistical test, 'the *P*-value is the probability, calculated under the null hypothesis H_0, of obtaining a test result as extreme as that observed in the sample . . . . If this probability is regarded as small, H_0 should be rejected; otherwise it should not be rejected' (Gibbons, 2005).

A.5　Maximum likelihood method

In 'classical' statistics maximum likelihood estimators are frequently used to estimate a parameter (or vector of parameters) θ in a given density function $f(x_1, \ldots, x_n, \theta)$. They are based on the likelihood function, i.e. the likelihood of the data given the parameters

$$L(x_1, \ldots, x_n, \theta) = f(x_1, \ldots, x_n, \theta),$$

considered as a function of θ, where $x_1, \ldots, x_n$ are the observations. For computational reasons the logarithm of the likelihood, i.e. the log-likelihood function $l(x_1, \ldots, x_n, \theta)$, is often considered instead. The parameter θ is typically estimated by numerical maximisation or by solving the likelihood equations

$$\frac{\partial}{\partial \theta_s} l(x_1, \ldots, x_n, \theta) = 0 \qquad \text{for } s = 1, 2, \ldots, p, \tag{A.2}$$

for $\theta = (\theta_1, \ldots, \theta_p)$ often using a numerical algorithm. An important by-product of this approach are the asymptotic variances of the maximum likelihood estimators for the θ_j, which may be obtained from the diagonal of the Hessian matrix (the matrix of second derivatives of the log-likelihood function) evaluated for the solution of (A.2). These may be used to calculate confidence intervals for the estimated parameters, providing information on their precision.

A.6　Kernel estimators

Kernel estimators are commonly used in classical statistics for the estimation of probability densities and appear in several contexts throughout this book. The following briefly explains the fundamental idea as applied to probability density estimation to facilitate the understanding of similar approaches in the context of estimation of point process characteristics.

Let $f(x)$ be an unknown p.d.f. to be estimated, given a sample $x_1, \ldots, x_n$. Further, let $k(x)$ be another p.d.f., the kernel function, usually taken to be symmetric, i.e.

$$k(x) = k(-x).$$

Then

$$\hat{f}(x) = \frac{1}{n} \sum_{i=1}^{n} k(x - x_i)$$

is an estimator for $f(x)$. There are many possible kernel functions to choose from. Common choices are the simple *box kernel*,

$$k(x) = \begin{cases} \frac{1}{2h} & \text{for } -h \leq x \leq h, \\ 0 & \text{otherwise,} \end{cases}$$

and the slightly more complicated *Epanechnikov kernel*,

$$k(x) = \begin{cases} \frac{3}{4h}\left(1 - \frac{x^2}{h^2}\right) & \text{for } -h \leq x \leq h, \\ 0 & \text{otherwise,} \end{cases}$$

where h is the bandwidth (see below).

Note that the choice of the Epanechnikov kernel in probability density estimation is based on certain optimisation considerations. Experience shows that the specific choice of $k(x)$ is not as important as the choice of the appropriate *bandwidth h*. Large values of h result in smooth estimated density functions $\hat{f}(x)$, but sometimes important local distributional details are lost. On the other hand, small values of h result in rather 'wiggly' functions and the true global structure of the distribution might be obscured. Choosing h appropriately is a difficult issue, frequently discussed in the literature. In a large study, the behaviour of the estimator with a range of different bandwidths should be assessed.

Also note the following problem. The p.d.f. to be estimated often has the property that

$$f(x) = 0 \qquad \text{for } x < a,$$

where a is some known constant. In particular, in many applications the data are positive, i.e. $a = 0$.

If a symmetric kernel function $k(x)$ is applied in this situation, it is not unlikely that $\hat{f}(x) > 0$ for $a - h < x < a$. A viable means of ensuring that $\hat{f}(x) = 0$ for $x < a$, and that the probability mass below a is not lost, is to use the so-called *reflection method:*

$$\hat{f}(x) := \begin{cases} \hat{f}(x) + \hat{f}(a - x) & \text{for } x \geq a, \\ 0 & \text{otherwise.} \end{cases}$$

As noted, kernel estimators are frequently used in point process statistics and throughout this book and principles similar to the above also hold. Note, however, that the methods developed for choosing the optimal bandwidth for p.d.f. estimation only apply in this specific context and not in most of the more general contexts addressed in this book.

Appendix B

Geometrical characteristics of sets

Set-geometrical ideas are frequently used throughout this book, and this appendix provides fundamental definitions and important formulas. For more details, refer to Stoyan et al. (1995) and Soille (1999).

B.1 Minkowski addition

For two subsets A and B of $\mathbb{R}^d$, the Minkowski addition is defined as

$$A \oplus B = \{x + y : x \in A \text{ and } y \in B\}.$$

Minkowski addition causes both a translation and enlargement of the original sets A and B.

Consider the special case where B is a set with a single element, i.e. a singleton $\{x\}$. Then

$$A \oplus \{x\} = A + x = A_x = \{x + y : y \in A\}.$$

The set A_x is the set A translated by the vector x.

Statistical Analysis and Modelling of Spatial Point Patterns   J. Illian, A. Penttinen, H. Stoyan and D. Stoyan
© 2008 John Wiley & Sons, Ltd

B.2   Calculation of $\nu(W \cap W_x)$

In many situations, the area (volume) of the intersection of the set W and the translated set W_x, i.e. $\nu(W \cap W_x)$, has to be determined. This is a standard problem in computational geometry and is used in this book in the context of stationary estimators of second-order summary characteristics. For some simple sets W, formulas are known. For a disc of radius R,

$$\nu(W \cap W_x) = 2R^2 \arccos\left(\frac{r}{2R}\right) - \frac{r}{2}\sqrt{4R^2 - r^2} \qquad \text{for } r \le 2R,$$

and for a sphere of radius R,

$$\nu(W \cap W_x) = \frac{4}{3}\pi R^3 \left(1 - \frac{3r}{4R} + \frac{r^3}{16R^3}\right) \qquad \text{for } r \le 2R.$$

In both these cases r is the length $\|x\|$ of x. For a rectangle with sides a and b,

$$\nu(W \cap W_x) = (a - \|\xi\|)(b - \|\eta\|),$$

where $x = (\xi, \eta)$, $a > \xi$ and $b > \eta$, and for a rectangular parallelepiped with sides a, b and c,

$$\nu(W \cap W_x) = (a - \|\xi\|)(b - \|\eta\|)(c - \|\zeta\|),$$

where $x = (\xi, \eta, \zeta)$ and $0 \le \xi \le a$, $0 \le \eta \le b$, $0 \le \zeta \le c$. Of course, it is always assumed that x is such that $W \cap W_x \neq \emptyset$.

If W has a more general shape, several efficient algorithms can be applied. Two basic algorithms for the planar case are the following.

1. *Application of the discrete version of Green's theorem.* Assume that W is a polygon and suppose that $e_1, \ldots, e_n$ are the edges of W indexed in anticlockwise order. Without loss of generality assume that W lies in the positive quadrant $(x > 0, y > 0)$. Let h_i be the orthogonal projection of e_i onto the x-axis. Let T_i be the quadrilateral bounded by e_i and h_i and two vertical lines. Then the indicator function $\mathbf{1}_W(x)$ of W can be decomposed as

$$\mathbf{1}_W(x) = \sum_{i=1}^{n} s_i \mathbf{1}_{T_i}(x),$$

where s_i is ± 1. Representing both $\mathbf{1}_W$ and $\mathbf{1}_{W_x}$ in this form, multiplying these and then integrating the product, one gets a representation for the area $\nu(W \cap W_x)$ as a sum over all pairs of edges of W.

2. *Application of the Fast Fourier Transform.* Consider a set W represented by a binary pixel image. $\nu(W \cap W_x)$ can be computed by taking the Fourier transform of the indicator function $\mathbf{1}_W(x)$, evaluating the squared modulus, and then taking the inverse Fourier transform; see Ohser and Schladitz (2007).

The library `spatstat` for the software package R (Baddeley and Turner, 2005, 2006) contains implementations of both algorithms. Algorithm 1 is called `overlap.owin` and algorithm 2 is `setcov`.

B.3 Set covariance and isotropised set covariance

The isotropised set covariance $\overline{\gamma}_W(r)$ is a function assigned to the subset W of $\mathbb{R}^d$, which is relevant in the context of many formulas and estimators in this book. Consider the intersection of W and the translated set $W + ru$, and its area (volume), the *set covariance* or *geometric covariogram* $\gamma_W(r, u)$, i.e.

$$\gamma_W(r, u) = \nu(W \cap W_{ru}) \qquad \text{for } r \geq 0,$$

where

$$W_{ru} = W + ru \qquad \text{with } u \in S^{d-1},$$

i.e. u is a unit vector and r a non-negative number, the length of the translation vector ru.

The *isotropised set covariance* is the rotation average of the set covariance $\gamma_W(r, u)$, i.e.

$$\overline{\gamma}_W(r) = \frac{1}{db_d} \int_{S^{d-1}} \gamma_W(r, u) du,$$

where S^{d-1} is the $(d-1)$-dimensional sphere.

For the cases of a rectangular, parallelepipedal, circular and spherical window W formulas for $\overline{\gamma}_W(r)$ are known. For a disc of radius R,

$$\overline{\gamma}_W(r) = 2R^2 \arccos\left(\frac{r}{2R}\right) - \frac{r}{2}\sqrt{4R^2 - r^2} \qquad \text{for } r \leq 2R,$$

and for a sphere of radius R,

$$\overline{\gamma}_W(r) = \frac{4}{3}\pi R^3 \left(1 - \frac{3r}{4R} + \frac{r^3}{16R^3}\right) \qquad \text{for } r \leq 2R.$$

These two formulas coincide with those for $\nu(W \cap W_x)$, since for both disc and sphere $\gamma_W(r, u)$ do not depend on u. For a rectangle with sides a and b ($a \leq b$),

$$\overline{\gamma}_W(r) = \begin{cases} ab - \dfrac{2r}{\pi}(a+b) + \dfrac{r^2}{\pi} & \text{for } r \le a, \\[4mm] \dfrac{ab}{\pi}\left(2\arcsin\left(\dfrac{a}{r}\right) - \dfrac{a}{b} - 2\left(\dfrac{r}{a} - \sqrt{\dfrac{r^2}{a^2} - 1}\right)\right) & \text{for } a < r \le b \end{cases}$$

(see Stoyan and Stoyan, 1994, p. 123, for the case $r > b$). For a rectangular parallelepiped with sides a, b and c $(a \le b \le c)$,

$$\overline{\gamma}_W(r) = abc - \frac{1}{2}(ab + ac + bc)r + \frac{2}{3\pi}(a+b+c)r^2 - \frac{1}{4\pi}r^3 \qquad \text{for } r \le a.$$

For the calculation of the full $\overline{\gamma}_W(r)$ a program is given in Ohser and Mücklich (2000, p. 357). (Note that there $M\_2\_PI = \frac{2}{\pi}$, $M\_1\_PI = \frac{1}{\pi}$ and $M\_PI\_2 = \frac{\pi}{2}$.)

B.4 Calculation of $v(W \cap b(x, r))$

The quantity $v(W \cap b(x, r))$ for $x \in W$ is used in this book in the context of the estimation of second-order summary characteristics. Its calculation is simple for $W = b(o, R)$, $x \in W$ and $r < R$. In the planar case,

$$v(W \cap b(x, r)) = \begin{cases} \pi r^2 & \text{for } d \le R - r, \\[4mm] R^2 \arccos\left(\dfrac{R^2 - r^2 + d^2}{2dR}\right) + r^2\left(\pi - \arccos\left(\dfrac{R^2 - r^2 - d^2}{2dr}\right)\right) \\[4mm] \quad -d\sqrt{r^2 - \left(\dfrac{R^2 - r^2 - d^2}{2d}\right)^2} & \text{otherwise.} \end{cases}$$

In the spatial case,

$$v(W \cap b(x, r)) = \begin{cases} \dfrac{4}{3}\pi r^3 & \text{for } r \le R \text{ and } \|x\| \le R - r, \\[4mm] \dfrac{\pi}{12d}(R + r - d)^2(d^2 - 3(R-r)^2 + 2d(R+r)) & \text{otherwise,} \end{cases}$$

with $d = \|x\|$.

If W is a rectangle of sides a and b $(a \le b)$, the following program shows the calculation, where ra[0] is $v(W \cap b(x, r))$. This is elementary mathematics but complicated because of many case distinctions.

```
%
double ra[2]={0.0,0.0};
void f(double d1, double d2, double r, double *ra)
{
double r2=r*r,d3,w;
if (d1<r)
```

```
{
d3=sqrt(r2-d1*d1);
if (d3<d2)
{
w=atan(d2/d1)-atan(d3/d1);
ra[0]+=0.5*(d1*d3+r2*w);
ra[1]+=w*r;
}
else ra[0]+=0.5*d1*d2;
}
else
{
w=atan(d2/d1);
ra[0]+=0.5*r2*w;
ra[1]+=w*r;
}
}


double *rect(double a, double b, double x, double y, double r)
/* returns the vector (ra[0],ra[1])
*/
{
f(a-x,b-y,r,ra);
f(a-x,y,r,ra);
f(y,a-x,r,ra);
f(y,x,r,ra);
f(x,y,r,ra);
f(x,b-y,r,ra);
f(b-y,x,r,ra);
f(b-y,a-x,r,ra);
return ra;
}
```

For the case of a parallelepipedal window W formulas may be found in Baddeley et al. (1993).

B.5   Calculation of $\nu_{d-1}(W \cap \partial b(x, r))$

The surface area of $b(x, r)$ in W, $\nu_{d-1}(W \cap \partial b(x, r))$, for $x \in W$ is used in this book in the context of the estimation of second-order characteristics. In particular,

$$w(x_1, x_2) = \nu_{d-1}(W \cap \partial b(x, r))$$

with $x = x_1$ and $r = \|x_1 - x_2\|$, see p. 229. The calculation of $\nu_{d-1}(W \cap \partial b(x, r))$ is again simple for $W = b(o, R)$, $x \in W$ and $r < R$. In the planar case,

$$\nu_1(W \cap \partial b(x, r)) = \begin{cases} 2\pi r & \text{for } d \leq R - r, \\ 2r\left(\pi - \arccos\left(\dfrac{R^2 - r^2 - d^2}{2dr}\right)\right) & \text{otherwise.} \end{cases}$$

In the spatial case,

$$\nu_1(W \cap \partial b(x, r)) = \begin{cases} 4\pi r^2 & \text{for } d \leq R - r, \\ 2\pi rc & \text{otherwise,} \end{cases}$$

with $d = \|x\|$ and $c = \dfrac{R^2 - (r - d)^2}{2d}$. In the planar case where W is a rectangle of sides a and b ($a \leq b$), the program above also covers the calculation of $\nu_1(W \cap \partial b(x, r))$, which is denoted by ra[1]. Explicit formulas are given in Goreaud and Pélissier (1999).

For the case of a parallelepipedal window W the formulas may be found in Baddeley et al. (1993).

B.6 $K_{\text{fin}}(r)$ for a rectangular window W

The general definition of $K_{\text{fin}}(r)$ for a binomial process is

$$K_{\text{fin}}(r) = \nu(W \oplus W \cap b(o, r));$$

see p. 130. The side lengths of the window are a and b, with $a \leq b$. Here the radii are $r_1 = a$ and $r_2 = \sqrt{a^2 + b^2}$. Then $W \oplus \check{W}$ is the rectangle of side lengths $2a$ and $2b$ with centre at the origin o, and

$$K_{\text{fin}}(r) = \begin{cases} \pi r^2 & \text{for } r \leq a, \\ 2ac + 8r^2\left(\dfrac{\pi}{2} - \arctan\left(\dfrac{a}{c}\right)\right) & \text{for } a < r \leq b, \\ 4ab, & \text{for } a^2 + b^2 \leq r^2, \\ 2bf + 2ac + 8r^2\left(\dfrac{\pi}{2} - \left(\arctan\left(\dfrac{b}{f}\right) + \arctan\left(\dfrac{a}{c}\right)\right)\right) & \text{otherwise,} \end{cases}$$

with $c = \sqrt{r^2 - a^2}$ and $f = \sqrt{r^2 - b^2}$.

Appendix C

Fundamentals of geostatistics

In some applications of point process statistics, random fields and the statistical methods for these, termed *geostatistics,* have an important role. The corresponding theory and applications are discussed in Cressie (1993), Chilès and Delfiner (1999) and Wackernagel (2003). This appendix briefly covers some of the basic ideas.

A *random field* is a family of random variables $Z(x)$ with x in $\mathbb{R}^d$. For example, $Z(x)$ may denote a parameter describing soil quality or altitude at location x. A random field is a regionalised variable since a value $Z(x)$ is associated with every x. Note that random fields differ from marked point processes, as the marks $m(x)$ are only defined for points x of the processes. Sometimes a random field is written as $\{Z(x)\}_{x \in \mathbb{R}^d}$, in order to emphasise that the random field is the totality of all $Z(x)$; a simpler symbol is $\{Z(x)\}$.

In many applications, the random field is assumed to be stationary (homogeneous) and isotropic. These concepts resemble the analogous concepts as defined for point processes: the distribution of the random field is invariant under translation and rotation. Similarly, stationary random fields may be described with first- and second-order moments.

The first-order moment of a random field is the *mean value function* $m(x)$ given by

$$\mathbf{E}Z(x) = m(x),$$

where **E** denotes the expectation operator. For a stationary random field, $m(x)$ does not depend on x and is denoted by m. m is estimated based on values of $Z(y_i)$ at measurement locations y_i which are independent of the random field; the estimator of m is simply the arithmetic mean of these values.

The classical second-order characteristic considered for random fields is the *(semi-)variogram* $\gamma(r)$, which is defined as the mean squared difference of the field values at x and $x + \mathbf{r}$ divided by 2, where $\mathbf{r}$ is a difference vector of length r:

$$\gamma(\mathbf{r}) = \frac{1}{2} \mathbf{E} \left(Z(x) - Z(x + \mathbf{r}) \right)^2 .$$

Assuming stationarity, $\gamma(\mathbf{r})$ does not depend on x; and if isotropy is also assumed, $\gamma(\mathbf{r})$ depends only on the length r of $\mathbf{r}$ or the distance r between x and $x + \mathbf{r}$. Therefore, the simplified notation $\gamma(r)$ is used whenever stationarity and isotropy are assumed, as in the following.

A variogram $\gamma(r)$ of a stationary random field has the properties

$$\gamma(0) = 0, \qquad \gamma(\infty) = \sigma^2 ,$$

where σ^2 is the field variance, i.e. the variance of $Z(x)$ (which does not depend on x in the stationary case). Furthermore, the variogram is conditionally negative definite, which means that it satisfies an analytical condition which is not discussed here.

The rate of convergence of $\gamma(r)$ towards σ^2 for $r \to \infty$ characterises the range of correlation. If there is a finite r_0 with $\gamma(r) = \sigma^2$ for $r \geqslant r_0$, r_0 is called the *range of correlation*. Otherwise the smallest value r_p with $\gamma(r_p) = (1 - p)\sigma^2$ is used for some p. The specific value $r_{0.05}$ is often called the *practical range*.

It is possible that

$$\lim_{r \to 0} \gamma(h) = \kappa^2 > 0.$$

If this is the case, there is a 'nugget effect' and κ^2 is called the *nugget variance*. The nugget effect indicates short-range irregularities caused by very small structures or measurement errors.

Estimating variograms based on values of the random field $Z(y_i)$ observed at observation points y_i yields so-called sample variograms or *empirical variograms* $\hat{\gamma}(r)$.

The geostatistical variogram estimator is

$$\hat{\gamma}(r) = \frac{1}{2n_r} \sum_{\|y_\beta - y_\alpha\| = r} \left(Z(y_\beta) - Z(y_\alpha) \right)^2 ,$$

where the sum is extended over the n_r pairs (y_β, y_α) of measurement points of inter-point distance r; see Chilès and Delfiner (1999). The points of measurement y_i

have to be independent of the random field. They are either points chosen without knowledge of the values of the field or based on points in a point process which is independent of the random field. If the y_i are not independent (e.g., they tend to be located at places where the field has large values) the resulting estimators are biased. In these cases even the estimation of the mean m can be difficult.

Several theoretical variogram models have been developed in geostatistics. These models may be used to interpret the spatial variability of random fields, and are based on variogram parameters that may be usefully interpreted. A simple example is the exponential variogram given by

$$\gamma(r) = \sigma^2 \left(1 - \exp(-\alpha r)\right) ,$$

where α is a model parameter and σ^2 the field variance.

Several approaches to the simulation of random fields have been developed. Usually samples of Gaussian random fields are generated, i.e. random fields where $Z(x)$ and random vectors of the form $\{Z(x_1), Z(x_2), \ldots, Z(x_n)\}$ are Gaussian random variables. These simulations can be performed with the `RandomFields` package (Schlather, 2001b) available for the statistical computing environment R. Since Gaussian random fields may have negative values, often log-Gaussian fields are used that are calculated from Gaussian fields by exponentiation, $\exp(Z(x))$. Conditional simulation of (Gaussian) random fields, where the observed values $Z(y_i)$ are retained at the observation points v_i, is considered by Lantuéjoul (2002). This means simulation where the observed values $Z(y_i)$ are retained at the points y_i of observation.

One of the main aims in geostatistics is to find spatial interpolations, i.e. to predict the value $Z(x)$ at location x different from the measurement locations $y_1, y_2, \ldots,$ y_n using the $Z(y_i)$. The geostatistical standard method in this context is kriging, which yields unbiased predictions based on least-squares methods.

References

Abbe, E. (1879) Über Blutkörperzählung. *Jena Z. Med. Naturwiss.* **13** (Neue Serie), 98–105 (in German).

Adler, F.R. (1996) A model of self-thinning through local competition. *Proc. Natl. Acad. Sci. USA* **93**, 9980–9984.

Aguirre, O., Hui, G.Y., Gadow, K. von, and Jimenez, J. (2003) An analysis of spatial forest structure using neighbourhood-based variables. *Forest Ecol. Manag.* **183**, 137–145.

Akaike, H. (1974) A new look at the statistical model identification. *IEEE Trans. Autom. Control* **AC-19**, 716–723.

Anikeenko, A.V., Medvedev, N.N., and Gavrilova, M.L. (2006) Application of Procrustes distance to shape analysis of Delaunay simplexes. In M.L. Gavrilova (ed.), *The 3rd International Symposium on Voronoi Diagrams in Science and Engineering*, pp. 148–152. IEEE Computer Society, Los Alamitos, CA.

Armitage, P., Berry, G., and Matthews, J.N.S. (2001) *Statatistical Methods in Medical Research*. Blackwell Science, Malden, MA.

Armstrong, P. (1991) Species patterning in the heath vegetation of the northern sandplain. Honours thesis, University of Western Australia.

Babalievski, F. (1998) Cluster counting: The Hoshen–Kopelman algorithm vs. spanning tree approaches. *Int. J. Modern Phys. C* **9**, 43–60.

Babu, G.J., and Feigelson, E.D. (1996) *Astrostatistics*. Chapman & Hall, London.

Baccelli, F., and Blaszczyszyn, B. (2001) On a coverage process ranging from the Boolean model to the Poisson–Voronoi tessellation. *Adv. Appl. Probab.* **33**, 293–323.

Baccelli, F., and Bordenave, C. (2007) The radial spanning tree of a Poisson process. *Ann. Appl. Probab.* **17**, 305–359.

Baccelli, F., Klein, M., Lebourges, M., and Zuyev, S. (1997) Stochastic geometry and architecture of communication networks. *Telecomm. Syst.* **7**, 209–227.

Baddeley, A.J. (1999) Spatial sampling and censoring. In O.E. Barndorff-Nielsen, W.S. Kendall and M.N.M. van Lieshout (eds), *Stochastic Geometry, Likelihood and Computation*. Chapman & Hall/CRC, Boca Raton, FL, pp. 37–78.

Baddeley, A.J., and Gill, R.D. (1997) Kaplan–Meier estimators of distance distributions for spatial point processes. *Ann. Statist.* **25**, 263–292.

Baddeley, A.J., and Lieshout, M.N.M. van (1995) Area-interaction point processes. *Ann. Inst. Statist. Math.* **46**, 601–619.

Baddeley, A.J., and Silverman, B.W. (1984) A cautionary example for the use of second-order methods for analysing point patterns. *Biometrics* **40**, 1089–1094.

Baddeley, A.J., and Turner, R. (2000) Practical maximum pseudolikelihood for spatial point patterns. *Austral. New Zealand J. Statist.* **42**, 283–322.

Baddeley, A.J. and Turner, R. (2005) spatstat: An R package for analyzing spatial point patterns. *J. Statist. Software* **12**(6), 1–42.

Baddeley, A.J. and Turner, R. (2006) Modelling spatial point patterns in R. In A. Baddeley, P. Gregori, J. Mateu, R. Stoica and D. Stoyan (eds), *Case Studies in Spatial Point Process Modeling*, Lecture Notes in Statistics 185. Springer-Verlag, New York, pp. 23–76.

Baddeley, A.J., Moyeed, R.A., Howard, C.V., and Boyde, A. (1993) Analysis of a three-dimensional point pattern with replication. *Appl. Statist.* **42**, 641–668.

Baddeley, A.J., Møller, J., and Waagepetersen, R. (2000) Non- and semi-parametric estimation of interaction in inhomogeneous point patterns. *Statist. Neerland.* **54**, 329–350.

Baddeley, A.J., Turner, R., Møller, J., and Hazelton, M. (2005) Residual analysis for spatial point processes. *J. Roy. Statist. Soc. B* **67**, 617-666.

Baddeley, A.J., Møller, J., and Pakes, A.G. (2007) Properties of residuals for spatial point processes. *Ann. Inst. Stat. Math.* To appear.

Bai, Y., Walsworth, N., Roddan, B., Hill, D.A., Broersma, K., and Thompson, D. (2005) Quantifying treecover in the forest-grassland ecotone of British Columbia using crown delineation and pattern detection. *Forest Ecol. Manag.* **212**, 92–100.

Ballani, F. (2006) On modelling of refractory castables by marked Gibbs and Gibbsian-like processes. In A. Baddelely, P. Gregori, J. Mateu, R. Stoica and D. Stoyan (eds), *Case Studies in Spatial Point Process Modeling*, Lecture Notes in Statistics 185. Spinger-Verlag, New York, pp. 153–167.

Ballani, F., Daley, D.J., and Stoyan, D. (2005) Modelling the microstructure of concrete with spherical grains. *Comput. Materials Sci.* **35**, 399–407.

Banerjee, S., Carlin, P., and Gelfand, A. (2004) *Hierarchical Modeling and Analysis for Spatial Data*, 2nd edition. Chapman & Hall/CRC, Boca Raton, FL.

Bargiel, M., and Moscinski, J. (1991) C language program for the irregular packing of hard spheres. *Comput. Phys. Comm.* **64**, 183–192.

Barot, S., and Gignoux, J. (2003) Neighbourhood analysis in the savanna palm *Borassus aethiopum*: Interplay of intraspecific competition and soil patchiness. *J. Veget. Sci.* **14**, 79–88.

Barot, S., Gignoux, J., and Menaut, J.-C. (1999) Demography of a savanna palm tree: Predictions from comprehensive spatial pattern analyses. *Ecology* **80**, 1987–2005.

Bartlett, M.S. (1964) The spectral analysis of two-dimensional point processes. *Biometrika* **51**, 299–311.

Bedford, T., and Berg, J. van den (1997) A remark on the Van Lieshout and Baddeley J-function for point processes. *Adv. Appl. Probab.* **29**, 19–25.

Bella, I.E. (1971) A new competition model for individual trees. *Forest Sci.* **17**, 364–372.

Beneš, V. (1957) Fluctuations of telephone traffic. *Bell System Tech. J.* **36**, 965–973.

Beneš, V., Bodlak, K., Møller, J., and Waagepetersen, R.P. (2005) A case study on point process modelling in disease mapping. *Image Anal. Stereol.* **24**, 159–168.

Berger, U., and Hildenbrandt, H. (2000) A new approach to spatially explicit modelling of forest dynamics: Spacing, ageing and neighbourhood competition. *Ecol. Modelling* **132**, 287–302.

Berman, M. (1986) Testing for spatial association between a point process and another stochastic process. *Appl. Statist.* **35**, 54–62.

Berman, M., and Turner, R. (1992) Approximating point process likelihoods with GLIM. *Appl. Statist.* **41**, 31–38.

Bernadeau, F., and Weygaert, R. van de (1996) A new method for accurate estimation of velocity field statistics. *Mon. Not. Roy. Astron. Soc.* **279**, 693–711.

Berndt, S., and Stoyan, D. (1997) Automatic determination of dendritic arm spacing in directionally solidified matters. *Z. Metallkunde* **88**, 758–763.

Besag, J.E. (1975) Statistical analysis of non-lattice data. *The Statistician* **24**, 179–195.

Besag, J.E. (1977) Contribution to the discussion of Dr. Ripley's paper. *J. Roy. Statist. Soc. B* **39**, 193–195.

Besag, J.E. (1978) Some methods of statistical analysis for spatial data. *Bull. Int. Statist. Inst.* **47**, 77–92.

Bezrukov, A., and Stoyan, D. (2006) Simulation and statistical analysis of random packings of ellipsoids. *Part. Part. Syst. Charact.* **23**, 388–398.

Bezrukov, A., Bargiel, M., and Stoyan, D. (2002) Statistical analysis of simulated random packings of spheres. *Part. Part. Syst. Charact.* **19**, 111–118.

Bitterlich, W. (1947) Die Winkelzahlmessung. *Allg. Forst. Holzwirtsch. Ztg.* **58**, 94–96 (in German).

Bitterlich, W. (1984) *The Relascope Idea. Relative Measurements in Forestry.* Commonwealth Agricultural Bureaux, Norwich.

Blackwell, P.G. (2001) Bayesian inference for a random tessellation process. *Biometrics,* **57**, 502–507.

Blackwell, P.G., and Møller, J. (2003) Bayesian analysis of deformed tessellation models. *Adv. Appl. Probab.* **35**, 4–26.

Brix, A. (1999) Generalized gamma measures and shot-noise Cox processes. *Adv. Appl. Probab.* **31**, 929–953.

Brix, A., and Diggle, P.J. (2001) Spatio-temporal prediction for log-Gaussian Cox processes. *J. Roy. Statist. Soc. B* **63**, 823–841.

Brix, A., and Kendall, W.S. (2002) Simulation of cluster point processes without edge effects. *Adv. Appl. Probab.* **34**, 267–280.

Brix, A., and Møller, J. (2001) Space–time multi-type log Gaussian Cox processes with a view to modelling weeds. *Scand. J. Statist.* **28**, 471–488.

Brodatzki, U., and Mecke, K. (2002) Simulating stochastic geometries: morphology of overlapping grains. *Comput. Phys. Comm.* **147**, 218–221.

Buckland, S.T., Anderson, D.R., Burnham, K.P., Laake, J.L., Borchers, D.L., and Thomas, L. (2001) *Introduction to Distance Sampling.* Oxford University Press, Oxford.

Bugmann, H. (2001) A review of forest gap models. *Climatic Change* **51**, 259–305.

Buryak, O., and Doroshkevich, A. (1996) Correlation functions as a measure of the structure. *Astron. Astrophys.* **306**, 1–8.

Byers, S.D. and Raftery, A.E. (2002) Bayesian estimation and segmentation of spatial point processes using Voronoi tilings. In A.B. Lawson and D.G.T. Denison (eds), *Spatial Cluster Modelling.* Chapman & Hall/CRC, Boca Raton, FL, pp. 109–121.

Byth, K. (1981) θ-stationary point processes and their second-order analysis. *J. Appl. Probab.* **18**, 864–878.

Byth, K., and Ripley, B.D. (1980) On sampling spatial patterns by distance methods. *Biometrics* **36**, 279–284.

Canham, C.D., Cole, J.J., and Lauenroth, W.K. (2003) The role of modeling in ecosystem science. In: C.D. Canham, J.J. Cole and W.K. Lauenroth (eds), *Models in Ecosystem Science*, Princeton University Press, Princeton, NJ.

Capobianco, R., and Renshaw, E. (1998) The autocovariance function for marked point processes: A comparison between two different approaches. *Biometrical J.* **40**, 431–446.

Carroll, R.J., and Lombard, F. (1985) A note on *n* estimators for the binomial distribution. *J. Amer. Statist. Assoc.* **80**, 423–426.

Chave, J. (1999) Study of structural, successional and spatial patterns in tropical rain forests using TROLL, a spatially explicit forest model. *Ecol. Modelling.* **124**, 233–254.

Chilès, J.P., and Delfiner, P. (1999) *Geostatistics. Modelling Spatial Uncertainity.* John Wiley & Sons, Inc., New York.

Chiu, S.N. (2003) Spatial point pattern analysis by using Voronoi diagrams and Delaunay tessellations – a comparative study. *Biometrical J.* **45**, 367–376.

Chiu, S.N. (2007) Correction to Koen's critical values in testing spatial randomness. *J. Statist. Comput. Simul.* **77**, 1001–1004.

Chiu, S.N., and Molchanov, I.S. (2003) A new graph related to the direction of nearest neighbours in a point process. *Adv. Appl. Probab.* **35**, 47-55.

Chiu, S.N., and Stoyan, D. (1998) Estimators of distance distributions for spatial patterns. *Statist. Neerland.* **52**, 239–246.

Clapham, A.R. (1936) Over-dispersion in grassland communities and the use of statistical methods in plant ecology. *J. Ecology* **24**, 232–251.

Clark, J., and Gelfand, A. (2006) *Hierarchical Modelling for the Environmental Sciences.* Oxford University Press, Oxford.

Clark, P.J., and Evans, F.C. (1954) Distance to nearest neighbour as a measure of spatial relationships in populations. *Ecology* **35**, 445–453.

Cliff, A.D., and Ord, J.K. (1981) *Spatial Processes. Models and Applications.* Pion, London.

Coles, P., and Jones, B. (1991) A lognormal model for the cosmological mass distribution. *Mon. Not. Roy. Astron. Soc.* **248**, 1–13.

Comas, C., and Mateu, J. (2007) Modelling forest dynamics: A perspective form point process methods. *Biometrical J.* **49**, 176–196.

Conover, W.J. (1999) *Practical Nonparametric Statistics,* 3rd edition. John Wiley & Sons, Inc., New York.

Corral-Rivas, J.J. (2006) *Models of Tree Growth and Spatial Structure for Multi-Species, Uneven-aged Forests in Durango (Mexico).* Doctoral thesis, University of Göttingen.

Cottam, G., Curtis, J.T., and Hale, B.W. (1953) Some sampling characteristics of a population of randomly dispersed individuals. *Ecology* **34**, 741–757.

Cowling, A., Hall, P., and Phillips, M.J. (1996) Bootstrap confidence regions for the intensity of a Poisson point process. *J. Amer. Statist. Assoc.* **91**, 1514–1524.

Cox, D.R. (1955) Some statistical models related with series of events. *J. Roy. Statist. Soc. B* **17**, 129–164.

Cox, D.R., and Isham, V. (1980) *Point Processes.* Chapman & Hall, London.

Cressie, N. (1993) *Statistics for Spatial Data,* revised edition. John Wiley & Sons, Inc., New York.

Cressie, N., and Collins, L.B. (2001) Analysis of spatial point patterns using bundles of product density LISA functions. *J. Agric. Biol. Envir. Statist.* **6**, 118–135.

Dale, M.R.T. (1999) *Spatial Pattern Analysis in Plant Ecology.* Cambridge Universiy Press, Cambridge.

Dale, M.R.T., Dixon, P., Fortin, M.-J., Legendre, P., Myers, D.E., and Rosenberg, M.S. (2002) Conceptual and mathematical relationships among methods for spatial analysis. *Ecography* **25**, 558–577.

Daley, D.J., and Vere-Jones, D. (1988) *An Introduction to the Theory of Point Processes.* Springer-Verlag, New York.

Daley, D.J., and Vere-Jones, D. (2003) *An Introduction to the Theory of Point Processes. Volume I: Elementary Theory and Methods,* 2nd edition. Springer-Verlag, New York.

Daley, D.J., and Vere-Jones, D. (2008) *An Introduction to the Theory of Point Processes. Volume II: General Theory and Structure,* 2nd edition. Springer-Verlag, New York.

Davison, A.C., and Hinkley, D.V. (1997) *Bootstrap Methods and their Applications.* Cambridge University Press, Cambridge.

Dean, C. (2003) Calculation of wood volume and stem taper using terrestrial single-image close-range photogrammetry and contemporary software tools. *Silva Fennica* **37**, 359–380.

Degenhardt, A. (1999) Description of tree distribution patterns and their development through marked Gibbs processes. *Biometrical J.* **41**, 457–470.

Devroye, L. (1986) *Non-uniform Random Variate Generation.* Springer-Verlag, Berlin.

Diggle, P.J. (1979) On parameter estimation and goodness-of-fit testing for spatial point-patterns. *Biometrics* **35**, 87–101.

Diggle, P.J. (1982) Some statistical aspects of spatial distribution models for plants and trees. *Studia Forest. Suec.* no. 162.

Diggle, P.J. (1983) *Statistical Analysis of Spatial Point Patterns.* Academic Press, London.

Diggle, P.J. (1985) A kernel method for smoothing point process data. *Appl. Statist.* **34**, 138–147.

Diggle, P.J. (2003) *Statistical Analysis of Spatial Point Patterns.* Arnold, London.

Diggle, P.J. (2007) Spatio-temporal point processes: methods and applications. In B. Finkenstad, L. Held and V. Isham (eds), *Statistical Methods for Spatio-temporal Systems.* Chapman & Hall/CRC Press, Boca Raton, FL, pp. 1–45.

Diggle, P.J., Gates, D.J., and Stibbard, A. (1987) A nonparametric estimator for pairwise interaction point processes. *Int. Statist. Rev.* **62**, 763–770.

Diggle, P.J., Lange, N., and Beneš, F. (1991) Analysis of variance for replicated spatial point patterns in clinical neuroanatomy. *J. Amer. Statist. Assoc.* **86**, 618–625.

Diggle, P.J., Fiksel, T., Grabarnik, P., Ogata, Y., Stoyan, D., and Tanemura, M. (1994) On parameter estimation for pairwise interaction point proceses. *Int. Statist. Rev.* **62**, 99–117.

Diggle, P.J., Chetwynd, A.G., Haggkvist, R., and Morris, S. (1995) Second-order analysis of space–time clustering. *Statist. Meth. Medical Res.* **4**, 124–136.

Diggle, P.J., Mateu, J., and Clough, H.E. (2000) A comparison between parametric and non-parametric approaches to the analysis of replicated spatial point patterns. *Adv. Appl. Probab.* **32**, 331–343.

Diggle, P.J., Eglen, S.J., and Troy, J.B. (2006) Modelling the bivariate spatial distribution of amacrine cells. In A. Baddelely, P. Gregori, J. Mateu, R. Stoica and D. Stoyan (eds), *Case Studies in Spatial Point Process Modeling,* Lecture Notes in Statistics 185. Spinger-Verlag, New York, pp. 215–233.

Diggle, P.J., Gómez-Rubio, V., Brown, P.E., Chetwynd, A.G., and Gooding, S. (2007) Second-order analysis of inhomogeneous spatial point processes using case-control data. *Biometrics* **63**, 550–557.

Dimov, L.D., Chambers, J.L., and Lockhart, B.R. (2005) Spatial continuity of tree attributes in bottomland hardwood forests in the southeastern United States. *Forest Sci.* **51**, 532–540.

Döge, G. (2001) Perfect simulation for random sequential adsorption of d-dimensional spheres with random radii. *J. Statist. Comput. Simul.* **69**, 141–156.

Döge, G., Mecke, K., Møller, J., Stoyan, D., and Waagepetersen, R. (2004) Grand canonical simulations of hard-disk systems by simulated annealing. *Int. J. Modern Phys. C* **15**, 129–147.

Doguwa, S.I. (1989) On second order neighbourhood analysis of mapped point patterns. *Biometrical J.* **31**, 451–457.

Doguwa, S.I., and Upton, G.J.G. (1989) Edge-corrected estimators for the reduced second moment measure of point processes. *Biometrical J.* **31**, 563–675.

Dubé, P., Fortin, M.-J., Canham, C.D., and Marceau, D.J. (2001) Quantifying gap dynamics and spatio-temporal structures in spatially-explicit models of temperate forest ecosystems. *Ecol. Modelling* **142**, 39–60.

Edelsbrunner, H. (1995) The union of balls and its dual shape. *Discrete Comput. Geom.* **13**, 415–440.

Edelsbrunner, H., and Mücke, E.P. (1994) Three-dimensional alpha shapes. *ACM Trans. Graphs* **13**, 43–72.

Efron, B., and Tibshirani, R.J. (1993) *An Introduction to the Bootstrap.* Chapman & Hall, London.

Eisenstein, D.J., and Hut, P. (1998) HOP: A new group-finding algorithm for N-body simulations. *Astrophys. J.* **498**, 137–142.

Elliott, P., Wakefield, J.C., Best, N.G., and Briggs, D.J. (eds) (2000) *Spatial Epidemiology.* Oxford University Press, Oxford.

Evans, J.W. (1993) Random and cooperative adsorption. *Rev. Modern Phys.* **65**, 1281–1329.

Everitt, B.S., Landau, S., and Leese, N. (2001) *Cluster Analysis.* Arnold, London.

Felsenstein, J. (1975) A pain in the torus: Some difficulties with models of isolation by distance. *Amer. Nat.* **109**, 359–368.

Ferreira, J.T.A.S. (2002) Partition modelling. In A.B. Lawson and D.G.T. Denison (eds), *Spatial Cluster Modelling.* Chapman & Hall/CRC, Boca Raton, FL, pp. 125–146.

Fiksel, T. (1984) Estimation of parametrized pair potentials of marked and non-marked Gibbsian point processes. *Elektron. Informationsverarb. Kybernet.* **20**, 270–278.

Fiksel, T. (1986) Estimation of interaction potentials of Gibbsian point processes. *Statistics* **19**, 77–86.

Fisher, N.I., Lewis, T. and Embleton, B.I.J. (1987) *Statistical Analysis of Spherical Data.* Cambridge University Press, Cambridge.

Fisher, R.A., Thornton, H.G., and Mackenzie, W.A. (1922) The accuracy of the plating method of estimating the density of bacterial populations, with particular reference to the use of Thornton's agar medium with soil samples. *Ann. Appl. Biol.* **9**, 325–359.

Fishman, G.S. (1996) *Monte Carlo. Concepts, Algorithms, and Applications.* Springer-Verlag, New York.

Fleischer, E., Eckel, S., Schmidt, I., Schmidt, V., and Kazda, M. (2006) Point process modelling of root distribution in pure stands of *Fagus sylvatica* and *Picea abies. Can. J. Forest Res.* **36**, 227–237.

Floresroux, E.M., and Stein, M.L. (1996) A new method of edge correction for estimating the nearest neighbor distribution. *J. Statist. Plann. Infer.* **50**, 353–371.

Fotheringham, A.S., and Zhan, F.B. (1996) A comparison of three exploratory methods for cluster detection in spatial point processes. *Geogr. Anal.* **28**, 200–218.

Franklin, J., Michaelsen, J., and Strahler, A.H. (1985) Spatial analysis of density dependent pattern in coniferous forest stands. *Vegetatio* **64**, 29–36.

Freudenthal, A.M. (1950) *The Inelastic Behaviour of Engineering Materials and Structures.* J. Wiley & Sons, Chichester.

Füldner, K. (1995) *Strukturbeschreibung von Buchen-Edellaubholz-Mischwäldern.* Doctoral dissertation, University of Göttingen. Cuvillier-Verlag, Göttingen (in German).

Gadow, K. von, Hui, G.Y., and Albert, M. (1998) Das Winkelmaß – ein Strukturparameter zur Beschreibung der Individualverteilung in Waldbeständen. *Forstwiss. Centralbl.* **115**, 1–9 (in German).

Gadow, K. von, Hui, G.Y., Chen, B.W. and Albert, M. (2003) Beziehungen zwischen Winkelmaß und Baumabständen. *Forstwiss. Centralbl.* **122**, 127–137 (in German).

Gardi, J.E., Nyengaard, J.R., and Gundersen, H.J.G. (2006) Using biased image analysis for improving unbiased stereological number estimation – a pilot simulation study of the smooth fractionator. *J. Microsc.* **222**, 242–250.

Gates, D.J., and Westcott, M. (1986) Clustering estimates in spatial point processes with stable potentials. *Ann. Inst. Statist. Math.* **38**, 123–135.

Gelman, A., Carlin, J.B., Stern, H., and Rubin, B.D. (2004) *Bayesian Data Analysis,* 2nd edition. Chapman & Hall/CRC, Boca Raton, FL.

Gentle, J.E. (2003) *Random Number Generation and Monte Carlo Methods,* 2nd edition. Springer-Verlag, New York.

Georgii, H.-O. (1976) Canonical and grand canonical Gibbs states for continuum systems. *Comm. Math. Phys.* **48**, 31–51.

Getis, A., and Franklin, J. (1987) Second-order neighbourhood analysis of mapped point patterns. *Ecology* **68**, 473–477.

Geyer, C.J. (1999) Likelihood inference for spatial point processes. In O.E. Barndorff-Nielsen, W.S. Kendall and M.N.M. van Lieshout (eds), *Stochastic Geometry, Likelihood and Computation.* Chapman & Hall/CRC, Boca Raton, FL, pp. 79–140.

Geyer, C.J., and Møller, J. (1994) Simulation procedures and likelihood inference for spatial point processes. *Scand. J. Statist.* **21**, 359–373.

Geyer, C.J., and Thompson, E.A. (1992) Constrained Monte Carlo maximum likelihood for dependent data. *J. Roy. Statist. Soc. B* **54**, 657–699.

Ghorbani, H., Möller, H.J., and Stoyan, D. (2006) Using Pareto and Weibull distributions in the modelling of growth processes. *South African Statist. J.* **40**, 75–98.

Gibbons, J.D. (2005) *P*-values. In S. Kotz (ed.), *Encyclopedia of Statistical Sciences,* 2nd edition. John Wiley & Sons, Inc., Hoboken, NJ p. 6636.

Gignoux, J., Duby, C., and Barot, S. (1999) Comparing the performance of Diggle's tests of spatial randomness for small samples with and without edge-effect correction: Application to ecological data. *Biometrics* **55**, 156–164.

Gilks, W.R., Richardson, S., and Spiegelhalter, D.J. (1996) *Markov Chain Monte Carlo in Practice.* Chapman & Hall, London.

Givens, G.H., and Hoeting, J.A. (2005) *Computational Statistics.* John Wiley & Sons, Inc., Hoboken, NJ.

Glasbey, C., and Roberts, I.M. (1997) Statistical analysis of the distribution of gold particles over antigen sites after immunogold labelling. *J. Microsc.* **186**, 258–262.

Glass, L., and Tobler, W.R. (1971) Uniform distribution of objects in a homogeneous field: Cities on a plain. *Nature* **233**(5314), 67–68.

Goes, S., Spakman, W., and Bijwaard, H. (1999) A lower mantle source for central Europaean volcanism. *Science* **286**, 1928–1931.

Goreaud, F., and Pélissier, R. (1999) On explicit formulas of edge effect correction for Ripley's *K*-function. *J. Veget. Sci.* **10**, 433–438.

Goreaud, F., and Pélissier, R. (2003) Avoiding misinterpretation of biotic interactions with the intertype K_{12}-function: Population independence vs. random labelling hypotheses. *J. Veget. Sci.* **14**, 681–692.

Goulard, M., Särkkä, A., and Grabarnik, P. (1996) Parameter estimation for marked Gibbs point processes through the maximum pseudo-likelihood method. *Scand. J. Statist.* **23**, 365–379.

Grabarnik, P., and Chiu, S.N. (2002) Goodness-of-fit test for complete spatial randomness against mixtures of regular and clustered spatial point processes. *Biometrika* **89**, 411–412.

Greig-Smith, P. (1964) *Quantitative Plant Ecology*. Butterworths, London.

Guan, Y. (2007) A least-squares cross-validation band width selection approach in pair correlation function estimation. *Statist. Probab. Lett.* **18**.

Gubner, J.A., Chang, W.-B. and Hayat, M.M. (2000) Performance analysis of hypothesis testing for sparse pairwise interaction point processes. *IEEE Trans. Inform. Theory* **46**, 1357–1365.

Guillot, G., Kan-King-Yu, D., Michelin, J., and Huet, P. (2006) Inference of a hidden spatial tessellation from multivariate data: application to the delineation of homogeneous regions in an agricultural field. *Appl. Statist.* **55**, 407–430.

Gundersen, H.J.G. (1986) Stereology of arbitrary particles. *J. Microsc.* **143**, 3–45.

Gundersen, H.J.G. (2002) The smooth fractionator. *J. Microsc.* **207**, 191–210.

Günel, E., and Chilko, D. (1989) Estimation of parameter n of the binomial distribution. *Comm. Statist. Simul. Comput.* **18**, 537–555.

Haase, P. (2001) Can isotropy vs. anisotropy in the spatial association of plant species reveal physical vs. biotic facilitation? *J. Veget. Sci.* **12**, 127–136.

Hahn, U., Micheletti, A., Pohlink, R., Stoyan, D., and Wendrock, H. (1999) Stereological analysis and modelling of gradient structures. *J. Microsc.* **195**, 113–124.

Hahn, U., Jensen, E.B.V., Lieshout, M.N.M. van, and Nielsen, L.S. (2003) Inhomogeneous spatial point processes by location dependent scaling. *Adv. Appl. Probab.* **35**, 319–336.

Hamilton, A.J.S. (1993) Toward better ways to measure the galaxy correlation function. *Astrophys. J.* **417**, 19–35.

Hanisch, K.-H. (1982) On inversion formulas for n-fold Palm distributions of point processes in LCS spaces. *Math. Nachr.* **106**, 171–179.

Hanisch, K.-H. (1983) Reduction of n-th moment measures and the special case of the third moment measure of stationary and isotropic planar point processes. *Math. Operationsforsch. Statist. Ser. Statistics* **14**, 421–435.

Hanisch, K.-H. (1984) Some remarks on estimators of the distribution function of nearest neighbour distance in stationary spatial point patterns. *Math. Operationsforsch. Statist. Ser. Statistics* **15**, 409–412.

Hanisch, K.-H., and Stoyan, D. (1983) Remarks on statistical inference and prediction for a hard-core clustering model. *Math. Operationsf. Statist., Ser. Statistics* **14**, 559–567.

Hanisch, K.-H., and Stoyan, D. (1984) Once more on orientations in point processes. *J. Inform. Proc. Cybernet.* **20**, 279–284.

Hansen, J.-P., McDonald, I.R. (1986) *Theory of Simple Liquids*. Academic Press, London.

Hansen, M.B., Baddeley, A., and Gill, R. (1999) First contact distributions for spatial patterns: Regularity and estimation. *Adv. Appl. Probab.* **31**, 15–33.

Hastings, W. (1970) Monte Carlo sampling methods using Markov chains and their application. *Biometrika* **57**, 97–109.

Hegyi, F. (1974) A simulation model for managing jack-pine stands. In J. Fries (ed.), *Growth Models for Tree and Stand Simulation*. Royal Coll. of Forestry, Stockholm, pp. 74–76.

Heikkinen, J., and Arjas, E. (1998) Non-parametric Bayesian estimation of a spatial Poisson intensity. *Scand. J. Statist.* **25**, 435–450.

Heikkinen, J., and Penttinen, A. (1999) Bayesian smoothing in the estimation of the pair potential function of Gibbs point processes. *Bernoulli* **5**, 1119–1136.

Heinrich, L. (1986) Asymptotic normality of a random point field characteristic. *Statistics* **17**, 453–460.

Heinrich, L. (1988) Asymptotic Gaussianity of some estimators for reduced factorial moment measures and product densities of stationary Poisson cluster processes. *Statistics* **19**, 87–106.

Heinrich, L. (1991) Goodness-of-fit tests for the second-order moment function of a stationary multidimensional Poisson process. *Statistics* **22**, 245–268.

Heinrich, L. (1992) Minimum contrast estimators for parameters of spatial ergodic point processes. In *Transactions of the 11th Prague Conference on Random Processes, Information Theory and Statistical Decision Functions.* Academia, Prague, pp. 479–492.

Heinrich, L. (1993) Asymptotic properties of minimum contrast estimators for parameters of Boolean models. *Metrika* **40**, 67–94.

Heinrich, L., and Prokešová, M. (2006) On estimating the asymptotic variance of stationary point processes. Thiele Center Research Report no. 17.

Heinrich, L., and Schmidt, V. (1985) Normal convergence of multidimensional shot-noise processes and rates of this convergence. *Adv. Appl. Probab.* **17**, 709–730.

Hickernell, F.J., Lemieux, C., and Owen, A.B. (2005) Control variates for quasi-Monte Carlo. *Statist. Sci.* **20**, 1–31.

Ho, L.P., and Chiu, S.N. (2006) Testing the complete spatial randomness by Diggle's test without an arbitrary upper limit. *J. Statist. Comput. Simul.* **76**, 585–591.

Ho, L.P., and Chiu, S.N. (2007a) Testing uniformity of a spatial point pattern. *J. Comput. Graph. Statist.* **16**, 378–398.

Ho, L.P., and Chiu, S.N. (2007b) Using weight functions in spatial point pattern analysis with application to plant ecology data. Preprint, Hong Kong Baptist University.

Ho, L.P., and Stoyan, D. (2008) Modelling marked point patterns by intensity-marked Cox processes. *Statist. Probab. Lett.* To appear

Hoel, P.G. (1943) On indices of dispersion. *Ann. Math. Statist.* **14**, 155–162.

Hoffman, C., Holroyd, A.E., and Peres, Y. (2006) A stable marriage of Poisson and Lebesgue. *Ann. Probab.* **34**, 1241–1272.

Hoffman, R., and Jain, A.K. (1983) A test of randomness based on the minimal spanning tree. *Pattern Recogn. Lett.* **1**, 175–180.

Holgate, P. (1965) Some new tests of randomness. *J. Ecology* **53**, 261–266.

Holroyd, A.E., and Peres, Y. (2005) Extra heads and invariant allocations. *Ann. Probab.* **33**, 31–52.

Hua, L.K., and Wang, Y. (1981) *Applications of Number Theory to Numerical Analysis.* Springer-Verlag, Berlin.

Huang, F., and Ogata, Y. (1999) Improvements of the maximum pseudo-likelihood estimators in various spatial statistical models. *J. Comput. Graph. Statist.* **8**, 510–530.

Hubalková, J., and Stoyan, D. (2003) On a qualitative relationship between degree of inhomogenity amd cold crushing strength of refractory castables. *Cement Concrete Res.* **33**, 747–753.

Hui, G.Y., Albert, M., and Gadow, K. von (1998) Das Umgebungsmaß als Parameter zur Nachbildung von Bestandesstrukturen. *Forstw. Centralblatt* **117**, 258–266 (in German).

Ickstadt, K., and Wolpert, R.L. (1997) Multiresolution assessment of forest inhomogeneity. In C. Gatsonis, J.S. Hodges, R.E. Kass, R. McCulloch, P. Rossi and N.D. Singpurwalla (eds), *Case Studies in Bayesian Statistics, Volume III.* Springer-Verlag, New York, pp. 371–386.

Illian, J.B. (2006) Spatial point process modelling of a biodiverse plant community. PhD thesis, University of Abertay Dundee.

Illian, J.B., Benson, E., Crawford, J., and Staines, H. (2006) Principal component analysis for spatial point processes – assessing the appropriateness of the approach in an ecological context. In A. Baddeley, P. Gregori, J. Mateu, R. Stoica and D. Stoyan (eds.), *Case Studies in Spatial Point Process Modeling,* Lecture Notes in Statistics 185. Springer-Verlag, New York, pp. 135–150.

Illian, J.B., Møller, J., and Waagepetersen, R.P. (2008) Hierarchical spatial point process analysis for a plant community with high biodiversity. *Environm. Ecol. Statist.* To appear.

Isham, V. (1987) Marked point processes and their correlations. In F. Droesbeke (ed.), *Spatial Processes and Spatial Time Series Analysis,* Proceedings of the 6th Franco-Belgian Meeting of Statisticians, 1985. Publications des Facultés Universitaires Saint-Louis, Brussels, pp. 63–75.

Ivanoff, G. (1982) Central limit theorems for point processes. *Stoch. Proc. Appl.* **12**, 171–186.

Jammalamadaka, S.R., and Penrose, M.D. (2000) Poisson limits for pairwise and area interaction point processes. *Adv. Appl. Probab.* **32**, 75–85.

Jensen, E.B.V., and Nielsen, L.S. (2000) Inhomogeneous Markov point processes by transformation. *Bernoulli* **6**, 761–782.

Jodrey, W.S., and Tory, E.M. (1979) Simulation of random packing of spheres. *J. Simulation* **32**, 1–12.

Jodrey, W.S., and Tory, E.M. (1985) Computer simulation of close random packing of equal spheres. *Phys. Rev. A* **32**, 2347–2351.

Johnson, N.L., Kemp, A.W., and Kotz, S. (2005) *Univariate Discrete Distributions.* John Wiley & Sons, Inc., Hoboken, NJ.

Jolivet, E. (1986) Parametric estimation of the covariance density for a stationary point process on $\mathbb{R}^d$. *Stoch. Proc. Appl.* **22**, 111–119.

Jones, B.J.T., Martínez V.J., Saar, E., and Trimble, V. (2004) Scaling laws in the distribution of galaxies. *Rev. Modern Phys.* **76**, 1211–1266.

Kallenberg, O. (1983) *Random Measures,* 3rd edition, revised and enlarged. Akademie-Verlag, Berlin, and Academic Press, London.

Kallenberg, O. (2002) *Foundations of Modern Probability,* 2nd edition. Springer-Verlag, New York

Kelly, F.P., and Ripley, B.D. (1976) A note on Strauss' model for clustering. *Biometrika* **63**, 357–360.

Kendall, W.S., and Thönnes, E. (1999) Perfect simulation in stochastic geometry. *Pattern Recogn.* **32**, 1569–1586.

Kendall, W.S., Lieshout, M.N.M. van, and Baddeley, J. (1999) Quermass-interaction processes: Conditions for stability. *Adv. Appl. Probab.* **31**, 315–342.

Kerle, N., Janssen, L.L.F., and Huurneman, G.C. (2004) *Principles of Remote Sensing. An Introductory Textbook.* International Institute for Geo-information Science and Earth Observation, Enschede (NL).

Kingman, J.F.C. (1993) *Poisson Processes.* Clarendon Press, Oxford.

Kint, V., Meirvenne, M. van, Nachergale, L., Geudens, G., and Lust, N. (2003) Spatial methods for quantifying forest stand development: A comparison between nearest neighbor indices and variogram analysis. *Forest Sci.* **49**, 36–49.

Konecny, G. (2003) *Geoinformation: Remote Sensing, Photogrammetry and Geographic Information Systems.* Taylor and Francis, London.

Krebs, C.J. (1998) *Ecological Methodology,* 2nd edition. Addison Wesley Longman, New York.

Kühlmann-Berenzon, S., Heikkinen, J., and Särkkä, A. (2005) An additive edge correction for the influence potential of trees. *Biometrical J.* **47**, 517–526.

Kuuluvainen, T., and Pukkala, T. (1989) Effect of Scots pine seed trees on the density of ground vegetation and tree seedlings. *Silva Fennica* **23**, 159–167.

Kvarnström, M., and Glasbey, C.A. (2007) Estimation of centres and radial intensity profiles of spherical nano-particles in digital microscopy. *Biometrical J.* **49**, 300–311.

Lachmanovich, E., Shvartsman, D.E., Malka, Y., Botvin, C., Henis, Y.I., and Weiss, A.M. (2003) Co-localization analysis of complex formation among membrane proteins by computerized fluorescence microscopy: Application to immunofluorescence co-patching studies. *J. Microsc.* **212**, 122–131.

Lancaster, J. (2006) Using neutral landscapes to identify patterns of aggregations across resource points. *Ecography* **29**, 385–396.

Lancaster, J., and Downes, B.J. (2004) Spatial point pattern analysis of available and exploited resources. *Ecography* **27**, 94–102.

Landy, S.L., and Szalay, A.S. (1993) Bias and variance of angular correlation functions. *Astrophys. J.* **412**, 64–71.

Lantuéjoul, C. (2002) *Geostatistical Simulation. Models and Algorithms.* Springer-Verlag, Berlin.

Last, G., and Schassberger, R. (2001) On the second derivative of the spherical contact distribution function of smooth grain models. *Probab. Theory Related Fields* **121**, 49–72.

Lawson, A.B., and Denison, D.G.T. (eds) (2002) *Spatial Cluster Modelling.* Chapman & Hall/CRC, Boca Raton, FL.

Lepš, J., and Kindlmann, P. (1987) Models of the development of the spatial pattern of an even-aged plant population over time. *Ecol. Modelling* **39**, 45–57.

Lewis, P.A.W., and Shedler, G.S. (1979) Simulation of non-homogeneous Poisson processes by thinning. *Naval Res. Log. Quart.* **26**, 403–413.

Liebhold, A.M., and Gurevitch, J. (2002) Integrating the statistical analysis of spatial data in ecology. *Ecography* **25**, 553–557.

Liemant, A., Matthes, K., and Wakolbinger, A. (1988) *Equilibrium Distributions of Branching Processes.* Kluwer, Dortrecht.

Lieshout, M.N.M. van (1995) *Stochastic Geometry Models in Image Analysis and Spatial Statistics.* CWI Tracts 108, Amsterdam.

Lieshout, M.N.M. van (2000) *Markov Point Processes and their Applications.* Imperial College Press, London.

Lieshout, M.N.M. van (2006a) A *J*-function for marked point patterns. *Ann. Inst. Statist. Math.* **58**, 235–259.

Lieshout, M.N.M. van (2006b) Markovianity in space and time. In D. Denteneer, F. den Hollander and E. Verbitskiy (eds), *Dynamics and Stochastics: Festschrift in Honor of M.S. Keane,* IMS Lecture Notes – Monograph Series 48. Institute of Mathematical Statistics, Beachwood, OH, pp. 154–168.

Lieshout, M.N.M. van (2006c) Maximum-likelihood estimation for random sequential adsorption. *Adv. Appl. Probab.* **38**, 889–898.

Lieshout, M.N.M. van, and Baddeley, A.J. (1995) Markov chain Monte Carlo methods for clustering of image features. In *Fifth International Conference on Image Processing and its Applications.* IEE, London, pp. 241–245.

Lieshout, M.N.M. van, and Baddeley, A.J. (1996) A nonparametric measure of spatial interaction in point patterns. *Statist. Neerland.* **50**, 344–361.

Lieshout, M.N.M. van, and Baddeley, A.J. (2002) Extrapolating and interpolating spatial patterns. In A.B. Lawson and D. Denison (eds), *Spatial Cluster Modelling.* Chapman & Hall/CRC, Boca Raton, FL, pp. 61–86.

Lieshout, M.N.M. van, and Stoica, R.S. (2006) Perfect simulation for marked point processes. *Comput. Statist. Data. Anal.* **51**, 679–698.

Lochmann, K., Anikeenko, A., Elsner, A., Medvedev, N., and Stoyan, D. (2006a) Statistical verification of crystallization in hard sphere packings under densification. *Eur. Phys. J. B* **53**, 67–76.

Lochmann, K., Oger, L., and Stoyan, D. (2006b) Statistical analysis of random sphere packings with variable radius distribution. *Solid State Sci.* **8**, 1397–1413.

Loosmore, N.B., and Ford, E.D. (2006) Statistical inference using the G or K point pattern spatial statistics. *Ecology* **87**, 1925–1931.

Low, R.J. (2002) Measuring order and biaxiality. *Eur. J. Phys.* **23**, 111–117.

Löwen, H. (2000) Fun with hard spheres. In K.R. Mecke and D. Stoyan (eds), *Statistical Physics and Spatial Statistics,* Springer Lecture Notes in Physics 554. Springer-Verlag, Berlin, pp. 295–331.

Lubachevsky, B.D., and Stillinger, F.H. (1990) Geometric properties of random disk packings. *J. Statist. Phys.* **60**, 561–583.

Lund, J., and Rudemo, M. (2000) Models for point processes observed with noise. *Biometrika* **87**, 235–249.

Mandallaz, D. (2000) Estimation of the spatial covariance in universal kriging: Application to forest inventory. *Environm. Ecol. Statist.* **7**, 263–284.

Manly, B.F.J. (2004) *Multivariate Statistical Methods: A Primer,* 3rd edition. Chapman & Hall/CRC, Boca Raton, FL.

Manly, B.F.J. (2006) *Randomization, Bootstrap and Monte Carlo Methods in Biology,* 3rd edition. Chapman & Hall/CRC, Boca Raton, FL.

Månsson M., and Rudemo, M. (2002) Random patterns of nonoverlapping convex grains. *Adv. Appl. Probab.* **34**, 718–738.

Marchette, D.J. (2004) *Random Graphs for Statistical Pattern Recognition* J. Wiley & Sons, Inc., Hoboken, NJ.

Mardia, K.V., Kent, J.T., and Bibby, J.M. (1989) *Multivariate Analysis,* 7th printing. Academic Press, London.

Martínez, V.J., and Saar, E. (2002) *Statistics of the Galaxy Distribution.* Chapman & Hall/CRC, Boca Raton, FL.

Martínez, V.J., Starck, J.L., Saar, E., Donoho, D.L., Reynolds, S.C., de la Cruz, P., and Paredes, S. (2005) Morphology of the galaxy distribution from wavelet denoising. *Astrophys. J.* **634**, 744–755.

Mase, S. (1996) The threshold method for estimating total rainfall. *Ann. Inst. Statist. Math.* **48**, 201–213.

Mase, S., Møller, J., Stoyan, D., Waagepetersen, R., and Döge, G. (2001) Packing densities and simulated tempering for hardcore Gibbs point processes. *Ann. Inst. Statist. Math.* **53**, 661–680.

Matérn, B. (1960) Spatial variation: Stochastic models and their applications to problems in forest surveys and other sampling investigations. *Meddelanden från Statens Skogsforskningsinstitut*, **49**(5).

Matérn, B. (1986) *Spatial Variation*, Lecture Notes in Statistics 36. Springer-Verlag, Berlin.

Mateu, J., and Montes, F. (2001) Likelihood inference for Gibbs processes in the analysis of spatial point processes. *Int. Statist. Rev.* **69**, 81–104.

Matheron, G. (1968) Schema booléen séquentiel de partition aléatoire. N-83 CMM, Paris School of Mines Publication.

Mattern, N., Sakowski, J., Macht, M.-P., Jovari, P., and Jiang, J. (2003) Structural behavior of $Pd_{40}Cu_{30}Ni_{10}P_{20}$ bulk metallic glass below and above the glass transition. *Appl. Phys. Lett.* **82**, 2589–2591.

Mattfeldt, T. (2005) Explorative statistical analysis of planar point processes in microscopy. *J. Microsc.* **220**, 131–139.

Mecke, J. (1967) Stationäre zufällige Maße auf lokalkompakten Abelschen Gruppen. *Z. Wahrscheinlichkeitsth. verw. Geb.* **9**, 36–58 (in German).

Mecke, K.R. (1998) Integral geometry and statistical physics. *Int. J. Modern Phys. B* **12**, 861–899.

Mecke, K.R. (2000) Additivity, convexity, and beyond: Applications of Minkowski functionals in statistical physics. In K.R. Mecke and D. Stoyan (eds.), *Statistical Physics and Spatial Statistics,* Lecture Notes in Physics 554. Springer-Verlag, Berlin, pp. 111–184.

Mecke, K.R., and Stoyan, D. (2005) Morphological characterisation of point patterns. *Biometrical J.* **47**, 473–488.

Mecke, K.R., Buchert, T., and Wagner, H. (1994) Robust morphological measures for large-scale structure in the universe. *Astron. Astrophys.* **288**, 697–704.

Medvedev, N.N. (2000) *The Voronoi–Delaunay Method for Non-Crystalline Systems*. Nauka, Novosibirsk (in Russian).

Medvedev, N.N., and Naberukhin, Yu.I. (1987) Shape of the Delaunay simplices in dense random packings of hard and soft spheres. *J. Non-Crystalline Solids* **94**, 402–406.

Metropolis, N., Rosenbluth, A.W., Rosenbluth, M.N., Teller, A.H., and Teller, E. (1953) Equation of state calculations by fast computing machines. *J. Chem. Phys.* **21**, 1087–1092.

Miina, J., and Pukkala, T. (2002) Application of ecological field theory in distance-dependent growth modelling. *Forest Ecol. Manag.* **161**, 101–107.

Moeur, M. (1993) Characterizing spatial patterns of trees using stem-mapped data. *Forest Sci.* **39**, 756–775.

Molchanov, I. (2005) *Theory of Random Sets*. Springer-Verlag, London.

Møller, J. (2001) A review of perfect simulation in stochastic geometry. In: I.V. Basawa, C.C. Heyde and R.L. Taylor (eds), *Selected Proceedings of the Symposium on Inference for Stochastic Processes*. IMS Lecture Notes – Monograph Series 37. Institute of Mathematical Statistics, Beachwood, OH, pp. 333–355.

Møller, J. (2003) Shot noise Cox processes. *Adv. Appl. Probab.* **35**, 614–640.

Møller, J., and Torrisi, G.L. (2005) Generalized shot noise Cox processes. *Adv. Appl. Probab.* **37**, 48–74.

Møller, J., and Waagepetersen, R.P. (2002) Statistical inference for Cox processes. In A.B. Lawson and D.G.T. Denison (eds), *Spatial Cluster Modelling*. Chapman & Hall/CRC, Boca Raton, FL, pp. 37–60.

Møller, J., and Waagepetersen, R.P. (2004) *Statistical Inference and Simulation for Spatial Point Processes*. Chapman & Hall/CRC, Boca Raton, FL.

Møller, J., and Waagepetersen, R.P. (2007) Modern statistics for spatial point processes. *Scand. J. Statist.* **34**, 643–684, discussion 685–711.

Møller, J., Syversveen, A.R., and Waagepetersen, R.P. (1998) Log Gaussan Cox processes. *Scand. J. Statist.* **25**, 451–482.

Møller, J., Pettitt, A.N., Berthelsen, K.K., and Reeves R.W. (2006) An efficient Markov chain Monte Carlo method for distributions with intractable normalising constants. *Biometrika* **93**, 451–458.

Moran, P.A.P. (1976) Another quasi-Poisson planar point process. *Probab. Theory Related Fields* **33**, 269–272.

Moravie, M.-A., Durand, M., and Houllier, F. (1999) Ecological meaning and predictive ability of social status, vigour and competition indices in a tropical rain forest (India). *Forest Ecol. Manag.* **117**, 221–240.

Moscinski, J., Bargiel, M., Rycerz, Z.A., and Jacobs, P.W.M. (1989) The force biased algorithmm for the irregular close packing of equal hard spheres. *Mol. Simul.* **3**, 201–212.

Mrkvicka, T., and Molchanov, I. (2005) Optimisation of linear unbiased intensity estimators. *Ann. Inst. Statist. Math.* **57**, 71–82.

Muche, L., Rother, P., Friesenegger, A., and Geupel, M. (2000) Evaluation of inhomogeneities in histological structures (cartilage, retina). *Image Anal. Stereol.* **19**, 119–124.

Mugglestone, M.A., and Renshaw, E. (1996a) A practical guide to the spectral analysis of spatial point processes. *Comput. Statist. Data Anal.* **21**, 43–65.

Mugglestone, M.A., and Renshaw, E. (1996b) The exploratory analysis of bivariate spatial point patterns using cross-spectra. *Environmetrics* **7**, 361–377.

Mugglestone, M.A., and Renshaw, E. (2001) Spectral tests for randomness for spatial point patterns. *Environm. Ecol. Statist.* **8**, 237–251.

Myles, J.P., Flenley, E.C., Fieller, N.R.J., Alkinson, H.V., and Jones, H. (1995) Statistical tests for clustering of second phases in composite materials. *Philos. Mag.* **72**, 515–528.

Myllymäki, M., and Penttinen, A. (2007) Conditionally heteroscedastic intensity-dependent marking of log Gaussian Cox processes. University of Jyväskylä, Department of Mathematics and Statistics, Preprint 357.

Naberukhin, Yu.I., Voloshin, V.P., and Medvedev, N.N. (1991) Geometrical analysis of the structure of simple liquids: Percolation approach. *Mol. Phys.* **73**, 917–936.

Näther, W., and Wälder, K. (2003) Experimental design and statistical inference for cluster point processes – with applications to the fruit dispersion of anemochorus forest trees. *Biometrical J.* **45**, 1006–1022.

Neyman, J., and Scott, E.L. (1952) A theory for the spatial distribution of galaxis. *Astrophys. J.* **116**, 144–163.

Neyman, J., and Scott, E.L. (1958) Statisical approach to problems of cosmology (with discussion). *J. Roy. Statist. Soc. B* **20**, 1–43.

Neyman, J., and Scott, E.L. (1972) Processes of clustering and applications. In P.A.W. Lewis (ed.), *Stochastic Point Processes*. J. Wiley & Sons, Inc., New York, pp. 646–681.

Neyman, J., Scott, E.L., and Shane, C.D. (1953) On the spatial distribution of galaxies: a specific model. *Astrophys. J.*, **117**, 92–133.

Neyrinck, M.C., Gnedin, N.Y., and Hamilton, A.J.S. (2005) VOBOZ: an almost-parameter-free halo-finding algorithm. *Mon. Not. Roy. Astron. Soc.* **356**, 1222–1232.

Niederreiter, H. (1992) *Random Number Generation and Quasi-Monte Carlo Methods*. SIAM, Philadelphia.

Nielsen, L.S. (2000) Modelling the position of cell profiles allowing for both inhomogeneity and interaction. *Image Anal. Stereol.* **19**, 183–187.

Ogata, Y. (1998) Space–time point process models for earthquake occurrences. *Ann. Inst. Statist. Math.* **50**, 379–402.

Ogata, Y., and Tanemura, M. (1981) Estimation of interaction potentials of spatial point patterns through the maximum likelihood procedure. *Ann. Inst. Statist. Math.* **33**, 315–338.

Ogata, Y., and Tanemura, M. (1984) Likelihood analysis of spatial point patterns. *J. Roy. Statist. Soc. B* **46**, 496–518.

Ogata, Y., and Tanemura, M. (1985) Estimation of interaction potentials of marked spatial point patterns through the maximum likelihood method. *Biometrics* **41**, 421–433.

Ogata, Y., and Tanemura, M. (1986) Likelihood estimation of interaction potentials and external fields of inhomogeneous spatial point patterns. In I.S. Francis, B.F.J. Mauly and F.C. Lam (eds), *Proceedings of the Pacific Statistical Congress*. North-Holland, Amsterdam, pp. 150–154.

Ohser, J., and Mücklich, F. (2000) *Statistical Analysis of Microstructures in Materials Science*. John Wiley & Sons, Ltd, Chichester.

Ohser, J., and Schladitz, K. (2008) *3D Images of Materials Structures – Processing and Analysis*. Wiley-VCH, Weinheim.

Ohser, J., and Stoyan, D. (1981) On the second-order and orientation analysis of planar stationary point processes. *Biometrical J.* **23**, 523–533.

Okabe, A., Boots, B.N., Sugihara, K, and Chiu, S.N. (2000) *Spatial Tessellations. Concepts and Applications of Voronoi Diagrams*. John Wiley & Sons, Ltd, Chichester.

Olkin, I., Petkau, A.J., and Židek, J.V. (1981) A comparison of n estimators for the binomial distribution. *J. Amer. Statist. Assoc.* **76**, 637–642.

Ovryn, B., and Izen, S. (2000) Imaging of transparent spheres through a planar interface using a high-numerical-aperture optical microscope. *J. Optical Soc. Amer. A* **17**, 1202–1213.

Pacala, S.W., Canham, C.D., and Silander, J.A. (1993) Forest models defined by field measurements: I. The design of a northeastern forest simulator. *Can. J. Forest Res.* **23**, 1980–1988.

Paulo, M.J., Stein, A., and Tomé, M. (2002) A spatial statistical analysis of cork oak competition in two Portuguese silvapastoral systems. *Can. J. Forest Res.* **32**, 1893–1903.

Peebles, P.J.E. (1980) *The Large-Scale Structure of the Universe*. Princeton University Press, Princeton, NJ.

Pélissier, R., and Goreaud, F. (2001) A practical approach to the study of spatial structure in simple cases of heterogeneous vegetation. *J. Veget. Sci.* **12**, 99–108.

Penrose, M.D. (2003) *Random Geometric Graphs*. Oxford University Press, Oxford.

Penrose, M.D. (2005) Multivariate spatial central limit theorems with applications to percolation and spatial graphs. *Ann. Probab.* **33**, 1945–1991.

Penrose, M.D., and Yukich, J.E. (2001) Central limit theorems for some graph in computational geometry. *Ann. Appl. Probab.* **11**, 1005–1041.

Penttinen, A. (1984) Modelling interaction in spatial point patterns: Parameter estimation by the maximum likelihood method. *Jyväskylä Stud. Comput Sci. Econom. Statist.* **7**.

Penttinen, A., and Niemi, A. (2007) On statistical inference for the random set generated Cox process with set-marking. *Biometrical J.* **49**, 197–213.

Penttinen, A., and Stoyan, D. (1989) Statistical analysis for a class of line segment processes. *Scand. J. Statist.* **16**, 153–161.

Penttinen, A., Stoyan, D., and Henttonen, H. (1992) Marked point processes in forest statistics. *Forest Sci.* **38**, 806–824.

Pfeifer, D., Bäumer, H.-P., and Schleier, U. (1996) The 'minimum area' problem in ecology: a spatial Poisson process approach. *Comput. Statist.* **11**, 415–428.

Pielou, E.C. (1959) The use of point-to-plant distances in the study of the pattern of plant populations. *J. Ecology* **48**, 575–584.

Pielou, E.C. (1977) *Mathematical Ecology.* John Wiley & Sons, Inc., New York.

Pommerening, A. (2002) Approaches to quantifying forest structures. *Forestry* **75**, 305–324.

Pommerening, A. (2006) Evaluating structural indices by reversing forest structural analysis. *Forest Ecol. Manag.* **224**, 266–277.

Pommerening, A., and Stoyan, D. (2006) Edge-correction needs in estimating indices of spatial forest structure. *Can. J. Forest Res.* **36**, 1723–1739.

Pretzsch, H. (2002) *Grundlagen der Waldwachstumsforschung.* Parey Buchverlag, Berlin (in German).

Pretzsch, H., Biber, P., and Dursky, J. (2002) The single tree based stand simulator Silva. Construction, application and evaluation. *Forest Ecol. Manag.* **162**, 3–21.

Prokešová, M., Hahn, U., and Jensen, E.B.V. (2006) Statistics for locally scaled point processes. In A. Baddeley, P. Gregori, J. Mateu, R. Stoica and D. Stoyan (eds.), *Case Studies in Spatial Point Process Modeling,* Lecture Notes in Statistics 185. Springer-Verlag, Berlin, pp. 99–123.

Provatas, N., Haataja, M., Asikainen, J., Majaniemi, S., Alava, M., and Ala-Nissila, T. (2000) Fiber deposition models in two and three spatial dimensions. *Colloids and Surfaces A* **165**, 209–229.

Quine, M.P., and Watson, D.F. (1984) Radial simulation of n-dimensional Poisson processes. *J. Appl. Probab.* **21**, 548–557.

R Development Core Team (2007) *R: A Language and Environment for Statistical Computing.* R Foundation for Statistical Computing, Vienna.

Ramsay, J.O., and Silverman, B.W. (2002) *Applied Functional Data Analysis.* Springer-Verlag, New York.

Ramsay, J.O., and Silverman, B.W. (2005) *Functional Data Analysis,* 2nd edition. Springer-Verlag, New York.

Rathbun, S.L. (1996) Estimation of Poisson intensity using partially observed concomitant variables. *Biometrics* **52**, 226–242.

Renshaw, E. (1997) Spectral techniques in spatial analysis. *Forest Ecol. Manag.* **94**, 165–174.

Renshaw, E. (2002) Two-dimensional spectral analysis for marked point processes. *Biometrical J.* **44**, 718–745.

Renshaw, E., Mateu, J., and Saura, F. (2007) Disentangling mark/point interaction in marked-point processes. *Comput. Statist. Data Anal.* **51**, 3123–3144.

Rényi, A. (1967) Remarks on the Poisson process. *Studia Sci. Math. Hung.* **2**, 119–123.

Ribbens, E., Silander, J.A. Jr., and Pacala, S.W. (1994) Seedling recruitment in forests: calibrating models to predict patterns of tree seedlings dispersion. *Ecology* **75** (6), 1794–1806.

Ripley, B.D. (1976) The second-order analysis of stationary point processes. *J. Appl. Probab.* **13**, 255–266.

Ripley, B.D. (1977) Modeling spatial patterns (with discussion). *J. Roy. Statist. Soc. B* **39**, 172–212.

Ripley, B.D. (1979) Tests of 'randomness' for spatial point patterns. *J. Roy. Statist. Soc. B* **41**, 368–374.

Ripley, B.D. (1981) *Spatial Statistics.* John Wiley & Sons, Inc., New York.

Ripley, B.D. (1987) *Stochastic Simulation.* John Wiley & Sons, Inc., New York.

Ripley, B.D. (1988) *Statistical Inference for Spatial Processes.* Cambridge University Press, Cambridge.

Ripley, B.D., and Kelly, F.P. (1977) Markov point processes. *J. London Math. Soc.* **15**, 188–192.

Ripley, B.D., and Silverman, B.W. (1978) Quick tests for spatial interaction. *Biometrika* **65**, 641–642.

Robert, C.P., and Casella, G. (2005) *Monte Carlo Statistical Methods,* 2nd edition. Springer-Verlag, New York.

Roberts, I.M. (1994) Factors effecting the efficiency of immunogold labelling of plant virus antigens in thin sections. *J. Virolog. Methods* **50**, 155–166.

Sakai, H., Stillinger, F.H. and Torquato, S. (2002) Equi-$g(r)$ sequence of systems derived from the square-well potential. *J. Chem. Phys.* **117**, 297–307.

Särkkä, A. (1993) Pseudo-likelihood approach for pair potential estimation of Gibbs processes. *Jyväskylä Stud. Comp. Sci. Econ. Statist.* **22**.

Särkkä, A., and Renshaw, E. (2006) The analysis of marked point patterns evolving through space and time. *Comput. Statist. Data Anal.* **51**, 1698–1718.

Särkkä, A., and Tomppo, E. (1998) Modelling interactions between trees by means of field observations. *Forest Ecol. Manag.* **108**, 57–62.

Saunders, R., Kryscio, R.J. and Funk, G.M. (1982) Poisson limits for a hard-core clustering model. *Stoch. Proc. Appl.* **12**, 97–106.

Schladitz, K., and Baddeley, A.J. (2000) A third order point process characteristic. *Scand. J. Statist.* **27**, 657–671.

Schladitz, K., Särkkä, A., Pavenstädt, I., Haferkamp, O., and Mattfeldt, T. (2003) Statistical analysis of intramembranous particles using freeze fracture specimens. *J. Microsc.* **211**, 137–153.

Schlather, M. (2001a) On the second-order characteristics of marked point processes. *Bernoulli* **7**, 99–117.

Schlather, M. (2001b) Simulation and analysis of random fields. *R News* **1**(2), 18–20.

Schlather, M., Ribeiro, P.J., and Diggle, P.J. (2004) Detecting dependence between marks and locations of marked point processes. *J. Roy. Statist. Soc. B* **66**, 79–93.

Schmidt, V. (1985) POISSON bounds for moments of shot noise processes. *Statistics* **16**, 253–262.

Schneider, M.K., Law, R., and Illian, J.B. (2006) Quantification of neighbourhood-dependent plant growth by Bayesian hierarchical modelling. *J. Ecology* **94**, 310–321.

Seneta, E. (1983) Modern probabilistic concepts in the work of E. Abbe and A. de Moivre. *Math. Sci.* **8**, 75–80.

Shepilov, M.P., Kalmykov, A.E., and Sycheva, G.A. (2007) Liquid-liquid phase separation in sodium borosilicate glass: Ordering phenomena in particle arrangement. *J. Non-Crystalline Solids* **353**, 2413–2430.

Sheth, R.K., and Saslaw, W. (1994) Synthesizing the observed distribution of galaxies. *Astrophys. J.* **437**, 35–55.

Shimatani, K. (2001) Multivariate point processes and spatial variation of species diversity. *Forest Ecol. Manag.* **142**, 215–229.

Shimatani, K. (2002) Point processes for fine-scale spatial genetics and molecular ecology. *Biometrical J.* **44**, 325–352.

Shimatani, K., and Kubota, Y. (2004) Quantitative assessment of multispecies pattern with high species diversity. *Ecol. Res.* **19**, 149–164.

Silverman, B.W. (1986) *Density Estimation for Statistics and Data Analysis.* Chapman & Hall, London.

Skare, Ø., Møller, J., and Jensen, E.B.V. (2007) Bayesian analysis of spatial point porcesses in the neighbourhood of Voronoi networks. *Statist. Comput.* **17**, 369–379.

Snethlage, M. (1999) Is bootstrap really helpful in point process statistics? *Metrika* **49**, 245–255.

Snethlage, M., Martínez, V.J., Stoyan, D., and Saar, E. (2002) Point field models for the galaxy point pattern. Modelling the singularity of the two-point correlation function. *Astron. Astrophys.* **388**, 758–765.

Soille, P. (1999) *Morphological Image Analysis.* Springer-Verlag, Berlin.

Stein, A., and Georgiadis, N. (2006) Spatial marked point patterns for herd dispersion in a savanna wildlife herbivore community in Kenya. In A. Baddeley, P. Gregori, J. Mateu, R. Stoica and D. Stoyan (eds), *Case Studies in Spatial Point Process Modeling,* Lecture Notes in Statistics 185. Springer-Verlag, Berlin, pp. 261–273.

Stillinger, F.H., DiMarzio, E.A., and Kornegay, R.L. (1964) Systematic approach to explanation of the rigid disk phase transition. *J. Chem. Phys.* **40**, 1564–1576.

Stillinger, F.H., Stillinger, D.K., Torquato, S., Truskett, T.M., and Debenedetti, P.G. (2000) Triangle distribution and equation of state for classical rigid disks. *J. Statist. Phys.* **100**, 49–72.

Stoica, R.S., Descombes, X., and Zerubia, J. (2004) A Gibbs point process for road extraction in remotely sensed images. *Int. J. Comput. Vision* **52**, 121–136.

Stoica, R.S., Martinez, V.J., Mateu, J., and Saar, D. (2005) Detection of cosmic filaments using the Candy model. *Astron. Astroph.* **434**, 423–432.

Stoica, R.S., Gay, E., and Kretzschmar, A. (2007) Cluster pattern detection in spatial data dased on Monte Carlo inference. *Biometrical J.* **49**, 508–519.

Stoyan, D. (1984) On correlations of marked point processes. *Math. Nachr.* **116**, 197–207.

Stoyan, D. (1988) Thinnings of point processes and their use in the statistical analysis of a settlement pattern with deserted villages. *Statistics* **19**, 45–56.

Stoyan, D. (2006) On estimators of the nearest neighbour distance distribution function for stationary point processes. *Metrika* **64**, 139–150.

Stoyan, D., and Beneš, V. (1991): Anisotropy analysis for particle systems. *J. Microsc.* **164**, 159–168.

Stoyan, D., and Grabarnik, P. (1991a) Statistics for the stationary Strauss model by the cusp point method. *Statistics* **22**, 283–289.

Stoyan, D., and Grabarnik, P. (1991b) Second-order chracteristics for stochastic structures connected with Gibbs point processes. *Math. Nachr.* **151**, 95–100.

Stoyan, D., and Mecke, K. (2005) The Boolean model: from Matheron till today. In M. Bilodeau, F. Meyer and M. Schmitt (eds), *Space, Structure, and Randomness,* Springer Lecture Notes in Statistics 183. Springer-Verlag, New York, pp. 151–181.

Stoyan, D., and Ohser, J. (1982) Correlations between random structuress, with an ecological application. *Biometrical J.* **24**, 631–647.

Stoyan, D., and Ohser, J. (1984) Cross-correlation measure of weighted random measures and their estimation. *Teor. Veroyatn. Primen.* **29**, 338–347.

Stoyan, D., and Penttinen, A. (2000) Recent applications of point process methods in forestry statistics. *Statist. Sci.* **15**, 61–78.

Stoyan, D., and Schlather, M. (2000) Random sequential adsorption: Relationship to dead leaves and characterization of variability. *J. Statist. Phys.* **100**, 969–979.

Stoyan, D., and Schnabel, H.-D. (1990) Description of relations between spatial variability of microstructure and mechanical strength of aluminia ceramics. *Ceramics Int.* **16**, 11–18.

Stoyan, D., and Stoyan, H. (1985) On one of Matern's hard-core point process models. *Math. Nachr.* **122**, 205–214.

Stoyan, D., and Stoyan, H. (1992) *Fraktale – Formen – Punktfelder.* Akademie Verlag, Berlin (in German).

Stoyan, D., and Stoyan, H. (1994) *Fractals, Random Shapes and Point Fields.* John Wiley & Sons, Ltd, Chichester.

Stoyan, D., and Stoyan, H. (1996) Estimating pair correlation functions of planar cluster processes. *Biometrical J.* **38**, 259–271.

Stoyan, D., and Stoyan, H. (1998) Non-homogeneous Gibbs process models for forestry – a case study. *Biometrical J.* **40**, 521–531.

Stoyan, D., and Stoyan, H. (2000) Improving ratio estimators of second order point process characteristics. *Scand. J. Statist.* **27**, 641–656.

Stoyan, D., and Wagner, S. (2001) Estimating the fruit dispersion of anemochorous forest trees. *Ecol. Modelling* **145**, 35–47.

Stoyan, D., Bertram, U., and Wendrock, H. (1993) Estimation variances for estimators of product densities and pair-correlation functions of planar point processes. *Ann. Inst. Statist. Math.* **45**, 211–221.

Stoyan, D., Kendall, W.S., and Mecke, J. (1995) *Stochastic Geometry and its Applications,* 2nd edition. John Wiley & Sons, Ltd, Chichester.

Strauss, D.J. (1975) A model for clustering. *Biometrika* **62**, 467–475.

Stroeven, P., and Stroeven, M. (1999) Assessment of packing characteristings by computer simulations. *Cement Concrete Res.* **29**, 1201–1206.

Stroeven, P., and Stroeven, M. (2001) Reconstructions by SPACE of the interfacial transition zone. *Cement Concrete Compos.* **23**, 189–200.

Svedberg, T. (1922) Ett bidrag till de statistiska metodernas användning inom växtbiologien. *Svensk Botanik Tidskrift* **16**, 1–8 (in Swedish).

Szapudi, I., Pan, J., Prunet, S. and Budavári, T. (2005) Fast edge-corrected measurement of the two-point correlation function and the power spectrum. *Astrophys. J. Lett.* **631**, L1–L4.

Takacs, R. (1986) Estimator for the pair-potential of a Gibbsian point process. *Statistics* **17**, 429–433.

Tanaka, U., Ogata, Y., and Stoyan, D. (2008) Model selection and estimation of the Neyman–Scott type spatial cluster models. *Biometrical J.* **50**. To appear.

Taylor, C.C., Dryden, I.L., and Franoosh, R. (2001) The K-function for nearly regular point processes. *Biometrics* **57**, 224–231.

Thomas, M. (1949) A generalization of Poisson's binomial limit for use in ecology. *Biometrika* **36**, 18–25.

Thönnes, E., and Lieshout, M.N.M. van (1999) A comparative study on the power of Van Lieshout and Baddeley's J-function. *Biometrical J.* **41**, 712–734.

Thorisson, H. (2000) *Coupling, Stationarity, and Regeneration.* Springer-Verlag, New York.

Tijms, H.C. (2003) *A First Course in Stochastic Models.* John Wiley & Sons, Ltd, Chichester.

Tomppo, E. (1986) Models and methods for analyzing spatial patterns of trees. *Comm. Inst. Forestal. Fenniae* **38**, Helsinki.

Torquato, S. (2002) *Random Heterogeneous Materials. Microstructure and Macroscopic Properties.* Springer-Verlag, New York.

Totsuji, H., and Kihara, T. (1969) The correlation function for the distribution of galaxies. *Publ. Astron. Soc. Japan* **21**, 221–229.

Tscheschel, A. (2001) Reconstruction of random porous media using a simulated annealing method. Diploma thesis, TU Bergakademie Freiberg.

Tscheschel, A., and Stoyan, D. (2006) Statistical reconstruction of random point patterns. *Comput. Statist. Data Anal.* **51**, 859–871.

Uche, O.U., Stillinger, F.H., and Torquato, S. (2005) On the realizability of pair correlation functions. *Physica A* **360**, 21–36.

Upton, G., and Fingleton, B. (1985) *Spatial Data Analysis by Example. Vol. 1: Point Pattern and Quantitative Data.* John Wiley & Sons, Ltd, Chichester.

Upton, G., and Fingleton, B. (1989) *Spatial Data Analysis by Example. Vol. 2: Categorial and Directional Data.* John Wiley & Sons, Ltd, Chichester.

Uriarte, M., Condit, R., Canham, C.D., and Hubbell, S.P. (2004) A spatially-explicit model of sapling growth in a tropical forest: Does the identity of neighbours matter? *J. Ecol.* **92**, 348–360.

Vere-Jones, D. (2009) Some models and procedures for space–time point processes. *Environm. Ecolog. Statist.* **16**. To appear.

Vincent, L., and Soille, P. (1991) Watersheds in digital spaces: An efficient algorithm based on immersion simulation. *IEEE Trans. Pattern Anal. Mach. Intell.* **13**, 583–598.

Vio, R., D'Odorico, V., Stoyan, H., and Stoyan, D. (2007) Ly-α forest: efficient unbiased estimation of second-order properties with missing data. *Astron. Astrophys.* **466**, 403–411.

Waagepetersen, R. (2007) An estimating function approach to inference for inhomogeneous Neyman–Scott processes. *Biometrics* **63**, 252–258.

Wackernagel, H. (2003) *Multivariate Geostatistics,* 3rd edition. Spinger-Verlag, Berlin.

Wagner, S. (1997) A model describing the fruit dispersal of ash (*Fraxinus excelsior*) taking into account directionality. *Allg. Forst- u. Jagdztg.* **168**, 149–155 (in German, with English abstract).

Wagner, S., Wälder, K., Ribbens, E., and Zeibig, A. (2004) Directionality in fruit dispersal models for anemochorus forest trees. *Ecol. Modelling* **179**, 487–498.

Wälder, O., and Stoyan, D. (1996) On variograms in point process statistics. *Biometrical J.* **38**, 895–905.

Warren, W.G. (1972) Point processes in forestry. In P.S.W. Lewis (ed.), *Stochastic Point Processes.* John Wiley & Sons, Inc., New York, pp. 801–816.

Wartenberg, D. (1990) Exploratory spatial analyses: outliers, leverage points, and influence functions. In: D. Griffith (ed.), *Spatial Statistics: Past, Present and Future.* Institute of Mathematical Geography, Ann Arbor, MI, pp. 133–162.

Weber, H., Marx, D., and Binder, K. (1995) Melting transition in two dimension: A finite-size scaling analysis of bond-orientational order in hard disks. *Phys. Rev. B* **51**, 14636–14651.

Weygaert, R. van de (1994) Fragmenting the universe III. The construction and statistics of 3-D Voronoi tessellations. *Astron. Astrophys.* **283**, 361–406.

Weygaert, R. van de, and Schaap, W. (2008) The cosmic web: Geometric analysis and modelling. In V. Martínez (ed.), *Data Analysis in Cosmology.* Springer-Verlag, Berlin.

Wiegand, T., and Moloney, K.A. (2004) Rings, circles, and null-models for point pattern analysis in ecology, *Oikos* **104**, 209–229.

Wiencek, K. (2000) A non-isotropic distribution of carbide dispersion in steel. In *Proc. VII Conference KomPlasTech'2000,* Krynica, Poland, pp. 133–139 (in Polish).

Wiencek, K., and Satora, K. (1999) Particle arrangement of a carbide dispersion in steel. In V. Beneš, J. Janáček and I. Saxl (eds), *Proc. S4G International Conference on Stereology, Spatial Statistics and Stochastic Geometry.* Union of Czech Mathematicians and Physicists, Prague, pp. 267–272.

Windhager, M. (1997) Die Berechnung des Ek und Monserud (1974) Konkurrenzindex für Randbäume nach unterschiedlichen Berechnugsmethoden. In G. Kenk (ed.), *Proceedings of the Jahrestagung of Deutscher Verband Forstlicher Versuchsanstalten, Sektion Ertragskunde, Freiburg*, pp. 74–86 (in German).

Winkler, G. (2003) *Image Analysis, Random Fields and Markov Chain Monte Carlo Methods*. Springer-Verlag, Berlin.

Winkler, P., and Gilli, M. (2004) Application of optimization heuristics to estimation and modelling problems. *Comput. Statist. Data Anal.* **47**, 211–223.

Wolpert, R.L., and Ickstadt, K. (1998) Poisson/gamma random field models for spatial statistics. *Biometrika* **85**, 251–267.

Wu, H.-I., Sharpe, P.J.H., Walker, J., and Penridge, L.K. (1985) Ecological field theory: A spatial analysis of resource interference among plants. *Ecol. Modelling* **29**, 215–243.

Zehavi, I., Weinberg, D.H., Zheng, Z., Berlind, A.A., Frieman, J.A., Scoccimarro, R., Sheth, R.K., Blanton, M.R., Tegmark, M., Mo, H.J., Bahcall, N.A., Brinkmann, J., Burless, S., Csabai, I., Fukugita, M., Gunn, J.E., Lamb, D.Q., Loveday, J., Lupton, R.H., Meiksin, A., Munn, J.A., Nichol, R.C., Schlegel, D., Schneider, D.P., Subba Rao, M., Szalay, A.S., Uomoto, A., and York, D.G. (2004) On departures from a power law in the galaxy correlation function. *Astrophys. J.* **608**, 16–24.

Zhuang, J., Ogata, Y., and Vere-Jones, D. (2002) Stochastic declustering of space–time earthquake occurences. *J. Amer. Statist. Assoc.* **97**, 369–380.

Zhuang, J., Ogata, Y., and Vere-Jones, D. (2006) Diagnostic analysis of space–time branching processes for earthquakes. In A. Baddeley, P. Gregori, J. Mateu, R. Stoica and D. Stoyan (eds.), *Case Studies in Spatial Point Process Modeling*, Lecture Notes in Statistics 185. Springer-Verlag, Berlin, pp. 275–292.

Zimmerman, D.L. (1993) A bivariate Cramér–von Mises type of test for spatial randomness. *Appl. Statist.* **42**, 43–54.

Zwicky, F. (1953) Frequencies of clusters of galaxies. *Publ. Astron. Soc. Pacific* **65**, 215–216.

Notation index

Abbreviations

d.f.	distribution function	
i.i.d.	independent and identically distributed	
p.d.f.	probability density function	
lu	length unit	6
mse	mean squared error	480
^	'hat', denotes an estimator; for example, $\hat{\theta}$ is an estimator of the parameter θ.	

Frequently used symbols

b_d	volume of unit sphere in $\mathbb{R}^d$	
$b(x, r)$	sphere (disc) of radius r centred at x	
CE	Clark–Evans index	196
$d(r)$	p.d.f. for $D(r)$	207
$d(x)$	distance from point x to its nearest neighbour	177
$D(r)$	nearest-neighbour distance d.f.	206
$D_{\text{fin}}(r)$	finite nearest-neighbour distance d.f.	126
$\mathbf{E}$	expectation	
$\mathbf{E}_o$	expectation with respect to Palm distribution	178
$f_{\mathcal{M}}(m)$	p.d.f. for $F_{\mathcal{M}}(m)$	302
$f_n(x_1, \dots, x_n)$	location density function	103
$F_{\mathcal{M}}(m)$	mark d.f.	300
$g(r)$	pair correlation function	218
$g_{ij}(r)$	partial pair correlation function	325
$G(t)$	point density d.f.	29
h	bandwidth	230, 482
$h_s(r)$	p.d.f. for $H_s(r)$	201
$H_s(r)$	spherical contact d.f.	200

Statistical Analysis and Modelling of Spatial Point Patterns   J. Illian, A. Penttinen, H. Stoyan and D. Stoyan
© 2008 John Wiley & Sons, Ltd

Author index

Statistical Analysis and Modelling of Spatial Point Patterns   J. Illian, A. Penttinen, H. Stoyan and D. Stoyan
© 2008 John Wiley & Sons, Ltd

Subject index

STATISTICS IN PRACTICE

Human and Biological Sciences

Berger – Selection Bias and Covariate Imbalances in Randomized Clinical Trials
Brown and Prescott – Applied Mixed Models in Medicine, Second Edition
Chevret (Ed) – Statistical Methods for Dose-Finding Experiments
Ellenberg, Fleming and DeMets – Data Monitoring Committees in Clinical Trials: A Practical
   Perspective
Hauschke, Steinijans & Pigeot – Bioequivalence Studies in Drug Development: Methods
   and Applications
Lawson, Browne and Vidal Rodeiro – Disease Mapping with WinBUGS and MLwiN
Lui – Statistical Estimation of Epidemiological Risk
Marubini and Valsecchi – Analysing Survival Data from Clinical Trials and Observation
   Studies
Molenberghs and Kenward – Missing Data in Clinical Studies
O'Hagan, Buck, Daneshkhah, Eiser, Garthwaite, Jenkinson, Oakley & Rakow – Uncertain
   Judgements: Eliciting Expert's Probabilities
Parmigiani – Modeling in Medical Decision Making: A Bayesian Approach
Pintilie – Competing Risks: A Practical Perspective
Senn – Cross-over Trials in Clinical Research, Second Edition
Senn – Statistical Issues in Drug Development, Second Edition
Spiegelhalter, Abrams and Myles – Bayesian Approaches to Clinical Trials and Health-Care
   Evaluation
Whitehead – Design and Analysis of Sequential Clinical Trials, Revised Second Edition
Whitehead – Meta-Analysis of Controlled Clinical Trials
Willan and Briggs – Statistical Analysis of Cost Effectiveness Data

Earth and Environmental Sciences

Buck, Cavanagh and Litton – Bayesian Approach to Interpreting Archaeological Data
Glasbey and Horgan – Image Analysis in the Biological Sciences
Helsel – Nondetects and Data Analysis: Statistics for Censored Environmental Data
Illian, Penttinen, Stoyan, H and Stoyan D – Statistical Analysis and Modelling of Spatial
   Point Patterns
McBride – Using Statistical Methods for Water Quality Management
Webster and Oliver – Geostatistics for Environmental Scientists, Second Edition
Wymer – Statistical Framework for Recreational Quality Criteria and Monitoring

Industry, Commerce and Finance

Aitken – Statistics and the Evaluation of Evidence for Forensic Scientists, Second Edition
Balding – Weight-of-evidence for Forensic DNA Profiles
Brandimarte – Numerical Methods in Finance and Economics: A MATLAB-Based Intro-
   duction, Second Edition

Printed and bound by CPI Group (UK) Ltd, Croydon, CR0 4YY

16/04/2025

14658497-0004